CONTENTS

TRANSONIC AERODYNAMICS

NORTH-HOLLAND SERIES IN

APPLIED MATHEMATICS AND MECHANICS

EDITORS:

J. D. ACHENBACH
Northwestern University

B. BUDIANSKY
Hardvard University

W. T. KOITER
University of Technology, Delft

H. A. LAUWERIER
University of Amsterdam

L. VAN WIJNGAARDEN
Twente University of Technology

VOLUME 30

NORTH-HOLLAND –AMSTERDAM · NEW YORK · OXFORD · TOKYO

TRANSONIC AERODYNAMICS

Julian D. COLE

Department of Mathematical Sciences
Rensselaer Polytechnic Institute
Troy, NY, U.S.A.

L. Pamela COOK

Department of Mathematics
University of Delaware
Newark, DE, U.S.A.

1986

NORTH-HOLLAND – AMSTERDAM · NEW YORK · OXFORD · TOKYO

ISBN: 0 444 87958 7

Publishers:
ELSEVIER SCIENCE PUBLISHERS B.V.
P.O. Box 1991
1000 BZ Amsterdam
The Netherlands

Sole distributors for the U.S.A. and Canada:
ELSEVIER SCIENCE PUBLISHING COMPANY, INC.
52 Vanderbilt Avenue
New York, N.Y. 10017
U.S.A.

PRINTED IN THE NETHERLANDS

The authors would like to thank the Air Force Office of Scientific Research for their support of this project under contract # F49620-79-C-0162. The second author would also like to thank the National Science Foundation for support under grants # MCS 80-02203 and DMS-8401738.

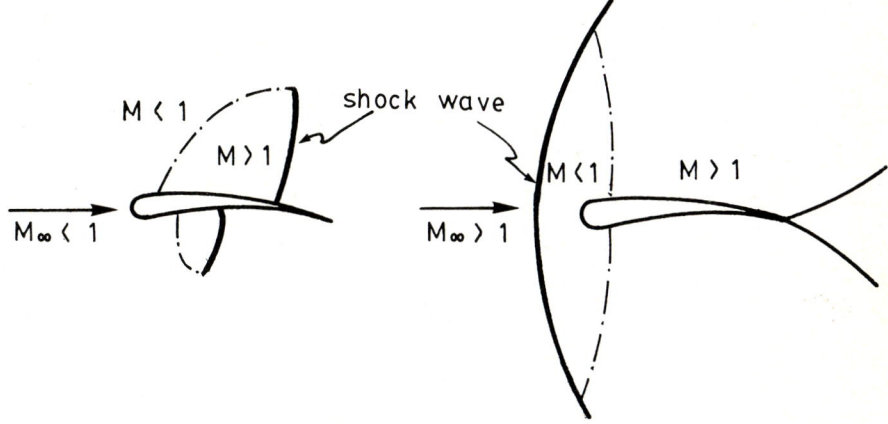

1. Introduction

Transonic flows are those in which the local flow speed is close to the local sonic speed. That is the local Mach number M, the ratio of flow speed q to sound speed a, is close to one: $M = \frac{q}{a} \doteq 1$. This means that the dynamic pressure $\frac{\rho q^2}{2}$ and the static pressure P are the same order of magnitude since $\frac{1}{2}\frac{\rho q^2}{P} \sim \frac{\gamma M^2}{2} \sim 0(1)$.* Since the local flow speed is approaching a critical value, we can expect some special phenomena to occur, in contrast to other flow regimes, and indeed they do. The qualitative features that dominate the situation are the existence of throats in streamtubes when the local Mach number is one and the possible occurrence of shock waves when the local Mach number is supersonic. In general when a local supersonic zone is formed in the flow around an airfoil a shock wave occurs. This also occurs when the flight speed is supersonic. (See Figure 1.1.1)

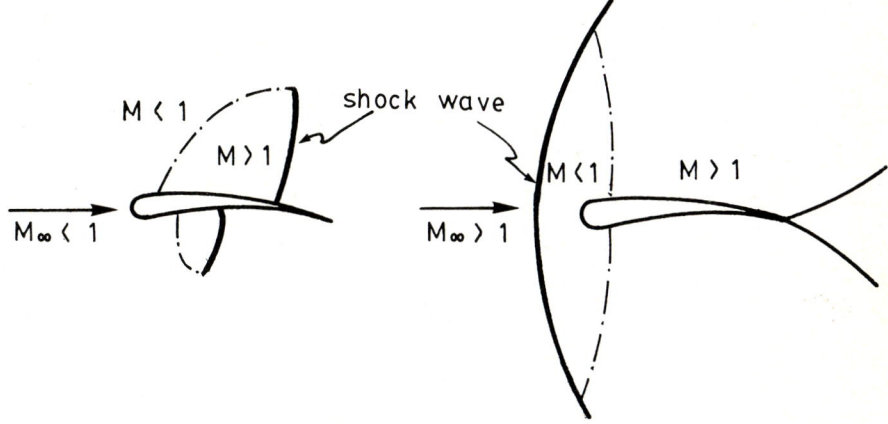

Figure 1.1.1

Typical Transonic Flow Patterns

In technical applications this type of flow occurs in the neighborhood of airplanes, such as the Boeing 727, 747 that fly close to the speed of sound. The next generation of transports might include a "boom-less" airplane which flies at supersonic speed at altitude but is subsonic with respect to sound speed at

---

* For an ideal gas $a^2 = \gamma RT = \frac{\gamma P}{\rho}$. At 10 Km altitude, $P/P_0 = 2.6153 \times 10^{-1}$, P_0 = ground pressure = $1.01325 \times 10^5 \mathrm{Newtons/m^3}$, $\rho = 4.153 \times 10^{-1}\,\mathrm{Kg/m^2}$, $a = 299.53\,\mathrm{m/sec}$, $T = 223.25°\mathrm{K}$.

the ground. Transonic flows also occur in compressors and turbines and around helicopter blades, in the throat regions of supersonic wind tunnels, in inlets and in rocket nozzles. Even at highly supersonic speeds a transonic region appears near the nose of a blunt body. "Quasi" transonic flows appear when an important component of the flow velocity is close to sonic, as for example when a wing is swept back close to the Mach angle $\theta_M = \sin^{-1} \frac{1}{M_\infty}$, then the component normal to the edge is sonic.

The aim of this book is to present relatively a self-contained treatment, based on an elementary knowledge of fluid mechanics.

1.1 Framework

The book covers mainly ideal inviscid flow theory (gasdynamics). The results, for external flows, are then applicable to flows at high Reynolds numbers past streamlined bodies. The viscous effects are assumed to be confined to the interior of very thin regions. These shapes are desirable for technical applications and experience shows that wide classes of engineering problems are amenable to this theory. For example, lift, drag, and moment for three-dimensional wings can be calculated.

Viscous effects and interactions may be important but in any case one must know how to calculate the inviscid flows.

1.2 Mathematics

Perturbation methods will be used to give a systematic discussion of transonic small disturbance theory. The use of this theory is justified by the fact the simplified equations exhibit all the essential features and provide in many cases a good numerical approximation to experimental results. Various important similarity rules appear which are not available for the exact equation. Further, since a systematic procedure is employed corrections to the first-order theory can be studied.

Some special problems for more exact equations will also be studied.

Significant mathematical areas which enter the discussion are:

— Partial Differential Equations of Mixed Type

— Weak Solutions (Shock Waves)

— Hodograph Transformations

— Similarity Solutions

— New Numerical Methods for Equations of Mixed Type

1.3 Historical Note

Transonic flows have been studied theoretically since the beginning of the century: (e.g.): S.A. Chaplygin "On Gas Jets", Moscow University Press (1902) (O Гazobix Cmpyax).

Shock waves as isolated phenomena have been known for a long time. Early U.S. experiments were done in the 1930's (NACA Briggs, Dryden, Stack). These were motivated by sonic effects near propeller tips.

Pioneering work in the field was done by Guderley (early 1950's) and Frankl, as evidenced by many references throughout this book. In more recent years a vast literature has accumulated on the subject of transonic flow. In particular many papers dealing with computations have appeared for both approximate and more exact equations.

It is not possible in this work to review all of these developments. We try here to give a detailed theoretical picture of the basis of transonic flow and some discussion of the ideas behind recent numerical approaches.

References

Useful books dealing especially with transonic flow are:

[1.1] Guderley, K. G., *Theorie Schallnahe Strömungen*, Springer-Verlag Berlin 1957. English translation: Addison-Wesley 1962.

[1.2] Bers, L., *Mathematical Aspects of Subsonic and Transonic Gas Dynamics*, John Wiley, N.Y. 1958.

[1.3] Ferrari, C. and Tricomi, F., *Transonic Aerodynamics*, Academic Press 1968.

REMARK: Equations of mixed type appear also in other contexts such as oceanography, elastic shell theory, viscoelastic fluids.

2. Linearized Theory – Transonic Breakdown

Airplanes and slender objects cause only a small disturbance to the ambient state on passage through the air. The theory of "Acoustics" describes the propagation of such small disturbances usually in a uniform medium at rest. Thus all of linearized aerodynamics (subsonic, supersonic, unsteady) is equivalent to acoustics. The solutions of acoustics are solutions to the classical wave equation. However, in aerodynamics new and typical boundary value problems appear.

For technical applications it would be very useful if linear theory gave a good approximation. Linear solutions are easy to compute and further very general problems can be formulated and solved. For example, R. T. Jones has shown, in linearized supersonic theory, how to distribute the lift on a wing of given span so as to obtain the minimum wave drag. Unfortunately, linearized theory cannot give the correct answer in the transonic range.

In order to understand the breakdown of linearized theory, we can consider the development of the acoustic field around a body flying at sonic speed. For this we need the equations of acoustics. The assumption of isentropy is adequate for the weak disturbances of acoustics. This point will be discussed in some detail later.

The framework of acoustics is that of an inviscid ideal gas. Viscous effects are supposed to be confined to thin layers, such as boundary layers adjacent to solid surfaces, vortex sheets, and the interior of "discontinuous" jumps in pressure (shock waves). The main interest here is the calculation of forces normal to solid surfaces and this can be done if flow separation does not occur. The fact that viscous effects may modify the downstream flow considerably does not affect the calculation of forces on the solid surfaces producing this flow. The ideal gas assumption is not necessary since only small disturbances appear and an arbitrary equation of state could be treated. However it is convenient and sufficiently accurate for most technical applications.

This same framework will also cover most of our considerations on transonic flows.

2.1 Equations of Acoustics

Let $\mathbf{q}\,(x, y, z, t)$ be the flow velocity in the rest frame. (cf Figure 2.1.1). For acoustics this is assumed to be small in some sense, e.g.,

$$\frac{|\mathbf{q}|}{a_\infty} \ll 1; \quad a_\infty = \text{sound speed at infinity} = \left(\frac{\gamma P_\infty}{\rho_\infty}\right)^{\frac{1}{2}},$$

$$\gamma = \frac{c_p}{c_v} = \frac{7}{5} \quad \text{for a diatomic gas,}$$

$$= \frac{5}{3} \quad \text{for a monotomic gas,}$$

$$= \text{ratio of specific heats.}$$

BASIC EQUATIONS:

continuity
$$\frac{\partial \rho}{\partial t} + \nabla \cdot \rho \mathbf{q} = 0, \quad \nabla \cdot \equiv \mathbf{div};$$

momentum
$$\rho \left\{ \frac{\partial \mathbf{q}}{\partial t} + \mathbf{q} \cdot \nabla \mathbf{q} \right\} = -\nabla P, \quad \nabla \equiv \mathbf{grad}; \quad (2.1.1)$$

isentropy
$$\frac{P}{\rho^\gamma} = \frac{P_\infty}{\rho_\infty^\gamma}.$$

The acoustic equations are derived by assuming small disturbances

$$\rho / \rho_\infty = 1 + s,$$

$$p, s \ll 1. \quad (2.1.2)$$

$$P / P_\infty = 1 + p,$$

These forms are substituted in the basic equations and squares and higher powers of small quantities are neglected. Isentropy gives

$$1 + p = (1 + s)^\gamma = 1 + \gamma s + \cdots$$

or

$$p = \gamma s. \quad (2.1.3)$$

Continuity and momentum are:

$$\frac{\partial s}{\partial t} + \nabla \cdot \mathbf{q} = 0, \quad (2.1.4)$$

$$\rho_\infty \frac{\partial \mathbf{q}}{\partial t} = -P_\infty \nabla p. \quad (2.1.5)$$

Nonlinear convective effects are thus neglected. The kinematic consequence follows from (2.1.5)

$$\frac{\partial \omega}{\partial t} = 0, \quad \omega = \nabla \times \mathbf{q} \equiv \mathbf{curl}\,\mathbf{q} = \text{vorticity.} \quad (2.1.6)$$

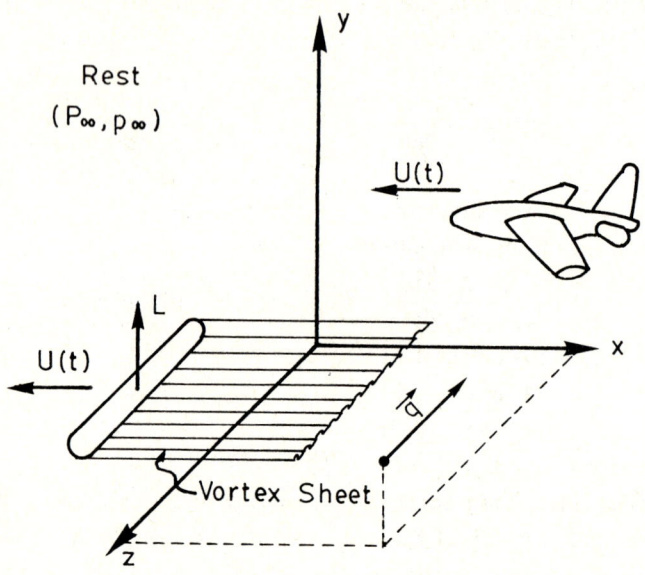

Figure 2.1.1

Coordinates for acoustics

The vorticity at any point in the flow cannot vary with time. Since we assume no distributed vorticity initially there can never be any distributed vorticity. In aerodynamics, however, a concentrated vortex sheet must appear behind a lifting wing. The vorticity so introduced obeys (2.1.6). In general, viscosity is the mechanism by which vorticity is introduced into an aerodynamic flow. For a subsonic trailing edge a Kutta condition is applied that the flow leave the trailing edge smoothly. This is an indirect expression of viscous effects as Reynolds numbers $Re \to \infty$ and makes possible the unique specification of the flow with a vortex sheet. At a supersonic trailing edge this condition of smooth exit is taken care of by a trailing edge wave system. In linearized theory, an edge can be classified as supersonic or subsonic according to the undisturbed component of flow normal to the edge.

Thus a perturbation velocity potential exists such that

$$\mathbf{q} = \nabla\phi, \quad \phi = \phi(x, y, z, t). \tag{2.1.7}$$

The momentum equation reads $\nabla\left(\rho_\infty \frac{\partial\phi}{\partial t} + P_\infty p\right) = 0$ so that integration yields

$\rho_\infty \frac{\partial \phi}{\partial t} + P_\infty p = f(t) = 0$, since disturbances vanish at infinity. Thus we have a linearized Bernoulli equation

$$p = \gamma s = -\frac{\rho_\infty}{P_\infty} \frac{\partial \phi}{\partial t} \quad \text{or} \quad \left\{ \begin{array}{l} P - P_\infty = -\rho \dfrac{\partial \phi}{\partial t}, \\[2mm] s = -\dfrac{1}{a_\infty^2} \dfrac{\partial \phi}{\partial t}, \end{array} \right\} \tag{2.1.8}$$

relating the pressure (and force) field to the potential. The equation for the potential comes from the continuity equation

$$-\frac{1}{a_\infty^2} \frac{\partial^2 \phi}{\partial t^2} + \nabla \cdot \nabla \phi = 0, \quad a_\infty^2 = \frac{\gamma P_\infty}{\rho_\infty}.$$

Thus we obtain the classical wave equation

$$\nabla^2 \phi - \frac{1}{a_\infty^2} \frac{\partial^2 \phi}{\partial t^2} = \left(\frac{\partial^2 \phi}{\partial x^2} + \frac{\partial^2 \phi}{\partial y^2} + \frac{\partial^2 \phi}{\partial z^2} \right) - \frac{1}{a_\infty^2} \frac{\partial^2 \phi}{\partial t^2} = 0. \tag{2.1.9}$$

The most typical property of solutions of the wave equation is that the effect of a concentrated disturbance spreads isotropically at a finite speed a_∞. This speed a_∞ is a property of the medium and is independent of the nature of the disturbance. A signal at the point P_0 at time $t = 0$ spreads to a distance $a_\infty t$ at time t (Figure 2.1-2). A basic solution illustrating this property is obtained from the spherically symmetric solution of (2.1.9) representing outgoing waves. In spherical coordinates $(R = \sqrt{x^2 + y^2 + x^2}, t)$ the wave equation is

$$\frac{\partial^2 \phi}{\partial R^2} + \frac{2}{R} \frac{\partial \phi}{\partial R} - \frac{1}{a_\infty^2} \frac{\partial^2 \phi}{\partial t^2} = 0. \tag{2.1.10}$$

This can, because of good luck, be written

$$\frac{\partial^2 (R\phi)}{\partial R^2} - \frac{1}{a_\infty^2} \frac{\partial^2 (R\phi)}{\partial t^2} = 0. \tag{2.1.11}$$

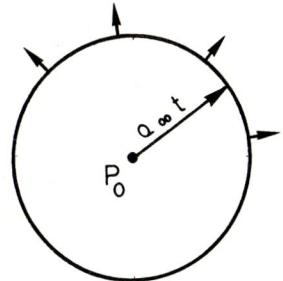

Figure 2.1.2

Spherically spreading disturbance

For only outgoing waves

$$\phi(R,t) = \frac{f(t - R/a_\infty)}{R}, \tag{2.1.12}$$

then the radial velocity $q_R = \frac{\partial \phi}{\partial R}$ has a "near" field and a "far" field

$$\frac{\partial \phi}{\partial R} = -\underbrace{\frac{f(t - R/a_\infty)}{R^2}}_{\text{"near field"}} - \underbrace{\frac{f'(t - R/a_\infty)}{a_\infty R}}_{\text{"far field"}}.$$

The solution can be considered to be produced by a source of fluid at the origin and the "near" field shows an essentially incompressible flow (why?). As $R \to 0$ the outward mass flux (units of ρ_∞) is

$$\lim_{R \to 0} 4\pi R^2 \frac{\partial \phi}{\partial R}(R,t) = -4\pi f(t) = Q(t) = \text{Source Strength} \tag{2.1.13}$$

For the special case of an impulsive source $Q(t) = \delta(t)$ we obtain the fundamental solution S_3, in 3-dimensional space, of the wave equation,

$$\phi(R,t) = S_3 = -\frac{1}{4\pi} \frac{\delta(t - R/a_\infty)}{R}. \tag{2.1.14}$$

This gives the potential at (R,t) due to unit source at $(0,0)$ and is the solution of

$$\frac{\partial^2 \phi}{\partial x^2} + \frac{\partial^2 \phi}{\partial y^2} + \frac{\partial^2 \phi}{\partial z^2} - \frac{1}{a_\infty^2} \frac{\partial^2 \phi}{\partial t^2} = \delta(x)\,\delta(y)\,\delta(z)\,\delta(t) \tag{2.1.15}$$

with $\phi = \phi_t = 0$ at $t = 0-$.

That is, (2.1.9) is a version of the continuity equation and the right hand side can be taken to represent the source strength.

(2.1.14) shows that indeed the propagation is sharp and that all of the disturbance potential is concentrated on $R = a_\infty t$. The corresponding pressure field

$$P - P_\infty = -\rho_\infty \frac{\partial \phi}{\partial t} = \frac{\rho_\infty}{4\pi} \frac{\delta'(t - R/a_\infty)}{R} \tag{2.1.16}$$

shows the arrival of a (singular) compression followed by an expansion (Figure 2.1.3). The superposition (valid because of linearity) of spherical fields can by envelope construction process produce, for example, cylindrical and plane waves (Figure 2.1-4). It is clear then that when a body travels steadily supersonically it out runs its signals. A moving point produces an envelope at the Mach angle θ_M (Figure 2.1.5).

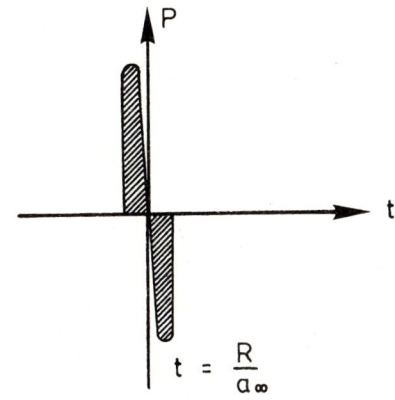

Figure 2.1.3

Pressure signal of an impulsive source

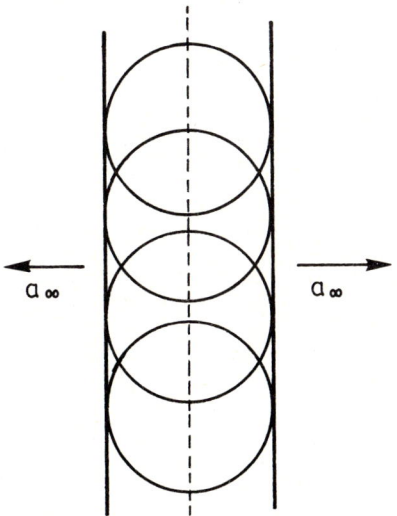

Figure 2.1.4

Envelope construction

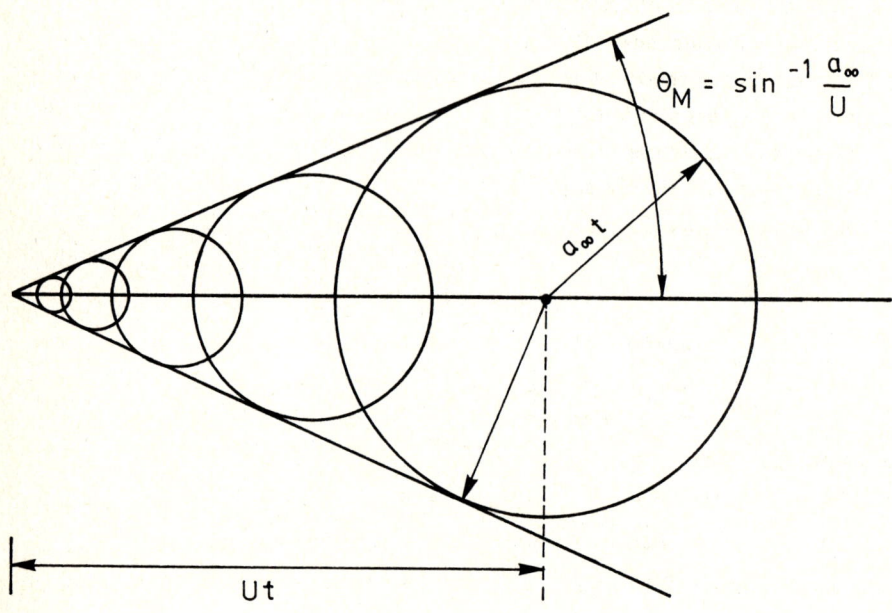

Figure 2.1.5

Supersonically moving disturbance

2.2 Galilean Transformation – Uniform Translation

The simplest solution of linearized theory is that of steady supersonic flow, past an airfoil of zero thickness. For example consider a flat plate at angle of attack $\alpha \ll 1$. In a linear theory it is always possible to decompose the flow about an airfoil into that part due to thickness and that part due to incidence (and camber) but having zero thickness. Thus this simple solution also gives a realistic estimate of the lift. Assume steady supersonic flow in a coordinate system fixed in the moving airfoil (x', y, t')

$$\left\{ \begin{array}{l} x' = x + Ut \\ t' = t \end{array} \right\}.$$

$$(2.2.1)$$

Then derivatives transform as

$$\frac{\partial}{\partial x} \rightarrow \frac{\partial}{\partial x'}, \quad \frac{\partial}{\partial t} \rightarrow \frac{\partial}{\partial t'} + U\frac{\partial}{\partial x'} \rightarrow U\frac{\partial}{\partial x'}$$

for steady flow, and

$$\frac{\partial^2}{\partial t^2} \rightarrow U^2 \frac{\partial^2}{\partial x'^2}, \qquad \frac{\partial^2}{\partial x^2} \rightarrow \frac{\partial^2}{\partial x'^2}.$$

The wave equation (2.1.9) for $\phi(x', y, t')$ thus reads, in the new coordinates,

$$(M_\infty^2 - 1)\frac{\partial^2 \phi}{\partial x'^2} - \frac{\partial^2 \phi}{\partial y^2} = 0.$$

$$(2.2.2)$$

ϕ is now the disturbance potential on a uniform stream.

The boundary conditions are:

(i) tangent flow

$$\frac{\partial \phi}{\partial y} = -\tan \alpha \left(U + \frac{\partial \phi}{\partial x'} \right) \qquad \text{on } y = -\tan \alpha \cdot x',$$

for $0 < x' < \ell$.

Approximating $\tan \alpha \doteq \alpha$, $\frac{\partial \phi}{\partial y}(x', -\alpha x') \doteq \frac{\partial \phi}{\partial y}(x', 0)$, and neglecting squared terms of small quantities a linearized boundary condition applied on the slit $y = 0, 0 < x' < \ell$ is found

$$\frac{\partial \phi}{\partial y}(x', 0) = -U\alpha, \quad 0 < x' < \ell$$

$$(2.2.3)$$

(ii) Radiation condition. No waves from infinity run into the airfoil - waves run downstream as in Figure (2.2.1).

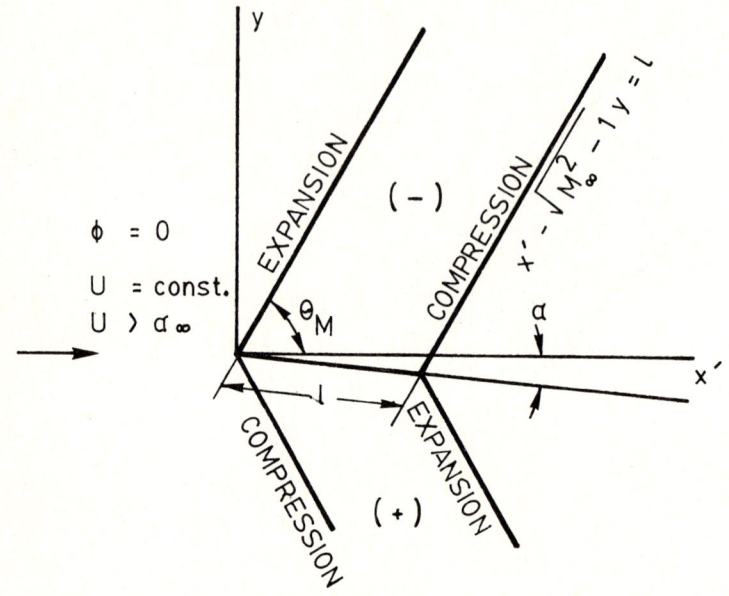

Figure 2.2.1

Wave pattern of supersonic airfoil

The general solution in the wave zone is thus

$$\phi(x', y) = f(x' - \sqrt{M_\infty^2 - 1}\, y) \quad \text{if} \quad y > 0, \quad 0 < x' - \sqrt{M_\infty^2 - 1}\, y < \ell$$
$$= g(x' + \sqrt{M_\infty^2 - 1}\, y) \quad \text{if} \quad y < 0, \quad 0 < x' + \sqrt{M_\infty^2 - 1}\, y < \ell \tag{2.2.4}$$

where f, g are arbitrary functions representing outgoing waves. f is found from the boundary condition

$$\frac{\partial \phi}{\partial y} = -\sqrt{M_\infty^2 - 1}\, f'(x' - \sqrt{M_\infty^2 - 1}\, y)$$

$$= -\sqrt{M_\infty^2 - 1}\, f'(x') \qquad y \to 0$$

$$= -U\alpha \quad \text{on} \quad y = 0,$$

hence,

$$f(x') = \frac{U\alpha}{\sqrt{M_\infty^2 - 1}} x'. \tag{2.2.5}$$

The solution for the wave system above the airfoil is thus

$$\phi = \frac{U\alpha}{\sqrt{M_\infty^2 - 1}} (x' - \sqrt{M_\infty^2 - 1}\, y). \tag{2.2.6}$$

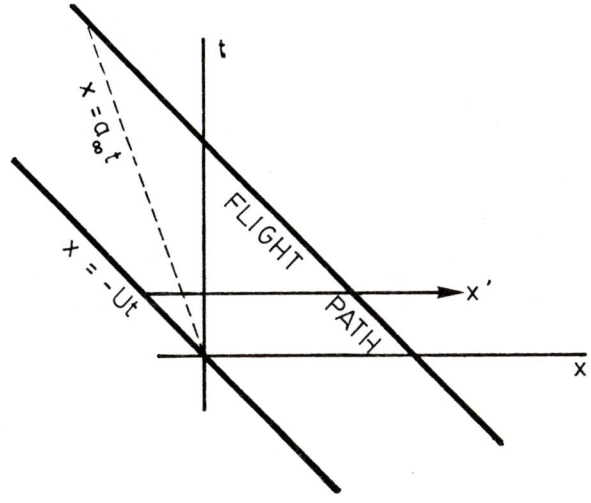

Figure 2.2.2

Representation of flight path

The pressure perturbation over the airfoil is found from (2.1.8) transformed

$$p = -\frac{\rho_\infty}{P_\infty}\frac{\partial\phi}{\partial t} \rightarrow -\frac{U\rho_\infty}{P_\infty}\frac{\partial\phi}{\partial x'} = -\frac{U^2\rho_\infty}{P_\infty}\frac{\alpha}{\sqrt{M_\infty^2 - 1}}$$

$$= \frac{\gamma M_\infty^2\,\alpha}{\sqrt{M_\infty^2 - 1}} < 0$$

(2.2.7)

There is an expansion for $y > 0$ and an equal compression for $y < 0$. The expansion wave at the nose and the compression wave at the tail are of equal intensity so that $\phi = $ constant downstream of the wave zone. The waves at nose and tail are "weak" solutions of the wave equation (2.2.2) in the sense that an integrated version of (2.2.2) holds across these waves. This concept will be discussed in detail later. Since these jumps in velocity and pressure are only finite, ϕ itself remains continuous.

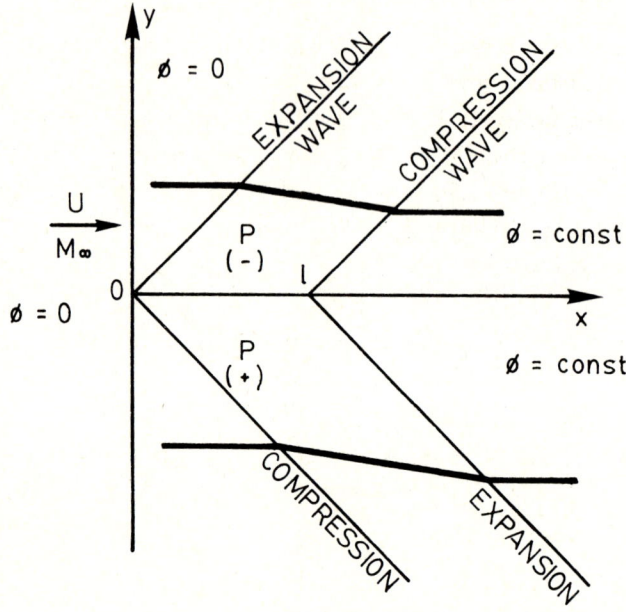

Figure 2.2.3

Supersonic flow past a flat plane

The transonic breakdown of the linearized approximations can be observed from (2.2.6, 7). As $M_\infty \to 1$ the disturbances which were assumed small tend to infinity. The perturbation pressure and lift are infinite. The difficulty is connected with the fact that if a body travels at the sonic speed then in this approximation all disturbances remain with the body. The steady state may not exist.

In order to investigate this point more carefully, we next study the waves produced by a slender body which accelerates to sonic speed and continues to fly exactly at sonic speed.

2.3 Slender Body Theory – Acoustics

The unsteady motion of a slender body along the x-axis can be represented by a distribution of acoustic sources along that axis. The sources flash on when the body passes a given location. The intensity of the sources can be related to the body geometry and motion (Figure 2.3.1). The sources are distributed over the flight path in (x, t) (Figure 2.3.2). Let $Q(x, t) =$ source intensity at x, t (units of ρ_∞). That is $Q =$ rate of mass addition per volume divided by ρ_∞. (cf. Equation 2.1.9). Then, from (2.1.14)

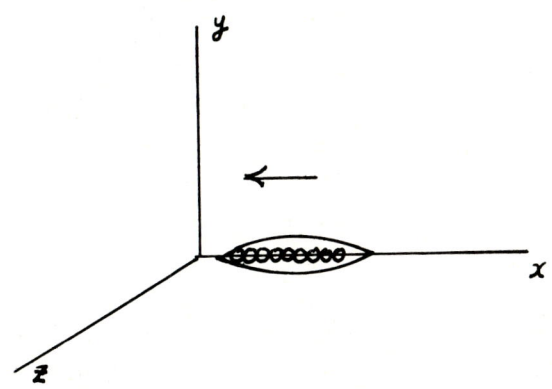

Figure 2.3.1

Slender Body in Motion

with $R = \sqrt{(x - \xi)^2 + y^2 + z^2}$,

$$-4\pi\phi(x, y, z, t) = \int_{-\infty}^{t} d\tau \int_{-\infty}^{\infty} \frac{Q(\xi, \tau)\delta\left(t - \tau - \dfrac{R}{a_\infty}\right)}{R} d\xi . \qquad (2.3.1)$$

Due to the sharp propagation of signals this becomes

$$-4\pi\phi(x,r,t) = \int_{-\infty}^{\infty} \frac{Q\left(\xi, t - \frac{\sqrt{(x-\xi)^2+r^2}}{a_\infty}\right)}{\sqrt{(x-\xi)^2+r^2}} \, d\xi \,, \qquad (2.3.2)$$

where

$$r = \sqrt{y^2 + z^2} \,.$$

This is the usual formula for the retarded potential. Axial-symmetry has been used. The only sources which contributed to the potential at a point (x, r, t) are those which flash on at the retarded time

$$\tau = t - \frac{R}{a_\infty}.$$

These lie on hyperbola

$$\tau = t - \frac{1}{a_\infty}\sqrt{(x-\xi)^2 + r^2}$$

on the flight path (cf. Figure 2.3.2). There are signals from the recent past (x_L, x_T) and there may be signals from the distant past.

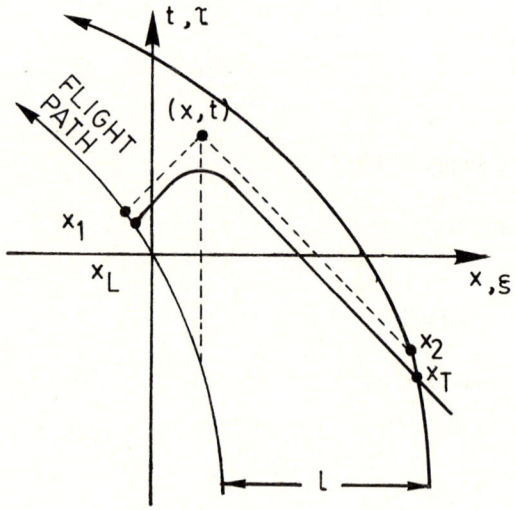

Figure 2.3.2

Dependence domains, accelerating body

For a slender body, a direct local relationship exists between the source strength $Q(x,t)$ and the body shape and motion. To see this the asymptotic expansion of (2.3.2) as $r \to 0$ can be constructed as follows: As $r \to 0$ the main contribution comes from near $x = \xi$. The integral can be broken into three parts.

$$\phi(x,r,t) = \phi_I + \phi_{II} + \phi_{III},$$

and these can be approximated:

$$-4\pi\phi_{II} = \int_{x-\epsilon(r)}^{x+\epsilon(r)} \frac{Q\left(\xi, t - \dfrac{1}{a_\infty}\sqrt{(x-\xi)^2 + r^2}\right)}{\sqrt{(x-\xi)^2 + r^2}} \, d\xi \,,$$

let $\xi = x + r \sinh \sigma$, then

$$-4\pi\phi_{II} = \int_{-\sigma_{II}}^{\sigma_{II}} Q\left(x + r\sinh\sigma \,,\, t - \frac{r}{a_\infty}\cosh\sigma\right) d\sigma \,,$$

$$= \int_{-\sigma_{II}}^{\sigma_{II}} \left\{ Q(x,t) + r\sinh\sigma \cdot Q_x(x,t) - \frac{r}{a_\infty}\cosh\sigma \cdot Q_t(x,t) + \cdots \right\} d\sigma$$

where $\sigma_{II} = \sinh^{-1}\left(\dfrac{\epsilon}{r}\right)$

choose $\epsilon(r)$ such that $\dfrac{\epsilon}{r} \to \infty$ as $r \to \infty$,

$$-4\pi\phi_{II} = 2Q(x,t)\sinh^{-1}\left(\frac{\epsilon}{r}\right) + O(\epsilon)$$

$$= 2Q(x,t)\log\left(\frac{\epsilon}{r} + \sqrt{\frac{\epsilon^2}{r^2} + 1}\right) + O(\epsilon)$$

$$-4\pi\phi_{II} = 2Q(x,t)\log\frac{2\epsilon}{r} + O\left(\epsilon, \frac{r^2}{\epsilon^2}\right) \qquad (2.3.3)$$

For ϕ_I, ϕ_{II} the singularity at $x = \xi$, $r = o$ is outside the interval, so

$$-4\pi\phi_I = \int_{x_L(x,r,t)}^{x-\epsilon(r)} \frac{Q\left(\xi, t - \dfrac{x-\xi}{a_\infty}\right)}{x-\xi} \, d\xi + \cdots$$

where $x_L(x,r,t)$, $x_T(x,r,t)$ are the intersections of the backward cone (hyperbola $t - \dfrac{R}{a_\infty} = \tau$) with the leading and trailing edges of the flight path. Partial integration shows

$$-4\pi\phi_I = Q\left(x_1, t - \frac{x-x_1}{a_\infty}\right)\log(x-x_1) - Q(x,t)\log\epsilon$$

$$- \int_{x_1}^{x} \log(x-\xi)\left\{ Q_x\left(\xi, t - \frac{x-\xi}{a_\infty}\right) + \frac{1}{a_\infty}Q_t\left(\xi, t - \frac{x-\xi}{a_\infty}\right) \right\} d\xi \,(2.3.4)$$

similarly

$$-4\pi\phi_{III} = \log(x_2 - x)Q\left(x_2, t - \frac{x_2 - x}{a_\infty}\right) - Q(x,t)\log\varepsilon$$

$$+ \int_x^{x_2} \log(\xi - x)\left\{Q_x\left(\xi, t - \frac{\xi - x}{a_\infty}\right) - \frac{1}{a_\infty}Q_t\left(\xi, t - \frac{\xi - x}{a_\infty}\right)\right\} d\xi,$$

where

$$x_{1,2}(x,t) = x_{L,T}(x,0,t) \qquad (2.3.5)$$

The final result is the slender body expansion for axial symmetry

$$-4\pi\phi(x,r,t) = 2Q(x,t)\log\frac{2}{r} + Q_1\log(x - x_1)$$

$$+ Q_2\log(x_2 - x) + \int_{x_1}^x \log(x - \xi)\left\{Q_x + \frac{1}{a_\infty}Q_t\right\}_{\text{ret}} d\xi$$

$$- \int_x^{x_2} \log(\xi - x)\left\{Q_x - \frac{1}{a_\infty}Q_t\right\}_{\text{ret}} d\xi$$

$$+ O(r^2\log r) \qquad (2.3.6)$$

where

$$Q_1 = Q\left(x_1, t - \frac{x - x_1}{a_\infty}\right), \quad Q_2 = Q\left(x_2, t - \frac{x_2 - x}{a_\infty}\right),$$

$$\{f\}_{\text{ret}} = f\left(\xi, t - \frac{|x - \xi|}{a_\infty}\right),$$

$$\left.\begin{array}{l} x_1(x,t) = x_L(x,0,t) \\ x_2(x,t) = x_T(x,0,t) \end{array}\right\} \qquad \begin{array}{l}\text{Intersection of backward cone} \\ \text{of signals and flight path.}\end{array}$$

Further terms in this asymptotic expansion are easily found by solving the wave equation (2.1.9) recursively

$$\frac{\partial^2\phi}{\partial r^2} + \frac{1}{r}\frac{\partial\phi}{\partial r} = \frac{1}{a_\infty^2}\frac{\partial^2\phi}{\partial t^2} - \frac{\partial^2\phi}{\partial x^2} \qquad (2.3.7)$$

using (2.3.6) in the right hand side. The first term $(\log r)$ of (2.3.6) represents the flow due to a source in a cross-section plane and the next terms a system of plane waves, a function of (x,t) produced at a distance from (x,r,t), near $r = 0$. This imcompressible source in a cross-plane is related to the near-field of the point source mentioned before.

Now the source strength $Q(x,t)$ can be related to body-geometry and motion by considering the boundary condition of tangent flow on the surface. If

$$B(x,r,t) = 0 = r - r_b(x,t) , \quad r = \sqrt{y^2 + z^2} \qquad (2.3.8)$$

is the body shape the general boundary condition reads

$$\frac{\partial B}{\partial t} + \mathbf{q} \cdot \nabla B = 0 \quad \text{on} \quad B = 0 . \qquad (2.3.9)$$

Thus

$$-\frac{\partial r_b}{\partial t} - \phi_x \frac{\partial r_b}{\partial x} + \phi_r = 0 \quad \text{on} \quad r = r_b$$

but neglecting the small terms we have

$$\frac{\partial \phi}{\partial r}(x, r_b, t) = \frac{\partial r_b}{\partial t} \qquad (2.3.10)$$

From (2.3.6) thus (valid since $r_b \to 0$)

$$\frac{1}{2\pi} \frac{Q(x,t)}{r_b} = \frac{\partial r_b}{\partial t}$$

or

$$\boxed{Q(x,t) = \frac{\partial A_b(x,t)}{\partial t}} \qquad (2.3.11)$$

where $A_b(x,t)$ = cross-section area of body. Note that for a body of fixed geometry in variable motion

$$A_b(x,t) = A_b(X) \qquad 0 < X < \ell$$

where

$$X = x + \int_0^t U(\tau)\, d\tau = \text{coordinate fixed in body} .$$

Thus

$$\frac{\partial A_b}{\partial t} = A_b'(X)\frac{\partial X}{\partial t} = U(t)A_b'(X) . \qquad (2.3.12)$$

2.3.1 Instantaneous Acceleration to Sonic.

In order to see how waves accumulate we can calculate the pressure field near a body which starts to move at sonic speed at $t = 0$ and then continues to fly steadily at that speed.

In this case, applying the formulas of the previous section

$$x_2 = x + a_\infty t = X \quad \text{(say)} = \text{coordinate fixed in body}$$
$$x_1 = \frac{x - a_\infty t}{2} \qquad (2.3.13)$$

(See Figure 2.3.3). For this problem x_2 is no longer located at the tail but comes from the initial time $t = 0$.

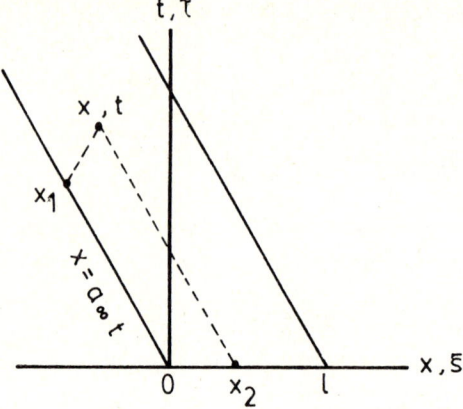

Figure 2.3.3

Domains of dependence–sudden start at sonic speed.

$$A_b(x,t) = A_b(X) ,$$

$$Q(x,t) = \frac{\partial A_b}{\partial t} = a_\infty A_b'(X) , \qquad (2.3.15)$$

$$Q_x + \frac{1}{a_\infty} Q_t = 2a_\infty A_b''(X) ,$$

$$Q_x - \frac{1}{a_\infty} Q_t = 0 .$$

Note also that for a pointed body

$$Q_1 = a_\infty A_b'(0) = 0$$

Thus the basic slender body formula for the potential (2.3.6) becomes, (valid for t larger than x/a_∞)

$$\phi(x,r,t) = \frac{a_\infty}{2\pi} A_b'(X) \log \frac{r}{2} - \frac{a_\infty}{4\pi} A_b'(X) \log(a_\infty t)$$

$$+ \frac{a_\infty}{2\pi} \int_{\frac{x-a_\infty t}{2}}^{x} \log(x - \xi) A_b''(2\xi + a_\infty t - x) \, d\xi + \cdots . \quad (2.3.16)$$

The integral in (2.3.16) can be transformed by letting

$$\sigma = 2\xi + a_\infty t - x , \quad \xi = \frac{\sigma}{2} + \frac{x - a_\infty t}{2} , \quad x - \xi = \frac{X - \sigma}{2} .$$

This shows that the integral depends only on X. Then

$$\phi(x,r,t) = \frac{a_\infty}{2\pi} A_b'(X) \log \frac{r}{2} - \frac{a_\infty}{4\pi} A_b'(X) \log(a_\infty t)$$

$$+ \frac{a_\infty}{4\pi} \int_0^X \log\left(\frac{X-\sigma}{2}\right) A_b''(\sigma)\, d\sigma \ ^*, \quad X = x + a_\infty t. \quad (2.3.17)$$

From (2.3.17) we can see that both the potential and the pressure disturbance grow logarithmically at $t \to \infty$ following the body (X fixed). In fact

$$P - P_\infty \to \frac{\rho_\infty a_\infty^2}{4\pi} A_b''(X) \log(a_\infty t) + \cdots . \quad (2.3.18)$$

We see, that due to the accumulation of waves for a long time the potential, which was assumed small, becomes logarithmically infinite. A steady state is never reached, and the basic assumptions of acoustics are violated.

This is true even though the drag may be calculated to be finite. The result here is not special and does not depend on the sudden start at sonic speed. In fact the same result occurs for a body which accelerates smoothly to sonic speed and then continues to fly steadily at sonic speed.

Naturally then, acoustic theory cannot give a good description of the flow in some neighborhood of flight near sonic speed. A more accurate theory is needed which takes into account how the speed of waves changes when the local state changes. This could be found from the solution of a second-order acoustic equation, one in which all quadratic terms are included[**]:

$$\frac{1}{a_\infty^2}\left(1 + (\gamma - 1)\frac{\phi_t}{a_\infty^2}\right)\phi_{tt} + \frac{1}{a_\infty^2}\frac{\partial}{\partial t}(\nabla\phi)^2 - \nabla^2\phi = 0. \quad (2.3.19)$$

The solution of this equation is enormously more complicated than that of the wave equation. Transonic theory, to be discussed later, provides a systematic simplification of (2.3.19).

---

[*]Note that this formula can be written

$$\phi(x,r,t) = \frac{a_\infty}{4\pi} \int_0^X \log\left(\frac{r^2}{a_\infty t(X - \sigma)}\right) A_b''(\sigma)\, d\sigma$$

in accordance with dimensional reasoning.

[**] J. D. Cole "Acceleration of Slender Bodies of Revolution Through Sonic Velocity" J. of Appl. Physics V. 26, No. 3, pp. 322-327, Mar. 1955.

In order to understand the validity of linear theory near $M_\infty = 1$, we can study second-order corrections for steady flow. This is done after the exact equations are studied in the next section.

Problem P 2.3.1

Show that for a two-dimensional airfoil which accelerates to sonic $p \sim \sqrt{t}$ as $t \to \infty$.

  Note: This problem can be worked for an airfoil which has a symmetric thickness distribution by a distribution of acoustic sources in the (x,t) plane, $y = 0$. The source potential for two-dimensions is

$$S_2 = \frac{1}{2\pi} \frac{1}{\sqrt{t^2 - \frac{x^2+y^2}{a_\infty^2}}} \quad \text{for } t > \frac{\sqrt{x^2+y^2}}{a_\infty}$$

$$= 0 \ \text{ elsewhere.}$$

2.4 Exact Equations of Planar Flow; Shock Waves and Entropy Jump.

The mathematical expansion procedure for linearised and second order theory will be discussed in the next section and following that the transonic expansion method (Section 3). For both these it is useful to have an exact equation of motion for a potential Φ (at least to a certain order in vorticity). The derivation of such an equation is given here starting from the basic principles of conservation of mass, momentum, and energy. For simplicity steady flow in an (x, y) plane is considered but it is easy to generalize the results to three dimensions.

In conservation form the basic equations are

continuity
$$\frac{\partial}{\partial x}(\rho q_x) + \frac{\partial}{\partial y}(\rho q_y) = 0 \quad \text{or} \quad \nabla \bullet \rho \mathbf{q} = 0 \,,$$

$$
\left.
\begin{aligned}
x - \text{momentum} \quad & \frac{\partial}{\partial x}(\rho q_x^2 + P) + \frac{\partial}{\partial y}(\rho q_x q_y) = 0 \\[2mm]
y - \text{momentum} \quad & \frac{\partial}{\partial x}(\rho q_x q_y) + \frac{\partial}{\partial y}(\rho q_y^2 + P) = 0
\end{aligned}
\right\} \quad \text{or} \quad \text{div}(\rho \mathbf{q} \bullet \mathbf{q} + PI) = 0 \,,
$$

$$(2.4.1)$$

energy
$$
\frac{\partial}{\partial x}\left\{ \left(\frac{1}{2}\rho q^2 + \frac{P}{\gamma - 1} \right) q_x + P q_x \right\}
$$
$$
+ \frac{\partial}{\partial y}\left\{ \left(\frac{1}{2}\rho q^2 + \frac{P}{\gamma - 1} \right) q_y + P q_y \right\} = 0 \,.
$$

$$\mathbf{q} = \mathbf{q}(x, y) = (q_x, q_y), \qquad q^2 = q_x^2 + q_y^2 \,,$$

$$\rho e = \text{internal energy per unit volume} = \rho c_v T = \frac{c_v}{R} P = \frac{P}{\gamma - 1} \,,$$

$$
\gamma = \frac{c_p}{c_v} = \text{ratio of specific heats} \quad \gamma = \frac{7}{5} \quad \text{diatomic} \,,
$$
$$
= \frac{5}{3} \quad \text{monatomic} \,,
$$

$$\frac{P}{\gamma - 1} + \frac{1}{2}\rho q^2 = \text{total gas energy per volume} \,.$$

Shock jump conditions are the integrated form of the conservation laws (2.4.1). That is, locally, the jumps in pressure, density and velocity are given by (see

Figure 2.4.1).

$$[\rho q_x]dy_s - [\rho q_y]dx_s = 0 \,,$$
$$[\rho q_x^2 + P]dy_s - [\rho q_y q_x]dx_s = 0 \,,$$
$$[\rho q_x q_y]dy_s - [\rho q_y^2 + P]dx_s = 0 \,, \qquad (2.4.2)$$
$$\left[\frac{1}{2}\rho q^2 + \frac{P}{\gamma - 1}q_x + Pq_x\right]dy_s - \left[\frac{1}{2}\rho q^2 + \frac{P}{\gamma - 1}q_y + Pq_y\right]dx_s = 0 \,.$$

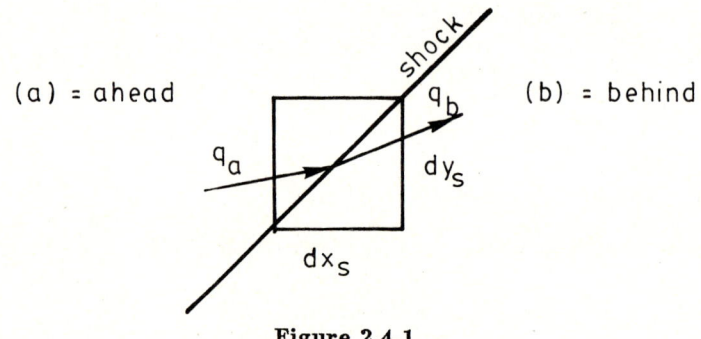

(a) = ahead   (b) = behind

Figure 2.4.1
Local shock element

The notation used here is:

$(\quad)_a$ = value ahead of shock

$(\quad)_b$ = value behind shock

$[\quad]$ = jump in quantity; e.g. $[\rho q_x] = (\rho q_x)_b - (\rho q_x)_a$

In this sense the shock jump conditions are contained in the differential equations (2.4.1).

The discontinuity of inviscid flow theory which represents a shock is an idealization of a very thin (compared to the characteristic size of the problem) zone in which dissipative processes such as viscosity and thermal conductivity are important. The flow passes through this thin zone from one uniform zone to another. If the viscous stresses and thermal fluxes are added to (2.4.1) they appear as divergences. Their contribution drops out on integration across the zone from (a) to (b). The memory of the dissipative process remains in the increase of entropy discussed below.

Certain kinematic forms are also useful. The momentum equations can be written

$$\rho\left(q_x\frac{\partial q_x}{\partial x} + q_y\frac{\partial q_x}{\partial y}\right) = -\frac{\partial P}{\partial x}\ ,$$

$$\rho\left(q_x\frac{\partial q_y}{\partial x} + q_y\frac{\partial q_y}{\partial y}\right) = -\frac{\partial P}{\partial y}\ ,$$

or

$$\rho\mathbf{q}\bullet\nabla\mathbf{q} = -\nabla P\ . \tag{2.4.3}$$

In vector invariant form we have

$$\rho\left(\nabla\left(\frac{q^2}{2}\right) - \mathbf{q}\times\omega\right) = -\nabla P\ ,$$

where

$$\omega = \operatorname{curl}\mathbf{q} = \nabla\times\mathbf{q} = \text{vorticity}\ .$$

The derivative along a streamline $\psi = $ constant is (see Figure 2.4.2),

$$\mathbf{q}\bullet\nabla = q_x\frac{\partial}{\partial x} + q_y\frac{\partial}{\partial y}\ .$$

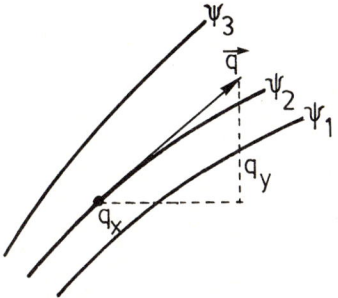

Figure 2.4.2

Streamlines

Thus, using the continuity equation it is easy to show that the specific total enthalpy is conserved along a streamline. The energy equation of (2.4.1) can be rewritten, as

$$\frac{\partial}{\partial x}\left\{\rho q_x\left(\frac{1}{2}q^2 + \frac{\frac{\gamma P}{\rho}}{\gamma - 1}\right)\right\} + \frac{\partial}{\partial y}\left\{\rho q_y\left(\frac{1}{2}q^2 + \frac{\frac{\gamma P}{\rho}}{\gamma - 1}\right)\right\} = 0 \tag{2.4.4}$$

or defining

$$h = \text{total specific enthalpy} = \frac{1}{2}q^2 + \frac{a^2}{\gamma - 1} \; ;$$

$$a^2 = \gamma\frac{P}{\rho} = \gamma RT = (\text{sound speed})^2 \; ;$$

$$\frac{\partial}{\partial x}\{\rho q_x h\} + \frac{\partial}{\partial y}\{\rho q_y h\} = 0 \; , \tag{2.4.5}$$

or

$$\nabla\bullet(\rho\mathbf{q}h) = 0 \; .$$

The kinematic form is (using $\nabla\bullet\rho\mathbf{q} = 0$)

$$q_x\frac{\partial h}{\partial x} + q_y\frac{\partial h}{\partial y} \equiv \mathbf{q}\bullet\nabla h = 0 \; . \tag{2.4.6}$$

This shows that $h = $ constant along a streamline. We can also see how h varies across a shock wave from the conservation form (2.4.5)

$$[\rho q_x h]dy_s - [\rho q_y h]dx_s = 0 \; . \tag{2.4.7}$$

The jumps in (2.4.7) can be expanded by using some simple rules from the calculus of jumps. If f, g are quantities with jumps then it is easy to verify that

$$[fg] = [f]\langle g\rangle + \langle f\rangle[g] \; ,$$
$$\langle fg\rangle = \langle f\rangle\langle g\rangle + \frac{1}{4}[f][g] \; , \tag{2.4.8}$$
$$[f^2] = 2\langle f\rangle[f] \; .$$

The jump in fgh can be calculated by repeated application of these formulas and so on. Here

$$[f] = f_b - f_a \; , \langle f\rangle = \text{average of} \quad f = \frac{1}{2}\{f_b + f_a\} \; .$$

Applying this to (2.4.7), we see that

$$\{[\rho q_x]\langle h\rangle + \langle\rho q_x\rangle[h]\}dy_s - \{[\rho q_y]\langle h\rangle + \langle\rho q_y\rangle[h]\} \, dx_s = 0 \; .$$

Using the jumps for the continuity equation in (2.4.2) it follows that

$$(\langle\rho q_x\rangle dy_s - \langle\rho q_y\rangle dx_s)[h] = 0 \; ,$$

or that

$$[h] = 0 \, . \tag{2.4.9}$$

Since h is constant along a streamline and does not jump across a shock we have the integral $h =$ constant along a streamline, even across shock waves. The constant can be evaluated if we assume a uniform state somewhere, for example at upstream infinity

$$\mathbf{q} = U\mathbf{i}_x \, , P = P_\infty \, , \rho = \rho_\infty \, ,$$

then the total enthalpy integral is

$$\boxed{\frac{q^2}{2} + \frac{a^2}{\gamma - 1} = \frac{U^2}{2} + \frac{a_\infty^2}{\gamma - 1}} \, , \quad a_\infty^2 = \frac{\gamma P_\infty}{\rho_\infty} = \gamma R T_\infty \, . \tag{2.4.10}$$

Since the entropy is intimately connected to vorticity in the flow (see below) it is useful also to see how the entropy varies along a streamline and across shock waves. From (2.4.6) we have

$$\mathbf{q} \cdot \nabla \left(\frac{q^2}{2} + \frac{\frac{\gamma P}{\rho}}{\gamma - 1} \right) = 0 \, ,$$

and from the momentum equation (2.4.3) it follows that

$$\mathbf{q} \cdot \nabla \left(\frac{q^2}{2} \right) = -\frac{\mathbf{q} \cdot \nabla P}{\rho} \, , \quad \text{since} \quad \mathbf{q} \cdot (\mathbf{q} \times \omega) = 0 \, ,$$

thus

$$-\frac{\mathbf{q} \cdot \nabla P}{\rho} + \mathbf{q} \cdot \nabla \left(\frac{\frac{\gamma P}{\rho}}{\gamma - 1} \right) = 0 \, ,$$

or

$$\frac{\mathbf{q} \cdot \nabla P}{P} - \frac{\gamma \mathbf{q} \cdot \nabla \rho}{\rho} = \mathbf{q} \cdot \nabla \log \left(\frac{P}{\rho^\gamma} \right) = 0 \, .$$

Since the specific entropy S for a perfect gas is given by

$$S - S_\infty = c_v \log \frac{P}{P_\infty} \left(\frac{\rho_\infty}{\rho} \right)^\gamma , \text{or} \quad \frac{P}{\rho^\gamma} = \frac{P_\infty}{\rho_\infty^\gamma} e^{\frac{S - S_\infty}{c_v}} \, , \tag{2.4.11}$$

we see that

$$\mathbf{q} \cdot \nabla S = 0 \, . \tag{2.4.12}$$

S is also constant along a streamline, but we do not have an overall conservation for S. We know, that due to dissipative processes in the interior of the shock wave, the entropy of the gas must increase on passing through a shock wave,

$$[S]_s > 0 . \tag{2.4.13}$$

In general the shock waves occurring in transonic flow are relatively weak. We now calculate the entropy jump across a shock in terms of the shock strength

$$\epsilon = \frac{q_a - q_b}{q_a} = -\frac{[q]}{q_a} , \quad \text{where} \quad q_x \equiv q . \tag{2.4.14}$$

It is sufficient to consider a normal shock wave (Figure 2.4-3) since for any shock the velocity can be resolved into normal and tangential components. The tangential component is preserved through a shock. The normal component measures the shock strength and entropy rise.

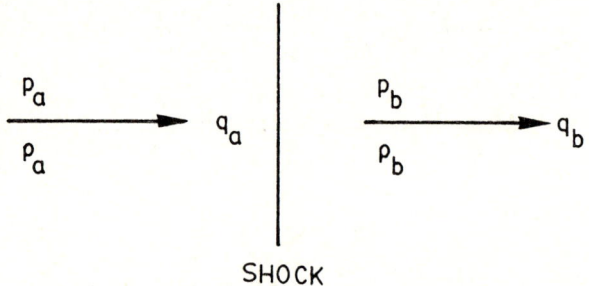

SHOCK

Figure 2.4.3

Shock

Since

$$[\rho q] = 0 , \quad \text{then} \quad \rho_b q_b = \rho_a q_a ,$$

or

$$\frac{\rho_a}{\rho_b} = 1 - \epsilon . \tag{2.4.15}$$

Momentum conservation (2.4.2) is

$$[P + \rho q^2] = 0 = P_b + \rho_b q_b^2 - P_a - \rho_a q_a^2 ,$$

or

$$P_b - P_a = \rho_a q_a (q_a - q_b) = \rho_a q_a^2 \epsilon . \tag{2.4.16}$$

Since

$$[h] = 0$$

$$\frac{1}{2}q_a^2 + \frac{a_a^2}{\gamma - 1} = \frac{1}{2}q_b^2 + \frac{a_b^2}{\gamma - 1} \ .$$

Now,

$$a_b^2 = \frac{\gamma P_b}{\rho_b} = \frac{\gamma(P_a + \rho_a q_a^2 \varepsilon)}{\rho_a}(1 - \varepsilon) = (a_a^2 + \gamma q_a^2 \varepsilon)(1 - \varepsilon) \ .$$

Thus, the total enthalpy conservation implies

$$\frac{1}{2}q_a^2 + \frac{a_a^2}{\gamma - 1} = \frac{1}{2}q_a^2(1 - \varepsilon)^2 + \frac{1}{\gamma - 1}(a_a^2 + \gamma q_a^2 \varepsilon)(1 - \varepsilon) \ ,$$

or

$$0 = q_a^2\left(-\varepsilon + \frac{\varepsilon^2}{2}\right) + \frac{1}{\gamma - 1}\{-\varepsilon a_a^2 + \gamma q_a^2(\varepsilon - \varepsilon^2)\} \ ,$$

$$0 = M_a^2\left\{\frac{\gamma}{\gamma - 1}(1 - \varepsilon) - 1 + \frac{\varepsilon}{2}\right\} - \frac{1}{\gamma - 1} \ ,$$

where

$$M_a = \frac{q_a}{a_a} = \text{shock (normal) Mach number.}$$

Finally

$$M_a^2 = \frac{1}{1 - \frac{\gamma + 1}{2}\varepsilon} \ . \tag{2.4.17}$$

Thus the range of ε is $0 \le \varepsilon \le 2/(\gamma + 1)$. For $\varepsilon > 0$, this shows that $M_a > 1$, the shock wave travels faster than a sound wave relative to the flow ahead. The fastest shock possible occurs when

$$\varepsilon = \frac{2}{\gamma + 1} \ , \quad M_a \to \infty \ , \quad \frac{\rho_b}{\rho_a} = \frac{\gamma + 1}{\gamma - 1} \ .$$

The expression for the entropy jump is now easily found

$$\frac{1}{c_v}[S] = \log\frac{P_b}{P_a}\left(\frac{\rho_a}{\rho_b}\right)^\gamma \ . \tag{2.4.18}$$

But, in terms of (cf. 2.4.16) (2.4.17)

$$\frac{P_b}{P_a} = 1 + \frac{\rho_a q_a^2}{P_a}\varepsilon = 1 + \gamma M_a^2 \varepsilon = 1 + \frac{\gamma\varepsilon}{1 - \frac{\gamma + 1}{2}\varepsilon} = \frac{1 + \frac{\gamma - 1}{2}\varepsilon}{1 - \frac{\gamma + 1}{2}\varepsilon} \ . \tag{2.4.19}$$

Thus

$$\frac{P_b}{P_a}\left(\frac{\rho_a}{\rho_b}\right)^\gamma = \frac{1 + \dfrac{\gamma - 1}{2}\varepsilon}{1 - \dfrac{\gamma + 1}{2}\varepsilon}(1 - \varepsilon)^\gamma \ . \tag{2.4.20}$$

This is an exact expression since we have not yet required ε to be small. We now wish to obtain the approximation to the entropy jump for small values of ε.

$$\frac{1}{c_v}[S] = \log\left(1 + \frac{\gamma - 1}{2}\varepsilon\right) - \log\left(1 - \frac{\gamma + 1}{2}\varepsilon\right) + \gamma\log(1 - \varepsilon) \tag{2.4.21}$$

and as $\varepsilon \to 0$

$$\frac{1}{c_v}[S] = \frac{\gamma - 1}{2}\varepsilon - \frac{1}{2}\left(\frac{\gamma - 1}{2}\varepsilon\right)^2 + \frac{1}{3}\left(\frac{\gamma - 1}{2}\varepsilon\right)^3 - \cdots$$

$$- \left\{-\frac{\gamma + 1}{2}\varepsilon - \frac{1}{2}\left(\frac{\gamma + 1}{2}\varepsilon\right)^2 - \frac{1}{3}\left(\frac{\gamma + 1}{2}\varepsilon\right)^3 - \cdots\right\}$$

$$+ \gamma\left\{-\varepsilon - \frac{\varepsilon^2}{2} - \frac{\varepsilon^3}{3} - \cdots\right\} \ ,$$

$$\frac{1}{c_v}[S] = \frac{1}{2}\left\{-\frac{(\gamma - 1)^2}{4} + \frac{(\gamma + 1)^2}{4} - \gamma\right\}\varepsilon^2$$

$$+ \frac{1}{3}\left\{\frac{(\gamma - 1)^3}{8} + \frac{(\gamma + 1)^3}{8} - \gamma\right\}\varepsilon^3 + \cdots \ ,$$

$$\boxed{\frac{[S]}{c_v} = \frac{\gamma(\gamma^2 - 1)}{12}\varepsilon^3 + O(\varepsilon^4)} \ , \tag{2.4.22}$$

or

$$\frac{[S]}{c_v} = \frac{2}{3}\frac{\gamma(\gamma - 1)}{(\gamma + 1)^2}(M_a^2 - 1)^3 + \cdots \ .$$

For weak shocks the entropy increase is of third-order in the shock strength. This means that to first and second order in ε we can consider transonic flow isentropic

$$\frac{P}{\rho^\gamma} = \frac{P_\infty}{\rho_\infty^\gamma}(1 + O(\varepsilon^3)) \ . \tag{2.4.23}$$

A kinematic consequence follows from Crocco's vortex theorem which we now discuss. The momentum equation (2.4.3) in a form with vorticity explicitly shown is

$$\mathbf{q} \times \omega = \nabla\left(\frac{q^2}{2}\right) + \frac{1}{\rho}\nabla P \ , \qquad \omega = \text{vorticity} = \nabla \times \mathbf{q} \ . \tag{2.4.24}$$

If $h = $ constant everywhere $= \dfrac{q^2}{2} + \dfrac{\gamma\left(\dfrac{P}{\rho}\right)}{\gamma - 1}$,

$$\nabla h = 0 = \nabla\frac{q^2}{2} + \frac{\gamma}{\gamma - 1}\nabla\left(\frac{P}{\rho}\right) .$$

Thus

$$\mathbf{q} \times \omega = \frac{1}{\rho}\nabla P - \frac{\gamma}{\gamma - 1}\nabla\left(\frac{P}{\rho}\right) .$$

$$= \frac{1}{\gamma - 1}\frac{\nabla P}{\rho} + \frac{\gamma}{\gamma - 1}\frac{P}{\rho^2}\nabla\rho = \frac{P}{\rho(\gamma - 1)}\left(\frac{\nabla P}{P} - \gamma\frac{\nabla\rho}{\rho}\right)$$

$$= \frac{RT}{\gamma - 1}\frac{\nabla S}{c_v} , \qquad \gamma = \frac{c_p}{c_v} , \qquad R = c_p - c_v$$

or

$$\boxed{\mathbf{q} \times \omega = -T\nabla S} . \tag{2.4.25}$$

It follows from (2.4.25) that

> (i)   If $\nabla S \neq 0$ then vorticity $\omega \neq 0$,
>
> (ii)   If $\omega = 0$ then $\nabla S = 0$, irrotational flow,

and

> (iii)   If $\nabla S = O(\varepsilon^3)$ then $\omega = O(\varepsilon^3)$.

In general when a shock wave appears in a flow $\nabla S \neq 0$ and distributed vorticity is introduced downstream of the shock waves. The only exceptions are shocks of uniform strength such as plane or conical shocks.

We now derive the equations of motion based on the idea that the vorticity is zero (no shocks) or negligibly small (weak shocks) so that the flow is isentropic (or approximately so). Then a potential $\Phi(x, y, z)$ exists.

$$\omega = \nabla \times \mathbf{q} = \text{curl}\,\mathbf{q} = 0 , \qquad \mathbf{q} = \nabla\Phi . \tag{2.4.26}$$

For basic equations, to order ε^3, or exactly if there are no shocks, we have:

continuity : $\quad \text{div}\,\rho\mathbf{q} \equiv \rho\nabla\bullet\mathbf{q} + \mathbf{q}\bullet\nabla\rho = 0$,

enthalpy integral : $\quad \dfrac{q^2}{2} + \dfrac{a^2}{\gamma - 1} = \dfrac{U^2}{2} + \dfrac{a_\infty^2}{\gamma - 1}$.

entropy constant : $\quad \dfrac{P}{\rho^\gamma} = \dfrac{P_\infty}{\rho_\infty^\gamma} = $ constant.

Note that:

$$a^2 = \frac{\gamma P}{\rho} = \gamma RT$$

so that

$$2\frac{da}{a} = (\gamma - 1)\frac{d\rho}{\rho} = \frac{da^2}{a^2} \ .$$

The continuity equation is thus

$$\mathbf{q} \cdot \nabla(a^2) + (\gamma - 1)a^2 \nabla \cdot \mathbf{q} = 0 \ .$$

But from the enthalpy integral

$$\nabla(a^2) = -\frac{\gamma - 1}{2}\nabla q^2 \ .$$

In terms of flow velocities and sound speeds, we can thus rewrite the basic system:

$$\boxed{\begin{aligned} &a^2 \nabla \cdot \mathbf{q} = \mathbf{q} \cdot \nabla\left(\frac{q^2}{2}\right) \ , \qquad \mathbf{q} = \nabla\Phi \ , \\ &\frac{q^2}{2} + \frac{a^2}{\gamma - 1} = \frac{U^2}{2} + \frac{a_\infty^2}{\gamma - 1} \ . \end{aligned}}$$

(2.4.27)

This formulation is, of course, valid in 2 or 3 dimensions. Note how as $a \to \infty$ the incompressible equation $\nabla^2 \Phi = 0$ is retrieved.

In two dimensions

$$\mathbf{q} \cdot \nabla = \Phi_x \frac{\partial}{\partial x} + \Phi_y \frac{\partial}{\partial y} \ .$$

Thus (2.4.27) becomes

$$\begin{aligned} a^2(\Phi_{xx} + \Phi_{yy}) &= \frac{1}{2}\left(\Phi_x \frac{\partial}{\partial x} + \Phi_y \frac{\partial}{\partial y}\right)(\Phi_x^2 + \Phi_y^2) \\ &= \Phi_x^2 \Phi_{xx} + 2\Phi_x \Phi_y \Phi_{xy} + \Phi_y^2 \Phi_{yy} \ . \end{aligned}$$

In summary, the basic system for $\Phi(x, y)$ is

$$\boxed{\begin{aligned} &\left(a^2 - \Phi_x^2\right)\Phi_{xx} - 2\Phi_x \Phi_y \Phi_{xy} + \left(a^2 - \Phi_y^2\right)\Phi_{yy} = 0 \ , \\ &\frac{1}{2}(\Phi_x^2 + \Phi_y^2) + \frac{a^2}{\gamma - 1} = \frac{1}{2}U^2 + \frac{a_\infty^2}{\gamma - 1} \ . \end{aligned}}$$

(2.4.28)

The system is exact for subsonic flow and whenever there are no shocks. It is adequate also for the basis of a transonic theory with shocks.

Once $\Phi(x, y)$ is found, $\mathbf{q}$, a, P, ρ follow. The basic equation is a version of the continuity equation and could be written as a divergence

$$\nabla \cdot \left(\frac{\rho}{\rho_\infty} \mathbf{q} \right) = 0 = \nabla \cdot \left(\left(\frac{a}{a_\infty} \right)^{\frac{2}{\gamma - 1}} \nabla \Phi \right) = 0 \qquad (2.4.29)$$

We now discuss briefly some properties of (2.4.28). The equation is quasi-linear (really non-linear) and its local type (elliptic or hyperbolic) depends on the local velocity $|\mathbf{q}| = \sqrt{\Phi_x^2 + \Phi_y^2}$. A critical speed a^* is the sound speed when $|\mathbf{q}| = a$, or

$$\left(\frac{1}{2} + \frac{1}{\gamma - 1} \right) a^{*^2} = \frac{\gamma + 1}{2(\gamma - 1)} a^{*^2} = \frac{U^2}{2} + \frac{a_\infty^2}{\gamma - 1} \,.$$

A critical speed q_v is the maximum possible flow speed which occurs when the flow expands to vacuum $a = 0$.

$$\frac{q_v^2}{2} = \frac{U^2}{2} + \frac{a_\infty^2}{\gamma - 1} \,.$$

For $q^2 < a^{*^2}$, then $q^2 < a^2$ and the flow is locally subsonic. For $q_v^2 > q^2 > a^{*^2}$, then $q^2 > a^2$ and the flow is locally supersonic. The regions are indicated in the hodograph (Figure 2.4.4). The type of equation is decided by the discriminant of the associated quadratic form.

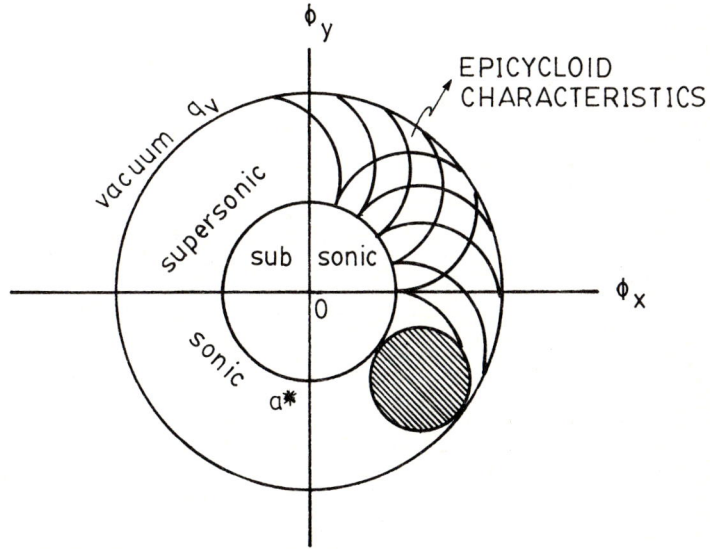

Figure 2.4.4

Schematic hodograph

$$\text{discriminant} = (2\Phi_x\Phi_y)^2 - 4(a^2 - \Phi_x^2)(a^2 - \Phi_y^2)$$
$$= 4a^2(\Phi_x^2 + \Phi_y^2 - a^2)$$
$$= 4a^2(q^2 - a^2).$$

Thus, the equation is of locally hyperbolic type and has real characteristics if the flow is supersonic $q^2 > a^2$, and is of elliptic type if $q^2 < a^2$. In the physical plane the characteristic curves are Mach lines, bisected by and at the Mach angle Θ_M to the streamline (Figure 2.4.5). The hodograph images are epicycloids (Figure 2.4.4).*

As a typical boundary value problem for the basic equation (2.4.28) we consider the problem of flow past an airfoil.

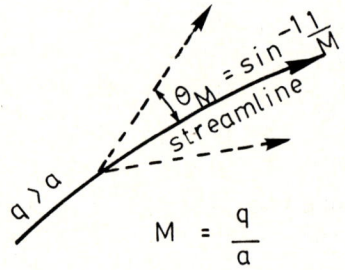

Figure 2.4.5

Local characteristics in supersonic flow

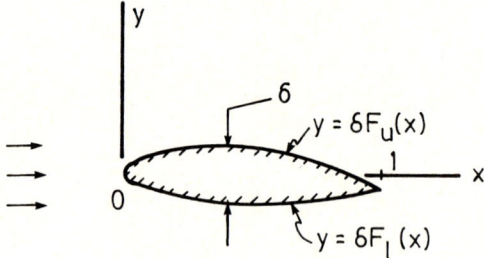

Figure 2.4.6

Flow past an airfoil

---

* See for example any good book on gas dynamics for details. e.g., Liepmann and Roshko "Elements of Gas Dynamics." John Wiley.

We choose the airfoil chord to be 1, the thickness ratio δ and represent the upper and lower surfaces by $F_{u,\ell}$

$$y = \delta F_{u,\ell}(x) \quad \text{for} \quad 0 < x < 1 . \tag{2.4.30}$$

(See Figure 2.4.6)

The boundary conditions are:

(i) Uniform flow along x-axis at infinity,

$$\Phi \to Ux \quad \text{or} \quad (\Phi_x \to U, \Phi_y \to 0) \quad \text{as} \quad x \to -\infty ; \tag{2.4.31}$$

(ii) Tangent flow at each point of the surface.

$$\frac{\Phi_y\big(x, \delta F_{u,\ell}(x)\big)}{\Phi_x\big(x, \delta F_{u,\ell}(x)\big)} = \delta F'_{u,\ell}(x) ; \tag{2.4.32}$$

(iii) Kutta Joukowski condition.

The assumption is made that for well streamlined shapes the flow leaves a locally subsonic trailing edge smoothly. (Figure 2.4.7) This condition is necessary to insure uniqueness. The hidden effects of viscosity are supposed to be the operative mechanism. At a supersonic trailing edge the flow can turn suddenly through an expansion fan or shock wave. The pressure and flow direction of the flow just aft of the trailing edge can be matched.

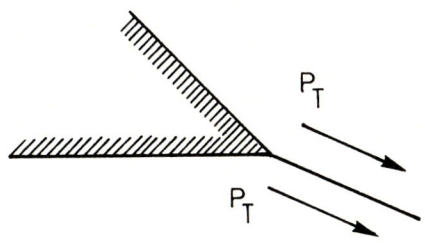

Figure 2.4.7

Trailing Edge Flow

This is exactly how the Kutta condition in general is to be formulated: Just aft of the trailing edge the flow direction and pressure must match for the flow coming from the upper and lower surfaces.

Finally, shock jump conditions such as (2.4.2) must be appended to (2.4.28), with the additional proviso that $[S] > 0$ for the shock waves. This rules out expansion shocks.

With this statement of the problem the solution presumably exists and is unique. Due to the nonlinearity of the problem the possible occurrence of shock waves and their location is not known in advance of the solutions. A similar remark applies to the slipstream or vortex wake which can appear behind a three- dimensional lifting wing. The problem is extremely difficult and essentially analytically intractable. There is no analytic solution known which incorporates a shock and some non-uniform flow, although it is generally true that shock waves occur whenever the flow is supersonic. Certain exceptions are shock-free airfoil flows and purely accelerating flows such as those in nozzles and jets.

Hence there has always been a great interest in approximate methods for this problem and related problems. In many practical problems an approximation based on small disturbances produced by small flow deflections is adequate. This standard procedure as it applies to thin airfoils is discussed in the next section.

2.5 Linearized Theory for Thin Airfoils

In this section, we study linearized theory and its higher corrections from a mathematical point of view in order to understand more precisely the transonic breakdown of linearized theory.

For simplicity consider plane flow past an airfoil of thickness ratio δ. A family of flows is considered, past similar airfoil shapes as $\delta \to 0$. (cf. Figure 2.5.1). An asymptotic expansion is constructed for the exact potential Φ. The various terms in this asymptotic expansion are found by applying the limit: $\delta \to 0$ with (M_∞, x, y) fixed.

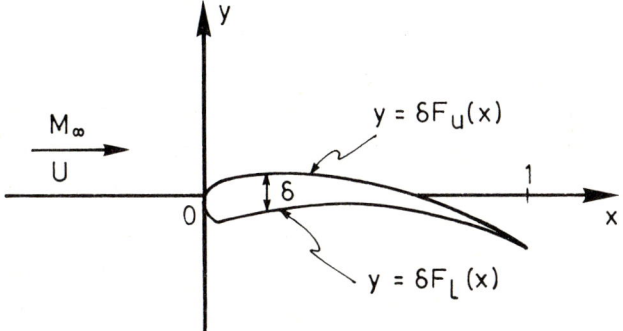

Figure 2.5.1

Family of airfoil shapes

The form is

$$\Phi(x, y; M_\infty, \delta) = U\left\{x + \delta\phi_1(x, y, M_\infty) + \delta^2\phi_2(x, y, M_\infty) + \cdots\right\}. \qquad (2.5.1)$$

To this order, it can again be shown that in the event of supersonic flow vorticity and entropy jumps introduced by shock waves are of no consequence to a first or second order solution. That is, we assume here an airfoil with a sharp leading edge operating so that the bow shock wave is attached to the airfoil. In any case shocks occur in this theory only for $M_\infty > 1$ and enter in a special way (see below).

The form of expansion (2.5.1) consists of the free stream plus small disturbances. The order δ of the first disturbance term is fixed by the order of the upwash Φ_y on the airfoil. The order δ^2 follows naturally if a non-trivial problem is to be found for ϕ_2.

The expansion is substituted into the basic equations (2.4.28) to obtain the succesive approximation equations. We have

$$\frac{1}{U}\Phi_x = 1 + \delta\phi_{1x} + \delta^2\phi_{2x} + \cdots$$

$$\frac{1}{U}\Phi_y = \delta\phi_{1y} + \delta^2\phi_{2y} + \cdots .$$

$$(2.5.2)$$

The total enthalpy integral is

$$\frac{a^2}{U^2} = \frac{1}{M_\infty^2} + \frac{\gamma-1}{2}\left\{1 - \frac{\phi_x^2 + \phi_y^2}{U^2}\right\} \quad , \qquad (2.5.3)$$

or

$$\frac{a^2}{U^2} = \frac{1}{M_\infty^2} - \delta\big\{(\gamma-1)\phi_{1x}\big\} - \delta^2\left\{(\gamma-1)\phi_{2x} + \frac{\gamma-1}{2}\left(\phi_{1x}^2 + \phi_{1y}^2\right)\right\} + \cdots \quad .$$

$$(2.5.4)$$

Thus, the equation for Φ becomes,

$$\left\{\frac{1}{M_\infty^2} - \delta(\gamma-1)\phi_{1x} - \cdots - (1 + 2\delta\phi_{1x}) - \cdots\right\}\left\{\delta\phi_{1xx} + \delta^2\phi_{2xx}\right\}$$

$$- 2\delta\phi_{1y}\delta\phi_{1xy} + \cdots$$

$$+ \left(\frac{1}{M_\infty^2} - \delta(\gamma-1)\phi_{1x} + \cdots\right)\left(\delta\phi_{1yy} + \delta^2\phi_{2yy} + \cdots\right) = 0 \quad .$$

Collecting terms of various orders, we have the first and second approximations.

$$\left(1 - M_\infty^2\right)\phi_{1xx} + \phi_{1yy} = 0 \qquad (2.5.5)$$

$$\left(1 - M_\infty^2\right)\phi_{2xx} + \phi_{2yy} = (\gamma-1)M_\infty^2\phi_{1x}\phi_{1xx}$$

$$+ (\gamma-1)M_\infty^2\phi_{1x}\phi_{1yy} + 2M_\infty^2\phi_{1y}\phi_{1xy}$$

$$= 2M_\infty^2\left\{\left(1 + \frac{\gamma-1}{2}M_\infty^2\right)\phi_{1x}\phi_{1xx} + \phi_{1y}\phi_{1xy}\right\} . \quad (2.5.6)$$

The first equation agrees with that obtained from acoustics by a Galilean transformation (2.2.2). The boundary condition on the airfoil to be attached to each equation comes from the expansion of the exact condition (2.4.32). For example, on the upper surface

$$\delta\phi_{1y}\left(x, \delta F_u\right) + \delta^2\phi_{2y}\left(x, \delta F_u\right) + \cdots$$

$$= \delta F_u'(x)\left\{1 + \delta\phi_{1x}\left(x, \delta F_u\right) = \cdots\right\} \quad \text{for} \quad 0 < x < 1 \quad .$$

Assuming the solution behaves in a regular way, we can further expand the boundary condition about $y = 0+$* to obtain

$$\phi_{1y}(x, 0+) = F'_u(x) \tag{2.5.7}$$

$$\phi_{2y}(x, 0+) = F'_u(x)\phi_{1x}(x, 0+) - F_u(x)\phi_{1yy}(x, 0+) . \tag{2.5.8}$$

A similar treatment can be applied to the lower surface so that the boundary condition is transferred to the slit $y = 0$, $0 < x < 1$. A further boundary condition is that the disturbances vanish at upstream infinity. Also the Kutta condition must be satisfied.

Note that the equation for ϕ_{1x} is always either elliptic $(M_\infty < 1)$ like the Laplace equation or hyperbolic $(M_\infty > 1)$ like a wave equation and that the equation for ϕ_2 has a similar structure. Since both equations are linear, shock waves in the ordinary sense do not appear. Further as long as $M_\infty < 1$ no discontinuities which might approximate shocks can appear since the equations are elliptic. Thus some of the qualitative features necessary for transonic flow are missing.

The defects of this approach can be seen most clearly by considering the solutions to a simple problem. The solution in general proceeds in steps. Once $\phi_1(x, y)$ is found the RHS of (2.5.6) and its boundary condition are known and ϕ_2 can be found.

The simplest problem is that of supersonic flow, when the flow on the upper and lower surfaces are independent. The general solution of (2.5.5) which has waves propagating only downstream is

$$\phi_1 = f\left(x - \sqrt{M_\infty^2 - 1}\, y\right) \quad \text{for} \quad y > 0 \quad , \tag{2.5.9}$$

where f is an arbitrary function (cf. Figure 2.5-2).

---

* This expansion of the boundary conditions is not strictly in accord with the limit process defining the asymptotic expansions. An inner limit valid near the airfoil should be used and matched to the outer flow, but the result is the same because of the smooth behavior of the potentials near $y = 0$.

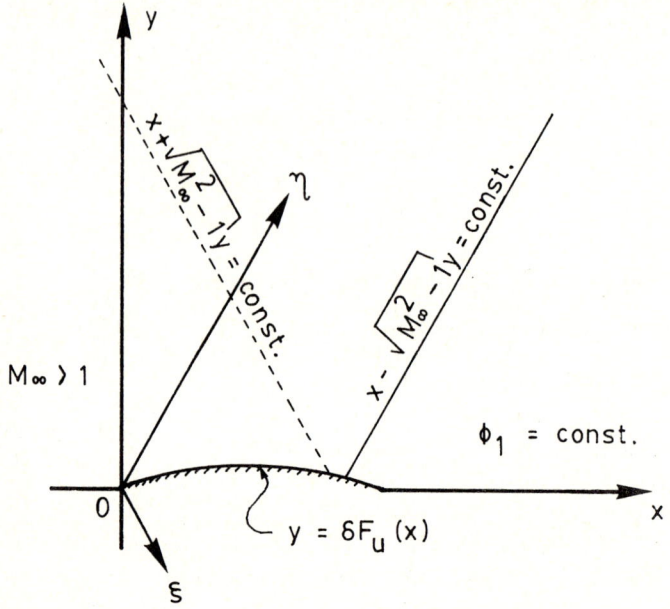

Figure 2.5.2

Linearized supersonic flow

We have that

$$\phi_1 = 0 \quad \text{for} \quad x < \sqrt{M_\infty^2 - 1}\, y$$

ahead of the wave from the nose. From (2.5.7) we find

$$\phi_{1y}(x, 0+) = -\sqrt{M_\infty^2 - 1}\, f'(x) = F_u'(x)$$

so that the solution is

$$\phi_1 = \frac{-F_u\left(x - \sqrt{M_\infty^2 - 1}\, y\right)}{\sqrt{M_\infty^2 - 1}} \qquad (2.5.10)$$

in the wave zone where

$$0 < x - \sqrt{M_\infty^2 - 1}\, y < 1 \quad .$$

Downstream of the tail we can take $\phi_1 = $ constant. Next in order to solve the problem for ϕ_2 it is convenient to introduce characteristic coordinates

$$\xi = x - \sqrt{M_\infty^2 - 1}\, y$$
$$\eta = x + \sqrt{M_\infty^2 - 1}\, y \quad . \tag{2.5.11}$$

We have

$$\phi_x = \phi_\xi + \phi_\eta \quad , \phi_y = -\sqrt{M_\infty^2 - 1}\, \phi_\xi + \sqrt{M_\infty^2 - 1}\, \phi_\eta \quad ,$$
$$\phi_{xx} = \phi_{\xi\xi} + 2\phi_{\xi\eta} + \phi_{\eta\eta} \quad ,$$
$$\phi_{yy} = \left(M_\infty^2 - 1\right)\left(\phi_{\xi\xi} - 2\phi_{\xi\eta} + \phi_{\eta\eta}\right) \quad .$$

Further, the solution ϕ_1 depends only on one characteristic

$$\phi_1 = -\frac{F_u(\xi)}{\sqrt{M_\infty^2 - 1}} \qquad \text{for} \quad 0 < \xi < 1 \quad . \tag{2.5.12}$$

Thus

$$-\left(M_\infty^2 - 1\right)\phi_{2xx} + \phi_{2yy} = -4\left(M_\infty^2 - 1\right)\phi_{2\xi\eta} \quad ,$$

and (2.5.6) becomes

$$\phi_{2\xi\eta} = -\frac{(\gamma + 1)M_\infty^4}{4\left(M_\infty^2 - 1\right)^2} F_u'(\xi)F''(\xi) \quad . \tag{2.5.13}$$

In this form the equation for ϕ_2 can be integrated directly

$$\phi_{2\eta} = g_2'(\eta) - \frac{(\gamma + 1)M_\infty^4}{8\left(M_\infty^2 - 1\right)^2} F_u'^2(\xi) \, ,$$

and

$$\phi_2(\xi, \eta) = f_2(\xi) + g_2(\eta) - \frac{(\gamma + 1)M_\infty^4}{8(M_\infty^2 - 1)^2} F_u'^2(\xi) \, . \tag{2.5.14}$$

The arbitrary functions f_2, g_2 can be determined from the boundary conditions. On the wave from the leading edge ($\xi = 0$) we must have $\phi_2(0, \eta) = 0$ so that

$$g_2(\eta) = \frac{(\gamma + 1)M_\infty^4}{8\left(M_\infty^2 - 1\right)^2}\eta F_u'^2(0) \quad , \tag{2.5.15}$$

if $f_2(0)$ is chosen zero. We assume here that $F'_u(0)$ is finite (pointed airfoil). Thus

$$\phi_2(\xi, \eta) = f_2(\xi) + \frac{(\gamma+1)^2 M_\infty^4}{8\left(M_\infty^2 - 1\right)^2} \eta \left\{ F''^2_u(0) - F'^2_u(\xi) \right\} \tag{2.5.16}$$

in the wave zone. Further, the boundary condition (2.5.8) becomes

$$\phi_{2y}(x, 0+) = \frac{(\gamma+1) M_\infty^4}{8\left(M_\infty^2 - 1\right)^{3/2}} \left\{ F'^2_u(0) - F'^2_u(x) \right\}$$

$$- \sqrt{M_\infty^2 - 1}\, f'_2(x) + \frac{(\gamma+1) M_\infty^4}{4\left(M_\infty^2 - 1\right)^{3/2}} x F'_u F''_u(x)$$

$$= -\frac{F'^2_u(x)}{\sqrt{M_\infty^2 - 1}} - \sqrt{M_\infty^2 - 1}\, F_u(x) F''_u(x) \quad . \tag{2.5.17}$$

Without solving explicitly it can be seen that in its dominant dependence on Mach number $f_2(\xi) \sim \dfrac{1}{(M_\infty^2 - 1)}$ also as $M_\infty \to 1$. Thus, the general form that the linearized expansion (2.5.1) takes ($M_\infty > 1$) is

$$\Phi = U \left\{ x - \delta \frac{F_u(\xi)}{\sqrt{M_\infty^2 - 1}} + \delta^2 \left\{ \frac{(\gamma+1) M_\infty^4}{8\left(M_\infty^2 - 1\right)^2} \eta \left(F'^2_u(0) - F'^2_u(\xi) \right) + \cdots \right\} \right.$$

$$\left. + O(\delta^3) + \cdots \right\} \quad .$$

Difficulties with this asymptotic expansion can be seen when the term that is supposed to be of $O(\delta^2)$ becomes comparable to the term of $O(\delta)$. When $M_\infty \approx 1$ the ratio of these terms, for fixed (x, y), is

$$\frac{\delta}{\left(M_\infty^2 - 1\right)^{\frac{3}{2}}} \quad .$$

This expansion is thus only valid for

$$\frac{\delta}{\left(M_\infty^2 - 1\right)^{\frac{3}{2}}} \ll 1 \quad . \tag{2.5.19}$$

This inequality marks out in a more precise way the boundaries of the transonic region (cf. Figure 2.5-3).

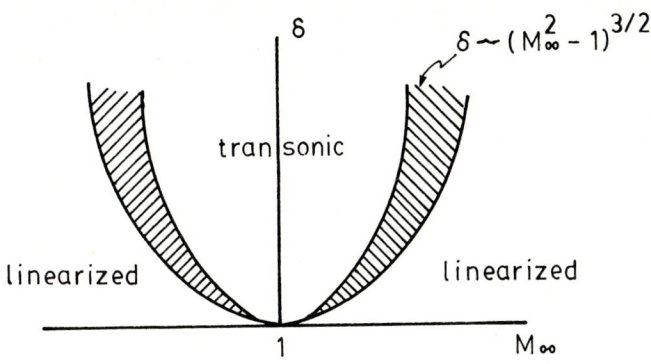

Figure 2.5.3

Approximate flow regimes

A study for $M_\infty < 1$ shows analogously that

$$\delta \ll \left(1 - M_\infty^2\right)^{\frac{3}{2}}$$

is necessary for linearized theory to be satisfactory.

We also note that shock waves $(M_\infty > 1)$ are approximated in this theory by velocity jumps across the characteristics $(\xi = 0)$. A non-uniformity of the expansion also appears near infinity in the wave zone $(\eta \to \infty, 0 < \xi < 1)$. This is a cumulative effect since the shock angle is slightly incorrect and the shock position at infinity is far off.

In the next chapter, we discuss the different expansion procedure that must be used for thin airfoils and bodies in order that it remain valid in the transonic range. In linearized theory it becomes clear that the validity of the expansion involves a relation of (M_∞, δ). A different relation is used below, to construct an expansion valid in the transonic regime.

3. Transonic Expansion Procedures; Simple Solutions, Integral Relations

In this chapter, small disturbance procedures that can be used to derive approximate equations valid in the transonic range are outlined. The expansion procedures are, at first, all based on the fact that a characteristic thickness ratio or deflection angle δ is very small. The linearized theory expansions, which fail in the transonic range, have the free stream Mach number M_∞ fixed as $\delta \to 0$. In order to obtain an expansion valid in the transonic range, we need to consider the simultaneous limit $\delta \to 0, M_\infty \to 1$.

After the expansion procedures for some simple cases are presented some simple solutions will be found. To conclude, in this chapter integral theorems for lift, drag and moment will be derived.

3.1 Expansion Procedure For Steady Flow Past Airfoils.

The simplest starting point is the full potential equation derived in Section 2.4, equation (2.4.28):

$$\left(a^2 - \Phi_x^2\right)\Phi_{xx} - 2\Phi_x\Phi_y\Phi_{xy} + \left(a^2 - \Phi_y^2\right)\Phi_{yy} = 0$$
$$\frac{1}{2}\left(\Phi_x^2 + \Phi_y^2\right) + \frac{a^2}{\gamma - 1} = \frac{1}{2}U^2 + \frac{a_\infty^2}{\gamma - 1} \tag{3.1.1}$$

The problem of flow past an airfoil was also outlined in (2.4) and we follow here the notation in Figure 3.1.1. The coordinates (x, y) are dimensionless with the characteristic length chosen as the airfoil chord.

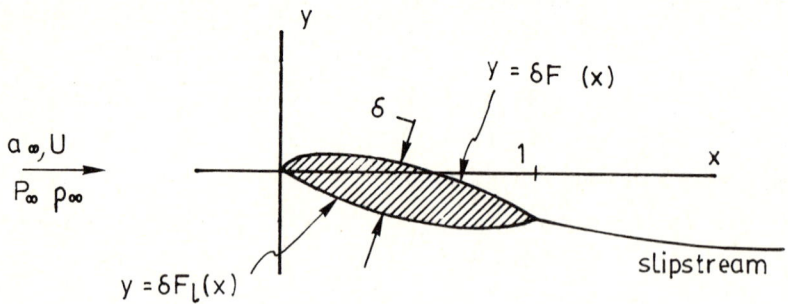

Figure 3.1.1

Airfoil problem

If we wish to specify a family of airfoil shapes with a specific thickness, camberline, and angle of attack, we may write the shape functions defining upper and lower surfaces as

$$\delta F_{u,\ell}(x) = \delta\left\{c(x) \pm t(x) - \frac{\alpha}{\delta}x\right\} \quad , \qquad 0 < x < 1 \quad , \tag{3.1.2}$$

where

$$c(x) = \text{camberline function,}$$

$$t(x) = \text{thickness distribution function,}$$

$$\alpha = \text{angle of attack (in the limit} \quad \delta \to 0, \quad A = \frac{\alpha}{\delta} \quad \text{is fixed).}$$

As discussed in Section 2.4, the full potential equation is valid exactly when the vorticity is zero and approximately valid to a certain order when vorticity is introduced due to the occurence of shock waves. According to (2.4.22), the entropy jump is

$$\frac{1}{c_v}[S] = \frac{\gamma(\gamma^2 - 1)}{12}\epsilon^3 + \cdots \quad = \frac{2}{3}\frac{\gamma(\gamma - 1)}{(\gamma + 1)^2}(M_a^2 - 1)^3 + \cdots \quad , \tag{3.1.3}$$

where M_a = normal Mach number ahead of the shock. An estimate for this quantity [S] and the corresponding vorticity produced will be given when the orders of magnitude for various quantities in transonic flow are established. For this reason, it is not necessary to discuss the shock relations separately at first, and it will be shown below how they are contained in the approximate equations.

The boundary conditions were also discussed in section (2.4) and are summarized here:

(i) uniform flow at upstream infinity; $\Phi_x \to U, \Phi_y \to 0$, (BC1)

(ii) tangent flow;

$$\Phi_y(x, \delta F_{u,\ell}(x)) = \delta F'_{u,\ell}(x)\Phi_x(x, \delta F_{u,\ell}(x)), \text{ (BC2)}$$

(iii) Kutta-Joukowski Condition; (KJ)

Flow direction and pressure match at the tail. Note that due to possible entropy changes a slipstream (where the same pressure corresponds to different velocities) can come from the trailing edge.

In order that the potential be a one-valued function in general it is necessary to consider the (x, y) plane as cut and allow the potential to jump. A convenient

place for the cut is along the slipstream. Of course no appreciable slip can appear
without the idea of a potential being invalid, but the idea of a cut remains due
to the circulation which develops when lift is carrried on the airfoil. The ideas
in general are analogous to those for incompressible flow and play an essential
role in the solutions only for subsonic free-stream Mach numbers. Details appear
below.

As the Mach number $M_\infty \to 1$ linearized theory shows that the disturbance
field has a greater and greater lateral extent, that is, $\partial/\partial y \ll \partial/\partial x$. A formal
mathematical way to express this fact is to use a stretched coordinate

$$\tilde{y} = \beta(\delta)y \tag{3.1.4}$$

where $\beta \to 0$ as $\delta \to 0$ so that $\dfrac{\partial}{\partial y} \to \beta \dfrac{\partial}{\partial \tilde{y}}$. The use of this coordinate is necessary
if a non-trivial transonic equation is to be obtained. Thus we consider a family
of flows which represent small disturbances on a uniform stream and for which

$$\delta \to 0$$

$$M_\infty^2 = 1 - K\mu(\delta) \to 1, \quad \mu(\delta) \to 0$$

and for which

$$(x, \tilde{y}, K)$$

are fixed. K measures the rate at which $M_\infty \to 1$; $K = 0$ corresponds to $M_\infty = 1$.

The form of the expansion is

$$\Phi(x, y; M_\infty, \delta) = U\{x + \varepsilon(\delta)\phi(x, \tilde{y}; K) + \cdots \} \ . \tag{3.1.5}$$

The orders (ϵ, μ, β) must be chosen to give a non-trivial equation which can
describe a transonic flow. The resulting equation must have the possibility of
describing flow which is locally subsonic or locally supersonic. That is, the equation must be able to change from elliptic type, locally resembling the Laplace
equation, to hyperbolic type locally resembling a wave equation. Thus the equation must be non-linear since the regions of local supersonic or local subsonic
flow can not be prescribed in advance. The non-linearity is also necessary to
allow the occurence of shock waves which can approximate the true shock jumps.
Finally we can hope that the equation will be valid in the entire flow field from
the airfoil surface to infinity. However, we must expect a local breakdown of the
approximation near a stagnation point, just as in linearized subsonic theory.

Now we note

$$\frac{\Phi_x}{U} = 1 + \varepsilon\phi_x \quad , \qquad \frac{\Phi_y}{U} = \varepsilon\beta\phi_{\tilde{y}} \ . \tag{3.1.6}$$

Thus (BC2) becomes

$$\varepsilon\beta\phi_{\tilde{y}}\big(x, \beta\delta F_{u,\ell}(x)\big) = \delta F'_{u,\ell}(x)\big\{1 + \varepsilon\phi_x\big\} \quad .$$

As $\delta \to 0$ we can apply this B.C. at $\tilde{y} = 0$ and obtain

$$\phi_{\tilde{y}}(x, 0\pm) = F'_{u,\ell}(x) \ , \qquad 0 < x < 1 \ , \tag{3.1.7}$$

if we choose

$$\boxed{\varepsilon\beta = \delta \ .} \tag{3.1.8}$$

Strictly speaking, the limit process just used does not fit in with our expansion at the boundary, and it should be replaced with an inner limit. This procedure is useful in the axially symmetric case but can be dispensed with here because of the expected smooth behavior of $\phi_{\tilde{y}}$ as $\tilde{y} \to 0$.

Note further

$$\frac{1}{U^2}\Phi_x^2 = 1 + 2\varepsilon\phi_x \ , \tag{3.1.9}$$

$$\frac{1}{U^2}\Phi_y^2 = \delta^2\phi_{\tilde{y}}^2 + \cdots \quad , \tag{3.1.10}$$

$$\frac{1}{U^2}\big(\Phi_x^2 + \Phi_y^2\big) = 1 + 2\varepsilon\phi_x + \cdots \quad , \tag{3.1.11}$$

$$\frac{a^2}{U^2} = \frac{1}{M_\infty^2} + \frac{\gamma - 1}{2}\left(1 - \frac{\Phi_x^2 + \Phi_y^2}{U^2}\right) = 1 + K\mu(\delta) - (\gamma - 1)\varepsilon\phi_x + \cdots \ , \tag{3.1.12}$$

from (3.1.1). Thus the basic full potential equation (3.1.1) becomes

$$\big(1 + K\mu - (\gamma - 1)\varepsilon\phi_x + \cdots \quad - 1 - 2\varepsilon\phi_x\big)\big(\varepsilon\phi_{xx} + \cdots\big) - 2(1 + \cdots)\delta^2\phi_{\tilde{y}}\phi_{x\tilde{y}}$$
$$+ (1 + \cdots)\big(\delta\beta\phi_{\tilde{y}\tilde{y}} + \cdots\big) = 0 \quad .$$

The dominant orders in this equation are $\mu\varepsilon$, ε^2, $\delta\beta$, and the only equation which has a chance to meet the requirements outlined above is obtained when all these orders are equal. This is also a distinguished limit that results in a definite order for $(\mu, \varepsilon, \beta)$.

$$\mu\varepsilon = \varepsilon^2 = \delta\beta \ ,$$
$$\mu = \varepsilon \ , \quad \varepsilon^2 = \delta^2/\varepsilon \ ,$$

so that

$$\boxed{\varepsilon = \delta^{\frac{2}{5}} , \quad \mu = \delta^{\frac{2}{5}} , \quad \beta = \delta^{\frac{1}{5}} .} \tag{3.1.13}$$

In summary then, the transonic expansion takes the form

$$\Phi(x, y; M_\infty, \delta) = U\left\{ x + \delta^{\frac{2}{5}} \phi(x, \tilde{y}; K) + \cdots \right\} , \tag{3.1.14}$$

where

$$\boxed{K = \frac{1 - M_\infty^2}{\delta^{\frac{2}{5}}} = \text{transonic similarity parameter} ,}$$

$$\tilde{y} = \delta^{\frac{1}{5}} y$$

$$\frac{\Phi_x}{U} = 1 + \delta^{\frac{2}{5}} \phi_x , \quad \frac{\Phi_y}{U} = \delta \phi_{\tilde{y}} , \quad \frac{a^2}{U^2} = 1 + \delta^{\frac{2}{5}} \left(K - (\gamma - 1)\phi_x \right) \quad .$$

It is not suprising that the velocity perturbation in the freestream is larger in order of magnitude $O(\delta^{2/3})$ than the transverse perturbation $O(\delta)$ The same feature is observed on the epicycloid characteristics of Figure 2.4-4 where near sonic $\Phi_y \sim (\Phi_x - a^*)^{\frac{3}{2}}$. This would be the behaviour in a near-sonic exact simple wave.

Note that

$$\frac{a^2}{a_\infty^2} = 1 + \frac{\gamma - 1}{2} M_\infty^2 \left(1 - \frac{\Phi_x^2 + \Phi_y^2}{U^2} \right) = 1 - \delta^{\frac{2}{5}}(\gamma - 1)\phi_x + \cdots = \frac{T}{T_\infty} ,$$

so that

$$\frac{\rho}{\rho_\infty} = \left(\frac{a^2}{a_\infty^2} \right)^{\frac{1}{\gamma - 1}} = 1 - \delta^{\frac{2}{5}} \phi_x + \cdots ,$$

and

$$\frac{P}{P_\infty} = \left(\frac{\rho}{\rho_\infty} \right)^\gamma = 1 - \delta^{\frac{2}{5}} \gamma \phi_x + \cdots ,$$

$$\boxed{\begin{aligned} &\frac{P}{P_\infty} - 1 = -\gamma \delta^{\frac{2}{5}} \phi_x , \\ &c_p = \frac{P - P_\infty}{\left(\dfrac{\rho_\infty U^2}{2} \right)} = -2\delta^{\frac{2}{5}} \phi_x \quad \text{pressure coefficient} . \end{aligned}} \tag{3.1.15}$$

That is the pressure disturbance depends on the perturbation in the x-direction, just as in linearized theory.

The resulting equation for the disturbance potential is

$$\boxed{\left(K - (\gamma + 1)\phi_x\right)\phi_{xx} + \phi_{\tilde{y}\tilde{y}} = 0 \, ,} \qquad \text{K-G equation} \qquad (3.1.16)$$

which is to be solved in the $(x, \tilde{y})$ plane with the boundary conditions

$$\phi_x, \phi_{\hat{y}} \to 0 \quad \text{at upstream infinity} \quad , \qquad\qquad \text{(BC1)}$$

$$\phi_{\hat{y}}(x, 0\pm) = F'_{u,\ell}(x) \quad \text{tangent flow} \, . \qquad\qquad \text{(BC2)}$$

The K-J condition is approximately realized with no pressure loading at the subsonic trailing edge

$$\phi_x(1, 0+) - \phi_x(1, 0-) \equiv [\phi_x]_{\text{TE}} = 0 \, . \qquad\qquad \text{(K-J)}$$

Consequences of this condition will be discussed later when the boundary value problems for an airfoil are studied.

This equation in its various forms is commonly referred to as the K-G equation named after Karman and Guderley who are among the early workers on the subject [3.1.1], [3.1.2]. Frankl in the USSR also gave an early independent derivation.

The equation is of changing type. it is

locally elliptic , subsonic flow, if $K - (\gamma + 1)\phi_x > 1$,

locally hyperbolic , supersonic flow, if $K - (\gamma + 1)\phi_x < 0$.

The sonic line occurs in the flow where $K = (\gamma + 1)\phi_x$. In the hyperbolic region the characteristics are real and have the directions

$$\frac{d\tilde{y}}{dx} = \pm \frac{1}{\sqrt{(\gamma + 1)\phi_x - K}} \, . \qquad\qquad (3.1.17)$$

The approximate streamlines $\tilde{y} = $ constant bisect these characteristics. The local Mach number M_ℓ is given by

$$M_\ell^2 = \frac{\Phi_x^2 + \Phi_y^2}{a^2} = \frac{1 + 2\delta^{\frac{2}{3}}\phi_x + \cdots}{1 + \delta^{\frac{2}{3}}\left(K - (\gamma - 1)\phi_x\right)} = 1 - \left(K - (\gamma + 1)\phi_x\right)\delta^{\frac{2}{3}} \, ,$$

$$\frac{1 - M_\ell^2}{\delta^{\frac{2}{3}}} = K - (\gamma + 1)\phi_x \, .$$

$$(3.1.18)$$

Thus the K-G equation is analogous to the equation of linearized theory locally (cf. 2.5.5). The characteristic directions are along the local Mach waves also (cf. 3.1.17). If we compare with the derivation of second order theory (equation 2.5.6) we see that some of the non-linear terms have become large $(\phi_x \phi_{xx})$ and appear in the first approximation equation as $M_\infty \to 1$. This non-linear term (uu_x) is responsible in the classical Riemann theory of unsteady waves for the steepening of wave fronts and the formation of shock waves. The same happens here. The problem is solved for a fixed value of K which is the transonic similarity parameter. Fixed K can correspond to varying sets of values of (M_∞, δ) so that there is a definite rule of correspondence for airfoils of affinely similar shapes flying at related Mach numbers. Some rules are written out in detail later (See 3.4).

Now it is useful to note that the K-G equation (3.1.16) is also a conservation equation. That is, it can be written in a divergence form and corresponds to the conservation of a physical quantity. We have

$$\boxed{\left(K\phi_x - \frac{\gamma+1}{2}\phi_x^2 \right)_x + (\phi_{\hat{y}})_{\hat{y}} = 0 \,,}$$ conservation form K-G equation.

$$(3.1.19)$$

In order to show that this is a physical conservation law we need expressions for the density and mass flux. It follows from isentropy (to a sufficient order) that

$$\frac{\rho}{\rho_\infty} = \left(\frac{a^2}{a_\infty^2} \right)^{\frac{1}{\gamma-1}} \,.$$

Carrying out an expansion of the total enthalpy integral to a higher order we find (cf. 3.1.1), (3.1.12 ff)

$$\frac{a^2}{a_\infty^2} = 1 + \frac{\gamma-1}{2}M_\infty^2 \left(1 - \frac{\Phi_x^2 + \Phi_y^2}{U^2} \right)$$

$$= 1 + \frac{\gamma-1}{2}\left(1 - K\delta^{\frac{2}{3}}\right)\left(-2\delta^{\frac{2}{3}}\phi_x - \delta^{\frac{4}{3}}\phi_x^2\right) + \cdots \,,$$

$$\frac{a^2}{a_\infty^2} = 1 - (\gamma-1)\delta^{\frac{2}{3}}\phi_x - (\gamma-1)\delta^{\frac{4}{3}}\left\{ \frac{\phi_x^2}{2} - K\phi_x \right\} + \cdots \,. \qquad (3.1.20)$$

Thus

$$\frac{\rho}{\rho_\infty} = \left\{ 1 - (\gamma - 1)\delta^{\frac{2}{3}}\phi_x - (\gamma - 1)\delta^{\frac{4}{3}}\left(\frac{\phi_x^2}{2} - K\phi_x\right) + \cdots \right\}^{\frac{1}{\gamma - 1}},$$

$$\frac{\rho}{\rho_\infty} = 1 - \delta^{\frac{2}{3}}\phi_x - \delta^{\frac{4}{3}}\left(\frac{\gamma - 1}{2}\phi_x^2 - K\phi_x\right) + \cdots. \tag{3.1.21}$$

In the calculations above it is not necessary to include the second order potential $(\delta^{4/3}\phi_{2x})$ because its effect on the mass flux drops out to order $\delta^{4/3}$. The mass flux components are

$$\frac{\rho\Phi_x}{\rho_\infty U} = \left(1 - \delta^{\frac{2}{3}}\phi_x + \delta^{\frac{4}{3}}\left(K\phi_x - \frac{\gamma - 1}{2}\phi_x^2\right) + \cdots \right)\left(1 + \delta^{\frac{2}{3}}\phi_x\right) + \cdots$$

$$\boxed{\frac{\rho\Phi_x}{\rho_\infty U} = 1 + \delta^{\frac{4}{3}}\left(K\phi_x - \frac{\gamma + 1}{2}\phi_x^2\right) + \cdots,} \tag{3.1.22}$$

$$\frac{\rho\Phi_y}{\rho_\infty U} = \left(1 - \delta^{\frac{2}{3}}\phi_x + \cdots \right)\left(\delta\phi_{\tilde{y}} + \cdots \right)$$

$$\boxed{\frac{\rho\Phi_y}{\rho_\infty U} = \delta\phi_{\tilde{y}} + \cdots.} \tag{3.1.23}$$

Thus we can see that (3.1.19) is the divergence of the mass flux vector, with div $= (\partial/\partial x, \delta^{1/3}\partial/\partial\tilde{y})$. We note that the mass flux perturbation in the x-direction

$$\frac{\left(\dfrac{\rho\phi_x}{\rho_\infty U} - 1\right)}{\delta^{\frac{4}{3}}} = K\phi_x - \frac{\gamma + 1}{2}\phi_x^2, \tag{3.1.24}$$

has a local maximum when $K = (\gamma + 1)\phi_x$, (see Figure 3.1.2) that is, at the local sonic speed. This behaviour is the same as that of an exact ideal gas near local sonic speed. As long as we are away from sonic it may be adequate to approximate the mass flux curve by a tangent as in linearized theory. But, near local sonic a second order approximation by a parabola is necessary. This expresses the well known fact that stream tubes have a throat at local sonic speed.

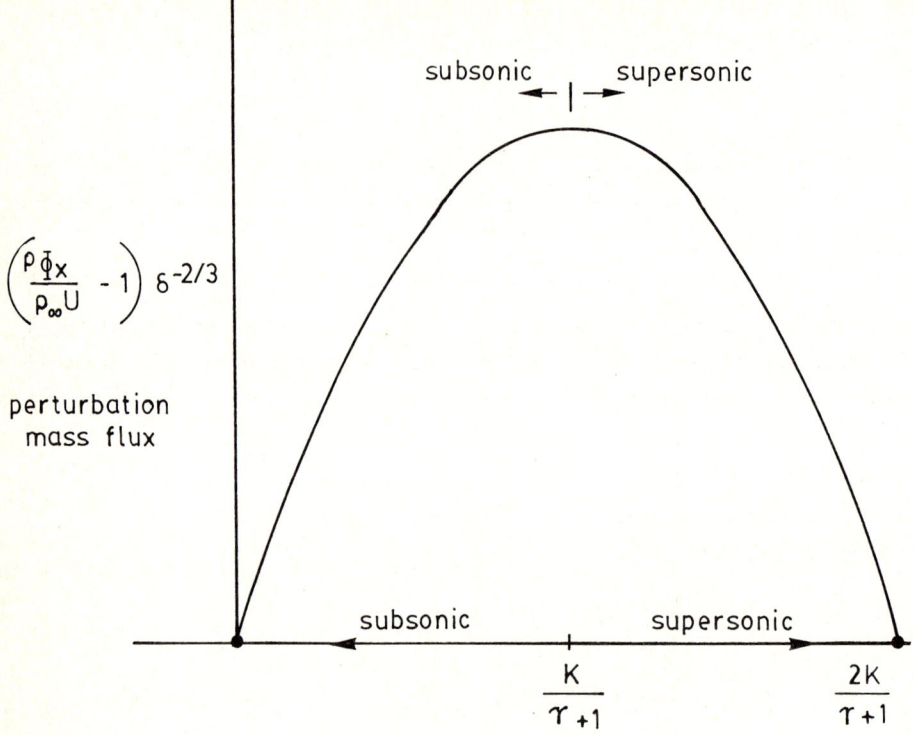

subsonic | supersonic

$\left(\dfrac{\rho\,\phi_x}{\rho_\infty U} - 1\right)\delta^{-2/3}$

perturbation
mass flux

subsonic   supersonic

$\dfrac{K}{\gamma+1}$ $\dfrac{2K}{\gamma+1}$

Figure 3.1.2

Transonic mass flux

Thus the (K-G) equation in form (3.1.19) is a form of physical conservation equation, the continuity equation. Shock waves must be allowed in the solution of the (K-G) equation. The shock jump conditions can be derived directly from the conservation form (3.1.19). Since (3.1.19) represents a physical conservation law its integral form holds across the shock jumps. With the notation of Figure 3.1.3 we have the following. Integration across the jumps yields

$$\left[K\phi_x - \frac{\gamma+1}{2}\phi_x^2\right]d\tilde{y}_s - [\phi_{\tilde{y}}]\,dx_s = 0\,. \qquad (3.1.25)$$

A second condition states that there is no jump in potential across the shock, or equivalently no jump in tangential velocity across the shock surface,

$$[\phi] = 0\,, \quad \text{or} \quad [\phi_x]\,dx_s + [\phi_{\tilde{y}}]\,d\tilde{y}_s = 0\,. \qquad (3.1.26)$$

The notation here is that the square bracket [   ] denotes a jump

$$[\quad] = (\quad)_b - (\quad)_a ,\tag{3.1.27}$$

that is, the value of a quantity behind the shock less the value ahead. A last condition to be imposed on shock waves is that the shock jumps should represent only compressions,

$$[\phi_x] < 0 .\tag{3.1.28}$$

As will be seen below this rules out one branch of possible solutions to (3.1.25, 26). The simplest expression for the entropy rise across a shock comes from the expressions for the density ratio cf (2.4.15), (3.1.21)

$$\varepsilon = \frac{-\rho_a}{\rho_b} + 1 = -\frac{1 - \delta^{\frac{2}{3}}\phi_{xa}}{1 - \delta^{\frac{2}{3}}\phi_{xb}} + 1 = -\delta^{\frac{2}{3}}[\phi_x] + \cdots \quad .\tag{3.1.29}$$

Thus (3.1.3) gives

$$\frac{1}{c_v}[S] = -\frac{\gamma(\gamma^2 - 1)}{12}\delta^2[\phi_x]^3 + \cdots ,\tag{3.1.30}$$

for any shock wave. This can also be checked by calculation of $(M_n^2 - 1)$ where M_n is the Mach number normal to the shock. It follows from Crocco's theorem (2.4.25) that the vorticity ω is at most $O(\delta^2)$. The flow is irrotational in $O(\delta^{2/3}, \delta^{4/3})$ and disturbance potentials $\delta^{1/3}\phi_1, \delta^{4/3}\phi_2$ can be introduced. The shock jump conditions for this system are as usual. If, for example, the state of the shock is known ahead $(\phi_{x_a}, \phi_{\tilde{y}_a})$ and the shock angle $(dx/d\tilde{y})_s$ is known then we have two equations (3.1.25, 26) for the two unknowns $(\phi_{x_b}, \phi_{\tilde{y}_b})$.

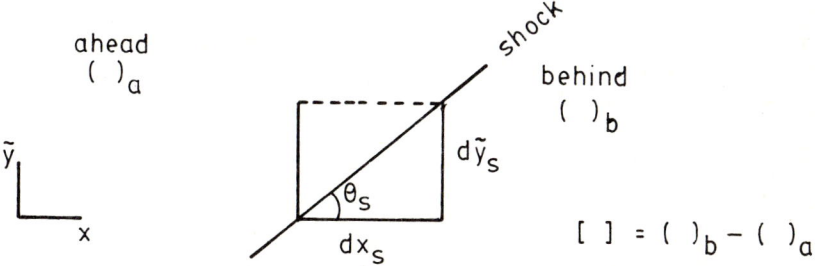

Figure 3.1.3

Transonic shock jumps

In summary the expansions constructed here result in an approximate equation which preserves all the essential features of a compressible nearly sonic flow such as local structure, shock waves, throats, in a mathematically consistent way.

In the next section the simple extension of this theory that is needed to describe flow around three dimensional wings is made.

3.1.1 Extension to Three Dimensional Wings

The first approximation of the expansion is easily carried out using the fact that disturbances spread in the same way in both lateral directions $(y,\ z)$. The three dimensional wing surface lying close to the plane $y = 0$ is represented by a family of forms:

$$S(x, y, z) = y - \delta F_{u,\ell}(x, \frac{z}{b}) = 0\ , \qquad (3.1.31)$$

for upper and lower surfaces respectively. (See Figure 3.1.4). The wing is scaled to have unit chord and semispan b, δ is the thickness ratio. Let the projections of the leading and trailing edges in the plane $y = 0$ be given by

$$x = x_{\text{LE}}\left(\frac{z}{b}\right) = x_{\text{LE}}\left(\frac{\tilde{z}}{B}\right)\ , \qquad \text{leading edge}\ ,$$

$$x = x_{\text{TE}}\left(\frac{z}{b}\right) = x_{\text{TE}}\left(\frac{\tilde{z}}{B}\right)\ , \qquad \text{trailing edge}\ , \qquad (3.1.32)$$

where $\tilde{z} = \delta^{1/3}z$ and $B = \delta^{1/3}b$.

The basic transonic expansion (3.1.5) now takes the form

$$\Phi(x, y, z;\ M_\infty, \delta, b) = U\left\{x + \delta^{\frac{2}{3}}\phi(x, \tilde{y}, z;\ K, B) + \cdots\right\}\ . \qquad (3.1.33)$$

Thus the boundary condition of tangent flow $\mathbf{q} \cdot \nabla S = 0$ on $S = 0$ takes the form

$$\Phi_y = \delta\frac{\partial F}{\partial x}\Phi_x + \frac{\delta}{b}\frac{\partial F}{\partial(2)}\Phi_z \quad \text{on} \quad y = \delta F\ . \quad \text{(BC2)} \qquad (3.1.34)$$

$\dfrac{\partial F}{\partial(2)}$ denotes the derivative of the shape function with respect to its second argument. In the transonic limit we must keep

$$B = \delta^{\frac{1}{3}}b \quad \text{fixed as} \quad \delta \to 0\ , \quad M_\infty \to 1\ .$$

The K-J condition is approximately realized with no pressure loading at the subsonic trailing edge

$$\phi_x(1, 0+) - \phi_x(1, 0-) \equiv [\phi_x]_{\text{TE}} = 0\ . \quad \text{(K-J)}$$

Consequences of this condition will be discussed later.

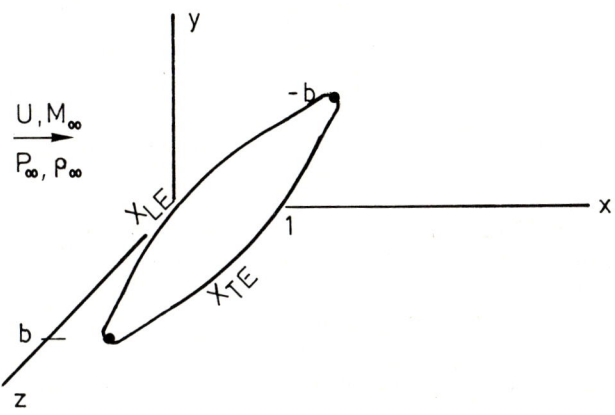

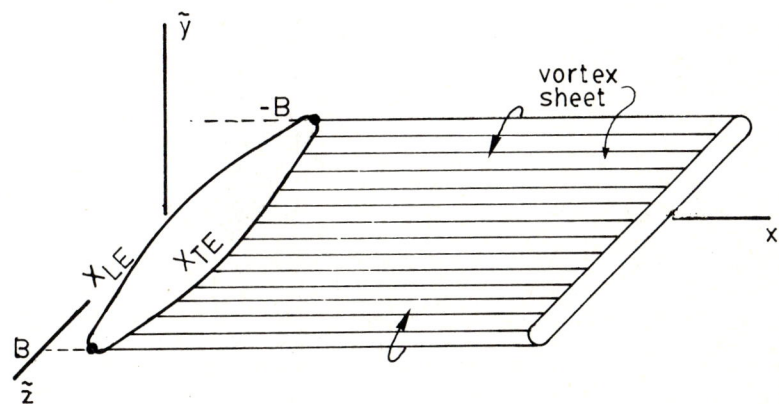

Figure 3.1.4

Three dimensional wing in physical space and in transonic space.

As before the boundary conditions of tangent flow can be applied approximately in the plane $y = 0$. Using the expansion (3.1.33).

$$\frac{\Phi_y}{U} = \delta\phi_{\tilde{y}} , \quad \frac{\Phi_z}{U} = \delta\phi_{\tilde{z}} , \quad \frac{\Phi_x}{U} = 1 + \delta^{\frac{2}{3}}\phi_x , \tag{3.1.35}$$

so that (BC2) becomes

$$\phi_{\tilde{y}}(x, 0\pm, \tilde{z}) = \frac{\partial F_{u,\ell}\left(x, \frac{\tilde{z}}{B}\right)}{\partial x} , \quad \text{(BC2)} \tag{3.1.36}$$

for

$$x_{\text{LE}}\left(\frac{\tilde{z}}{B}\right) < x < x_{\text{TE}}\left(\frac{\tilde{z}}{B}\right) .$$

The Kutta-Joukowski condition must be applied at the trailing edge just as in the two dimensional case and within the small-disturbance theory there is no ambiguity about the location of the trailing edge $x = x_{\text{TE}}$. The formula for pressure disturbance is also the same

$$[\phi_x]_{\text{TE}} = \phi_x(x_{\text{TE}}, 0+, \tilde{z}) - \phi_x(x_{\text{TE}}, 0-, \tilde{z}) = 0 , \quad \text{(K-J)} .$$

A vortex sheet wake trails downstream of a lifting wing and to a sufficient approximation, just as in linearized subsonic theory, it lies in the plane $\tilde{y} = 0$. This particular approximation is certainly not valid near $x = \infty$ since the vortex sheet moves down (and rolls up) under its own induction.

The approximate (K-G) equation follows from the full potential equation practically as before. In general the full potential equation is (cf. 2.4.27)

$$a^2 \nabla^2 \Phi = \mathbf{q} \cdot \nabla \left(\frac{(\nabla\Phi)^2}{2}\right) ,$$

$$\frac{a^2}{\gamma - 1} + \frac{(\nabla\Phi)^2}{2} = \frac{a_\infty^2}{\gamma - 1} + \frac{U^2}{2} . \tag{3.1.37}$$

Thus now

$$\nabla^2\phi \to \phi_{\tilde{y}\tilde{y}} + \phi_{\tilde{z}\tilde{z}} ,$$

and

$$\boxed{\left(K - (\gamma + 1)\phi_x\right)\phi_{xx} + \phi_{\tilde{y}\tilde{y}} + \phi_{\tilde{z}\tilde{z}} = 0 ,} \quad \text{K-G equation} \tag{3.1.38}$$

or in conservation form

$$\left(K\phi_x - \frac{\gamma + 1}{2}\phi_x^2\right)_x + (\phi_{\tilde{y}})_{\tilde{y}} + (\phi_{\tilde{z}})_{\tilde{z}} = 0 . \tag{3.1.39}$$

The shock jump condition of mass conservation can again be derived by integrating the divergence form (3.1.39) across the jump. The corresponding mass flux vector is

$$\frac{\rho \mathbf{q}}{\rho_\infty U} = \mathbf{i} \left\{ 1 + \delta^{\frac{4}{3}} \left(K\phi_x - \frac{\gamma+1}{2}\phi_x^2 \right) + \cdots \right\} + \delta \tilde{\nabla} \phi + \cdots \quad , \qquad (3.1.40)$$

where $\nabla = \mathbf{i}_{\tilde{y}} \frac{\partial}{\partial \tilde{y}} + \mathbf{i}_{\tilde{z}} \frac{\partial}{\partial \tilde{z}}$. The unit normal to the shock surface represented by $S(x, \tilde{y}, \tilde{z}) = 0$ is

$$\mathbf{n} = \frac{S_x \mathbf{i}_x + \delta^{\frac{1}{3}} \tilde{\nabla} S}{\sqrt{S_x^2 + \delta^{\frac{2}{3}} (\tilde{\nabla} S)^2}} = \mathbf{i}_x \left\{ 1 - \frac{\delta^{\frac{2}{3}}}{2} \frac{(\tilde{\nabla} S)^2}{S_x^2} + \cdots \right\}$$

$$+ \delta^{\frac{1}{3}} \frac{(\tilde{\nabla} S)}{S_x^2} + \cdots \qquad (3.1.41)$$

(cf. Figure 3.1.5).

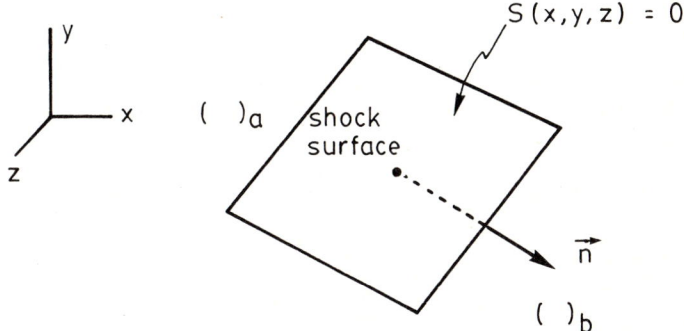

Figure 3.1.5

Shock element

Conservation of mass across the shocks $[\mathbf{q} \cdot \mathbf{n}]_S = 0$ is

$$\left[K\phi_x - \frac{\gamma+1}{2}\phi_x^2 \right] + [\nabla \tilde{\phi}] \cdot \frac{[\tilde{\nabla} S]}{S_x} = 0 \qquad (3.1.42)$$

The conservation of tangential component of velocity is expressed by $[\mathbf{q}] \times \mathbf{n} = 0$. thus,

$$\mathbf{i}_x [\phi_x] \times \frac{(\tilde{\nabla} S)}{S_x} + [\nabla \tilde{\phi}] \times \mathbf{i}_x = 0 . \qquad (3.1.43)$$

For given shock geometry (3.1.42,43) provide three relations for the components of velocity perturbation behind the shock in terms of those ahead. Equivalent to (3.1.41) is $[\phi] = 0$. These shock relations are analysed in more detail in Section 3.2.

References

[3.1.1] Karman, T. von: The similarity law of transonic flow, *Journal of Math. and Physics*, v.**26**, No. 3, October 1947.]

[3.1.2] Guderley G: Considerations of the Structure of Mixed Subsonic and Supersonic Flow Patterns, Wright Field Report, F-TR-2168-ND, October 1947.

3.2 Expansion Procedure Applied to the Basic System of Equations.

In this section the approximate transonic equation is derived in another way which serves as a check on the arguments of the previous section. This method can also be easily modified to account for a certain amount of vorticity in the upstream flow. The approach is to use the expected form of the expansion in the basic equations, continuity, momentum balance, and enthalpy conservation both in differential form and in integral form as applied for shock waves. The advantage of this approach is freedom from assumptions about isentropy and vorticity thus clarifying the notion of the approximate potential ϕ. A second advantage is that the shock relations are treated explicit. The disadvantage is that the expansion must be carried to a higher order to obtain the results; to second order to obtain a first order potential, to third order to obtain a second order potential, and to discuss the drag directly from the system.

The flow deflection is again characterized by the parameter δ, and in the limit process $M_\infty \to 1$ so that the transonic similarity parameter K is fixed

$$M_\infty^2 = 1 - K\delta^{\frac{2}{3}} . \tag{3.2.1}$$

The expansion is carried out in the system of coordinates $x, \tilde{y}, \tilde{z}$ fixed where as previously

$$\tilde{y} = \delta^{\frac{1}{3}} y , \quad \tilde{z} = \delta^{\frac{1}{3}} z .$$

A characteristic unit length in the x-direction is chosen. It is convenient to split the velocity into perturbations along the free-stream direction and transverse to it, denoted by $(\tilde{\ })$. Thus, in accord with the orders of magnitude discussed in the last section the flow quantities are considered to have the following asymptotic expansions

$$\frac{\mathbf{q}}{U} = \left(1 + \delta^{\frac{2}{3}} u_1(x, \tilde{y}, \tilde{z}; K) + \delta^{\frac{4}{3}} u_2 + \cdots\right)\mathbf{i}_x + \delta\tilde{\mathbf{v}}_1 + \delta^{\frac{5}{3}}\tilde{\mathbf{v}}_2 + \cdots , \tag{3.2.2a}$$

$$\frac{P}{P_\infty} = 1 + \delta^{\frac{2}{3}} p_1 + \delta^{\frac{4}{3}} p_2 + \cdots , \tag{3.2.2b}$$

$$\frac{\rho}{\rho_\infty} = 1 + \delta^{\frac{2}{3}} \sigma_1 + \delta^{\frac{4}{3}} \sigma_2 \cdots , \tag{3.2.2c}$$

where $\mathbf{i}_x$ = unit vector in x-direction , and

where $\tilde{\mathbf{v}}$ = transverse velocity perturbation = (v, w) in the cartesian system.

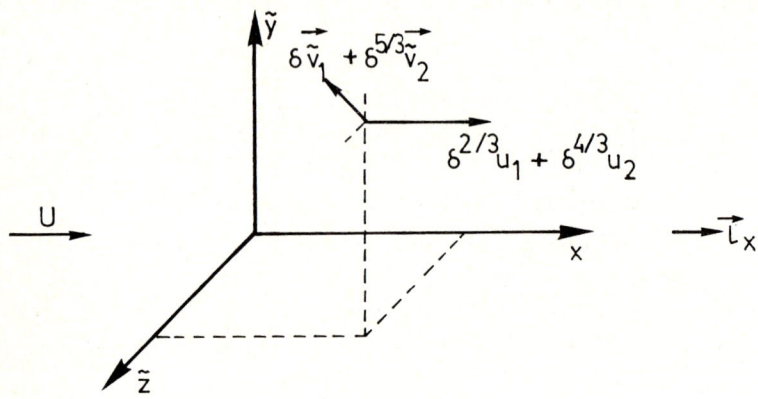

Figure 3.2.1

Coordinates for transonic expansion

First consider the integrals which arise from the total enthalpy integral (2.4.10) which holds throughout the flow field since it is conserved across shock waves.

$$\frac{a^2}{a_\infty^2} = 1 + \frac{\gamma - 1}{2}\left(1 - \frac{q^2}{U^2}\right). \tag{3.2.3}$$

The temperature ratio $\frac{T}{T_\infty}$ is, keeping only the necessary orders, equal to

$$\frac{a^2}{a_\infty^2} = \frac{P}{P_\infty} \cdot \frac{\rho_\infty}{\rho} = \frac{1 + \delta^{\frac{2}{3}}p_1 + \delta^{\frac{4}{3}}p_2}{1 + \delta^{\frac{2}{3}}\sigma_1 + \delta^{\frac{4}{3}}\sigma_2}$$

$$= \left(1 + \delta^{\frac{2}{3}}p_1 + \delta^{\frac{4}{3}}p_2\right)\left(1 - \delta^{\frac{2}{3}}\sigma_1 - \delta^{\frac{4}{3}}(\sigma_2 - \sigma_1^2)\right),$$

or

$$\frac{a^2}{a_\infty^2} = 1 + \delta^{\frac{2}{3}}(p_1 - \sigma_1) + \delta^{\frac{4}{3}}(p_2 - \sigma_2 + \sigma_1^2 - p_1\sigma_1) + \cdots, \tag{3.2.4}$$

Also,

$$\frac{q^2}{U^2} = 1 + \delta^{\frac{2}{3}}(2u_1) + \delta^{\frac{4}{3}}(2u_2 + u_1^2) + \cdots. \tag{3.2.5}$$

Hence the enthalpy integral (3.2.3) contains the following terms

$$1 + \delta^{\frac{2}{3}}(p_1 - \sigma_1) + \delta^{\frac{4}{3}}(p_2 - \sigma_2 - \sigma_1(p_1 - \sigma_1))$$

$$= 1 + \frac{\gamma - 1}{2}\left\{-\delta^{\frac{2}{3}}(2u_1) - \delta^{\frac{4}{3}}(2u_2 + u_1^2)\right\}\left\{1 - K\delta^{\frac{2}{3}}\right\}.$$

Thus, equating orders of magnitude we obtain the first and second order enthalpy integrals.

$$
\begin{aligned}
p_1 - \sigma_1 &= -(\gamma - 1)u_1 , \\
p_2 - \sigma_2 &= -(\gamma - 1)u_2 - (\gamma - 1)\left\{ \frac{u_1^2}{2} + \sigma_1 u_1 - K u_1 \right\} .
\end{aligned}
\tag{3.2.6}
$$

Next consider the continuity and momentum equation in differential form. The mass flux vector is

$$
\frac{\rho \mathbf{q}}{\rho_\infty U} = \left(1 + \delta^{\frac{2}{3}} \sigma_1 + \delta^{\frac{4}{3}} \sigma_2\right)\left(\left(1 + \delta^{\frac{2}{3}} u_1 + \delta^{\frac{4}{3}} u_2\right)\mathbf{i}_x + \delta \tilde{\mathbf{v}}_1 + \delta^{\frac{5}{3}} \tilde{\mathbf{v}}_2 \right)
$$

$$
\frac{\rho \mathbf{q}}{\rho_\infty U} = \left(1 + \delta^{\frac{2}{3}}(u_1 + \sigma_1) + \delta^{\frac{4}{3}}(u_2 + \sigma_2 + u_1\sigma_1)\right)\mathbf{i}_x + \delta \tilde{\mathbf{v}}_1 + \delta^{\frac{5}{3}}(\tilde{\mathbf{v}}_2 + \sigma_1\tilde{\mathbf{v}}_1) + \cdots
\tag{3.2.7}
$$

The operation of divergence can be decomposed

$$
\text{div} \equiv \nabla \cdot = \left(\mathbf{i}_x \frac{\partial}{\partial x} + \delta^{\frac{1}{3}} \tilde{\nabla} \right) ,
\tag{3.2.8}
$$

where $\tilde{\nabla} \equiv (\frac{\partial}{\partial \tilde{y}}, \frac{\partial}{\partial \tilde{z}})$ is the transverse gradient in transonic coordinates. Therefore from $\nabla \cdot (\rho \mathbf{q}) = 0$ the various orders give

$$
\begin{aligned}
O(\delta^{\frac{2}{3}}) \quad & (u_1 + \sigma_1)_x = 0 , \\
O(\delta^{\frac{4}{3}}) \quad & (u_2 + \sigma_2 + u_1\sigma_1)_x + \tilde{\nabla} \cdot \tilde{\mathbf{v}}_1 = 0 .
\end{aligned}
\qquad \text{continuity equation} \tag{3.2.9}
$$

Correspondingly, the Euler form of the momentum equation is expressed in terms of the derivative along a streamline,

$$
\mathbf{q} \cdot \nabla = U\left\{ (1 + \delta^{\frac{2}{3}} u_1 + \cdots)\mathbf{i}_x + \delta \tilde{\mathbf{v}}_1 + \cdots \right\} \cdot \left\{ \mathbf{i}_x \left(\frac{\partial}{\partial x} \right) + \delta^{\frac{1}{3}} \tilde{\nabla} \right\} ,
$$

$$
\mathbf{q} \cdot \nabla = U\left\{ (1 + \delta^{\frac{2}{3}} u_1) \frac{\partial}{\partial x} + \delta^{\frac{4}{3}} \tilde{\mathbf{v}}_1 \cdot \tilde{\nabla} \right\} .
\tag{3.2.10}
$$

The Euler equations are

$$
\left(\frac{\rho}{\rho_\infty} \right) \left(\frac{\rho_\infty U^2}{P_\infty} \right) \left\{ \frac{\mathbf{q}}{U} \cdot \nabla \frac{\mathbf{q}}{U} \right\} = -\nabla \frac{P}{P_\infty} .
\tag{3.2.11}
$$

Note that

$$
\frac{\rho_\infty U^2}{P_\infty} = \gamma M_\infty^2 = \gamma \left(1 - K \delta^{\frac{2}{3}}\right) .
\tag{3.2.12}
$$

The x-momentum equation becomes

$$\gamma\left(1 - K\delta^{\frac{2}{3}}\right)\left(1 + \delta^{\frac{2}{3}}\sigma_1\right)\left(\left(1 + \delta^{\frac{2}{3}}u_1\right)\frac{\partial}{\partial x} + \delta^{\frac{4}{3}}\tilde{\mathbf{v}}_1\cdot\tilde{\nabla}\right)\left(\delta^{\frac{2}{3}}u_1 + \delta^{\frac{4}{3}}u_2\right)$$

$$= -\delta^{\frac{2}{3}}p_{1_x} - \delta^{\frac{4}{3}}p_{2_x} ,$$

or

$$
\begin{array}{l}
O(\delta^{\frac{2}{3}}) \\[4pt]
O(\delta^{\frac{4}{3}})
\end{array}
\boxed{
\begin{array}{l}
\gamma u_{1x} = -P_{1x} \\[4pt]
\gamma u_{2x} + \gamma(\sigma_1 + u_1 - K)u_{1x} = -P_{2x}
\end{array}
}
\qquad x\text{-momentum equations}
$$

$$(3.2.13)$$

The transverse momentum is also useful. As above

$$\left(\gamma\left(1 + \delta^{\frac{2}{3}}(\sigma_1 + u_1 - K) + \cdots\right)\frac{\partial}{\partial x} + \gamma\delta^{\frac{4}{3}}\tilde{\mathbf{v}}_1\cdot\tilde{\nabla}\right)\left(\delta\tilde{\mathbf{v}}_1 + \delta^{\frac{5}{3}}\tilde{\mathbf{v}}_2 + \cdots\right) = -\delta\tilde{\nabla}p_2 ,$$

or

$$
\begin{array}{l}
O(\delta) \\[4pt]
O(\delta^{\frac{5}{3}})
\end{array}
\boxed{
\begin{array}{l}
\gamma\tilde{\mathbf{v}}_{1x} = -\tilde{\nabla}p_1 , \\[4pt]
\gamma\tilde{\mathbf{v}}_{2x} + \gamma(\sigma_1 + u_1 - K)\tilde{\mathbf{v}}_{1x} = -\tilde{\nabla}p_2 .
\end{array}
}
\qquad
\begin{array}{c}
\text{transverse momentum} \\
\text{equations}
\end{array}
$$

$$(3.2.14)$$

It is necessary to study the corresponding approximate forms of the shock jump relations. Let the shock be a surface

$$S(x, y, z) = 0 = x - g(\tilde{y}, \tilde{z}) - \cdots \quad . \qquad (3.2.15)$$

The orientation of the surface is characterized by its unit normal $\mathbf{n}$

$$\mathbf{n} = \frac{\nabla S}{|\nabla S|} = \frac{\mathbf{i} - \delta^{\frac{1}{3}}\tilde{\nabla}g}{\sqrt{1 + \delta^{\frac{2}{3}}(\tilde{\nabla}g)^2}} = \mathbf{i}_x\left(1 - \delta^{\frac{2}{3}}\frac{1}{2}(\tilde{\nabla}g)^2\right) - \delta^{\frac{1}{3}}\tilde{\nabla}g ,$$

(cf. Figure 3.2.2), $\mathbf{n}$ is thus decomposed into x and transverse components

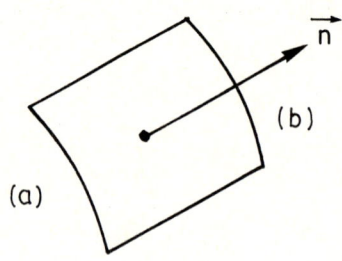

Figure 3.2.2

Shock geometry

$$\mathbf{n} = \mathbf{i}_x \left\{ 1 - \frac{\delta^{\frac{2}{3}}}{2} (\tilde{\mathbf{n}})^2 \right\} + \delta^{\frac{1}{3}} \tilde{\mathbf{n}} + \cdots , \qquad (3.2.16)$$

where $\tilde{\mathbf{n}} = -\tilde{\nabla} g$. The conditions of mass conservation can be expressed as zero jump in the normal mass flux

$$[\rho \mathbf{q} \cdot \mathbf{n}] = 0 . \qquad (3.2.17)$$

Using the expansion for the mass flux vector (3.2.7), this is

$$\left[\delta^{\frac{2}{3}} (u_1 + \sigma_1) + \delta^{\frac{4}{3}} (u_2 + \sigma_2 + u_1 \sigma_1) - \delta^{\frac{4}{3}} (u_1 + \sigma_1) \frac{1}{2} (\tilde{\mathbf{n}})^2 + \delta^{\frac{4}{3}} \tilde{\mathbf{v}}_1 \cdot \tilde{\mathbf{n}} \right] = 0 ,$$

or

$$\boxed{\begin{array}{l} [u_1 + \sigma_1] = 0 , \\[2mm] [u_2 + \sigma_2 + u_1 \sigma_1] + [\tilde{\mathbf{v}}_1] \cdot \tilde{\mathbf{n}} = 0 . \end{array}} \quad \text{shock mass conservation} \qquad (3.2.18)$$

Thus, we have a mass-flux integral valid everywhere. The first of the continuity equations (3.2.9) shows that

$$u_1 + \sigma_1 = fn(\tilde{y}, \tilde{z}) .$$

Since all disturbances are assumed to vanish at upstream infinity a non-zero right hand side can only be introduced by shock waves. But, (3.2.18) shows that shock waves do not change the value of $u_1 + \sigma_1$. Hence

$$\boxed{\sigma_1 + u_1 = 0 .} \quad \text{overall integral} \qquad (3.2.19)$$

It then also follows from the enthalpy integral that the entropy is constant to this order and that there is a simple relation between p_1, u_1,

$$\boxed{p_1 = \gamma \sigma_1 = -\gamma u_1 .} \qquad (3.2.20)$$

This also shows that the first of the x-momentum equations is included in these relations. Then the first of the transverse momentum equations shows that

$$\boxed{\tilde{\mathbf{v}}_{1_x} = \tilde{\nabla} u_1 ,} \qquad (3.2.21)$$

or that to this order no transverse components of vorticity can be introduced into the original vortex free flow by the presence of shock waves. Note that in general the vorticity ω

$$
\begin{aligned}
\frac{\omega}{U} = \nabla \times \frac{\mathbf{q}}{U} &= \left(\mathbf{i}_x \frac{\partial}{\partial x} + \delta^{\frac{1}{3}} \tilde{\nabla} \right) \\
&\quad \times \left(\left(\delta^{\frac{2}{3}} u_1 + \delta^{\frac{4}{3}} u_2 \right) \mathbf{i}_x + \delta \tilde{\mathbf{v}}_1 + \delta^{\frac{5}{3}} \tilde{\mathbf{v}}_2 \right) \\
&= \delta \left((v_{1x} - u_{1\hat{y}}) \mathbf{i}_z - (w_{1x} - u_{1\hat{z}}) \mathbf{i}_y \right) \\
&\quad + \delta^{\frac{5}{3}} \left((v_{2x} - u_{2\hat{y}}) \mathbf{i}_z + (w_{2x} - v_{2z}) \mathbf{i}_y \right) + \delta^{\frac{4}{3}} \left(\tilde{\nabla} \times \tilde{\mathbf{v}}_1 \right) + \cdots . \quad (3.2.22)
\end{aligned}
$$

In order to obtain a first order system of equations for $(u_1, \tilde{\mathbf{v}}_1)$ we need to be able to find an expression for $(u_2 + \sigma_2)_x$ in the second continuity equation (3.2.9). Now it follows from the second x-momentum equation (3.2.13) that

$$
p_{2_x} = -\gamma u_{2x} + \gamma K u_{1x} ,
$$

and from the second enthalpy integral (3.2.6) and $u_1 + \sigma_1 = 0$ that

$$
p_2 = \sigma_2 - (\gamma - 1) u_2 + (\gamma - 1) \left\{ \frac{u_1^2}{2} + K u_1 \right\} .
$$

Thus

$$
(\sigma_2 + u_2)_x = \left(K u_1 - \frac{\gamma - 1}{2} u_1^2 \right)_x . \quad (3.2.23)
$$

Using irrotationality (3.2.21) and the continuity equation (3.2.9) we have the basic differential system for $(u_1, \tilde{\mathbf{v}}_1)$.

$$
\boxed{
\begin{aligned}
\tilde{\mathbf{v}}_{1x} - \tilde{\nabla} u_1 &= 0 , \\
\left(K u_1 - \frac{\gamma + 1}{2} u_1^2 \right)_x + \tilde{\nabla} \cdot \tilde{\mathbf{v}}_1 &= 0 .
\end{aligned}
}
\qquad \text{Basic Transonic System} \qquad (3.2.24)
$$

The pressure perturbations are calculated for this system from (3.2.20)

$$
\boxed{ p_1 = -\gamma u_1 , \quad \sigma_1 = -u_1 . } \qquad (3.2.25)
$$

A first order perturbation potential then exists such that

$$
\tilde{\mathbf{v}}_1 = \tilde{\nabla} \phi_1 , \quad u_1 = \phi_{1x} . \qquad (3.2.26)
$$

The basic equation for the potential is the same as (3.1.38)

$$\left(K\phi_{1x} - \frac{\gamma+1}{2}\phi_{1x}^2 \right)_x + \tilde{\nabla}^2\phi_1 = 0 \ . \tag{3.2.27}$$

Next the shock jump conditions corresponding to the basic system (3.2.24) will be derived from the exact shock relations. First, there is no jump in the component of velocity tangential to the shock surface so that

$$\left[\frac{\mathbf{q}}{U} \times \mathbf{n} \right] = 0 \ ,$$

or

$$\left[\left(\delta^{\frac{2}{3}} u_1 + \delta^{\frac{4}{3}} u_2 + \cdots \right) \mathbf{i}_x + \delta\tilde{\mathbf{v}}_1 \right] \times \left(\mathbf{i}_x \left(1 - \delta^{\frac{2}{3}} \frac{(\tilde{\mathbf{n}})^2}{2} \right) + \delta^{\frac{1}{3}}\tilde{\mathbf{n}} \right)$$

$$= \delta[u_1]\mathbf{i}_x \times \tilde{\mathbf{n}} + \delta[\tilde{\mathbf{v}}] \times \mathbf{i}_x + \cdots = 0 \ .$$

Thus

$$[u_1]\tilde{\mathbf{n}} = [\tilde{\mathbf{v}}_1] \ . \tag{3.2.28}$$

This is a "wave angle" relationship which shows that the transverse perturbation jump is in the direction of the transverse normal.

We also consider the momentum flow $\mathcal{M}_n$ normal to the shock surface which gives exactly (cf. 2.4.2)

$$[\mathcal{M}_n] \equiv \left[\frac{P}{P_\infty} + \gamma M_\infty^2 \frac{\rho}{\rho_\infty} \left(\frac{\mathbf{q} \cdot \mathbf{n}}{U} \right)^2 \right] = 0 \ , \qquad \frac{\rho_\infty U^2}{P_\infty} = \gamma M_\infty^2 \ .$$

Note that $\gamma M_\infty^2 = \gamma(1 - K\delta^{\frac{2}{3}})$,

$$(1 - K\delta^{\frac{2}{3}})\frac{\rho}{\rho_\infty} = 1 + \delta^{\frac{2}{3}}(\sigma_1 - K) + \delta^{\frac{4}{3}}(\sigma_2 - K\sigma_1) + \cdots \ ,$$

$$\frac{\mathbf{q}}{U} \cdot \mathbf{n} = 1 + \delta^{\frac{2}{3}}u_1 + \delta^{\frac{4}{3}} \left(u_2 - u_1\frac{(\tilde{\mathbf{n}})^2}{2} + \tilde{\mathbf{v}}_1 \cdot \tilde{\mathbf{n}} \right) + \cdots \ ,$$

$$\left(\frac{\mathbf{q}}{U} \cdot \mathbf{n} \right)^2 = 1 + \delta^{\frac{2}{3}}(2u_1) + \delta^{\frac{4}{3}} \left(2u_2 - u_1(\tilde{\mathbf{n}})^2 + u_1^2 + 2\tilde{\mathbf{v}}_1 \cdot \tilde{\mathbf{n}} \right) + \cdots \ .$$

Therefore

$$\mathcal{M}_n = 1 + \delta^{\frac{2}{3}}p_1 + \delta^{\frac{4}{3}}p_2 + \cdots$$

$$+ \gamma \Big\{ \left(1 + \delta^{\frac{2}{3}}(\sigma_1 - K) + \delta^{\frac{4}{3}}(\sigma_2 - \sigma_1 K) \right)$$

$$\left(1 + \delta^{\frac{2}{3}}(2u_1) + \delta^{\frac{4}{3}} \left(2u_2 - u_1(\tilde{\mathbf{n}})^2 + u_1^2 + 2\tilde{\mathbf{v}}_1 \cdot \tilde{\mathbf{n}} \right) \right) \Big\}$$

$$= 1 + \delta^{\frac{2}{3}}p_1 + \delta^{\frac{4}{3}}p_2 + \gamma \Big\{ 1 + \delta^{\frac{2}{3}}(\sigma_1 + 2u_2 - K)$$

$$+ \delta^{\frac{4}{3}} \left(\sigma_2 + 2u_2 - K\sigma_1 - u_1(\tilde{\mathbf{n}})^2 + u_1^2 + 2\tilde{\mathbf{v}}_1 \cdot \tilde{\mathbf{n}} + 2u_1\sigma_1 - 2Ku_1 \right) \Big\} \ .$$

Thus for $O(\delta^{\frac{2}{3}})$,

$$\left[p_1 + \gamma(\sigma_1 + 2u_1)\right] = 0 \, ,$$

or using

$$\sigma_1 = -u_1 \, , \quad [p_1 - \gamma\sigma_1] = 0 \, ,$$

which is the already known result of approximate isentropy. The significant result comes from the $O(\delta^{\frac{4}{3}})$ terms $\mathcal{M}_{n_2}$,

$$\mathcal{M}_{n_2} = \delta^{\frac{4}{3}}\left(p_2 + \gamma\left(\sigma_2 + 2u_2 - Ku_1 - u_1^2 + 2\tilde{\mathbf{v}}_1\bullet\tilde{\mathbf{n}} - u_1(\tilde{\mathbf{n}})^2\right)\right) \, ,$$

or using the enthalpy integral

$$\mathcal{M}_{n_2} = \delta^{\frac{4}{3}}\left\{\sigma_2 - (\gamma - 1)u_2 + (\gamma - 1)\left(\frac{u_1^2}{2} + Ku_1\right)\right.$$
$$\left. + \gamma(\sigma_2 + 2u_2) - \gamma Ku_1 - \gamma u_1^2 + 2\gamma\tilde{\mathbf{v}}_1\bullet\tilde{\mathbf{n}} - \gamma u_1(\tilde{\mathbf{n}}_1)^2\right\} \, ,$$

$$\mathcal{M}_{n_2} = \delta^{\frac{4}{3}}\left\{(\gamma + 1)(\sigma_2 + u_2) - \frac{\gamma + 1}{2}u_1^2 - Ku_1 + 2\gamma\tilde{\mathbf{v}}_1\bullet\tilde{\mathbf{n}} - \gamma u_1(\tilde{\mathbf{n}})^2\right\} \, .$$

Now it follows from shock mass conservation (3.2.18) that

$$[\sigma_2 + u_2] = -[\tilde{\mathbf{v}}_1]\bullet\tilde{\mathbf{n}} + [u_1^2] \, ,$$

and

$$\left[\mathcal{M}_{n_2}\right] = 0 = \delta^{\frac{4}{3}}\left[\frac{\gamma + 1}{2}u_1^2 - Ku_1 + (\gamma - 1)\tilde{\mathbf{v}}_1\bullet\tilde{\mathbf{n}} - \gamma u_1(\tilde{\mathbf{n}})^2\right] \, .$$

Now using the kinematic relation (3.2.28) we have

$$\boxed{\left[\frac{\gamma + 1}{2}u_1^2 - Ku_1\right] - [v_1]\bullet\tilde{\mathbf{n}} = 0 \, .} \tag{3.2.29}$$

The shock jump conditions are equations (3.2.28,29) and correspond exactly to the integral forms of the basic system (3.2.24) taken across the shock surface.

The considerations of this section are confirmation of the approximate approach starting with the potential Φ used in section 3.1.

Also the approach used here can be extended to derive approximate equations when the upstream flow can be regarded as having given vorticity, for example, a sheared velocity profile in the x-direction and constant total enthalpy. The expansion is still that of (3.2.2), for example in planar flow:

$$\frac{\mathbf{q}}{U} = \left(1 + \delta^{\frac{2}{3}} u_1(x, \tilde{y}; K)\right)\mathbf{i}_x + \delta v_1(x, \tilde{y}; K)\mathbf{i}_{\tilde{y}} + \cdots \quad , \qquad (3.2.30a)$$

$$\frac{P}{P_\infty} = 1 + \delta^{\frac{2}{3}} p_1 + \cdots \quad . \qquad (3.2.30b)$$

Now as $x \to \infty$

$$u_1(x, \tilde{y}; K) \to u_\infty(\tilde{y}) , \quad v_1 \to 0 , \quad p_1 \to 0 , \qquad (3.2.31)$$
$$u_1(x, 0) = 0 \quad \text{(say)} ,$$

(see Figure 3.2.3).

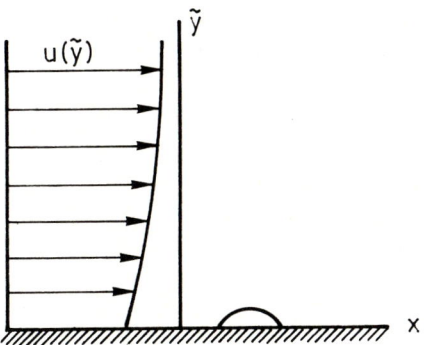

Figure 3.2.3

Transonic shear flow

The derivation follows as before but now there is $O(\delta)$ vorticity upstream which is conserved. The basic transonic system is thus

$$v_{1x} - u_{1\tilde{y}} = \frac{-du_\infty}{d\tilde{y}} = -u_\infty'(\tilde{y}) ,$$
$$\left(Ku_1 - \frac{(\gamma+1)}{2} u_1^2\right)_x + v_{1\tilde{y}} = 0 . \qquad (3.2.32)$$

This can be put in another form by letting

$$u_1 = u_\infty(\tilde{y}) + w(x, \tilde{y}) , \qquad (3.2.33)$$

so that

$$v_{1x} - w_{\tilde{y}} = 0 \,,$$

$$\left(Kw - (\gamma - 1)u_\infty w - \frac{\gamma + 1}{2}w^2 \right)_x + v_{\tilde{y}} = 0 \,.$$

Thus a perturbation potential $\varphi(x, \tilde{y})$ can be introduced

$$w = \varphi_x \,, \quad v_1 = \varphi_{\tilde{y}} \,, \tag{3.2.34}$$

and the continuity equation above becomes

$$\left(K - (\gamma + 1)u_\infty(\tilde{y}) \right)\varphi_{xx} - (\gamma + 1)\varphi_x\varphi_{xx} + \varphi_{\tilde{y}\tilde{y}} = 0 \,. \tag{3.2.35}$$

This is the same form as the K-G equation (3.1.14) with the similarity parameter there replaced by one depending on the upstream shear flow disturbance

$$K \to K_s(\tilde{y}) = K - (\gamma + 1)u_\infty(\tilde{y}) \,. \tag{3.2.36}$$

Note that the integral of x-momentum gives

$$p_1 = -\gamma w = -\gamma\varphi_x \,. \tag{3.2.37}$$

3.3 Expansion Procedures for Jet Flows

In transonic jet flows the half-width H for the planar case or radius of the jet for the axially symmetric case can be taken as the characteristic length in the transverse direction. (See Figure 3.3.1). This means that the transonic expansion has to be carried out with an x-coordinate of boundary-layer type to account for rapid changes in the x-direction. For the case of the subsonic jet the characteristic deflection δ can be specified by the geometry before the exit and M_∞ can be specified in terms of conditions far downstream.

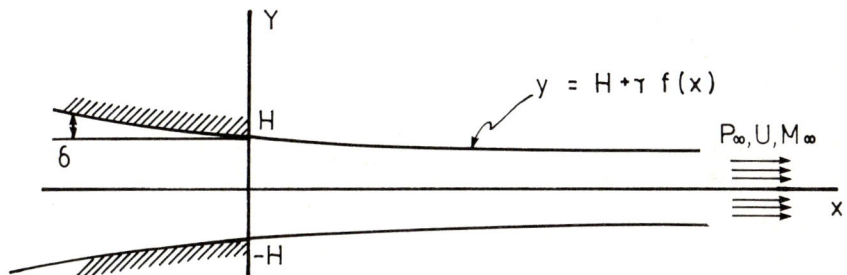

Figure 3.3.1

Subsonic jet

The expansion takes the form

$$\Phi(X,Y) = U\left\{X + \varepsilon(\delta)H\phi(x^*, y; K) + \dots\right\}$$

with $y = \dfrac{Y}{H}$, $\quad x^* = \dfrac{1}{\mu(\delta)}\dfrac{X}{H}$, $\quad K = \dfrac{1 - M_\infty^2}{\nu(\delta)}$,

ε, μ, ν are to be found. The velocity components are

$$\frac{\Phi_x}{U} = 1 + \frac{\varepsilon}{\mu}\phi_{x^*}, \quad \frac{\Phi_y}{U} = \varepsilon\phi_y. \tag{3.3.2}$$

For the flow to be $O(\delta)$,

$$\boxed{\varepsilon = \delta .} \tag{3.3.3}$$

The enthalpy integral of (3.1.1) has the expansion

$$\frac{a^2}{U^2} = 1 + \left(K - (\gamma - 1)\phi_{x^*}\right)\frac{\delta}{\mu} + \dots , \tag{3.3.4}$$

where

$$\nu = \frac{\varepsilon}{\mu} = \frac{\delta}{\mu} \,.$$

The equation for the potential (3.1.1) has the dominant terms

$$\left((K - (\gamma - 1)\phi_{x^*}) \frac{\delta}{\mu} - 2\frac{\delta}{\mu}\phi_{x^*} \right) \frac{\delta}{\mu^2}\phi_{x^* x^*} + \ldots + \delta\phi_{yy} = 0 \,,$$

so that the transonic equation results when $\frac{\delta^2}{\mu^3} = \delta$

$$\mu = \delta^{\frac{1}{3}} \,, \quad \text{and} \quad \nu = \delta^{\frac{2}{3}} \,.$$

Thus

$$\left(K - (\gamma + 1)\phi_{x^*} \right) \phi_{x^* x^*} + \phi_{yy} = 0 \,, \tag{3.3.5}$$

which is, of course, the usual transonic equation.

In summary the transonic expansion is

$$\Phi = U\{ X + \delta H \phi(x^*, y; K) + \ldots \} \,, \tag{3.3.6}$$

with $y = \dfrac{Y}{H}$, $\quad x^* = \delta^{-\frac{1}{3}}\dfrac{X}{H}$, $\quad K = \dfrac{1 - M_\infty^2}{\delta^{\frac{2}{3}}}$.

Consistent with this expansion the free streamline from the jet exit may be represented

$$Y = H + Tf(x^*) \,, \quad \text{or} \quad y = 1 + \tau(\delta)f(x^*) \,. \tag{3.3.7}$$

The requirement that the flow direction is tangent to the streamline shape is

$$\frac{\Phi_y}{U + \ldots} = \delta\phi_y(x^*, 1 + \ldots) = \frac{\tau(\delta)}{\mu(\delta)}f'(x^*) \,.$$

Thus, $\boxed{\tau = \delta^{\frac{4}{3}}}$ and the condition on the upper streamline is

$$\phi_y(x^*, 1) = f'(x^*) \,. \tag{3.3.8}$$

There is a corresponding condition on the lower streamline. These conditions serve to define the shape of the jet once the solutions are found. The actual boundary condition at the edge of the jet is that the pressure is constant. Since (cf. 3.2.20),

$$\frac{P}{P_\infty} = 1 - \gamma\delta^{\frac{2}{3}}\phi_{x^*} + \ldots \,,$$

the condition of constant pressure approximately at $y = \pm 1$ is

$$\phi_{x^*}(x^*, \pm 1) = 0 . \qquad (3.3.9)$$

It is also possible to consider problems for jets in which the velocity at the exit is supersonic. In that case the small parameter which is related to the order of the flow deflection, pressure perturbation, etc. can be expressed in terms of the pressure difference between the exit and the ambient medium. If

$$\varepsilon = \left(1 - \frac{P_e}{P_\infty}\right) \frac{1}{\gamma} , \qquad (3.3.10)$$

the expansion becomes

$$\Phi(X, Y) = U\left\{X + \varepsilon^{\frac{3}{2}} H \phi(x^*, y; K) + \ldots\right\} ,$$

where

$$K = \frac{1 - M_\infty^2}{\varepsilon} , \quad y = \frac{Y}{H} , \quad x^* = \varepsilon^{-\frac{3}{2}} \frac{X}{H} .$$

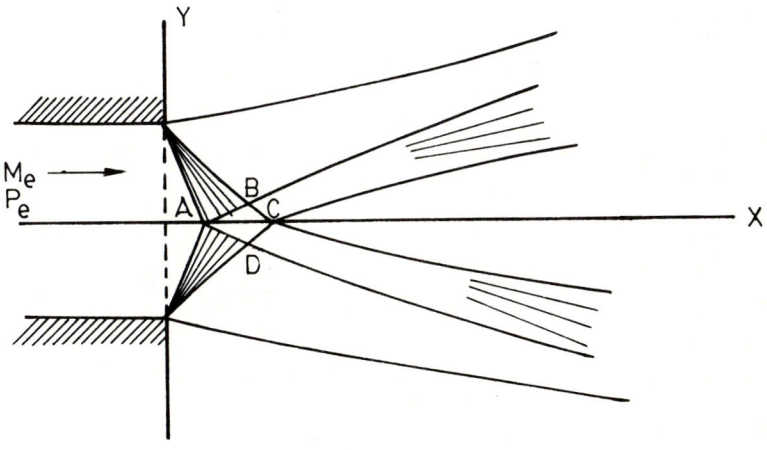

Figure 3.3.2

Supersonic jet

3.4 Transonic Similarity Rules

Transonic similarity rules are essentially contained in the expansion procedures of Sec. 3.1 or (3.2.3). The flows are calculated for fixed values of the similarity parameters (K, A, B for example). This enables a connection to be made between various airfoils or wings of similar shapes and planforms, with suitably chosen thickness (δ) and free stream Mach number (M_∞). There is a correspondence of local features of the flow, for example shock location and size, pressure at a point, as well as surface pressure distribution, lift, drag, and moment. It will be clear from the discussion here how more general rules can be devised for special circumstances.

In this section a brief discussion of these similarity rules is given. Consider first a family of three-dimensional wings as in Sec. 3.1.1. A more explicit representation of the shape can be used to indicate the angle of attack of a fixed wing shape. Let

$$F_{u,\ell}\left(x, \frac{z}{b}\right) = c\left(x, \frac{z}{b}\right) \pm t\left(x, \frac{z}{b}\right) - A . \qquad (3.4.1)$$

Here $A = \frac{\alpha}{\delta}$ is another similarity parameter, assumed held fixed as $\delta \to 0$, while α = angle of attack of the wing. $t(x, \frac{z}{b})$ gives the thickness distribution, $c(x, \frac{z}{b})$ the camber distribution. The planform is specified by the equations for the leading and trailing edges $x = x_{LE}(\frac{z}{b})$, $x = x_{TE}(\frac{z}{b})$. The transonic expansion, for a given family of wings (similar airfoil shapes and planforms) is thus

$$\Phi(x, y, z; \delta M_\infty, b, \alpha) = U\left\{x + \delta^{\frac{2}{3}} \phi(x, \tilde{y}, \tilde{z}; K, B, A) + \ldots\right\} , \qquad (3.4.2)$$

where the similarity parameters are

$$K = \frac{1 - M_\infty^2}{\delta^{\frac{2}{3}}} , \qquad B = b\delta^{\frac{1}{3}} , \qquad A = \frac{\alpha}{\delta}$$

and
$$\tilde{y} = \delta^{\frac{1}{3}} y , \qquad \tilde{z} = \delta^{\frac{1}{3}} z .$$

The pressure coefficient is

$$c_p = \frac{P - P_\infty}{\frac{\rho_\infty U^2}{2}} = -2\delta^{\frac{2}{3}} \phi_x(x, \tilde{y}, \tilde{z}; K, B, A) . \qquad (3.4.3)$$

Thus for two different values of M_∞, $M_{1\infty}$, $M_{2\infty}$,

$$\delta_2 = \delta_1 \left(\frac{1 - M_{2\infty}^2}{1 - M_{1\infty}^2}\right)^{\frac{3}{2}}$$

for similar flows in order to keep K constant. Similarly

$$b_2 = b_1 \left(\frac{\delta_1}{\delta_2}\right)^{\frac{1}{3}} = b_1 \left(\frac{1 - M_{1\infty}^2}{1 - M_{2\infty}^2}\right)^{\frac{1}{2}} \qquad \text{and} \qquad \alpha_2 = \alpha_1 \frac{\delta_2}{\delta_1} = \alpha_1 \left(\frac{1 - M_{2\infty}^2}{1 - M_{1\infty}^2}\right)^{\frac{3}{2}}$$

for corresponding flows. As the Mach number gets closer to one flows are kept similar for which the airfoil thickness is reduced, the angle α reduced and the span increased. At corresponding points in space $x_2 = x_1$, $y_2 = y_1 \left(\frac{\delta_2}{\delta_1}\right)^{\frac{1}{3}} = y_1 \left(\frac{1 - M_{2\infty}^2}{1 - M_{1\infty}^2}\right)^{\frac{1}{2}}$, $z_2 = z_1 \left(\frac{\delta_1}{\delta_2}\right)^{\frac{1}{3}}$ and at these corresponding points, which run away from the x-axis like the Mach angle in supersonic flow, there is a scaling of c_p

$$c_{p2} = c_{p1} \left(\frac{\delta_2}{\delta_1}\right)^{\frac{2}{3}} = c_{p1} \left(\frac{1 - M_{2\infty}^2}{1 - M_{1\infty}^2}\right)^{\frac{1}{2}}.$$

Shock location and strength scale accordingly. Note that $c_{p1} = \frac{P_1 - P_\infty}{\rho_{1\infty} \frac{U^2}{2}}$. The surface pressure distribution evaluated at $\tilde{y} = 0$ scales in exactly the same way. Since the lift L and drag D can be obtained by integration of the surface pressures similarity rules for the lift and drag follow.

$$L = c^2 \int_{-b}^{b} dz \int_{x_{LE}}^{x_{TE}} dx \, (P_\ell - P_u), \qquad \begin{array}{l}\text{where } c = \text{characteristic length} \\ \text{of the airfoil chord. } (c = 1 \text{ earlier),}\end{array}$$

$$\text{(3.4.4)}$$

where $P_{u,\ell}$ are the pressures on the upper and lower surfaces respectively. Thus

$$L = \rho_\infty \frac{U^2}{2} \delta^{-\frac{1}{3}} c^2 \int_{-B}^{B} d\tilde{z} \int_{x_{LE}}^{x_{TE}} dx \left(c_{p_\ell} - c_{p_u}\right),$$

or

$$L = \rho_\infty U^2 \delta^{\frac{1}{3}} c^2 \int_{-B}^{B} d\tilde{z} \int_{x_{LE}}^{x_{TE}} dx \left\{\phi_x(x, 0+, \tilde{z}; K, A, B) - \phi_x(x, 0-, \tilde{z}; K, A, B)\right\},$$

and

$$c_L = \frac{L}{\frac{\rho_\infty U^2}{2} (\text{wing area})} = \delta^{\frac{2}{3}} \text{fn}(K, A, B). \qquad \text{(3.4.5)}$$

if we use (wing area) $\sim bc$, $b/c = B/\delta^{1/3}$. In an analogous way for the drag

$$D = c^2 \delta \int_{-b}^{b} dz \int_{x_{LE}}^{x_{TE}} dx \left(P_u \frac{\partial F_u}{\partial x} - P_\ell \frac{\partial F_\ell}{\partial x}\right). \qquad \text{(3.4.6)}$$

Scaling in a similar way

$$D = \rho_\infty U^2 \delta^{\frac{4}{3}} c^2 \int_{-B}^{B} d\tilde{z} \int_{x_{LE}}^{x_{TE}} dx \left\{ \phi_x(0-)\frac{\partial F_\ell}{\partial x} - \phi_x(0+)\frac{\partial F_u}{\partial x} \right\} ,$$

$$c_D = \frac{D}{\frac{\rho_\infty u^2}{2}(\text{wing area})} = \delta^{\frac{5}{3}}\text{fn}(K, A, B) . \qquad (3.4.7)$$

One of the principal tests of transonic theory is a comparison of the similarity rules based on this theory with experiments on wings and airfoils of different (δ, M_∞). Local quantities such as the pressure can be scaled and the similarity of lift and drag can also be considered. Some examples of such comparisons are shown in Figs. (3.4.1, 3.4.2, 3.4.3). In general, the agreement is sufficiently good to reinforce our confidence in the theory. Fig. 3.4.1 is taken essentially from [3.4.1] in which Spreiter emphasizes the non-uniqueness of the transonic similarity parameter K.

The data of Fig. 3.4.1 falls more closely on a single curve, when a modified parameter

$$K_S = \frac{1 - M_\infty^2}{\left\{ (\gamma + 1)M_\infty^2 \delta \right\}^{\frac{2}{3}}}$$

is used. This question is taken up again by Hayes in [3.4.2], which discussed in great detail the second approximation to transonic flow, and the effect on it of the choice of similarity parameter. For example, Hayes suggests that the use of a parameter K_B based on Busseman's second order theory may be useful

$$K_B = \frac{1 - M_\infty^2}{\left\{ \left(\frac{\gamma+1}{2}M_\infty^4 + 2 - 2M_\infty^2 \right) \delta \right\}^{\frac{2}{3}}} .$$

Hayes also extends the previous results to gases with arbitrary equations of state (which is always possible for small-disturbance theories). There is no guarantee that the use of K_B or K_S is better in all problems, and probably more comparisons with experiments need to be made. Fig. 3.4.1 also contains theoretical results for the flow past a wedge at subsonic speeds ([3.4.4], [3.4.5]), sonic speed ([3.4.6]) and supersonic speed ([3.4.7]) all of which agree fairly well with experiment.

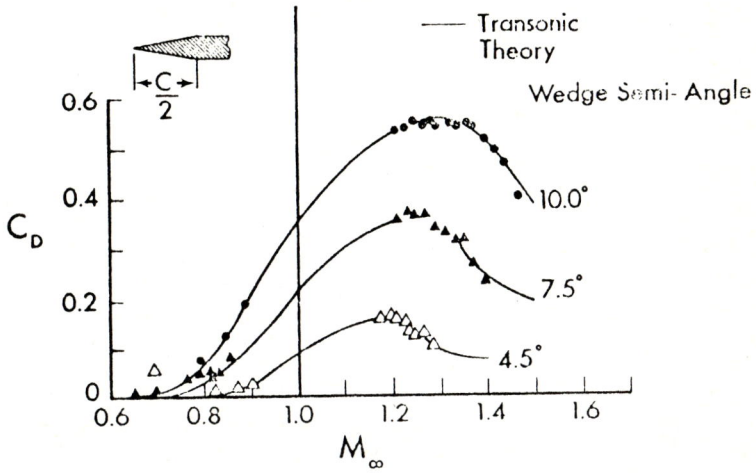

Figure 3.4.1a

Transonic Experiments; Drag Coefficient vs. Mach Number

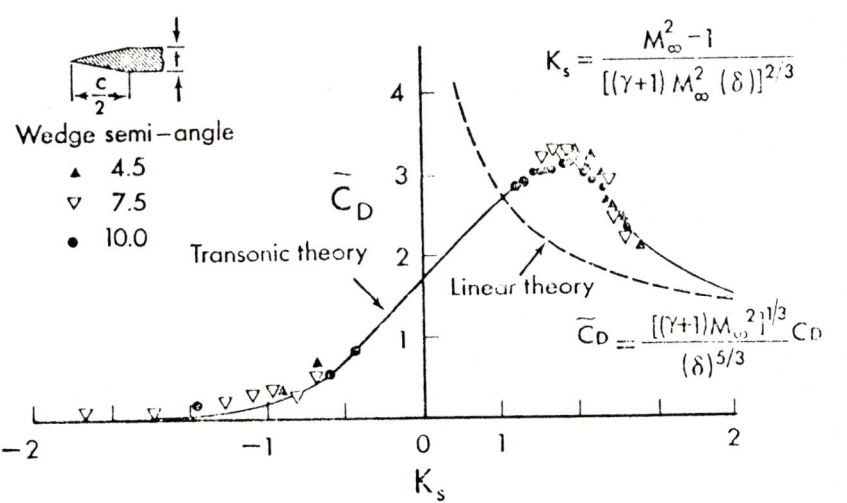

Figure 3.4.1b

Transonic Experiments; Similarity Form

Figs. 3.4.2, 3.4.3, based on the results of [3.4.3], show some of the effects of finite aspect ratio. There are corresponding experiments and similarity laws also for bodies of revolution and a variety of cases are discussed by Hayes (quasi-cylinder, cylinder, flat slender bodies, nonflat slender bodies).

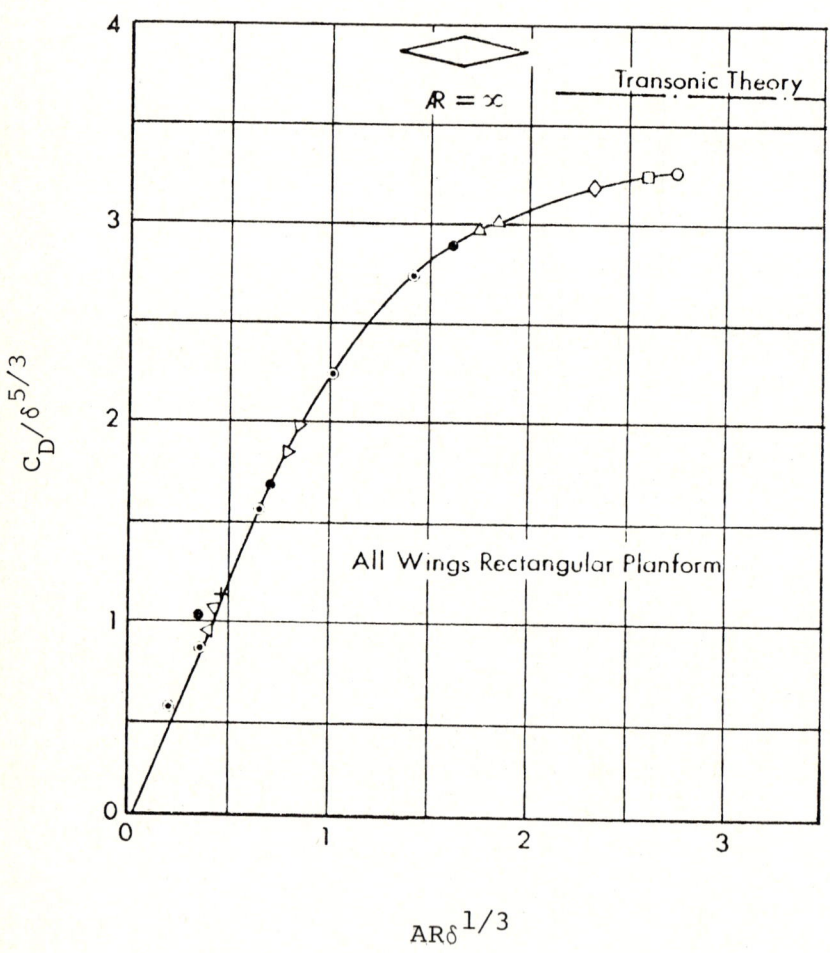

Figure 3.4.2

Transonic Experiments; Three Dimensional Similarity for Drag

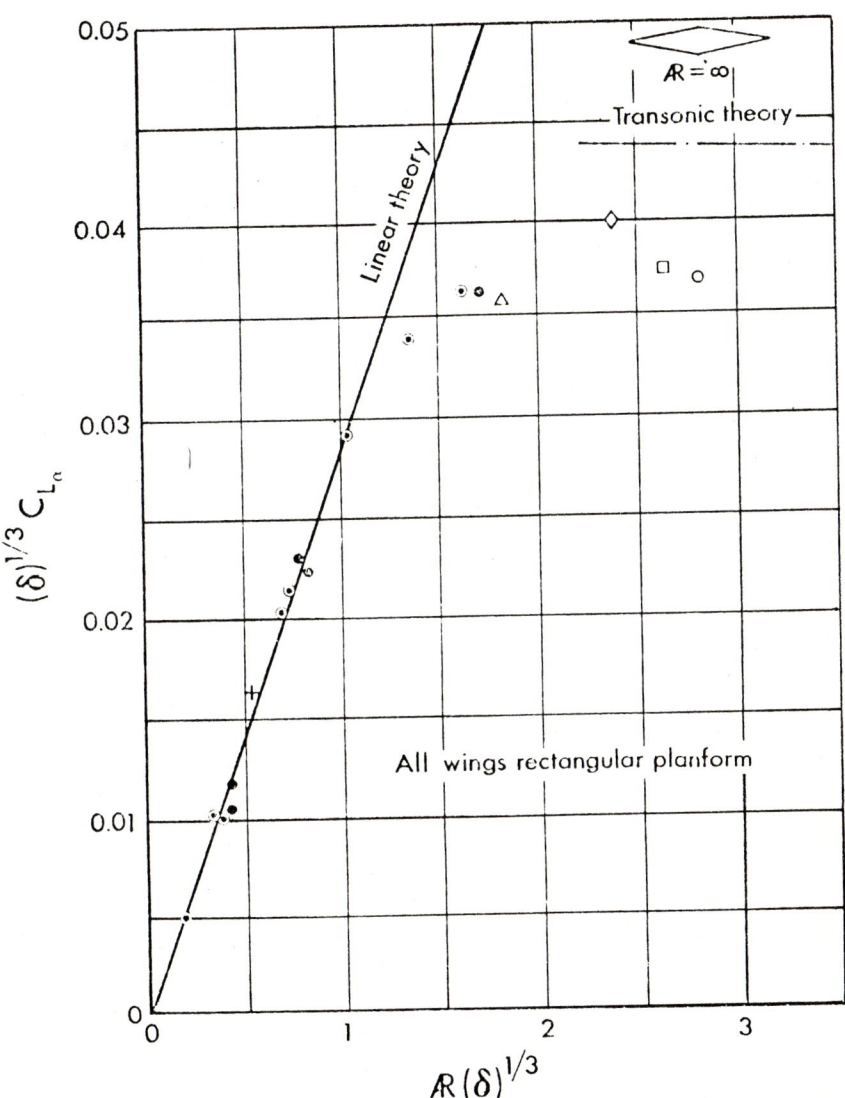

Figure 3.4.3

Transonic Experiments; Three Dimensional Similarity for Lift

A systematic study of the similarity parameter was undertaken by K. Kusunose [3.4.8]. For two-dimensional flow the aim was to use a higher order expansion of the similarity parameter

$$M_\infty^2 = 1 - K\delta^{\frac{2}{3}} + K_2(K)\delta^{\frac{4}{3}} + \ldots$$

together with a similar expansion for $\tilde{y}$, and second order transonic small disturbance theory to determine $K_2(K)$ etc. . Kusunose considered shock waves and simple waves and tried to determine $K_2(K)$ to make first order theory as good as possible. Unfortunately no simple results were achieved, so that empirical considerations for wide classes of problems remain the best method for fixing a practical form of K.

References

[3.4.1] Spreiter, J., On the Application of Transonic Similarity Rules to Wings of Finite Span, *NACA Tech. Rep.* 1153, 1953.

[3.4.2] Hayes, W. D., La seconde approximation pour les écoulements transoniques non visqueux, *Journal de Mechanique*, v. 5, n. 2, Juin 1966.

[3.4.3] Nelson, W. H. and McDevitt, J. B., The Transonic Characteristics of 17 Rectangular Symmetrical Wing Models of Varying Aspect Ratio and Thickness, *NACA Research Memo.* A51A12, 1951.

[3.4.4] Cole, J. D. Drag of Finite Wedge at High Subsonic Speeds, *J. Math. and Phys.*, **30**, No. 2, July 1951, pp. 79-93.

[3.4.5] Trilling, L. Transonic Flow Past a Wedge at Zero Angle of Attack, *ZAMP* **4**, No. 5, Sept. 1953.

[3.4.6] Guderley, G., and Yoshihara, H. The Flow over a Wedge Profile at Mach Number 1, *J. Aero. Sci.*, Vol. 17, No. 11, Nov. 1950, pp. 723-735.

[3.4.7] Vincenti, W. G., and Wagoner, C. B., Transonic Flow Past a Wedge Profile with Detached Bow Wave - General Analytic Method and Calculated Results, *NACA, TN* 2588, (1951).

[3.4.8] Kusunose, K. Two Dimensional Flow Past Convex and Concave Corners at Transonic Speed and Two Dimensional Flow around a Parabolic Nose at Subsonic and Transonic Speeds, PhD. Thesis, University of California Los Angeles 1979.

3.5 Hodograph equations for planar flow

The two dimensional small disturbance equation (3.1.14) and the irrotationality condition are equivalent respectively to the equations

$$ww_x - \vartheta_{\tilde{y}} = 0 \ ,$$
$$w_{\tilde{y}} - \vartheta_x = 0 \ ,$$

(3.5.1)

where

$$w = (\gamma + 1)\phi_x - K \ ,$$
$$\vartheta = (\gamma + 1)\phi_{\tilde{y}} \ .$$

(3.5.2)

Under certain conditions a local inversion of dependent and independent variables can be carried out on system (3.5.1). The resulting equations known as the hodograph equations have as independent variables the velocity components, w, ϑ and as dependent variables x, $\tilde{y}$. These equations are easier to deal with than (3.5.1) since they are linear, hence the characteristics and the location of the sonic line are known in advance in this hodograph (velocity) plane. However, in general the boundary conditions are difficult to handle in the hodograph plane, shocks in the physical plane correspond to gaps in the hodograph plane, and the inversion can often only be accomplished locally.

To carry out the inversion note that the increments of velocity dw, $d\vartheta$ are

$$dw = w_x dx + w_{\tilde{y}} d\tilde{y} \ ,$$
$$d\vartheta = \vartheta_x dx + \vartheta_{\tilde{y}} d\tilde{y} \ ,$$

(3.5.3)

and that the increments of position are

$$dx = x_w dw + x_\vartheta d\vartheta \ ,$$
$$dy = \tilde{y}_w dw + \tilde{y}_\vartheta d\vartheta \ .$$

(3.5.4)

Substitution of (3.5.3) into (3.5.4) gives

$$dx = (x_w w_x + x_\vartheta \vartheta_x)dx + (x_w w_{\tilde{y}} + x_\vartheta \vartheta_{\tilde{y}})d\tilde{y} \ ,$$
$$d\tilde{y} = (\tilde{y}_w w_x + y_\vartheta \vartheta_x)dx + (\tilde{y}_w w_{\tilde{y}} + \tilde{y}_\vartheta \vartheta_{\tilde{y}})d\tilde{y} \ .$$

Since dx, $d\tilde{y}$ are independent, these equations imply

$$x_w w_x + x_\vartheta \vartheta_x = 1 \ , \quad \tilde{y}_w w_x + \tilde{y}_\vartheta \vartheta_x = 0 \ ,$$
$$x_w w_{\tilde{y}} + x_\vartheta \vartheta_x = 0 \ , \quad \tilde{y}_w w_{\tilde{y}} + \tilde{y}_\vartheta \vartheta_{\tilde{y}} = 1 \ .$$

This system of equations can be solved uniquely for x_w, x_ϑ, $\tilde{y}_w$, $\tilde{y}_\vartheta$ in the neighborhood of those points $(x, \tilde{y})$ at which

$$j = \frac{\partial(w, \vartheta)}{\partial(x, \tilde{y})} = w_x \vartheta_{\tilde{y}} - \vartheta_x w_{\tilde{y}} \neq 0 \,. \tag{3.5.5}$$

That is, in the neighborhood of all points $(x, \tilde{y})$ such that the Jacobian $j \neq 0$, the inversion can be accomplished and it gives

$$\begin{aligned} x_w = \frac{1}{j} \vartheta_{\tilde{y}} \,, \qquad &\tilde{y}_w = -\frac{1}{j} \vartheta_x \,, \\ x_\vartheta = -\frac{1}{j} w_{\tilde{y}} \,, \qquad &\tilde{y}_\vartheta = \frac{1}{j} w_x \,. \end{aligned} \tag{3.5.6}$$

Subststitution of (3.5.6) into (3.5.1) gives the system of equations governing the flow in terms of the hodograph variables,

$$\begin{aligned} w\tilde{y}_\vartheta - x_w &= 0 \,, \\ \tilde{y}_w - x_\vartheta &= 0 \,. \end{aligned} \tag{3.5.7}$$

The equivalent second order partial differential equation is

$$w\tilde{y}_{\vartheta\vartheta} - \tilde{y}_{ww} = 0 \,, \tag{3.5.8}$$

known as the Tricomi equation. This variable coefficient equation changes type as $w \lessgtr 0$. If $w < 0$ the equation is elliptic, if $w > 0$ the equation is hyperbolic. The line $w = 0$, the ϑ axis in the hodograph plane, is the sonic line. This linear equation was obtained because the non-linear terms, namely j, cancelled. This cancellation does not occur in most cases, for example axially symmetric three dimensional flow, hence linearization can not be achieved in that case.

In a similar fashion the inversion from the hodograph plane back to the physical plane can be accomplished in the neighberhood of all points (w, ϑ) at which

$$J = \frac{\partial(x, \tilde{y})}{\partial(w, \vartheta)} = \frac{1}{j} = x_w \tilde{y}_\vartheta - x_\vartheta \tilde{y}_w \neq 0 \,. \tag{3.5.9}$$

The Jacobian j represents the directed ratio of the local area in the hodograph plane to that in the physical plane. This can be seen by calculating the vector product

$$\begin{aligned} d\mathbf{w} \times d\vartheta &= (w_x d\mathbf{x} + w_{\tilde{y}} d\tilde{\mathbf{y}}) \times (\vartheta_x d\mathbf{x} + \vartheta_{\tilde{y}} d\tilde{\mathbf{y}}) \\ &= j(d\mathbf{x} \times d\tilde{\mathbf{y}}) \,. \end{aligned}$$

Thus the map is

order reversing if $j < 0$,

order preserving if $j > 0$.

From (3.5.5) and (3.5.1),

$$j = w_x \vartheta_{\tilde{y}} - \vartheta_x w_{\tilde{y}} = w w_x^2 - w_{\tilde{y}}^2 , \qquad (3.5.10)$$

hence in subsonic zones, $w < 0$, the mapping, where defined, is order reversing $(j < 0)$, whereas in supersonic zones, $w > 0$, it may be either order preserving or order reversing.

From (3.5.10) it is clear that j can be zero at most at isolated points in the subsonic region, since if the solution of an elliptic equation is constant along a curve, then it is constant in the entire elliptic region. If $j = 0$ in a region of (supersonic) flow, then the region collapses in the hodograph, that is the image of this region in the hodograph is a point or a curve. Flows for which $j = 0$ along a curve in the physical plane (branch line) are considered in Section 3.7 as are also the unphysical flows for which $J = 0$ along a line in the hodograph plane (limit line). These latter are unphysical because the corresponding physical plane flow would consist of several sheets.

The characteristics in the physical plane are represented by

$$\frac{d\tilde{y}}{dx} = \pm \frac{1}{\sqrt{w}} . \qquad (3.5.11a)$$

To find the image of these characteristics in the hodograph plane note that along a curve $(x, \tilde{y}(x))$ in the physical plane we have

$$\frac{d\vartheta}{dw} = \frac{\vartheta_x + \vartheta_{\tilde{y}} \left(\frac{d\tilde{y}}{dx} \right)}{w_x + w_{\tilde{y}} \left(\frac{d\tilde{y}}{dx} \right)} ,$$

so that along a characteristic

$$\frac{d\vartheta}{dw} = \frac{\vartheta_x + w w_x \left(\pm \frac{1}{\sqrt{w}} \right)}{w_x + \vartheta_x \left(\pm \frac{1}{\sqrt{w}} \right)} ,$$

$$\frac{d\vartheta}{dw} = \pm \sqrt{w} \qquad (3.5.11b)$$

Integration of (3.5.11b) gives the explicit form of the curves,

$$\vartheta = \vartheta_0 \pm \frac{2}{3} w^{\frac{3}{2}} \qquad (3.5.12)$$

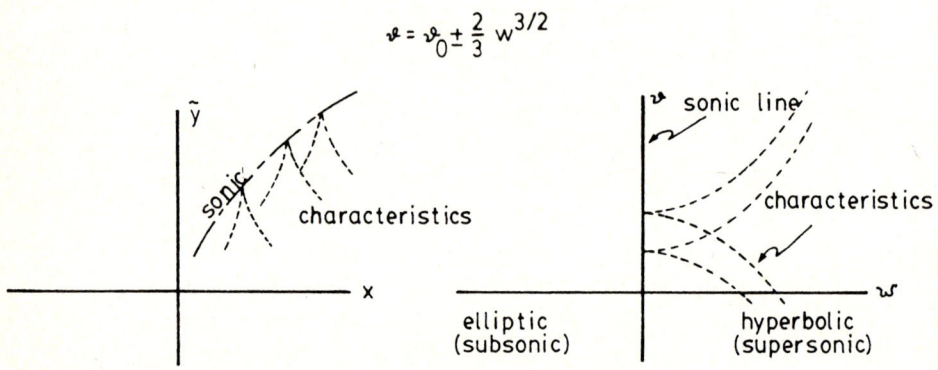

Figure 3.5.1

Characteristics in hodograph and physical planes

That the curves (3.5.12) are in fact the characteristics of system (3.5.7) can easily be checked. (See Figure 3.5.1).

Thus characteristics in the physical plane map onto characteristics in the hodograph plane, those of the plus family map to a direction orthogonal to the original minus family and vice versa.

As mentioned before although the equations are linear in the hodograph, and the location of the sonic line and of the characteristics is known in the hodograph, the location of the boundary curves are not known in advance except in a few special cases.

A direction must be set in the hodograph plane as time like for the supersonic $(w > 0)$ zone. In the physical plane the lines $\tilde{y} = $ constant are a first approximation to the streamlines. The streamlines bisect the characteristics (3.5.11a). In the physical plane the direction of flow along a streamline is taken as time like and the direction of the two characteristics at each point in the physical plane is taken as downstream. These two characteristics bound the region of influence, so a time-like direction is also fixed in the hodograph for the image of the streamline. Then the direction of the characteristics is known. This can best be illustrated by considering two elementary solutions of (3.5.7).

(1) Transonic Source Flow

Exact source flow has purely radial streamlines (cf. Figure 3.5.2).

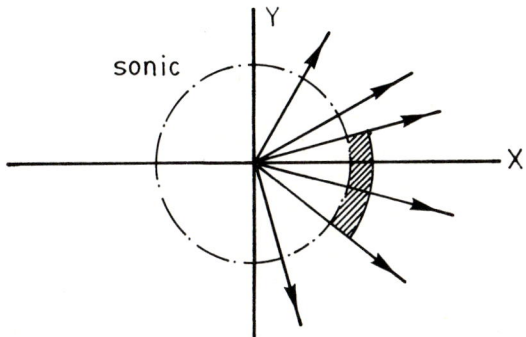

Figure 3.5.2

Exact source flow

The small disturbance flow is restricted to near sonic velocities and small deflections, the shaded region in Figure 3.5.2. The solution in the small disturbance approximation (3.5.7) is

$$\tilde{y} = c\vartheta$$
$$x = \frac{cw^2}{2} \, . \tag{3.5.13}$$

For this flow $\tilde{y}$ is constant along lines $\vartheta = $ constant, or streamlines carry constant deflection. With $c > 0$ the flow only exists in the physical plane for $x > 0$. (The map from the hodograph to the physical plane does not cover the full physical plane), and in fact the right half of the physical plane is doubly covered, once for $w > 0$, once for $w < 0$. Calculation of the Jacobian J gives $J = cw^2$ so, as expected, $J = 0$ if $w = 0$. The mapping to the physical plane can not be carried out past the sonic line. The sonic line is a limit line. This reflects the fact that the streamlines have no throat so that the flow can not pass through sonic. If the restriction is made to values of w such that $w > 0$, supersonic flow, then the map is well defined (1-1 and onto from $w > 0$ to $x > 0$). Inversion gives

$$\vartheta = \frac{1}{c}\tilde{y} ,$$

$$w = \sqrt{\frac{2x}{c}} .$$

The characterisics can be calculated explicitly in the physical plane,

$$\frac{d\tilde{y}}{dx} = \pm\frac{1}{\sqrt{w}} = \pm\left(\frac{c}{2x}\right)^{\frac{1}{4}} ,$$

$$\tilde{y} = \tilde{y}_0 \pm \frac{4}{3}\left(\frac{c}{2}\right)^{\frac{1}{4}} x^{\frac{3}{4}} .$$

The mapping is order preserving, streamlines in the hodograph plane bisect the expansion characteristics hence the name, source like-flow (Figure 3.5.2). This flow is supersonic accelerating from sonic. For the branch $w < 0$ we have a subsonic sink accelerating flow coming from infinity.

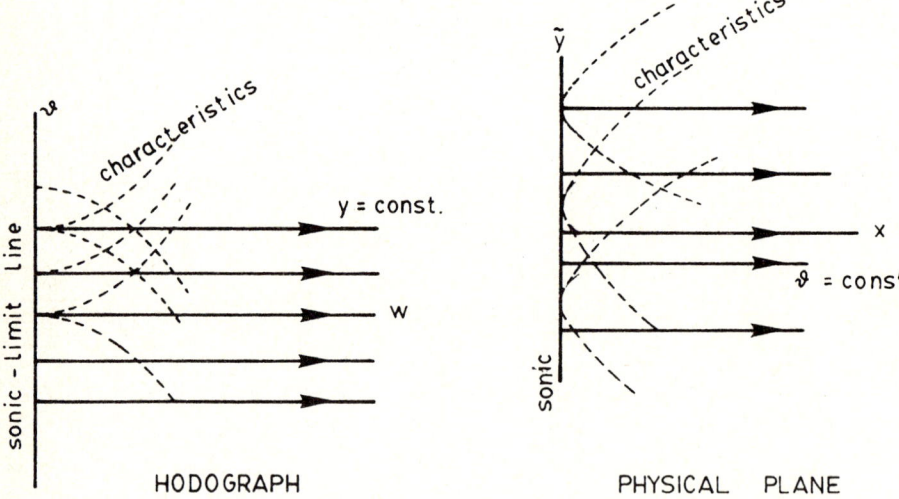

Figure 3.5.3

Transonic source flow

We can have a subsonic source accelerating from sonic and supersonic sink accelerating to sonic.

(2) Transonic Vortex Flow

For the exact vortex the streamlines are circles. For small disturbance flow this is expressed by another solution of (3.5.7) for which $\tilde{y}$ is constant on lines w = constant. It is given by

$$\tilde{y} = cw \, ,$$

$$x = c\vartheta \, ,$$

and applies to the shaded region on Figure 3.5.4. For this flow $J = -c^2$, hence the mapping between the hodograph and physical plane is order reversing and well defined on the entire plane. Inversion gives

$$w = \frac{\tilde{y}}{c} \, ,$$

$$\vartheta = \frac{x}{c} \, .$$

The characteristics in the physical plane are

$$\frac{d\tilde{y}}{dx} = \pm\sqrt{\frac{c}{\tilde{y}}} \, , \quad \text{or} \quad \frac{2}{3}\tilde{y}^{\frac{3}{2}} = \pm\sqrt{c}(x - x_0) \, .$$

In this flow the $\tilde{y}$ = constant streamlines bisect the characteristics one of which is an expansion, the other a compression. (Vortex-like flow). (Figure 3.5.5)

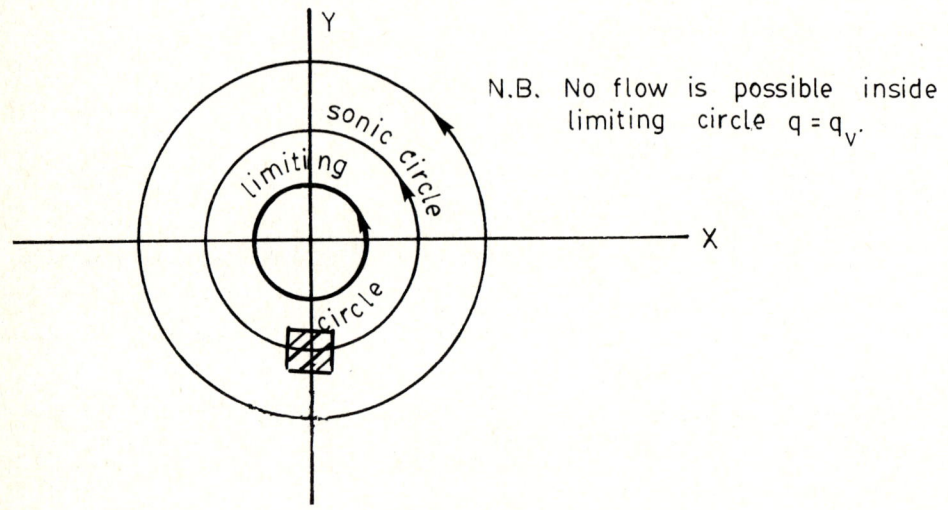

N.B. No flow is possible inside
limiting circle $q = q_v$.

Figure 3.5.4

Exact vortex flow

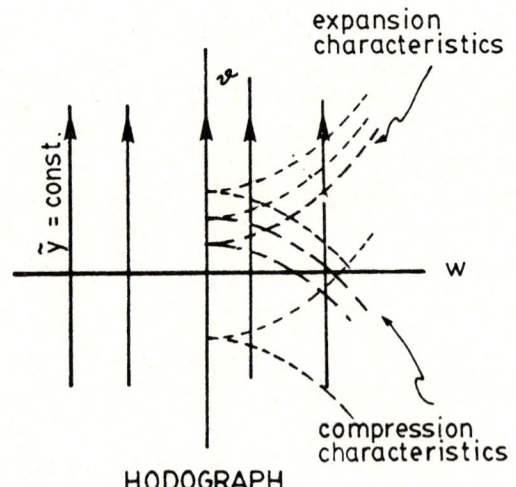

Figure 3.5.5

Transonic vortex flow (1)

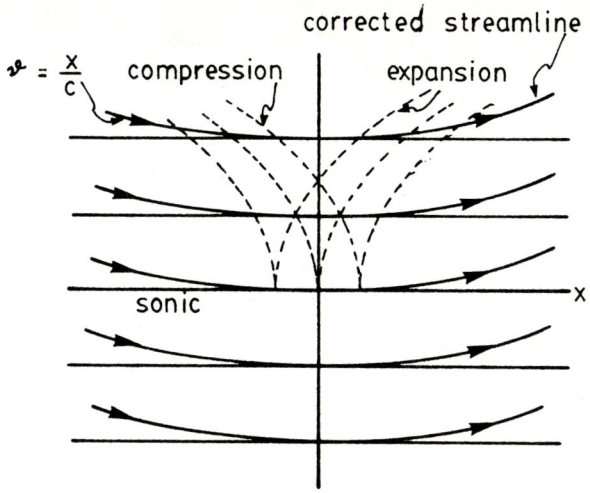

PHYSICAL   PLANE

Figure 3.5.5

Transonic vortex flow (2)

Due to the linearity of the hodograph system of equations it is possible to superimpose the hodograph flows of source and vortex to make a source-vortex flow of the general form

$$\begin{cases} \tilde{y} = c_1 \vartheta + c_2 w \\ x = \dfrac{c_1 w^2}{2} + c_2 \vartheta \end{cases} \tag{3.5.14}$$

The flow in the physical plane is evidently not merely a superposition of the two flows.

Several other exact solutions have been found, Many of these will be discussed elsewhere in the monograph. One set of exact solutions that will be mentioned here are those needed for jet flows. These solutions are obtained by direct separation of variables in (3.5.8).

Setting

$$\tilde{y}(w, \vartheta) = W(w)\Theta(\vartheta)$$

in (3.5.8) we find

$$\frac{\Theta''}{\Theta} = \frac{1}{w}\frac{W''}{W} = -\lambda ,$$

Thus,

$$\begin{cases} \Theta(\vartheta) = a_1 e^{i\sqrt{\lambda}\vartheta} + a_2 e^{-i\sqrt{\lambda}\vartheta} \\ W(w) = b_1 Ai(-\lambda w) + b_2 Bi(-\lambda w) \end{cases} \qquad (3.5.15)$$

where Ai, Bi are the Airy functions. They are related to the Bessel functions of order $1/3$ by

$$Ai(-z) = \sqrt{\frac{z}{3}} \left\{ J_{\frac{1}{3}}\left(\frac{2}{3}z^{\frac{3}{2}}\right) + J_{-\frac{1}{3}}\left(\frac{2}{3}z^{\frac{3}{2}}\right) \right\}$$

$$Bi(-z) = \sqrt{\frac{2}{3}} \left\{ J_{-\frac{1}{3}}\left(\frac{2}{3}z^{\frac{3}{2}}\right) - J_{\frac{1}{3}}\left(\frac{2}{3}z^{\frac{3}{2}}\right) \right\} .$$

Other properties of these functions are needed for later sections so they are summarized here [3.5.1].

For small values of z,

$$Ai(z) = c_1 f(z) - c_2 g(z) , \quad Bi(z) = \sqrt{3}\left(c_1 f(z) + c_2 g(z)\right) ,$$

where

$$f(z) = 1 + \frac{1}{3!}z^3 + O(z^6) ,$$

$$g(z) = z + O(z^4) ,$$

and

$$Ai(0) = c_1 = \frac{Bi(0)}{\sqrt{3}} = \left(3^{\frac{2}{3}}\Gamma(2/3)\right)^{-1} ,$$

$$Ai'(0) = -c_2 = \frac{-Bi'(0)}{\sqrt{3}} = \left(3^{\frac{1}{3}}\Gamma\left(\frac{1}{3}\right)\right)^{-1} .$$

For large values of z,

$$Ai(z) = \left\{ \left(2\sqrt{\pi}z^{\frac{1}{4}}\right)^{-1} + O\left(z^{-\frac{7}{4}}\right) \right\} e^{\frac{2}{3}z^{\frac{3}{2}}} \qquad \text{if } |\arg z| < \pi ,$$

$$Ai(-z) = \left(\sqrt{\pi}z^{\frac{1}{4}}\right)^{-1}\sin\left(\frac{2}{3}z^{\frac{3}{2}} + \frac{\pi}{4}\right) + O\left(z^{-\frac{7}{4}}\right) \quad \text{if } |\arg z| < \frac{2\pi}{3} ,$$

$$Bi(z) = \left\{ \frac{1}{\sqrt{\pi}z^{\frac{1}{4}}} + O\left(z^{-\frac{7}{4}}\right) \right\} e^{\frac{2}{3}z^{\frac{3}{2}}} \quad \text{if } |\arg z| < \frac{\pi}{3} ,$$

$$Bi(-z) = \frac{1}{\sqrt{\pi}z^{\frac{1}{4}}}\cos\left(\frac{2}{3}z^{\frac{3}{2}} + \frac{\pi}{4}\right) + O\left(z^{-\frac{7}{4}}\right) \quad \text{if } |\arg z| < \frac{2\pi}{3} .$$

The definitions are selected so that Bi grows exponentially for large (positive real part) argument and Ai decays exponentially. A graph of these functions appears below (Figure 3.5.6).

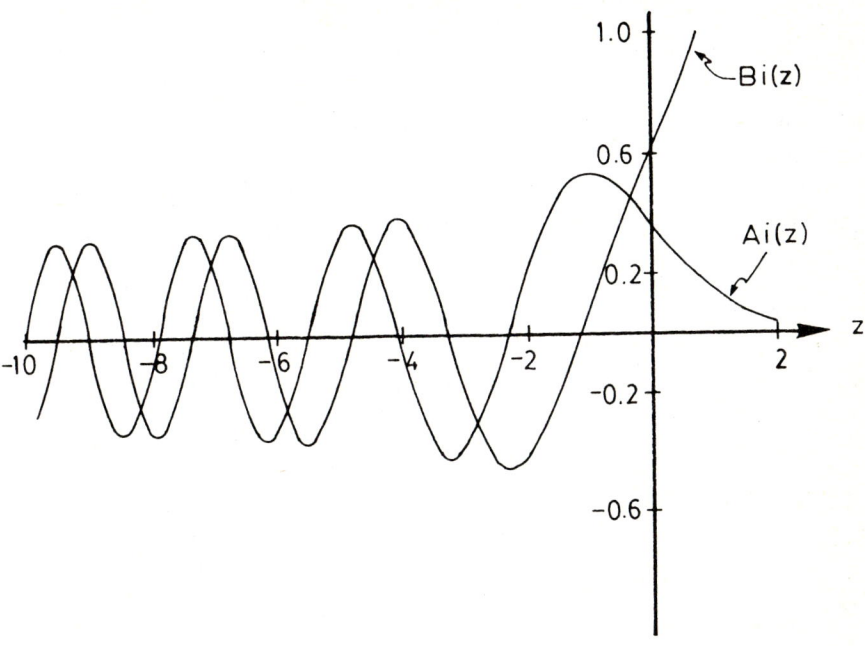

Figure 3.5.6

Airy function

Note that in the supersonic region $w > 0$ these special solutions, for real λ, are oscillatory both in ϑ and w; This mimics locally the behavior of the wave equation. But for the subsonic region, $w < 0$, the dependence on w is basically exponential while oscillatory in ϑ, just as a solution of the Laplace equation.

References

[3.5.1] M. Abramowitz and I. Stegun, *Handbook of Mathematical Functions*, Dover, New York 1965.

3.6. Simple Waves, Shock Waves, Detachment

The special solutions of this section are characterised by the property that they can appear both globally, that is over a finite region, and locally in the neighberhood of a point. Simple waves are supersonic flows in which only one family of characteristics carries a disturbance. That is the flow changes only when a streamline crosses a member of one family of characteristics (expansion *or* compression). Shock waves are compression jumps calculated according to the rules set out earlier (3.1.23,24).

According to a well-known theorem proved in Courant and Friedrichs [3.6.1], supersonic regions of uniform flow ($\vartheta = $ const, $w = $ const) must be bounded either by shock waves (of uniform strength) or by simple waves. Thus these solutions are important for building up more complex flow fields.

3.6.1. Simple Waves

Simple wave solutions for general hyperbolic systems can be sought by looking for solutions in which all the dependent variables depend on a single coordinate, the "phase" σ. In this case, in the transonic system (3.5.1)

$$\left\{ \begin{array}{l} w w_x - \vartheta_{\tilde{y}} = 0 \\ \vartheta_x - w_{\tilde{y}} = 0 \end{array} \right\} \tag{3.6.1}$$

let

$$w = w(\sigma) \,, \quad \vartheta = \vartheta(\sigma) \tag{3.6.2}$$

where $\sigma = \sigma(x, \tilde{y})$ is to be found. Then

$$\left\{ \begin{array}{l} w \dfrac{dw}{d\sigma} \sigma_x - \dfrac{d\vartheta}{d\sigma} \sigma_{\tilde{y}} = 0 \\ \dfrac{d\vartheta}{d\sigma} \sigma_x - \dfrac{dw}{d\sigma} \sigma_{\tilde{y}} = 0 \end{array} \right\} \tag{3.6.3}$$

No solutions for $(\sigma_x, \sigma_{\tilde{y}})$ exist unless the determinant vanishes

$$w \left(\frac{dw}{d\sigma} \right)^2 = \left(\frac{d\vartheta}{d\sigma} \right)^2 \tag{3.6.4}$$

(3.6.4) shows that:

(i) no solution exists unless the flow is locally supersonic $w > 0$,

(ii) as the phase σ changes, the hodograph image $\vartheta(w)$ of the flow lies on a single characteristic, obtained by integrating (3.6.5) (cf. 3.5.12)

$$\frac{d\vartheta}{d\sigma} = \pm\sqrt{w}\,\frac{dw}{d\sigma} \tag{3.6.5}$$

or

$$\vartheta(\sigma) = \vartheta_0 \pm \frac{2}{3}w^{\frac{3}{2}}(\sigma) . \tag{3.6.6}$$

Now, from (3.6.3) we can find the conditions which must hold on a given phase curve $\sigma = \text{constant}, 0 = d\sigma = \sigma_x + \sigma_{\tilde{y}}d\tilde{y}$,

$$\left(\frac{d\tilde{y}}{dx}\right)_{\sigma=\text{constant}} = -\frac{\sigma_x}{\sigma_{\tilde{y}}} = -\frac{\dfrac{d\vartheta}{d\sigma}}{w\dfrac{dw}{d\sigma}} = \mp\frac{1}{\sqrt{w(\sigma)}} \tag{3.6.7}$$

from (3.6.5).

This last relation shows that:

(i) the lines of constant phase σ in the physical plane are straight lines since $w(\sigma)$ is constant.

(ii) these straight lines have the characteristic direction in the physical plane (cf. 3.5.11), that is they are inclined at the local Mach angle to the approximate streamlines ($\tilde{y} = \text{constant}$).

$$\frac{d\tilde{y}}{dx} = \mp\frac{1}{\sqrt{w(\sigma)}} = \mp\frac{1}{\sqrt{(\gamma+1)\phi_x(\sigma) - K}} = \mp\frac{\delta^{\frac{1}{3}}}{\sqrt{M_\ell^2 - 1}}$$

as shown in (3.1.15,16).

(iii) these plane lines (straight characteristics) are orthogonal to the characteristic in the hodograph of slope $\left(\dfrac{d\vartheta}{dw}\right)$ onto which the flow is mapped.

Note that for this flow

$$j = \frac{\partial w}{\partial x}\frac{\partial \vartheta}{\partial \tilde{y}} - \frac{\partial \vartheta}{\partial x}\frac{\partial w}{\partial \tilde{y}} = \frac{dw}{d\vartheta}\frac{d\vartheta}{d\sigma}(\sigma_x\sigma_{\tilde{y}} - \sigma_x\sigma_{\tilde{y}}) = 0 \tag{3.6.8}$$

A finite area in the physical plane is mapped onto a curve (characteristic) in the hodograph.

As an application of the simple wave flow we consider the transition from one uniform state ($w = w_a = \text{constant}, \vartheta = \vartheta_a = 0$ say) to a final uniform state of higher speed and lower pressure, the acceleration of supersonic flow around a turn. The picture in physical coordinates appears in Figure 3.6.1.

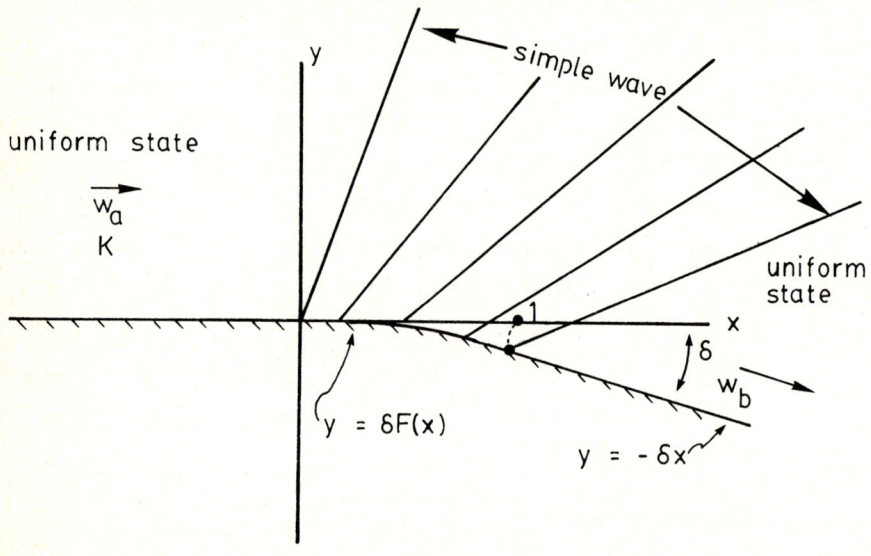

Figure 3.6.1

Simple Wave - Physical Plane

The parameter δ is used to characterize the turning angle of the flow, in accord with our previous derivation of the small-disturbance equations. The corresponding picture of this flow in the transonic and hodograph planes appears in Figure 3.6.2.

This formulation is completely general. Any function of a simple wave represents a possible flow. Uniform states ahead and behind are arbitrary.

The image of the entire flow in the simple wave lies along the characteristic in the hodograph which passes through the initial state

$$\vartheta + \frac{2}{3}w^{\frac{3}{2}} = \frac{2}{3}w_a^{\frac{3}{2}} \quad (\vartheta_a = 0) \tag{3.6.9}$$

This relation connects velocity (and pressure) to deflection angle for each point in the fan. The final uniform state is achieved for $\vartheta = \vartheta_b$

$$\frac{2}{3}w_b^{\frac{3}{2}} = \frac{2}{3}w_a^{\frac{3}{2}} - \vartheta_b = \frac{2}{3}w_a^{\frac{3}{2}} + (\gamma + 1) . \tag{3.6.10}$$

The formula (3.6.9) also gives the pressure on the surface at each point.

$$w(x,0) = \left(w_a^{\frac{3}{2}} - \frac{3}{2}\vartheta_f(x)\right)^{\frac{2}{3}} , \quad \vartheta_f(x) = +(\gamma + 1)F'(x) . \tag{3.6.11}$$

The family of straight characteristics each at the local Mach angle, are represented by

$$\tilde{y} = \frac{1}{\sqrt{w_f(\sigma)}}(x - \sigma) \tag{3.6.12}$$

where the phase σ has been chosen as x-coordinate when $\tilde{y} = 0$. Thus

$$w_f(\sigma) = \left(w_a^{\frac{3}{2}} - \frac{3}{2}\vartheta_f(\sigma)\right)^{\frac{2}{3}} . \tag{3.6.13}$$

The state is constant along each line in the fan; for each σ the values ϑ_f, w_f are known as well as the straight line along which these values are constant. The image of each streamline is along the segment of characteristic AB. Each point P_1, P_2, P_3 is the image of one entire ray σ_1, σ_2, σ_3. A given cross-characteristic, of the compression family, has its image all along the arc AB, but this family carries no disturbance.

A limiting case of the flow just discussed is a centered expansion fan in supersonic flow past a sharp corner, Figure 3.6.3.

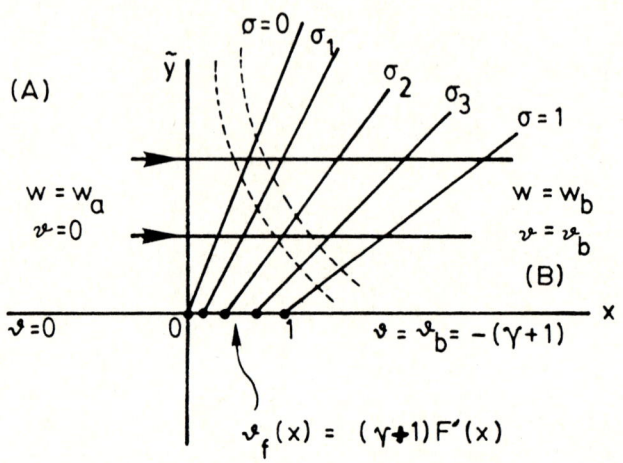

Figure 3.6.2a

Simple Wave, Transonic Plane

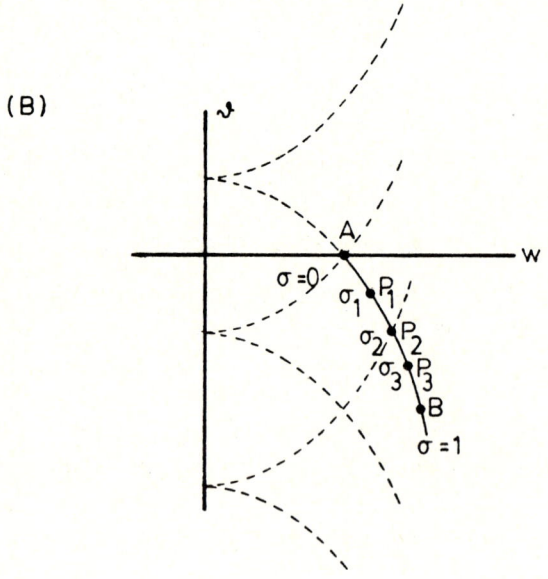

Figure 3.6.2b

Transonic Simple Wave, Hodograph Plane

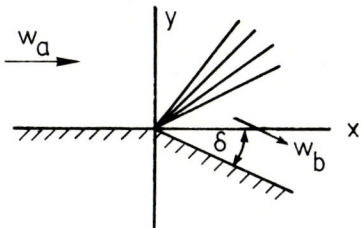

Physical Plane

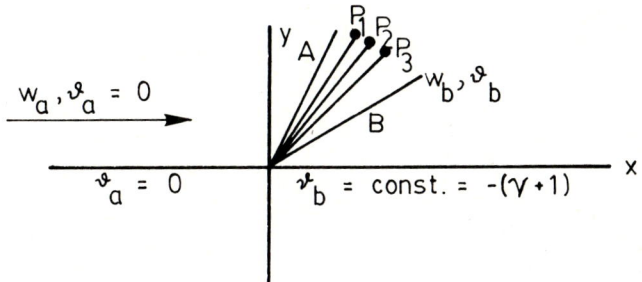

Transonic Plane

Figure 3.6.3

Centered Expansion Fan

The hodograph representation of the flow is the same as before, as is the connection between final and initial states. The rays are centered so that

$$\frac{\tilde{y}}{x} = \frac{1}{\sqrt{w}} = \frac{1}{\left(w_a^{\frac{3}{2}} - \frac{3}{2}\vartheta\right)^{\frac{1}{3}}} \qquad \vartheta_a \leq \vartheta < \vartheta_b \qquad (3.6.14)$$

ϑ itself can be used as the phase and (3.6.9) gives $w(\vartheta)$. The centered fan can also appear as a local solution when a non-uniform subsonic flow just reaches sonic velocity at a sharp corner. This occurs for example in subsonic flow towards a wedge airfoil.

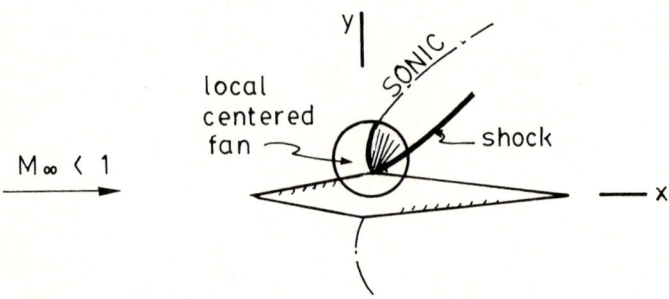

Figure 3.6.4

Local Centered Fan

In this flow the corner point $(x = \tilde{y} = 0)$ is mapped on the entire arc AB in the hodograph.

It is also possible to have simple flows of compression, but they are always limited in extent because the rays form envelopes. The flow pattern is illustrated in Figure 3.6.5.

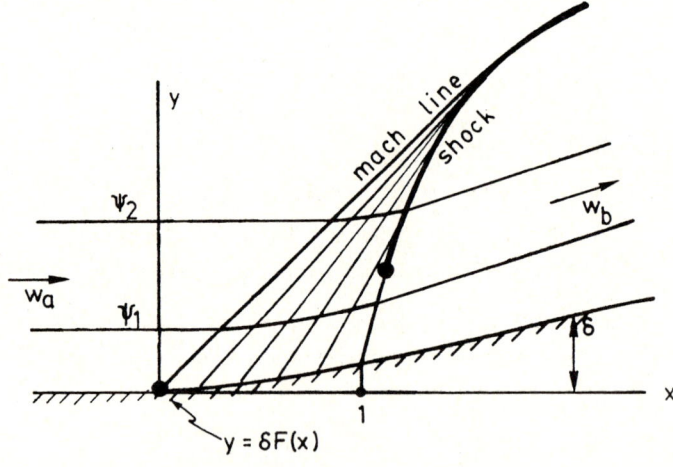

Figure 3.6.5

Envelope Formation at Compression Corner

The crossing of the characteristics (each of which carries constant w, ϑ) in the formation of an envelope implies the existence of an unphysical multiple valued region. A shock wave must form upstream of the location of the envelope of characteristics. For those streamlines which pass below the shock wave the flow is calculated as in a simple wave, but an interaction problem (simple wave and shock wave) must be solved to find the flow downstream of the shock.

The transonic and hodograph planes appear in Figure 3.6.6.

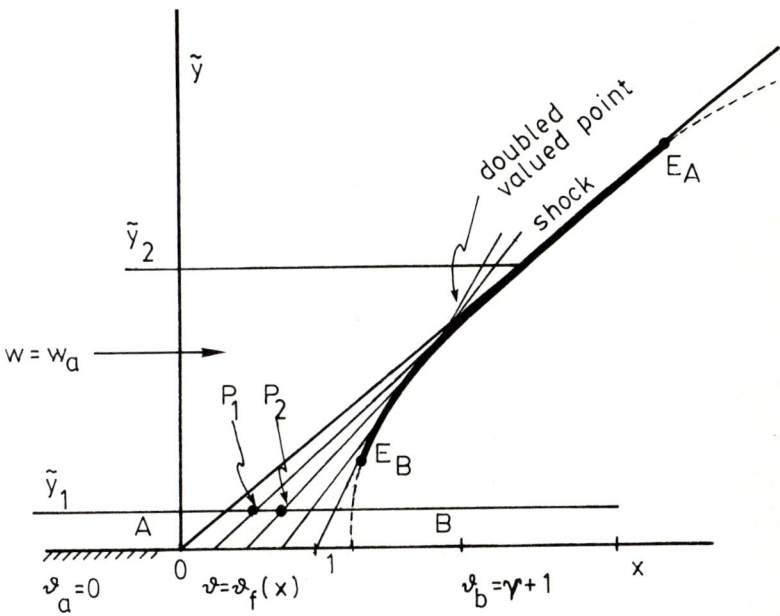

Figure 3.6.6a

Transonic Plane

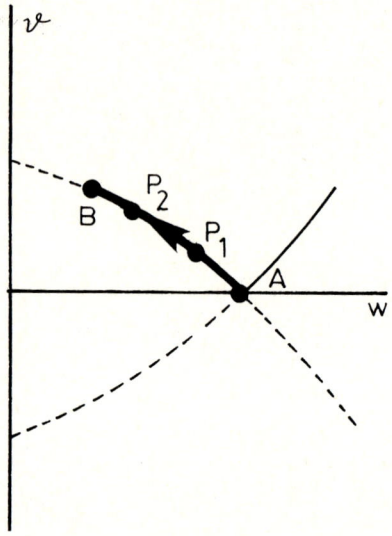

Figure 3.6.6b

Hodograph

A sample calculation of the envelope can now be given for the special shape of a parabola where

$$F(x) = \frac{x^2}{2}, \quad \vartheta_f(x) = (\gamma + 1)F'(x) = (\gamma + 1)x, \qquad 0 \le x \le 1. \qquad (3.6.15)$$

The family of straight rays is, from (3.6.12),

$$x - \sigma = \sqrt{w_f(\sigma)}\,\tilde{y} \qquad (3.6.16)$$

where by (3.6.13)

$$w_f(\sigma) = \left\{ w_a^{\frac{3}{2}} - \frac{3}{2}\vartheta_f(\sigma) \right\}^{\frac{2}{3}} = \left(w_a^{\frac{3}{2}} - \frac{3}{2}(\gamma + 1)\sigma \right)^{\frac{2}{3}} \qquad (3.6.17)$$

for this case. We must have

$$w_a^{\frac{3}{2}} \ge \frac{3}{2}(\gamma + 1)$$

to insure supersonic flow downstream; the flow is just compressed to sonic for the case of equality ($w_f = 0$). Writing (3.6.16) as

$$x - \sigma = \left(w_a^{\frac{3}{2}} - \frac{3}{2}(\gamma + 1)\sigma \right)^{\frac{1}{3}} \tilde{y} \qquad (3.6.18)$$

the envelope condition $\left(\frac{\partial}{\partial \sigma} = 0\right)$ is

$$0 = 1 - \frac{\gamma + 1}{2} \frac{\tilde{y}}{\left(w_a^{\frac{3}{2}} - \frac{3}{2}(\gamma + 1)\sigma\right)^{\frac{2}{3}}}$$

Thus the parametric representation of the envelope is

$$\left.\begin{cases} \tilde{y}_E(\sigma) = \dfrac{2}{\gamma + 1}\left(w_a^{\frac{3}{2}} - \dfrac{3}{2}(\gamma + 1)\sigma\right)^{\frac{2}{3}} \\[3mm] x_E(\sigma) = \sigma + \dfrac{2}{\gamma + 1}\left(w_a^{\frac{3}{2}} - \dfrac{3}{2}(\gamma + 1)\sigma\right) = \dfrac{2}{\gamma + 1}w_a^{\frac{3}{2}} - 2\sigma \,. \end{cases}\right\} \qquad (3.6.19)$$

Corresponding to $\sigma = 1$, the last ray, we have the point E_B with coordinates

$$x_{E_B} = \frac{2}{\gamma + 1}w_a^{\frac{3}{2}} - 2 > 1$$

$$\tilde{y}_{E_B} = \frac{2}{\gamma + 1}\left(w_a^{\frac{3}{2}} - \frac{3}{2}(\gamma + 1)\right)^{\frac{2}{3}} > 0$$

and on the first ray $\sigma = 0$

$$x_{E_A} = \frac{2}{\gamma + 1}w^{\frac{3}{2}}$$

$$\tilde{y}_{E_A} = \frac{2}{\gamma + 1}w_a \,.$$

The point E_B moves towards the surface as w_a decreases, and reaches ($x = 1$, $\tilde{y} = 0$) when the flow is just sonic behind the fan. By elimination of the parameter σ (3.6.19) gives the equation of the envelope

$$\tilde{y}_E = \frac{3^{\frac{2}{3}}}{2^{\frac{1}{3}}(\gamma + 1)^{\frac{1}{3}}}\left(x_E - \frac{2}{3}\frac{w_a^{\frac{3}{2}}}{\gamma + 1}\right)^{\frac{2}{3}} . \qquad (3.6.20)$$

In various situations interactions of simple waves may take place resulting in more complicated flows. For example, in the supersonic jet exit (Figure 3.3.2) the zone $ABCD$ is the interaction zone. Simple waves emanate from all four sides of the zone.

3.6.2. Shock Waves, Detachment.

A useful representation of the possibilities for shock waves is in the hodograph where the locus of all possible downstream states from a given upstream state can be studied. This locus is the shock polar. It is an essential part of discussing various boundary value problems and provides the details of shock wave attachment or detachment in supersonic flow towards a pointed wedge.

Here the shock wave relations are worked out in (w, ϑ). The results can be taken over to $(\phi_x, \phi_{\tilde{y}})$ for any particular problem. The results are essentially equations (3.1.25,26) for shock jumps and they can be derived here by integration of the conservative forms of (3.6.1) across the jumps (cf. Figure 3.6.7)

$$\left\{ \begin{array}{l} \left[\dfrac{w^2}{2} \right] d\tilde{y}_s + [\vartheta] dx_s = 0 \\[2mm] [\vartheta] d\tilde{y}_s + [w] dx_s = 0 \end{array} \right\} \text{ Shock Jump Conditions} \qquad (3.6.21)$$

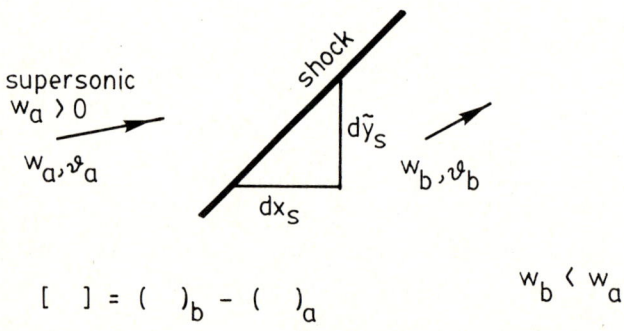

Figure 3.6.7

Shock Geometry

Elimination of the angle $\left(\dfrac{dx}{d\tilde{y}} \right)_s$ gives the basic equation for the shock polar,

$$\frac{1}{2}[w^2][w] = [\vartheta]^2 , \qquad (3.6.22)$$

or

$$\langle w \rangle [w]^2 = [\vartheta]^2 , \quad \text{where} \quad \langle \ \rangle = \text{Average}$$
$$= \frac{1}{2}\{(\ )_b + (\ )_a\} , \qquad (3.6.23)$$

using the usual jump rules.

When the jumps are known the local shock geometry follows from (3.6.21)

$$\left(\frac{dx}{d\tilde{y}}\right)_s = -\frac{[\vartheta]}{[w]} = \mp\sqrt{\langle w \rangle} \,, \qquad \langle w \rangle = \frac{1}{2}(w_b + w_a) \,. \qquad (3.6.24)$$

This provides us with an inequality showing the relationship of the slope of the characteristics (Mach waves) when there is supersonic flow ahead and behind the shock to the shock slope

$$\left(\frac{dx}{d\tilde{y}}\right)_{c_a} = \sqrt{w_a} > \left(\frac{dx}{d\tilde{y}}\right)_s = \sqrt{\frac{1}{2}(w_b + w_a)} > \left(\frac{dx}{d\tilde{y}}\right)_{c_b} = \sqrt{w_b} \qquad (3.6.25)$$

for the $+$ branch, as illustrated in Figure 3.6.8. The arrows indicate the downstream direction for propagation of disturbances.

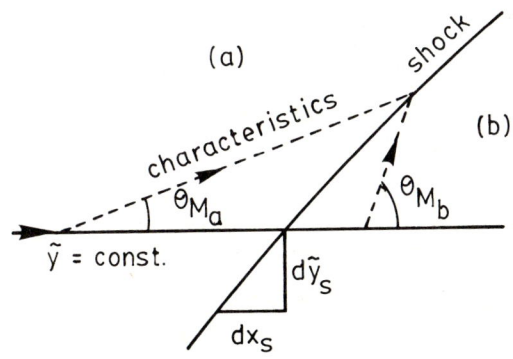

Figure 3.6.8

Characteristics and Shocks

It is sufficient to plot the polar curve (3.6.23) for $\vartheta_a = 0$ since the equations (3.6.1) and shock jump conditions (3.6.21) are invariant under translation in ϑ; ϑ_b can always be replaced by $\vartheta_b - \vartheta_a$ if $\vartheta_a \neq 0$. Then, written out, (3.6.22) is

$$\boxed{\vartheta_b^2 = \frac{1}{2}(w_b + w_a)(w_b - w_a)^2 \qquad \text{shock polar}} \qquad (3.6.26)$$

ϑ_b is cubic in w_b for given w_a. These are a family of curves for varying values of w_a as in figure 3.6.9, which give the locus of all possible shock jumps.

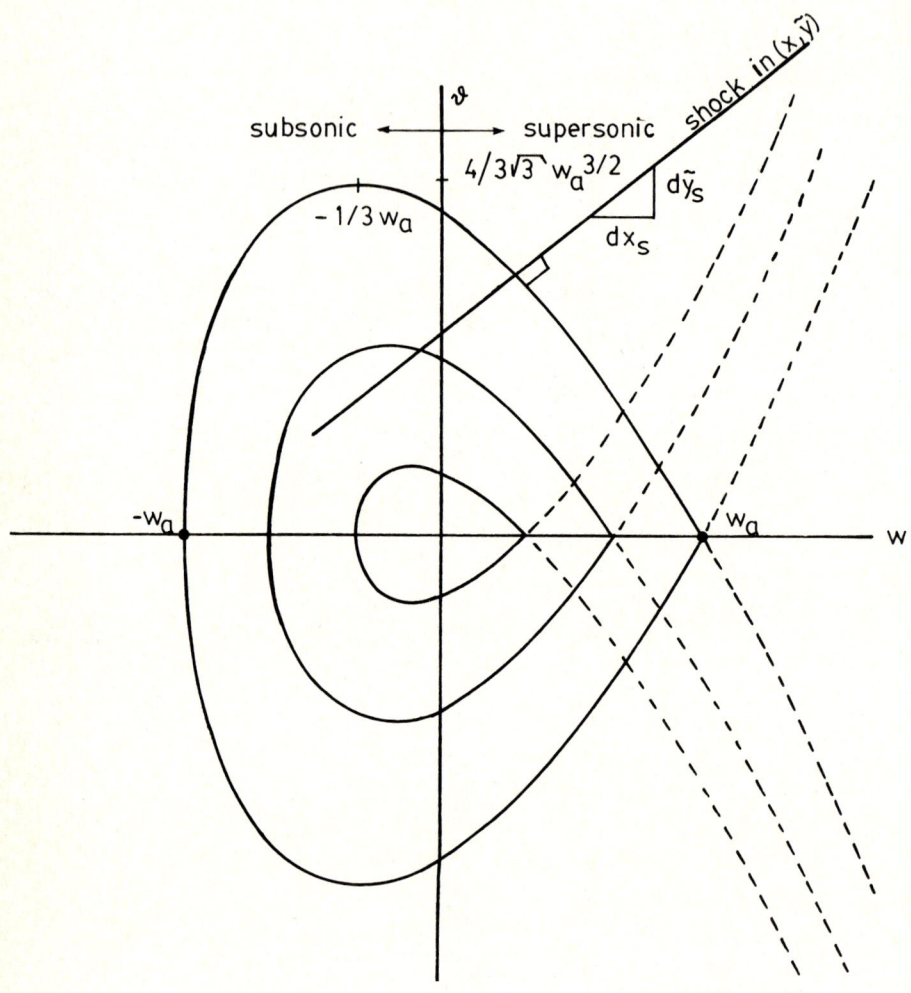

Figure 3.6.9.

Transonic Shock Polar Family

According to (3.6.24) and (3.6.26) the shock angle is in $(x, \tilde{y})$ locally orthogonal to the tangent to the curves $\left(\dfrac{d\tilde{y}}{dx}\right) = \pm \left(\dfrac{w_b - w_a}{\vartheta_b}\right)$. For a curved shock the locus lies along some arc of a shock polar. Notice that for the shock of vanishing strength $w_b - w_a \to 0$, $\langle w \rangle \to w_a$ and the shock angle $\left(\dfrac{dx}{d\tilde{y}}\right)_s \to \mp\sqrt{w_a} = \left(\dfrac{dx}{d\tilde{y}}\right)_c$ approaches the Mach angle. Shocks normal to the stream have $w_b = -w_a$, i.e. subsonic flow behind. It can thus be seen from the polar that a variety of downstream states exist. Subsonic and supersonic final states are accessible from a given initial state. The dotted portion of the curve is not however accessible $(w_a > w_b)$ for it represents an expansion shock. An elementary calculation shows that the maximum flow deflection is reached behind an oblique shock for

$$w_b = -\frac{1}{3}w_a , \quad |\vartheta_b| = \frac{4}{3\sqrt{3}}w_a^{\frac{3}{2}} , \qquad (3.6.27)$$

It is possible to bring all the shock polars to a single curve by the scaling

$$\vartheta_b = w_a^{\frac{3}{2}}\hat{\vartheta} , \quad w_b = w_a\hat{w} \qquad (3.6.28)$$

so that the polar reads

$$\boxed{\hat{\vartheta}^2 = \frac{1}{2}(\hat{w} + 1)(\hat{w} - 1)^2 \qquad \text{scaled polar}} \qquad (3.6.29)$$

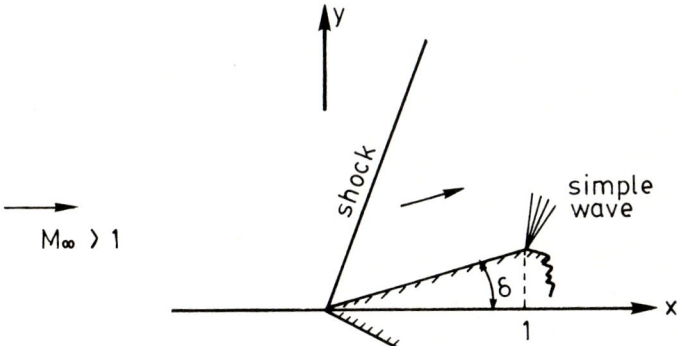

Figure 3.6.10

Supersonic Flow towards a Wedge

Consider now the supersonic flow towards the straight side wedge as in Figure 3.6.10. As long as the mach number M_∞ is sufficiently high and the deflection angle δ sufficiently small the shock is attached with uniform supersonic flow behind.

If in fact the flow is supersonic behind, then the length of the wedge is irrelevant for the flow near the surface since no signals concerning the corner can reach the shock (except by spreading out along the simple wave at the corner). Only when M_∞ decreases or δ increases sufficiently, so that the sonic or subsonic flow is reached behind the shock, will the effect of the corner be felt. The shock can still be attached but must have a curvature (the characteristic wedge length fixes the scale). Finally the shock must detach when the flow deflection exceeds the maximum possible. In order to find the flow in these latter two cases a boundary value problem must be solved since the shock is non-uniform. These problems are outside the scope of this section. For details see [3.1.1] or [3.6.2]. The sequence of flows just described is illustrated with the help of the shock polar (3.6.29) in Figure 3.6.13.

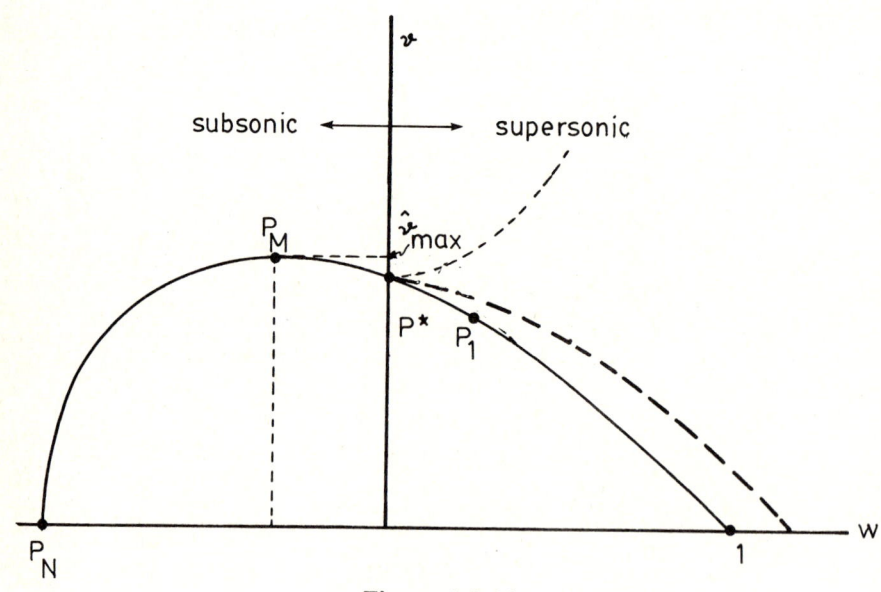

Figure 3.6.11

Scaled Shock Polar

The equation of the wedge surface is

$$F(x) = \delta x \quad \text{for} \quad 0 < x < 1\,, \tag{3.6.30}$$

so that the surface boundary condition is

$$\vartheta(x,0) = (\gamma + 1)\,. \tag{3.6.31}$$

The state ahead is characterized by the transonic similarity parameter

$$w_a = -K = |K| = \frac{M_\infty^2 - 1}{\delta^{\frac{2}{3}}}\,. \tag{3.6.32}$$

Thus in terms of the scaled variables

$$\hat{\vartheta}(x,0) = (\gamma + 1)|K|^{-\frac{3}{2}} = (\gamma + 1)\frac{\delta}{(M_\infty^2 - 1)^{\frac{3}{2}}}\,. \tag{3.6.33}$$

As long as $0 < \hat{\vartheta}(x,0) < \hat{\vartheta}^*$ (point P_1 in the figure) the shock is straight and attached, with uniform supersonic flow (w_b, ϑ_b) behind. When $\hat{\vartheta}$ increases to $\hat{\vartheta}^*$, $w_b = 0$, and the flow behind is just sonic. Thus

$$\hat{\vartheta}^* = \frac{1}{\sqrt{2}} = (\gamma + 1)|K|^{-\frac{3}{2}}\,. \tag{3.6.34}$$

For $\hat{\vartheta}^* < \hat{\vartheta} < \hat{\vartheta}_{\max}$ the shock is attached but curved, subsonic behind. As shown in [3.6.1], the curvature at the nose changes from finite to infinite for some critical $\hat{\vartheta}$ in this range. Note that

$$\hat{\vartheta}_{\max} = \frac{4}{3\sqrt{3}}\,.$$

For $\hat{\vartheta} > \vartheta_{\max}$ the shock must detach from the sharp point. Sketches of the flow for these cases appear below.

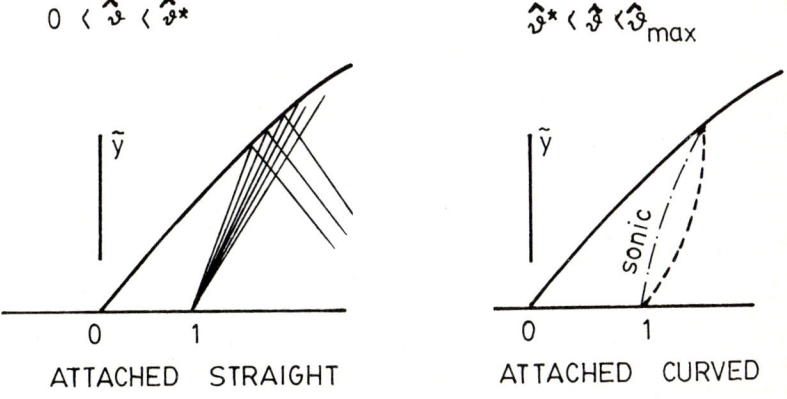

Figure 3.6.12a,b

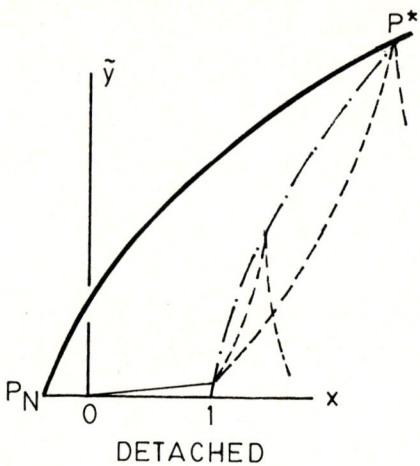

Figure 3.6.12c

Shock Patterns on Wedge

The simple wave at the corner is only local in the last two cases where the flow is subsonic behind the shock and reaches the corner. The sonic line runs from the corner to the shock. In the detached case the image of the shock in the hodograph is all the points $P_N \ldots P_M \ldots P^*$ etc.

The only case where the pressure on the wedge surface can be calculated simply is the supersonic case. The pressure coefficient is

$$c_p = -2\delta^{\frac{2}{3}} \phi_x = 2\delta^{\frac{2}{3}} \frac{|K| - \hat{w}_b}{\gamma + 1} \tag{3.6.35}$$

where w_b is the appropriate root of the cubic (3.6.24) representing the polar for $\hat{\vartheta}_b$.

For flows with shock waves in a uniform stream when boundary value problems are solved in the hodograph the Tricomi equations for $\tilde{y}(w, \vartheta)$ must be solved. It is possible to find a homogeneous boundary condition on the shock polar involving the partial derivatives $(\tilde{y}_w, \tilde{y}_\vartheta)$. This boundary condition is essential for the boundary value problem. Since

$$\left(\frac{d\vartheta}{dw}\right)_{\tilde{y} \,=\, \text{constant on shock}} = -\frac{(\tilde{y}_w)_s}{(\tilde{y}_\vartheta)_s}, \tag{3.6.36}$$

this boundary condition specifies the hodograph direction of the line $\tilde{y} = $ constant at the shock polar. The condition is derived as follows. Along the shock

$$\vartheta_b = (w_a - w_b)\sqrt{\frac{w_a + w_b}{2}} \qquad (+ \quad \text{branch}) \qquad (3.6.37)$$

from (3.6.21). Thus

$$\left(\frac{d\vartheta_b}{dw_b}\right) = -\sqrt{\frac{w_a + w_b}{2}} + \frac{w_a - w_b}{2\sqrt{2}\sqrt{w_a + w_b}} = \frac{-(3w_b + w_b)}{2\sqrt{2}\sqrt{w_a + w_b}} \cdot \qquad (3.6.38)$$

Further, behind the shock

$$\left(\frac{d\tilde{y}}{dx}\right)_s = \frac{w_a - w_b}{\vartheta_b} = \sqrt{\frac{2}{w_a + w_b}} = \frac{(\tilde{y}_w)_s dw_b + (\tilde{y}_w)_s d\vartheta_b}{(x_w)_s dw_b + (x_\vartheta)_s d\vartheta_b} \ ,$$

or

$$\sqrt{\frac{2}{w_a + w_b}} = -\frac{(\tilde{y}_w)_s + (y_\vartheta)_s \left(\dfrac{d\vartheta}{dw}\right)_b}{w_b(\tilde{y}_\vartheta)_s + (\tilde{y}_w)_s \left(\dfrac{d\vartheta}{dw}\right)_b} = \frac{(y_\vartheta)_s - \dfrac{3w_b + w_a}{2\sqrt{2}\sqrt{w_a + w_b}}(\tilde{y}_w)_s}{w_b(\tilde{y}_\vartheta)_s - \dfrac{3w_b + w_a}{2\sqrt{2}\sqrt{w_a + w_b}}(\tilde{y}_w)_s}$$

$$(3.6.39)$$

on using (3.6.38). (3.6.39) is the desired relation and on simplification reads

$$(5w_b + 3w_a)\left(\frac{\partial\tilde{y}}{\partial w}\right)_s - \sqrt{\frac{w_a + w_b}{2}}(7w_b + \vartheta_a)\left(\frac{\partial\tilde{y}}{\partial\vartheta}\right)_s = 0 \qquad (3.6.40)$$

or

$$\left(\frac{d\vartheta}{dw}\right)_{\tilde{y}=\text{constant}} = -\frac{7w_b + w_a}{5w_b + 3w_a}\sqrt{\frac{w_b + w_a}{2}} \qquad (3.6.41)$$

for the slope of $\tilde{y} = $ constant at the polar. A sketch of these elements appears in Figure 3.6.12. The pattern of streamlines for $\tilde{y} = $ constant is reminiscent of that produced by a singularity, imagined in this case to be inside the polar. This is the so-called porcupine due to Busemann.

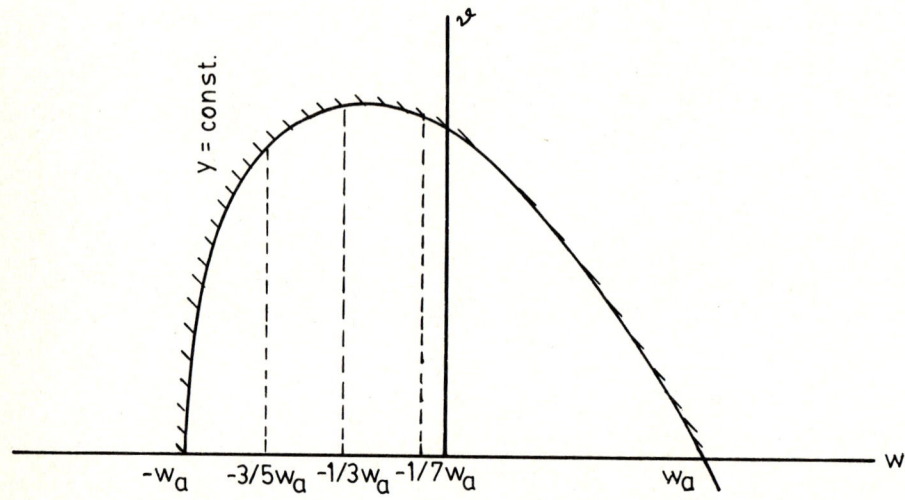

Figure 3.6.13

Shock Polar with Line Elements of $\tilde{y} = $ constant

References

[3.6.1] Courant, R. and Friedrichs, K. O., *Supersonic Flow and Shock Waves.* Interscience Publishers (J. Wiley and Sons) New York, 1948.

[3.6.2] Guderley, K. G., *The Theory of Transonic Flow*, Pergamon Press, London, 1962.

[3.6.3] Vincenti, W. G. and Wagoner, C. G., Transonic Flow past a Wedge Profile with Detached Bow Wave, *NACA TN* **2588**, 1951.

3.7 Nozzle Flow, Branch Lines, Limit Lines

Two flows are described in this section, each of which is an exact solution of the transonic small disturbance equations. The first flow, acceleration through sonic in a throat, exhibits a branch line. This flow corresponds to that in a De Laval nozzle [3.7.2]. The second flow is related to Ringleb's solution and the variants thereof obtained by separation of variables in the full potential equation (in $q^2 = \phi_x^2 + \phi_y^2$, $\vartheta = \tan^{-1}(\phi_y/\phi_x)$ coordinates) [3.7.1]. This flow, which is constructed in the hodograph, exhibits a limit line if the flow boundaries are not judiciously chosen. It is also a smooth transonic flow, that is, the streamlines pass from subsonic to supersonic and then back to subsonic flow regions with no shock.

The general properties of branch lines and limit lines are summarized here; they are then illustrated via the two simple flows.

A curve $x(\tilde{y})$ is a branch of a flow (a solution of (3.5.1)) if

$$j = \frac{\partial(w,\vartheta)}{\partial(x,\tilde{y})} = ww_x^2 - w_{\tilde{y}}^2 = 0 \tag{3.7.1}$$

along the curve, and $j \neq 0$ in a neighborhood of the curve. In Section 3.5 it was shown that under this condition the mapping from the $(x, \tilde{y})$ plane to the (w, ϑ) plane can not be uniquely determined. In fact this singular curve, the branch line, is a characteristic. The image of the branch line in the hodograph is therefore also a characteristic (see Section 3.5). All curves in the hodograph which hit this latter characteristic and are not tangent to it are cusped, as for example the members of the other family of characteristics. The streamlines $\tilde{y}(w, \vartheta) =$ constant are tangent to this characteristic. The hodograph plane folds along the image of the branch line and portions of the plane are doubly covered. A limit line is the $(x, \tilde{y})$ image of a curve $w(\vartheta)$ along which,

$$J = \frac{\partial(x,\tilde{y})}{\partial(w,\vartheta)} = w\tilde{y}_\vartheta^2 - \tilde{y}_w^2 = \left(\sqrt{w}\frac{\partial\tilde{y}}{\partial\vartheta} + \frac{\partial\tilde{y}}{\partial w}\right)\left(\sqrt{w}\frac{\partial\tilde{y}}{\partial\vartheta} - \frac{\partial\tilde{y}}{\partial w}\right) = 0 . \tag{3.7.2}$$

The streamlines $\tilde{y} =$ constant are tangent to the characteristics at the hodograph image of the limit line, but not to the same characteristic. In the physical plane these characteristics are cusped at the limit line and the $(x, \tilde{y})$ plane is doubly covered. The limit line is thus an envelope of one family of characteristics. It represents a fold in the plane, and the streamlines fold back on themselves at the limit line (see Figure 3.7.2). Clearly a flow which exhibits a limit line is unphysical.

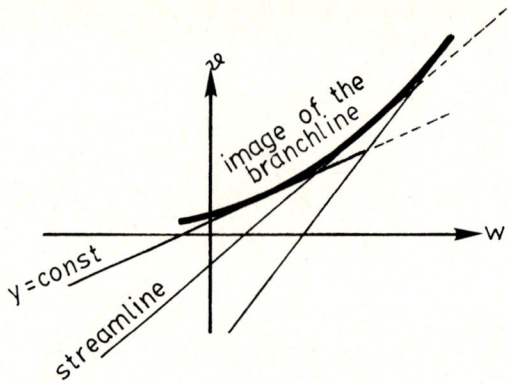

Figure 3.7.1

Hodograph Image of a Branchline

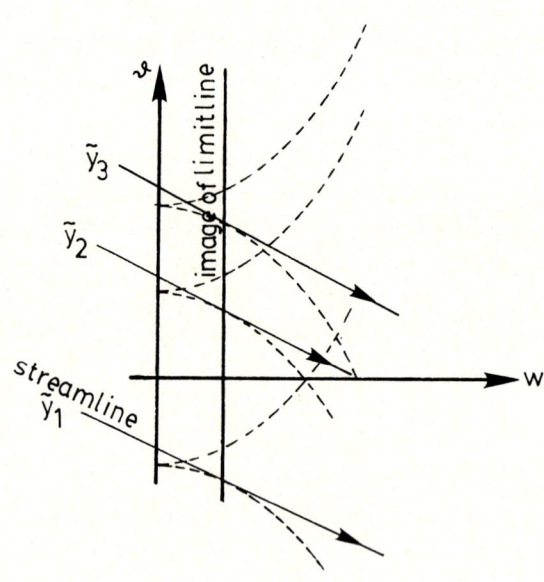

Figure 3.7.2a

Limit line - hodograph plane

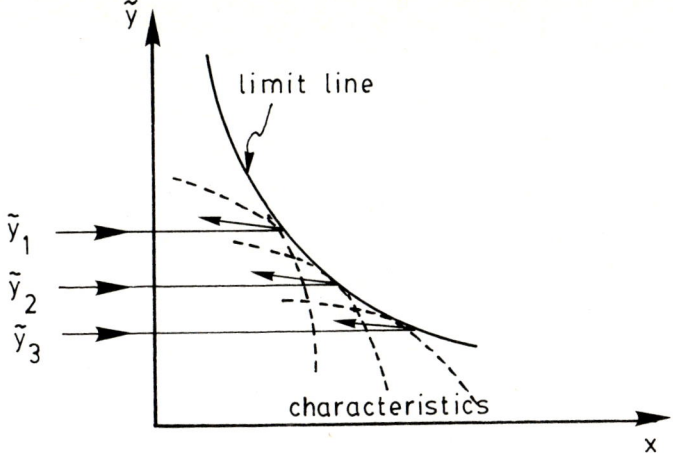

Figure 3.7.2b

Limit line - physical plane

Note that although the limit line and the branch line both represent the vanishing of the Jacobian of a map between the (w, ϑ) and the $(x, \tilde{y})$ plane, or vice versa, along a line, they have quite different properties. This is due to the fact that the governing equations are linear in the hodograph, nonlinear in the physical plane.

Nozzle Flow

The flow considered here is the acceleration through sonic in a slight throat. The half width of the throat is chosen as unity so that the shape of the throat is given by

$$y = \pm\bigl(1 + \delta F(x)\bigr) , \tag{3.7.3}$$

where

$$\delta \ll 1 ,$$

and

$$F'(0) = 0 . \tag{3.7.4}$$

The full potential equations describing the flow are then equations (3.1.1) with $U = a^*$ where a^* is the flow velocity at the speed of sound. The enthalpy integral is then

$$\frac{a^2}{\gamma - 1} + \frac{\Phi_x^2 + \Phi_y^2}{2} = \frac{a^{*2}(\gamma + 1)}{2(\gamma - 1)} , \tag{3.7.5}$$

and this is coupled with

$$(a^2 - \Phi_x^2)\Phi_{xx} - 2\Phi_x\Phi_y\Phi_{xy} + (a^2 - \Phi_y^2)\Phi_{yy} = 0 \ . \tag{3.7.6}$$

The flow must be tangent to the boundary, so

$$\frac{\Phi_y}{\Phi_x} = \pm\delta F'(x) \quad \text{on} \quad y = \pm\big(1 + \delta F(x)\big) \ . \tag{3.7.7}$$

(see Figure 3.7.3).

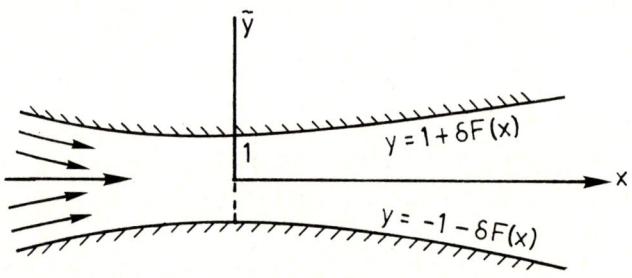

Figure 3.7.3

Nozzle Flow

The y scale is fixed at unity, so in order to make the rapid changes in the flow variables in the x direction near the throat comparable to those changes taking place in the y direction, x must be stretched. The small disturbance approximation is then obtained by holding

$$x^* - x_0^* = \frac{x}{\nu(\delta)} \quad \text{fixed as} \quad \delta \to 0 \ , \tag{3.7.8}$$

in the potential expansion

$$\Phi = a^*\big\{x + \varepsilon(\delta)\Phi(x^*, y) + \cdots\big\} \ , \tag{3.7.9}$$

where $\nu, \varepsilon \ll 1$ are to be determined as is the shift in origin, x_0^*.

Substitution of the derivatives

$$\frac{\Phi_x}{a^*} = 1 + \Big(\frac{\varepsilon}{\nu}\Big)\phi_{x^*} + \cdots \ , \qquad \frac{\Phi_y}{a^*} = \varepsilon\phi_y + \cdots$$

$$\frac{\Phi_{yy}}{a^*} = \varepsilon\phi_{yy} + \cdots \ , \qquad \frac{\Phi_{xx}}{a^*} = \frac{\varepsilon}{\nu^2}\phi_{x^*x^*} + \cdots$$

of (3.7.9) into (3.7.5) gives

$$\frac{a^2}{a^{*2}} = \frac{\gamma+1}{2} - \frac{\gamma-1}{2}\left(\frac{\Phi_x^2 + \Phi_y^2}{a^{*2}}\right)$$
$$= 1 - (\gamma - 1)(\varepsilon/\nu)\phi_{x^*} + \cdots , \tag{3.7.10}$$

and then substitution of the derivatives and (3.7.10) into (3.7.6) gives to leading order

$$-(\gamma+1)\frac{\varepsilon^2}{\nu^3}\phi_{x^*}\phi_{x^*x^*} + \varepsilon\phi_{yy} = 0 . \tag{3.7.11}$$

The distinguished limit of (3.7.11) is obtained for

$$\frac{\varepsilon}{\nu^3} = 1 . \tag{3.7.12}$$

In the small disturbance coordinates the boundary condition (3.7.7) becomes

$$\varepsilon\phi_y = \pm\delta\nu(x^* - x_0^*)F''(0) + \cdots ,$$

since $F'(0) = 0$. Thus to retain the shape in the limit $\delta \to 0$ we must have

$$\varepsilon = \delta\nu . \tag{3.7.13}$$

This coupled with (3.7.12) gives

$$\nu = \sqrt{\delta} , \quad \varepsilon = \delta^{3/2} .$$

Thus, the transonic small disturbance approximation for the nozzle flow is given by

$$\Phi = a^*\left\{x + \delta^{3/2}\phi(x^*, y) + \cdots\right\} , \tag{3.7.14}$$

with

$$x^* - x_0^* = x/\sqrt{\delta} \quad \text{held fixed} \quad \text{as} \quad \delta \to 0 , \tag{3.7.15}$$

where ϕ satisfies

$$(\gamma+1)\phi_{x^*}\phi_{x^*} - \phi_{yy} = 0 , \tag{3.7.16}$$

$$\phi_y|_{y=\pm1} = \pm(x^* - x_0^*)F''(0) . \tag{3.7.17}$$

Note that this is the same approximation as that for jet flows in Section 3.3 with the exception that now $K = 0$ (perturbation from sonic) and we have a translated origin (x_0^*). The physical meaning of δ can be made clear if it is considered with respect to the radius of curvature (Figure 3.7.4).

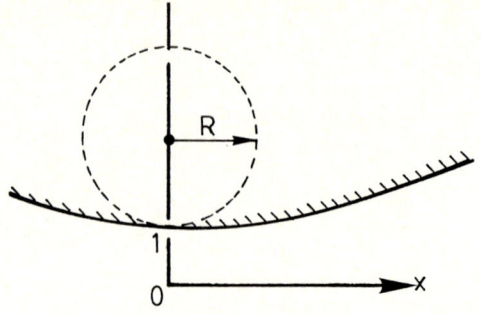

Figure 3.7.4

Nozzle Throat

At the throat the boundary curve is coincident with the circle

$$\left(y - (R+1)\right)^2 + x^2 = R^2 \, ,$$

hence

$$y = R + 1 - \sqrt{R^2 - x^2}$$

$$= R + 1 - R \left(1 - \frac{x^2}{2R^2} + \cdots \right) \, ,$$

$$= 1 + \frac{x^2}{2R} + \cdots \quad \text{for large} \quad R$$

Thus if

$$y = 1 + \delta F(x) \, ,$$

as in (3.7.7), then

$$\delta = \frac{1}{R} \, , \quad F = \frac{x^2}{2}$$

so

$$\delta = \text{half gap/radius of curvature at the throat,}$$

and then

$$F''(0) = 1 \, .$$

Note that in this formulation the small disturbance problem (3.7.16), (3.7.17) is

$$(\gamma + 1)\phi_{x^*} \phi_{x^* x^*} - \phi_{yy} = 0 \, , \tag{3.7.18}$$

$$\phi_y\big|_{y=\pm 1} = \pm(x^* - x_0^*) \, . \tag{3.7.19}$$

This system has, among other solutions, a fourth order polynomial solution. Making use of the symmetry in y, and condition (3.7.19), the form of ϕ is given by

$$\phi = ax^{*2} + bx^*y^2 + cy^4 . \tag{3.7.20}$$

In fact since

$$\phi_{x^*} = 2ax^* + by^2 , \quad \phi_{x^*x^*} = 2a , \quad \text{(uniform acceleration)}$$

$$\phi_{yy} = 12cy^2 + 2bx^* ,$$

(3.7.20) is an exact sulution of (3.7.18) iff

$$(\gamma + 1)2a(2ax^* + by^2) = 12cy^2 + 2bx^* ,$$

or

$$4a^2(\gamma + 1) = 2b , \quad 2ab(\gamma + 1) = 12c . \tag{3.7.21}$$

The final relation between the three constants a,b,c is given by the boundary condition (3.7.19) which implies

$$\pm 2bx^* \pm 4c = \pm x^* \mp x_0^* . \tag{3.7.22}$$

Thus

$$2b = 1 ,$$

and then from (3.7.21)

$$a^2 = \frac{1}{4(\gamma + 1)} , \quad c = \frac{(\gamma + 1)^{1/2}}{24} .$$

The constants a,b,c are now determined and (3.7.22) gives the shift of origin,

$$x_0^* = -\frac{(\gamma + 1)^{1/2}}{6} .$$

Summarizing, one exact solution of (3.7.18), (3.7.19) is

$$\phi(x^*,y) = \frac{1}{2\sqrt{\gamma + 1}}x^{*2} + \frac{1}{2}x^*y^2 + \frac{\sqrt{\gamma + 1}}{24}y^4 ,$$

with

$$w = (\gamma + 1)\phi_{x^*} = \sqrt{\gamma + 1}x^* + \frac{(\gamma + 1)y^2}{2} , \tag{3.7.23}$$

$$\vartheta = (\gamma + 1)\phi_y = (\gamma + 1)x^*y + \frac{(\gamma + 1)^{3/2}y^3}{6} \ . \qquad (3.7.24)$$

To find the details of this flow note that the sonic line is given by

$$w = (\gamma + 1)\phi_{x^*} = 0 \ , \quad \text{or}$$

$$x^* = -\sqrt{\gamma + 1}\frac{y^2}{2} \ , \quad \text{sonic line} \ . \qquad (3.7.25)$$

The Jacobian, $j = ww_{x^*}^2 - w_y^2$, is zero along

$$(\gamma + 1)^{\frac{3}{2}}x^* + (\gamma + 1)^2\frac{y^2}{2} = (\gamma + 1)^2y^2 \ ,$$

or

$$x^* = \sqrt{\gamma + 1}\frac{y^2}{2} \ , \quad \text{branch line} \ . \qquad (3.7.26)$$

The flow is

$$\text{order reversing,} \quad j < 0 \quad \text{if} \quad x^* < \sqrt{\gamma + 1}\frac{y^2}{2} \ ,$$

and changes as it passes through the branch line to

$$\text{order preserving,} \quad j > 0 \quad \text{if} \quad x^* > \sqrt{\gamma + 1}\frac{y^2}{2} \ .$$

The flow deflection is zero along $\vartheta = 0$ or if

$$x^* = -\sqrt{\gamma + 1}\frac{y^2}{6} \ , \qquad \text{zero deflection}$$

or

$$y = 0 \qquad \qquad , \qquad \text{zero deflection} \ .$$

The characteristics are given by

$$\frac{d\tilde{y}}{dx^*} = \pm\frac{1}{\sqrt{w}} = \pm\frac{1}{\sqrt{(\gamma + 1)^{1/2}x^* + \frac{\gamma+1}{2}y^2}} \ .$$

Note that along curves $x^*/y^2 = \lambda = \text{constant}$,

$$\frac{dy}{dx^*} = \pm\frac{1}{2\sqrt{\lambda x^*}} \ .$$

The characteristics therefore coincide with these curves if

$$\pm \frac{1}{2\sqrt{\lambda x^*}} = \pm \frac{1}{\sqrt{(\gamma+1)^{\frac{1}{2}} x^* + \frac{(\gamma+1)}{2}\frac{x^*}{\lambda}}} \; ,$$

or

$$2\sqrt{\lambda} = \sqrt{(\gamma+1)^{\frac{1}{2}} + \frac{(\gamma+1)}{2\lambda}} \; ,$$

$$4\lambda^2 - (\gamma+1)^{\frac{1}{2}}\lambda - \frac{(\gamma+1)}{2} = 0 \; ,$$

or

$$\lambda = \frac{\sqrt{\gamma+1}}{2} \; , \qquad -\frac{\sqrt{\gamma+1}}{4} \; .$$

The first of these curves,

$$x^* = \frac{\sqrt{\gamma+1}}{2}y^2 \; , \tag{3.7.27}$$

is the branch line (3.7.26). Thus, as was asserted earlier, the branch line is a characteristic. The other characteristic

$$x^* = \frac{-\sqrt{\gamma+1}}{4}y^2 \; , \tag{3.7.28}$$

is the limiting characteristic. Properties of the flow downstream of this characteristic can not influence the upstream regions. This is a compression characteristic, whereas along the branch line the flow expands. These flow properties are summarized in Figure 3.7.5.

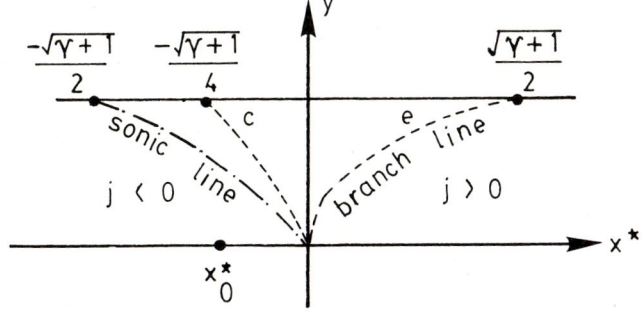

Figure 3.7.5

Branchline in Nozzle Flow

In order to sketch the hodograph we first obtain the mapping of the branch-line. Substituting (3.7.26) into (3.7.23) and (3.7.24) gives

$$w = (\gamma + 1)y^2$$

and

$$\vartheta = \frac{2}{3}(\gamma + 1)^{\frac{3}{2}} y^3 \ .$$

The first equation yields $y = \pm\sqrt{\frac{w}{\gamma+1}}$ and therefore, using the second equation, the description of the branchline in the hodograph is given by

$$\vartheta = \pm\frac{2}{3} w^{\frac{3}{2}} \ .$$

From section 3.5, these curves are characteristics. Following a similar procedure for the limiting characteristic (3.7.28b), we obtain

$$w = \frac{\gamma + 1}{4} y^2$$

and

$$\vartheta = -\frac{(\gamma + 1)^{\frac{3}{2}}}{12} y^3 \ .$$

Thus, $y = \pm 2\sqrt{\frac{w}{\gamma+1}}$ and the description of the limiting characteristic in the hodograph is given by

$$\vartheta = \mp\frac{2}{3} w^{\frac{3}{2}} \ .$$

In general, we note from (3.7.23) that

$$x^* = \frac{w - (\gamma + 1)\frac{y^2}{2}}{\sqrt{\gamma + 1}} \ .$$

Hence, using (3.7.24),

$$\vartheta = -\frac{(\gamma + 1)^{\frac{3}{2}}}{3} y^3 + (\gamma + 1)^{\frac{1}{2}} yw \ . \tag{3.7.29}$$

This cubic, if solved, would give $y(w, \vartheta)$. There can exist three distinct values of y which solve (3.7.29) for certain w, ϑ. Note that streamlines ($y = $ constant) appear in the hodograph as straight lines with slope $(\gamma + 1)^{1/2}y$ and ϑ-intercept $\frac{-(\gamma+1)^{3/2}}{3} y^3$.

From the above, the slope or image of the branch line is given by

$$\frac{d\vartheta}{dw} = \pm w^{\frac{1}{2}} .$$

Equating this to the slope of the image of the streamlines, there results

$$\pm w^{\frac{1}{2}} = (\gamma + 1)^{\frac{1}{2}} y .$$

Also, the point of intersection of the images of the branchlines and streamlines satisfies (using (3.7.29))

$$\pm \frac{2}{3} w^{\frac{3}{2}} = -\frac{(\gamma+1)^{\frac{3}{2}}}{3} y^3 + (\gamma+1)^{\frac{1}{2}} yw .$$

The first condition satisfies the second and, therefore, the image of the streamline is tangent to the image of the branchline at their point of intersection.

We can resolve the ambiguity in the $\pm$ sign by noting that $\vartheta = +\frac{2}{3} w^{3/2}$ serves as the branchline in the hodograph for $y > 0$ (with limiting characteristic given by $\vartheta = -\frac{2}{3} w^{3/2}$). Similarly, for $y < 0$, $\vartheta = -\frac{2}{3} w^{3/2}$ is the branch line in the hodograph and $\vartheta = +\frac{2}{3} w^{3/2}$ is the limiting characteristic. These results are presented in Figure 3.7.6a.

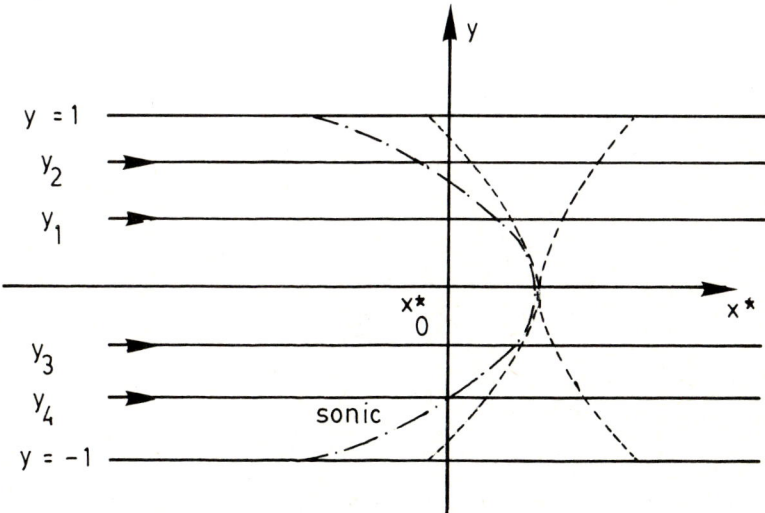

Figure 3.7.6a

Physical Plane of Nozzle Flow

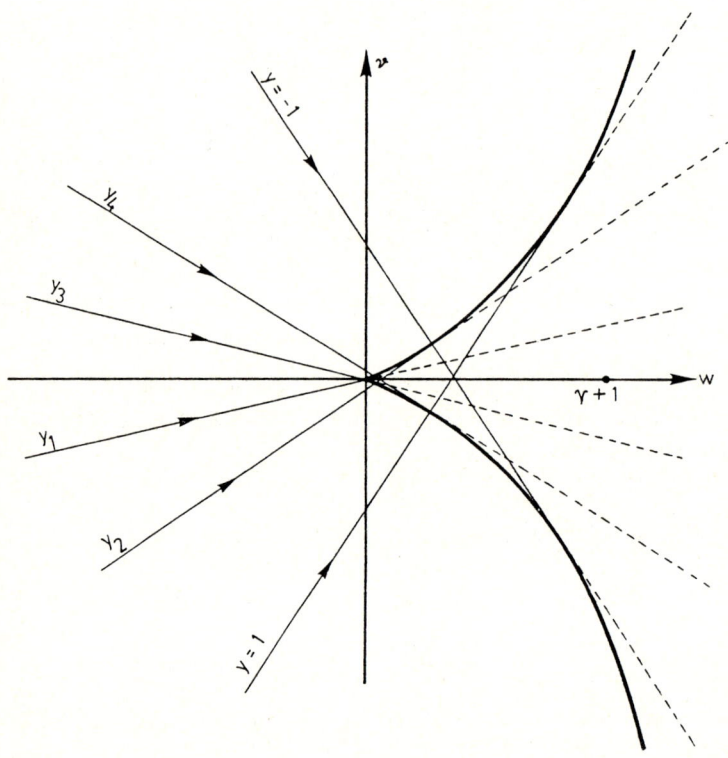

Figure 3.7.6b
Hodograph Plane of Nozzle Flow

A Smooth Transonic Flow

The flow discussed in this paragraph is a generalization to the transonic case of the incompressible flow around a half-plane. This was worked out for the full potential equation by Ringleb [3.7.3]. For incompressible flow the streamlines in the physical and hodograph planes appear as in Figure 3.7.7. For compressible flow the generalization is made starting in the hodograph where the equations are linear. It no longer represents flow around a half-plane.

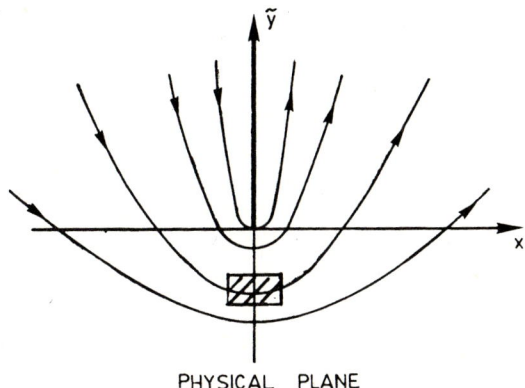

PHYSICAL PLANE

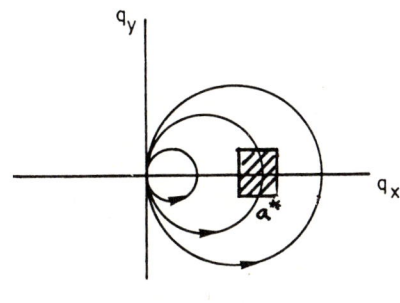

HODOGRAPH

Figure 3.7.7.

Flow Around a Half Plane

Since there is no characteristic length in the problem the scale in the physical plane is arbitrary. With some scale in mind a region of supersonic flow can be calculated in which the streamtubes behave qualitatively differently from incompressible flow. A transonic portion of this region where the flow deviates slightly from uniform sonic flow is shaded in the figure.

The stream function for incompressible flow in the shaded region is of the form $\psi \sim u'^2 + v'^2$ where $q_x = 1 + u'$, $q_y = v'$. A slightly more complicated polynomial provides the transonic solutions. The small disturbance hodograph equations (3.5.7) are

$$\left\{ \begin{array}{l} wy_\vartheta - x_w = 0 \\ \tilde{y}_w - x_\vartheta = 0 \end{array} \right\} \tag{3.5.7}$$

The solution of (3.5.7) of interest is

$$\left\{ \begin{array}{l} \tilde{y} = \dfrac{\vartheta^2}{2} + w\left(1 + \dfrac{w^2}{6}\right) \\ x = \vartheta\left(1 + \dfrac{w^2}{2}\right) \end{array} \right\} \tag{3.7.30)}$$

For this flow the Jacobian J is,

$$J = w\tilde{y}_\vartheta^2 - \tilde{y}_w^2 = w\vartheta^2 - \left(1 + \dfrac{w^2}{2}\right)^2 ,$$

hence $J = 0$ if

$$\vartheta^2 = \dfrac{1}{w}\left(1 + \dfrac{w^2}{2}\right)^2 . \tag{3.7.31}$$

The lines $\tilde{y} = $ constant correspond to

$$\tilde{y} = c = \dfrac{\vartheta^2}{2} + w\left(1 + \dfrac{w^2}{6}\right) . \tag{3.7.32}$$

A $\tilde{y} = $ constant streamline intersects (3.7.31) if

$$2c - 2w\left(1 + \dfrac{w^2}{6}\right) = \dfrac{1}{w}\left(1 + \dfrac{w^2}{2}\right)^2 .$$

That is, if $f(w) = 0$ where

$$f(w) = \dfrac{7}{12}w^4 + 3w^2 - 2cw + 1 .$$

In particular there is a $c_0 > 0$ such that the $\tilde{y} = c$ streamline never intersects the curve (3.7.31) if $c < c_0$, and intersects it four times if $c_0 = \sqrt{\frac{2}{7}}\left(\frac{10}{3}\right)$. To see that this is true note that

$$f'(w) = \dfrac{7}{3}w^3 + 6w - 2c ,$$
$$f''(w) = 7w^2 + 6 .$$

Thus f is always concave up. If $c \leq 0$ the minimum of f is at a $w_0 < 0$ $\left(f'(w_0) = 0\right)$, and $f(w_0) > 0$, hence f never crosses zero. If $c > 0$, $c \gg 1$, then the minimum of f is at a point $w_0 \to \left(\frac{6}{7}c\right)^{\frac{1}{3}}$ and $f(w_0) \to -\left(\frac{6}{7}\right)^{1/3}\left(\frac{3}{2}\right)c^{4/3} + \cdots < 0$.

In order to sketch the lines $\tilde{y} = c$ (3.7.32) in the hodograph note that for $w, \vartheta \gg 1$, the curves behave as

$$c = \frac{\vartheta^2}{2} + \frac{w^3}{6} \quad \text{or} \quad \vartheta \sim \pm \frac{(-w)^{\frac{3}{2}}}{3^{\frac{1}{2}}} ;$$

for $w \ll 1$

$$c = \frac{\vartheta^2}{2} + w , \qquad \vartheta = \pm\sqrt{2(c-w)} = \pm\sqrt{2c}\left\{1 - \frac{w}{2c} + \cdots\right\} ;$$

and for $\vartheta \ll 1$, $w = w_0 + w^*$, where $c = w_0\left(1 + \frac{w^2}{6}\right)$, $w^* \ll w_0$ and then

$$\vartheta = \pm\sqrt{w(c)(w_0 - w)} ;$$

where $w(c) = 2 + w_0^2$. Finally, on the streamlines (3.7.32)

$$\vartheta\frac{d\vartheta}{dw} = -\left(1 + \frac{w^2}{2}\right) .$$

The streamline intersects the image of the limit line (3.7.31) at the points (w_i, ϑ_i) where

$$\vartheta_i = \frac{1}{\sqrt{w_i}}\left(1 + \frac{w_i^2}{2}\right) , \qquad i = 1, 2, 3, 4 .$$

Thus,

$$\pm\frac{1}{\sqrt{w_i}}\left(1 + \frac{w_i^2}{2}\right)\frac{d\vartheta}{dw} = -\left(1 + \frac{w_i^2}{2}\right) , \qquad \text{or} \quad \frac{d\vartheta}{dw} = \mp\sqrt{w_i} .$$

Hence, from equation (3.5.11), the streamline is tangent to a characteristic at its intersection with the image of the limit line. (Figure 3.7.8).

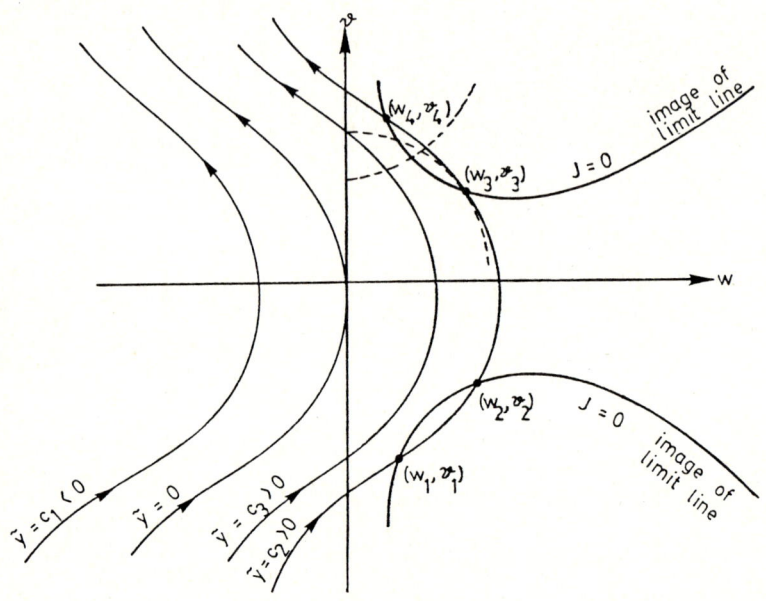

Figure 3.7.8

Hodograph plane streamlines and image of limit line

To recover the physical plane for the flow note first that the sonic line, $w = 0$, corresponds to the parabola,

$$\tilde{y} = \frac{x^2}{2} \, .$$

The limit line is the image of (3.7.31)

$$\vartheta^2 = \frac{1}{w} \left(1 + \frac{w^2}{2} \right)^2 ,$$

We can parametrize $x, \tilde{y}$ along the limit line as a function only of w by substituting (3.7.31) in (3.7.30), so

$$\left(\begin{array}{l} \tilde{y}_\ell = \dfrac{1}{2} \left\{ \left(\dfrac{1}{w} \right) \left(1 + \dfrac{w^2}{2} \right)^2 \right\} + w \left(1 + \dfrac{w^2}{6} \right) , \\[3mm] x_\ell = \pm \sqrt{\dfrac{1}{w}} \left(1 + \dfrac{w^2}{2} \right)^2 , \end{array} \right.$$

or

$$\left(\begin{aligned} \tilde{y}_\ell &= \left\{ \frac{1}{2w} + \frac{3w}{2} + 7\frac{w^3}{24} \right\}, \\ x_\ell &= \pm\sqrt{\frac{1}{w}}\left(1 + \frac{w^2}{2}\right)^2. \end{aligned} \right) \quad \begin{aligned} &\text{parametric representation of limit line} \\ &\text{in physical plane for } 0 < w < \infty \end{aligned}$$

For w large

$$\left(\begin{aligned} \tilde{y}_\ell &\sim \frac{7}{24}w^3 \\ x_\ell &\sim \pm\frac{w^{\frac{7}{2}}}{4} \end{aligned} \right),$$

so that

$$\tilde{y}_\ell \sim \frac{7}{24}x_\ell^{\frac{6}{7}}4^{\frac{6}{7}},$$

and for w small

$$\left(\begin{aligned} \tilde{y}_\ell &\sim \frac{1}{2w} \\ x_\ell &\sim \pm\sqrt{\frac{1}{w}} \end{aligned} \right)$$

so that

$$\tilde{y}_\ell \sim \frac{1}{2}x_\ell^2.$$

Also

$$\frac{dx_\ell}{dw} = 0 \quad \text{if} \quad w = \pm\sqrt{\frac{2}{7}},$$

$$\frac{d\tilde{y}_\ell}{dw} = 0 \quad \text{if} \quad w = \pm\sqrt{\frac{2}{7}},$$

and it is easily checked that $\dfrac{dx_\ell}{dw}$ and $\dfrac{d\tilde{y}_\ell}{dw}$ change sign at $w = +\sqrt{\dfrac{2}{7}}$. Thus, the limit line is cusped at $\left(x_\ell\left(+\sqrt{\dfrac{2}{7}}\right), \ \tilde{y}_\ell\left(+\sqrt{\dfrac{2}{7}}\right) \right) = \left(\pm4\sqrt{\dfrac{7}{2}\dfrac{64}{49}}, \ \sqrt{\dfrac{2}{7}\dfrac{10}{3}} \right)$ (see Figure 3.7.9).

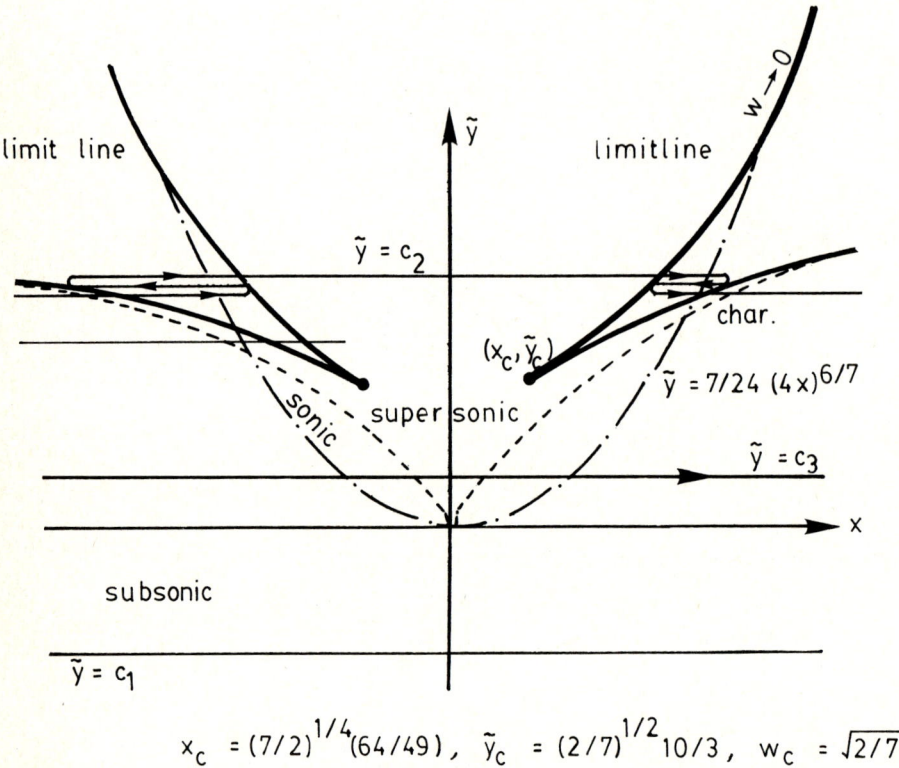

$$x_c = (7/2)^{1/4}(64/49), \quad \tilde{y}_c = (2/7)^{1/2}10/3, \quad w_c = \sqrt{2/7}$$

Figure 3.7.9
Physical Plane

Note that a realistic single valued solution to the small disturbance equations is obtained if the boundaries are located at

$$\tilde{y} = c_0 \equiv \sqrt{\frac{2}{7}\left(\frac{10}{3}\right)} \approx 1.78 \,,$$

for example at

$$\tilde{y} = \pm 1 \,.$$

Then in order to reconstruct the physical problem of which (3.7.30) is a small disturbance approximation, we must find ϑ on the boundaries, $\tilde{y} = \pm 1$, since this gives $F'(x)$ where $\vartheta_b(x) = (\gamma + 1)F'(x)$ and

$$\tilde{y} = \pm\left(1 + \delta F_{u,\ell}(x)\right), \qquad \begin{array}{l} F_u \text{ corresponds to } \tilde{y} = +1, \\ F_\ell \quad \text{to } \tilde{y} = -1 \end{array}$$

Now on $\tilde{y} \pm 1$,

$$\left\{ \begin{array}{c} \vartheta_b = \pm\sqrt{2\left(\pm 1 - w\left(1 + \dfrac{w^2}{6}\right)\right)}, \\[4mm] x_b = \vartheta_b\left(1 + \dfrac{w^2}{2}\right). \end{array} \right\} \tag{3.7.33}$$

Although it is not possible to solve explicitly for $\vartheta(x)$, (3.7.33) gives ϑ, x at various points as parametrized by w. It is easy to check that ϑ is monotone decreasing and furthermore that $|\vartheta(x; y_0)|$ at a given x is larger as y_0 increases. Hence the physical flow looks like channel flow (Figure 3.7.10).

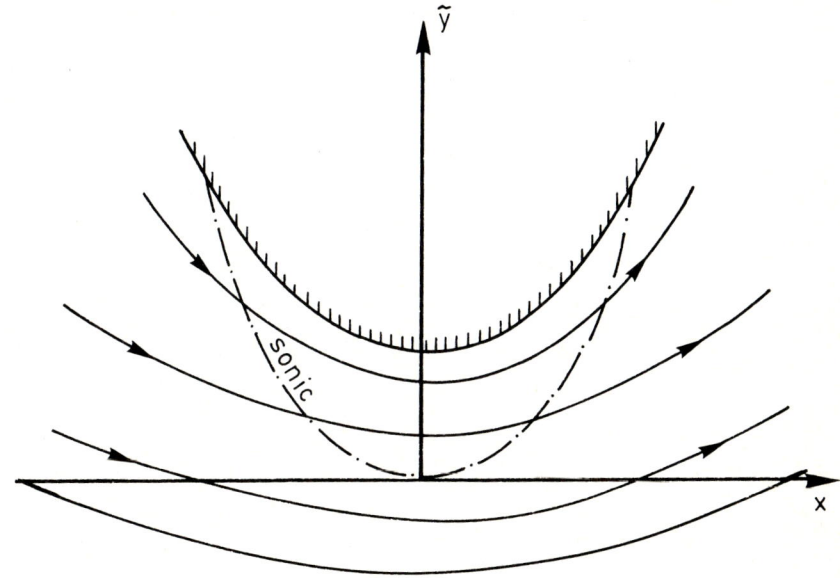

Figure 3.7.10

A Smooth Transonic Flow

Of course this flow is not realizable unless the entrance and exit conditions are adjusted properly, but precisely this could be achieved by phasing into a uniform flow.

References

[3.7.1] Courant, R. and Friedrichs, K. O., *Supersonic Flow and Shock Waves*, Interscience, New York 1948.

[3.7.2] Guderley, K. G., *The Theory of Transonic Flow*, Pergamon Press, London, 1962.

[3.7.3] Ringleb, F., Exacte Lösungen der Differentialgleichunsen eineradiabatischen Gasströmung, *ZAMM* **20** (1941), pp185-198.

3.8 Subsonic and Sonic Jets

The problem of the planar jet flow in which the final state is uniform subsonic or sonic flow admits an easy formulation in transonic hodograph variables. This is thus a simplified version of Chaplygin's original problem for the full potential equation (1906 On Gas Jets). The problem is simple because the boundaries appear as known curves in the hodograph; on the straight walls inside the jet the deflection ϕ_y is known and in the jet boundary the pressure is known ($\phi_{x^*} = 0$). The structure of the solution, its singularities etc. can be found by an examination of the streamline pattern in the hodograph. Further no shock waves appear in the entirely subsonic flow. The main information of interest is the shape, and the decay to the steady state.

The expansion procedure for the jet flow was discussed in Section 3.3, (cf. Figure 3.3.1). In summary

$$\Phi(X,Y) = U\left\{ X + \delta H \phi(x^*, y; K) + \cdots \right\}, \quad y = \frac{Y}{H}, \quad x^* = \delta^{-\frac{1}{3}} \frac{X}{H}, \quad (3.8.1)$$

$$K = \frac{1 - M_\infty^2}{\delta^{\frac{2}{3}}}$$

and

$$\boxed{\left(K - (\gamma + 1)\phi_{x^*} \right)\phi_{x^* x^*} + \phi_{yy} = 0} \qquad (3.8.3)$$

The boundary conditions in transonic coordinates are shown in Figure 3.8.1.

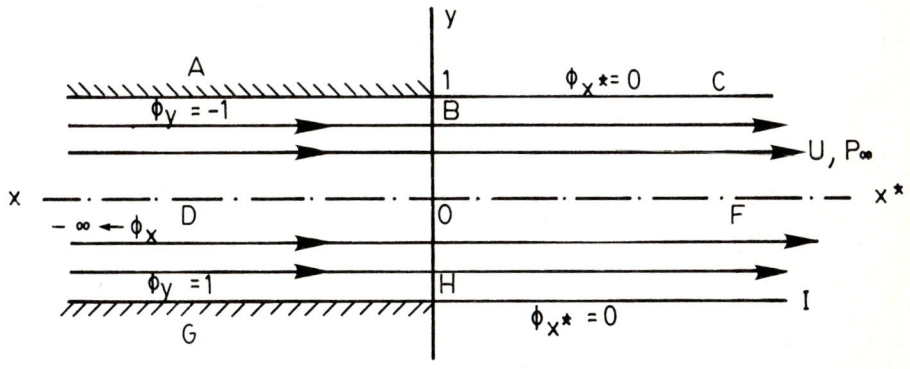

Figure 3.8.1

Jet in Transonic Coordinates

Transonic Aerodynamics

The flow comes out of a reservoir and accelerates to the low pressure outside. Within the small disturbance theory stagnation in the reservoir ($x^* \to -\infty$) appears as $\phi_{x^*} \to -\infty$ as is typical for small disturbance theories.

The K-G equation (3.8.2) is equivalent to the system

$$\left\{ \begin{array}{c} w w_{x^*} - \vartheta_y = 0 \\ w_y - \vartheta_x = 0 \end{array} \right\} \tag{3.8.4}$$

where

$$w = (\gamma + 1)\phi_{x^*} - K \,, \quad \vartheta = (\gamma + 1)\phi_y \,.$$

w measures the perturbations from local sonic flow, $w < 0$ for subsonic flow. This system can be expressed in the hodograph by considering $y(w, \vartheta)$, $x^*(w, \vartheta)$. (cf.section 3.5)

$$\left\{ \begin{array}{c} w y_\vartheta - x_w^* = 0 \\ x_\vartheta^* - y_w = 0 \end{array} \right\} \,, \tag{3.8.5}$$

or the approximate stream function y which satisfies the Tricomi equation.

$$\boxed{w y_{\vartheta\vartheta} - y_{ww} = 0} \,. \tag{3.8.6}$$

Once y is found, x^* can be found, for example from the second of (3.8.5). The boundary value problem as formulated in the hodograph (w, ϑ) appears in Figure 3.8.2.

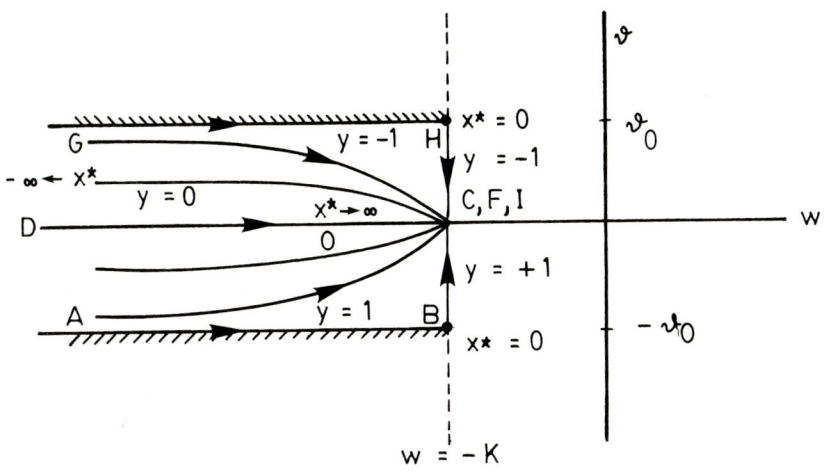

Figure 3.8.2

Hodograph Plane of Jet Flow

The boundary AB has the image

$$\vartheta = -\vartheta_0 , \quad \vartheta_0 = (\gamma + 1) \tag{3.8.7}$$

and the boundary GH has the image

$$\vartheta = +\vartheta_0 . \tag{3.8.8}$$

On the jet surface $y = \pm 1$, (BC and HI), The pressure is constant so that

$$w = -K . \tag{3.8.9}$$

The lines $y = $ constant with arrows indicating the directions of x^* increasing are shown in the figure.

$$x^* = 0 \quad \text{at} \quad \vartheta = \pm\vartheta_0 , \quad w = -K \quad \text{(Pts. BH)} \tag{3.8.10}$$

The uniform state at downstream infinity $(x^* \to \infty)$ is represented by the singular points $(w = -K, \quad \vartheta = 0)$ in the hodograph, since all the lines $y = $ constant run into this point. The local structure of this singular point is shown in Figure 3.8.3.

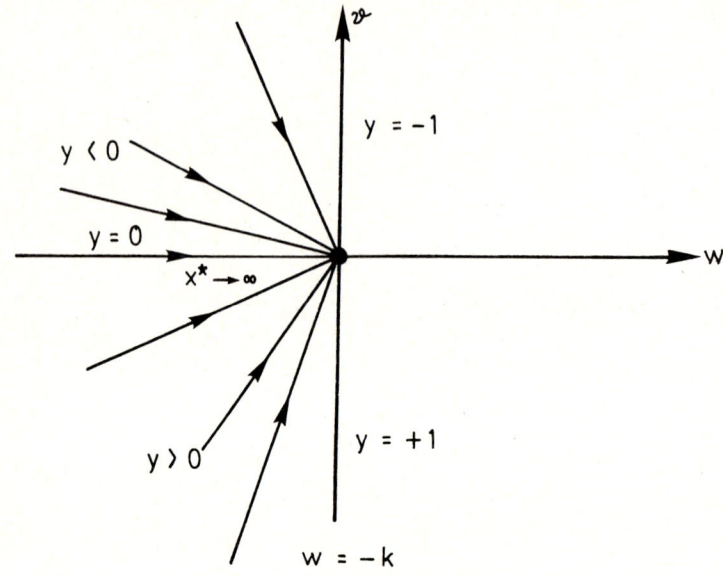

Figure 3.8.3

Local Free Stream Singularity

Locally, the Tricomi equation can be approximated by a (scaled) Laplace equation

$$K y_{\vartheta\vartheta} + y_{ww} + \cdots = 0 . \tag{3.8.11}$$

The stream lines are locally those of a sink flow or

$$y(w, \vartheta) = \frac{2}{\pi} \tan^{-1} \frac{\vartheta}{\sqrt{K}(w + K)} + \cdots ,$$

$$x^*(w, \vartheta) = -\frac{2}{\pi} \log \sqrt{\vartheta^2 + K(w + K)^2} + \cdots . \tag{3.8.12}$$

It is possible to find solutions of the Tricomi equation (3.8.6) which have the local singularity of (3.8.12) and to use the principle of reflections to take care of the boundary condition on $\vartheta = \pm \vartheta_0$. But it is probably simpler to construct an eigenfunction expansion using solutions generated by separation of variables. The nature of the local singularity can also be studied from the eigenfunction expansion. In order to use an eigenfunction expansion (or reflection) it is necessary to obtain homogeneous boundary conditions by subtracting the solutions at

$w \to -\infty$,

$$y = y_p = -\frac{\vartheta}{\vartheta_0} , \qquad x^* = x_p^* = -\frac{w^2}{2\vartheta_0} . \qquad (3.8.13)$$

Thus set

$$y(w, \vartheta) = \frac{-\vartheta}{\vartheta_0} + \sum_{n=1}^{\infty} \alpha_n \sin \frac{n\pi\vartheta}{\vartheta_0} \frac{Ai\left(-\left(\frac{n\pi}{\vartheta_0}\right)^{\frac{2}{3}} w \right)}{Ai\left(\left(\frac{n\pi}{\vartheta_0}\right)^{\frac{2}{3}} K \right)} \qquad (3.8.14)$$

(cf. section 3.5 for these solutions). The eigenfunctions $\sin\frac{m\pi\vartheta}{\vartheta_0}$ vanish at $\vartheta = \pm\vartheta_0$ and the Airy function $Ai(z)$ vanishes as $z \to \infty$. The Fourier coefficients α_n are found from the conditions on $w = -K$,

$$y(-K, \vartheta) = -\operatorname{sgn}\vartheta = \frac{-\vartheta}{\vartheta_0} + \sum_{n=1}^{\infty} \alpha_n \sin \frac{n\pi\vartheta}{\vartheta_0} , \qquad (3.8.15)$$

$$|\vartheta| < \vartheta_0 .$$

Thus

$$\alpha_n = \frac{1}{\vartheta_0} \int_{-\vartheta_0}^{\vartheta_0} \sin \frac{n\pi\vartheta}{\vartheta_0} \left\{ -\operatorname{sgn}\vartheta + \frac{\vartheta}{\vartheta_0} \right\} d\vartheta ,$$

$$= -2 \int_0^1 (1 - \beta)\sin n\pi\beta \, d\beta ,$$

$$\alpha_n = -\frac{2}{n\pi} , \qquad n = 1, 2, 3 \ldots . \qquad (3.8.16)$$

Thus (3.8.14) becomes

$$y(w, \vartheta) = -\frac{\vartheta}{\vartheta_0} - \frac{2}{\pi} \sum_{n=1}^{\infty} \frac{1}{n} \sin\left(n\pi \frac{\vartheta}{\vartheta_0} \right) \frac{Ai\left(-\left(\frac{n\pi}{\vartheta_0}\right)^{\frac{2}{3}} w \right)}{Ai\left(\left(\frac{n\pi}{\vartheta_0}\right)^{\frac{2}{3}} K \right)} \qquad (3.8.17a)$$

which is the eigenfunction representation of the solution. To find x^*

$$x_\vartheta^* = y_w = \frac{2}{\pi} \left(\frac{\pi}{\vartheta_0}\right)^{\frac{2}{3}} \sum_{n=1}^{\infty} \frac{1}{n^{\frac{1}{3}}} \sin\left(n\pi \frac{\vartheta}{\vartheta_0} \right) \frac{Ai'\left(\left(\frac{n\pi}{\vartheta_0}\right)^{\frac{2}{3}} w \right)}{Ai\left(\left(\frac{n\pi}{\vartheta_0}\right)^{\frac{2}{3}} K \right)} , \qquad (3.8.17b)$$

or

$$x^* = \text{fn}(w) - \frac{2}{\pi}\left(\frac{\pi}{\vartheta_0}\right)^{-\frac{1}{3}} \sum_{n=1}^{\infty} \frac{\cos\frac{n\pi\vartheta}{\vartheta_0}}{n^{\frac{4}{3}}} \frac{Ai'\left(-\left(\frac{n\pi}{\vartheta_0}\right)^{\frac{2}{3}} w\right)}{Ai\left(\left(\frac{n\pi}{\vartheta_0}\right)^{\frac{2}{3}} K\right)}.$$

Part of the function of integration is fixed from conditions at $(w = -\infty)$ so

$$x^* = -\frac{w^2}{2\vartheta_0} + x_0^* - \frac{2}{\pi}\left(\frac{\pi}{\vartheta_0}\right)^{-\frac{1}{3}} \sum_{n=1}^{\infty} \frac{\cos\frac{n\pi\vartheta}{\vartheta_0}}{n^{\frac{4}{3}}} \frac{Ai'\left(-\left(\frac{n\pi}{\vartheta_0}\right)^{\frac{2}{3}} w\right)}{Ai\left(\left(\frac{n\pi}{\vartheta_0}\right)^{\frac{2}{3}} K\right)}.$$

The constant x_0^* is found from the requirement that the jet exit

$$\vartheta = \vartheta_0, \quad w = -K, \quad x^* = 0,$$

$$0 = -\frac{K^2}{2\vartheta_0} + x_0^* - \frac{2}{\pi}\left(\frac{\pi}{\vartheta_0}\right)^{-\frac{1}{3}} \sum_{n=1}^{\infty} \frac{1}{n^{\frac{4}{3}}} \frac{Ai'\left(+\left(\frac{n\pi}{\vartheta_0}\right)^{\frac{2}{3}} K\right)}{Ai\left(\left(\frac{n\pi}{\vartheta_0}\right)^{\frac{2}{3}} K\right)}.$$

Thus

$$x^*(w, \vartheta) = \frac{K^2 - w^2}{2\vartheta_0}$$

$$- \frac{2}{\pi}\left(\frac{\pi}{\vartheta_0}\right)^{-\frac{1}{3}} \sum_{n=1}^{\infty} \frac{Ai'\left(-\left(\frac{n\pi}{\vartheta_0}\right)^{\frac{2}{3}} w\right)\cos\frac{n\pi\vartheta}{\vartheta_0} - (-)^n Ai'\left(\left(\frac{n\pi}{\vartheta_0}\right)^{\frac{2}{3}} K\right)}{Ai\left(\left(\frac{n\pi}{\vartheta_0}\right)^{\frac{2}{3}} K\right)}$$

$$\tag{3.8.18}$$

The nature of the singularities at $(w = -K, \vartheta = 0)$ in the sums (3.8.17, 3.8.18) can be investigated by studying the asymptotic behaviour of the terms (using the asymptotic formula for the Airy functions) and can be shown to be that of (3.8.12) for $K > 0$. The slope of the edge of the jet is given (implicitly) by

(3.8.18) $\left(x_J^*(\vartheta) \right)$ on $w = -K$

$$x_J^*(\vartheta) = x^*(-K, \vartheta) = -\frac{2}{\pi} \left(\frac{\pi}{\vartheta_0} \right)^{-\frac{1}{3}} \sum_{n=1}^{\infty} \frac{Ai'\left(\left(\frac{n\pi}{\vartheta_0} \right)^{\frac{2}{3}} K \right)}{n^{\frac{4}{3}} Ai\left(\left(\frac{n\pi}{\vartheta_0} \right)^{\frac{2}{3}} K \right)} \left\{ \cos \frac{n\pi\vartheta}{\vartheta_0} - (-)^n \right\}$$

$$(3.8.19)$$

Since

$$\frac{Ai'\left(\left(\frac{n\pi}{\vartheta_0} \right)^{\frac{2}{3}} K \right)}{Ai\left(\left(\frac{n\pi}{\vartheta_0} \right)^{\frac{2}{3}} K \right)} \rightarrow \left(\frac{n\pi}{\vartheta_0} \right)^{\frac{1}{3}} \sqrt{K} \quad \text{as} \quad n \rightarrow \infty \,,$$

we can see that $x_J^* \rightarrow \infty$ as $\vartheta \rightarrow 0$. Similar considerations apply to (3.8.17). The uniform state corresponding to the singular point at $(\vartheta = 0, w = -K)$ is reached asymptotically at downstream infinity.

It is worth remarking, however, that the nature of the singularity changes if the downstream Mach number is exactly equal to one

$$M_\infty = 1 \,, \quad K = 0 \,,$$

as $K \rightarrow 0$

$$Ai'\left(\left(\frac{n\pi}{\vartheta_0} \right)^{\frac{2}{3}} K \right) \rightarrow -c_2 \,, \quad Ai\left(\left(\frac{n\pi}{\vartheta_0} \right)^{\frac{2}{3}} K \right) \rightarrow c_1 \,. \qquad (3.8.20)$$

Thus as $\vartheta \rightarrow 0$ the sum (3.8.19) becomes

$$x_J^*(\vartheta) = x^*(0, \vartheta) = -\frac{2}{\pi} \left(\frac{\pi}{\vartheta_0} \right)^{-\frac{1}{3}} \sum_{n=0}^{\infty} \frac{1 - (-)^n}{n^{\frac{4}{3}}} \left(-\frac{c_2}{c_1} \right) = \mu^* \,, \qquad (3.8.21)$$

where

$$\frac{c_2}{c_1} = \frac{3^{\frac{2}{3}} \Gamma\left(\frac{2}{3} \right)}{3^{\frac{1}{3}} \Gamma\left(\frac{1}{3} \right)}$$

This shows that the uniform asymptotic state is reached at a finite distance from the exit. (See Figure 3.8.4)

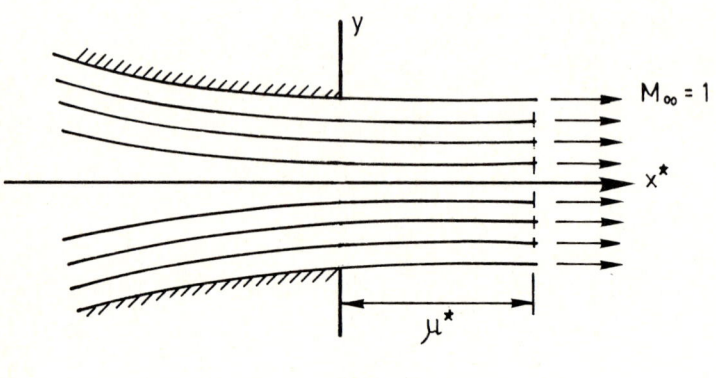

Figure 3.8.4
Sonic Jet

Downstream of $x^* = K^*$ we have a uniform sonic flow.

If the exit pressure is reduced a little from that for sonic flow the flow can accelerate to supersonic speeds. The sonic line starts from the corner and is close to the sonic line in Figure 3.8.3. The flow expands at the corner with a local simple wave. The expansion waves reflect from the sonic line and from the edge of the jet. The calculation of this flow is more complicated since it involves the solution of a mixed subsonic and supersonic flow and it is not discussed here. Numerical approaches to this type of problem are discussed later.

3.9 Transonic Slender Bodies: Expansion Procedure, Area Rule

Slender bodies are characterized by span and thickness of the same order $\delta \ll 1$, δ = thickness ratio. For example the body surface can be represented in cartesian coordinates as

$$B\left(x', \frac{y'}{\delta}, \frac{z}{\delta}\right) = \frac{y'}{\delta} - F_{u,\ell}\left(x', \frac{z}{\delta}\right) = 0 , \tag{3.9.1}$$

where $F_{u,\ell}$ = equations of upper and lower surfaces respectively, and x',y' are body fixed coordinates as in Figure 3.9.1.

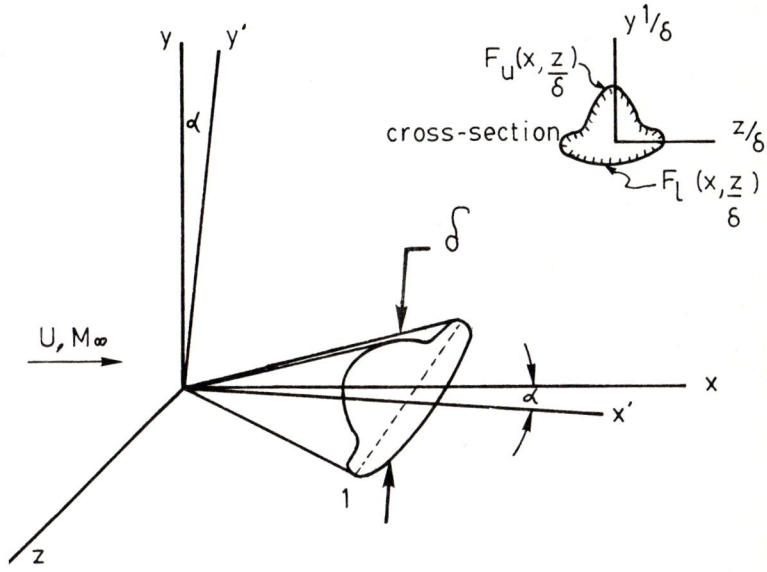

Figure 3.9.1

Slender Body

The length of the body is taken to be unity. The angle of attack is α such that $A = \dfrac{\alpha}{\delta} = O(1)$, and we can expect our theory to be valid uniformly as $A \to 0$. We note

$$y' = y\cos\alpha + x\sin\alpha = y\left(1 - \frac{\alpha^2}{2}\right) + x\left(\alpha - \frac{\alpha^3}{6}\right) + \cdots$$

$$x' = x\cos\alpha - y\sin\alpha = x\left(1 - \frac{\alpha^2}{2}\right) - y\left(\alpha - \frac{\alpha^3}{6}\right) + \cdots . \tag{3.9.2}$$

In the limit as $\delta \to 0$ a slender body shrinks to a line. Thus an inner expansion is needed to preserve the body geometry and represent the boundary conditions. The inner limit has $(x, y^* = \frac{y}{\delta}, z^* = \frac{z}{\delta})$ fixed as $\delta \to 0$, $M_\infty \to 1$. As shown below this leads to a series of unforced and forced Laplace equations in cross-section planes (x fixed). A different limit is needed to yield suitable non-linear equations of changing type. Analogous to Section 3.1 this limit has $\delta \to 0$ ($x, \tilde{y} = \delta y, \tilde{z} = \delta z$) fixed as well as $K = \frac{1-M_\infty^2}{\delta^2}$ fixed. These results were obtained systematically in [3.9.1], [3.9.2]. These inner and outer expansions match to first order but in fact, in general, do not match to higher order. This is perhaps not surprising since in the inner limit (y^*, z^* fixed) the representative y, z point runs to the axis with the body surface while in the outer limit ($\tilde{y}, \tilde{z}$) fixed, (y, z) runs away from the axis. This latter limit expresses the large lateral extent of the flow field. In this section it is shown that an intermediate limit and intermediate expansion exists (x, y, z) fixed which matches to both inner and outer expansions and provides the necessary connection between them.

In this section some details of this procedure are set out. Various switchback terms (with $\log \delta$, $\log^2 \delta \ldots$) occur in the expansion and these are anticipated here. A convenient starting point is the exact potential equation (2.4.28) which we write

$$(a^2 - \Phi_x^2)\Phi_{xx} + (a^2 - \Phi_y^2)\Phi_{yy} + (a^2 - \Phi_z^2)\Phi_{zz}$$
$$= 2\Phi_x\Phi_y\Phi_{xy} + 2\Phi_y\Phi_z\Phi_{yz} + 2\Phi_z\Phi_x\Phi_{zx} , \qquad (3.9.3)$$

$$\frac{a^2}{U^2} = \frac{1}{M_\infty^2} + \frac{\gamma - 1}{2}\left\{1 - \frac{q^2}{U^2}\right\} , \quad \text{where} \quad q^2 = \Phi_x^2 + \Phi_y^2 + \Phi_z^2 . \qquad (3.9.4)$$

For the inner expansion we have $(x, y^* = \frac{y}{\delta}, z^* = \frac{z}{\delta}, K = \frac{1-M_\infty^2}{\delta^2}, A = \frac{\alpha}{\delta})$ fixed as $\delta \to 0$. The form of the transonic similarity parameter K is verified when the outer expansion is considered below. (cf. [3.9.1], [3.9.2])

Inner Expansion:

$$\Phi = U\left\{x + \delta^2 \log \delta \left(2S_1(x)\right) + \delta^2\varphi_1(x, y^*, z^*)\right.$$
$$\left. + \delta^4 \log \delta \varphi_{21}(x, y^*, z^*) + \delta^4\varphi_2(x, y^*, z^*) + \cdots\right\} . \qquad (3.9.5)$$

We have

$$\frac{\Phi_x}{U} = 1 + \delta^2 \log \delta (2S_1') + \delta^2 \varphi_{1x} + \cdots ,$$

$$\frac{\Phi_{xx}}{U} = \delta^2 \log \delta 2S_1'' + \delta^2 \varphi_{1xx} + \cdots ,$$

$$\frac{\Phi_y}{U} = \delta \varphi_{1y^*} + \delta^3 \log \delta \varphi_{21y^*} + \delta^3 \varphi_{2y^*} + \cdots ,$$

$$\frac{\Phi_{yy}}{U} = \varphi_{1y^*y^*} + \delta^2 \log \delta \varphi_{21y^*y^*} + \delta^3 \varphi_{2y^*y^*} + \cdots ,$$

$$\frac{\Phi_{xy}}{U} = \delta \varphi_{1xy^*} + \cdots ,$$

where $(\ )' = \frac{d}{dx}$,

$$\frac{1}{M_\infty^2} = 1 + K\delta^2 + K\delta^4 + \cdots$$

$$\frac{a^2}{U^2} = 1 - \delta^2 \log \delta (2S_1') + \delta^2 \left(K - (\gamma - 1)\varphi_{1x} - \frac{\gamma - 1}{2}(\varphi_{1y^*}^2 + \varphi_{1z^*}^2) \right) + \cdots ,$$

$$\frac{a^2 - \Phi_x^2}{U^2} = -\delta^2 \log \delta \left((\gamma + 1)2S_1' \right)$$

$$+ \delta^2 \left(K - (\gamma + 1)\varphi_{1x} - \frac{\gamma - 1}{2}(\varphi_{1y^*}^2 + \varphi_{1z^*}^2) \right) + \cdots .$$

Thus (3.9.3) becomes

$$\left\{ 1 - \delta^2 \log \delta (\gamma - 1)2S_1' + \delta^2 \left(K - (\gamma - 1)\varphi_{1x} \right. \right.$$

$$\left. \left. - \frac{\gamma - 1}{2}(\varphi_{1y^*}^2 + \varphi_{1z^*}^2) - \delta^2 \varphi_{1y^*}^2 - \cdots \right\} \right.$$

$$\times \left\{ \varphi_{1y^*y^*} + \delta^2 \log \delta \varphi_{21y^*y^*} + \delta^2 \varphi_{2y^*y^*} + \cdots \right\}$$

$$+ \left\{ 1 - \delta^2 \log \delta (\gamma - 1)2S_1' + \delta^2 \left(K - (\gamma - 1)\varphi_{1x} - \frac{\gamma - 1}{2}(\varphi_{1y^*} + \varphi_{1z^*}) - \delta^2 \varphi_{1z^*}^2 \right\} \right.$$

$$\times \left\{ \varphi_{1z^*z^*} + \delta^2 \log \delta \varphi_{21z^*z^*} + \delta^2 \varphi_{2z^*z^*} + \cdots \right\} + O(\delta^4 \log^2 \delta)$$

$$= \delta^2 \cdot 2(1 + \cdots)\varphi_{1y^*}\varphi_{1xy^*} + \delta^2 \cdot 2\varphi_{1z^*}\varphi_{1xz^*} + \delta^2 \cdot 2\varphi_{1y^*}\varphi_{1z^*}\varphi_{1y^*z^*} + \cdots .$$

This yields the sequence of approximating equations

$$O(1): \quad \nabla^{*2}\varphi_1 \equiv \frac{\partial^2 \varphi_1}{\partial y^{*2}} + \frac{\partial \varphi_1}{\partial z^{*2}} = 0 , \tag{3.9.6}$$

$$O(\delta^2 \log \delta): \quad \nabla^{*2}\varphi_{21} = 0 , \tag{3.9.7}$$

$$O(\delta^2): \quad \nabla^{*2}\varphi_2 = \frac{\partial}{\partial x}(\varphi_{1y^*}^2 + \varphi_{1z^*}^2) + \varphi_{1y^*}^2\varphi_{1y^*y^*}$$

$$+ 2\varphi_{1y^*}\varphi_{1z^*}\varphi_{1y^*z^*} + \varphi_{1z^*}^2\varphi_{1z^*z^*} . \tag{3.9.8}$$

Next the inner boundary conditions of tangent flow will be derived. Each inner problem becomes a problem in a cross-section plane ($x = $ constant) as is typical of slender body theory. The far field behaviour ($r^* \to 0$) of the inner expression is crucial for matching.

Using (3.9.2) to expand the equation for the body surface, one finds

$$B^* = 0 = y^* + Ax - F_{u,\ell}(x, z^*) + \delta^2 H_{u,\ell}(x, z^*) + \cdots , \qquad (3.9.9)$$

where $H = \frac{A^3 x}{3} - \frac{A^2}{2}(xF)_x + AFF_x$, a correction term. The boundary condition of tangent flow $\nabla \Phi \bullet \nabla B = 0$ becomes

$$\Phi_x B_{x^*} + \Phi_y B_{y^*} \cdot \frac{1}{\delta} + \Phi_z B_{z^*}^* \cdot \frac{1}{\delta} = 0 \quad \text{on} \quad B^* = 0$$

or, with subscripts (u, ℓ) understood,

$$\left\{ 1 + \delta^2 \log \delta (2S_1') + \delta^2 \varphi_{1x}(x, F - Ax, z^*) \ldots \right\} \left\{ A - F_x + \delta^2 H_x + \cdots \right\}$$
$$+ \left\{ \delta \varphi_{1y^*}(x, F - Ax, z^*) - \delta^3 H \varphi_{1y^* y^*} + \delta^3 \log \delta \varphi_{21y^*} + \delta^3 \varphi_{2y^*} \right\} \frac{1}{\delta}$$
$$+ \left\{ \delta \varphi_{1z^*} - \delta^3 H \varphi_{1z^* y^*} + \delta^3 \log \delta \varphi_{21z^*} - \delta^3 \varphi_{2z^*} \right\} \frac{1}{\delta} \left\{ -F_{z^*} + \delta^2 H_{z^*} \right\} = 0 .$$
$$(3.9.10)$$

In (3.9.10) all derivatives of φ are evaluated on the basic surface $y^* = F_{u,\ell}(x, z^*) - Ax$. Thus the boundary conditions of various orders are obtained.

$$O(1) \qquad : \quad \varphi_{1y^*}(x, F - Ax, z^*) - F_{z^*} \varphi_{1z^*} = F_x - A \qquad (3.9.11)$$

$$O(\delta^2 \log \delta) \quad : \quad \varphi_{21y^*} - F_{z^*} \varphi_{21z^*} = 2S_1'(F_x - A) \qquad (3.9.12)$$

$$O(\delta^2) \qquad : \quad \varphi_{2y^*} - F_{z^*} \varphi_{2z^*} = -H_{z^*} \varphi_{1z^*} - H \varphi_{1z^* y^*} \cdot F_{z^*} + H \varphi_{1y^* y^*}$$
$$+ \varphi_{1x}(F_x - A) = J(x, z^*) \qquad \text{say.} \qquad (3.9.13)$$

The boundary conditions (3.9.11-3.9.13) are of the type where the normal derivative in a cross-section plane is prescribed on the surface $y^* = F_{u,\ell} - Ax$. Hence the far field of the solutions to (3.9.6-3.9.8) can be expressed by a superposition of singularities at the origin, including a source $S_1(x)$, constants of integration (which depend on x), and particular solutions. We know that $\varphi_1 \sim \log r^* + \cdots$ where $r^* = \sqrt{y^{*2} + z^{*2}}$ so that the dominant term in the right hand side of (3.9.8) as $r^* \to \infty$ is $\frac{\partial}{\partial x} \varphi_{1r^*}^2$. The particular solution φ_p we need is for

$$\frac{\partial^2 \varphi_p}{\partial r^{*2}} + \frac{1}{r^*} \frac{\partial \varphi_p}{\partial r^*} = 2S_1 S_1' \frac{1}{r^{*2}} + O\left(\frac{1}{r^{*3}}\right) .$$

Thus

$$\varphi_p = S_1 S_1' \log^2 r^* + O\left(\frac{1}{r^*}\right) . \qquad (3.9.14)$$

Further $\varphi_{21} = 2S_1'\varphi_1$ except for the constant of integration. In summary, the far field of the inner expansion is given by

$$\varphi_1 = S_1(x)\log r^* + g_1(x) + \frac{D_1(x)\cos\vartheta}{r^*} + \frac{E_1(x)\cos 2\theta}{r^{*2}} + \cdots \qquad (3.9.15)$$

$$\varphi_{21} = 2S_1 S_1' \log r^* + g_{21}(x) + 2S_2(x) + \cdots \qquad (3.9.16)$$

$$\varphi_2 = S_1 S_1' \log^2 r^* + S_2(x)\log r^* + g_2(x) + \cdots \qquad (3.9.17)$$

Here the source, doublet and quadrupole strength S_1, S_2, D_1, E_1 are known from the inner boundary value problems, while g_1, g_{21}, g_2 must be found from solving (numerically) outer problems. Downstream of the slender configuration ($x > 1$) only a vortex sheet remains so that $S_1 = S_2 = 0$, $E_1 = 0$ while $D_1 = \text{constant} = D_1(1)$.

Next the intermediate expansion is constructed in the form

Intermediate Expansion:

$$\Phi = U\Big\{ x + \delta^2 \log \delta S_1(x) + \delta^2 \bar\phi_1(x, y, z) + \delta^3 \bar\phi_2(x, y, z) + \delta^4 \log^2 \delta \bar\phi_{32}(x, y, z)$$

$$+ \delta^4 \log \delta \bar\phi_{31}(x, y, z) + \delta^4 \bar\phi_3(x, y, z) + \cdots \Big\} . \qquad (3.9.18)$$

Note that

$$\frac{\Phi_x}{U} = 1 + \delta^2 \log \delta S_1' + \delta^2 \bar\phi_{1x} , \qquad \frac{\Phi_{xx}}{U} = \delta^2 \log \delta S_1'' + \delta^2 \bar\phi_{1xx} + \cdots ,$$

$$\frac{\Phi_y}{U} = \delta^2 \bar\phi_{1y} + \delta^3 \bar\phi_{2y} + \delta^4 \log^2 \delta \bar\phi_{32y} + \delta^4 \log \delta \bar\phi_{31y} + \delta^4 \bar\phi_{3y} ,$$

$$\frac{\Phi_{yy}}{U} = \delta^2 \bar\phi_{1yy} + \delta^3 \bar\phi_{2yy} + \delta^4 \log^2 \delta \bar\phi_{32yy} + \delta^4 \log \delta \bar\phi_{31yy} + \delta^4 \bar\phi_{3yy} + \cdots ,$$

$$\frac{\Phi_{xy}}{U} = \delta^2 \bar\phi_{1xy} + \cdots \qquad \text{etc.,}$$

$$\frac{a^2}{U^2} = 1 - \delta^2 \log \delta(\gamma + 1)S_1' + \delta^2\big(K - (\gamma + 1)\bar\phi_{1x}\big) + \cdots ,$$

$$\frac{a^2 - \Phi_x^2}{U^2} = -\delta^2 \log \delta(\gamma + 1)S_1' + \delta^2\big(K - (\gamma + 1)\bar\phi_{1x}\big) .$$

Thus, the full potential equation (3.9.3) is

$$\left\{-\delta^2 \log \delta (\gamma + 1) S_1' + \delta^2 \left(K - (\gamma + 1)\bar{\phi}_{1x}\right) + \cdots\right\}\left\{\delta^2 \log \delta S_1'' + \delta^2 \bar{\phi}_{1xx} + \cdots\right\}$$

$$+ \left\{1 - \delta^2 \log \delta (\gamma - 1) S_1' + \delta^2 \left(K - (\gamma + 1)\bar{\phi}_{1x}\right) + \cdots\right\}$$

$$\times \left\{\delta^2 (\bar{\phi}_{1yy} + \bar{\phi}_{1zz}) + \delta^4 \log^2 \delta (\bar{\phi}_{32yy} + \bar{\phi}_{32zz})\right.$$

$$\left. + \delta^4 \log \delta (\bar{\phi}_{31yy} + \bar{\phi}_{31zz}) + \delta^4 (\bar{\phi}_{3yy} + \bar{\phi}_{3zz}) + \cdots\right\}$$

$$= \delta^4 \cdot 2(1 + \cdots)(\bar{\phi}_{1y}\bar{\phi}_{1xy} + \bar{\phi}_{1z}\bar{\phi}_{1xz}) \, .$$

Thus, for the various orders

$$O(\delta^2) \qquad \bar{\nabla}^2 \bar{\phi}_1 \equiv \frac{\partial^2 \bar{\phi}_1}{\partial y^2} + \frac{\partial^2 \bar{\phi}_1}{\partial z^2} = 0 \, , \tag{3.9.19}$$

$$O(\delta^3) \qquad \bar{\nabla}^2 \bar{\phi}_2 = 0 \, , \tag{3.9.20}$$

$$O(\delta^4 \log^2 \delta) \qquad \bar{\nabla}^2 \bar{\phi}_{32} = (\gamma + 1) S_1' S_1'' \, , \tag{3.9.21}$$

$$O(\delta^4 \log \delta) \qquad \bar{\nabla}^2 \bar{\phi}_{31} = -K S_1'' + (\gamma + 1)(S_1' \bar{\phi}_{1x})_x \, , \tag{3.9.22}$$

$$O(\delta^4) \qquad \bar{\nabla}^2 \bar{\phi}_3 = \frac{\partial}{\partial x}(\bar{\phi}_{1r}^2) - (K - (\gamma + 1)\bar{\phi}_{1x})\bar{\phi}_{1xx} \, , \tag{3.9.23}$$

$$\text{where} \qquad r = \sqrt{y^2 + z^2} \, .$$

The intermediate expansion must match to the inner as $r \to 0$ and to the outer as $r \to \infty$, and this, in a sense, provides the necessary boundary conditions for the intermediate equations. The right hand side of (3.9.23) contains the forcing term $\frac{\partial}{\partial x}(\bar{\phi}_{1r}^2)$ necessary to match the inner expansion and also the typically transonic forcing term $(\gamma + 1)\bar{\phi}_{1x}\bar{\phi}_{1xx}$ necessary to match to the outer. Equation (3.9.19) can be thought of as the usual Prandtl-Glauert linear equation (the small parameter is $\delta^2 \sim$ cross section area) with free stream Mach number $M_\infty \approx 1$. Given that

$$\bar{\phi}_1 = S_1(x) \log r + g_1(x) \tag{3.9.24}$$

for matching to the inner, the various particular solutions corresponding to the right hand sides of (3.9.21-23) can be worked out. We note

$$\frac{\partial^2 \phi_I}{\partial r^2} + \frac{1}{r}\frac{\partial \phi_I}{\partial r} = \log^2 r \, , \qquad \phi_I = \frac{r^2}{4} \log^2 r - \frac{r^2}{2} \log r + \frac{3}{8}r^2 \, ,$$

$$\frac{\partial^2 \phi_{II}}{\partial r^2} + \frac{1}{r}\frac{\partial \phi_{II}}{\partial r} = \log r \, , \qquad \phi_{II} = \frac{r^2}{4} \log r - \frac{r^2}{4} \, ,$$

and as before

$$\frac{\partial^2 \phi_{III}}{\partial r^2} + \frac{1}{r}\frac{\partial \phi_{III}}{\partial r} = \frac{1}{r^2} , \qquad \phi_{III} = \frac{1}{2}\log^2 r .$$

For matching with the inner expansion we add the necessary Laplace equation solutions to find

$$\bar{\phi}_1 = S_1(x)\log r + g_1(x) , \tag{3.9.25}$$

$$\bar{\phi}_2 = \frac{D_1(x)\cos\vartheta}{r} , \tag{3.9.26}$$

$$\bar{\phi}_{31} = g_{21}(x) + S_2(x) + \frac{\gamma+1}{4}(S_1'^2)'r^2 \log r$$
$$+ ((\gamma+1)(g_1'S_1')' - KS_1'' - (\gamma+1)(S_1'^2)')\frac{r^2}{4} , \tag{3.9.27}$$

$$\bar{\phi}_{32} = -S_1 S_1' + \frac{\gamma+1}{4}S_1'S_1''r^2 , \tag{3.9.28}$$

$$\bar{\phi}_3 = \frac{E_1\cos^2\theta}{r^2} + S_1 S_1'\log^2 r + S_2(x)\log r$$
$$+ g_2(x) + \frac{\gamma+1}{4}S_1'S_1''r^2 \log^2 r + T(x)r^2 \log r + V(x)r^2 , \tag{3.9.29}$$

where

$$T(x) \equiv \frac{(\gamma+1)(S_1'g_1')' - KS_1''}{4} - \frac{\gamma+1}{2}S_1'S_1'' ,$$

$$V(x) \equiv \frac{3}{8}(\gamma+1)S_1'S_1'' - \frac{(\gamma+1)(S_1'g_1')' - KS_1''}{4} - \frac{Kg_1''}{4} + \frac{\gamma+1}{4}g_1'g_1'' .$$

The matching to the inner expansion is carried out with the help of a matching limit $\delta \to 0$ ($r_\eta = \frac{r}{\eta(\delta)}$ fixed) where $\delta \ll \eta(\delta) \ll 1$ such that $y \to 0$, $\delta \to 0$, $\frac{\eta}{\delta} \to \infty$. Then

$$r = \eta r_\eta \to 0 , \qquad r^* = \frac{\eta r_\eta}{\delta} \to \infty .$$

$\eta(\delta)$ represents a whole order class of limits between the inner and intermediate. By considering the error terms more restrictive bounds on $\eta(\delta)$ for the existence of overlap can be found. Here only matching is demonstrated. The expansions must read the same to a certain order when expressed in r_η, under the matching limit. Note that

$$\log r^* = \log\frac{\eta r_\eta}{\delta} = \log \eta r_\eta - \log \delta ,$$

$$\log^2 r^* = \log^2 \eta r_\eta - 2\log \eta r_\eta \log \delta + \log^2 \delta ,$$

Thus writing both inner and intermediate expansions in terms of r_η (cf. 3.9.25-29, 3.9.15-17) and the expansions (3.9.5, 3.9.18):

Inner Expansion:

$$\delta^2 \log \delta 2S_1 + \delta^2 \left\{ S_1(\log \eta r_\eta - \log \delta) + g_1 + \frac{D_1 \cos \vartheta}{\frac{\eta r_\eta}{\delta}} + \frac{E_1 \cos 2\theta}{\left(\frac{\eta r_\eta}{\delta}\right)^2} + \cdots \right\}$$

$$+ \delta^4 \log \delta \left\{ 2S_1 S_1'(\log \eta r_\eta - \log \delta) + (g_{21} + 2S_2) + \cdots \right\}$$

$$+ \delta^4 \left(S_1 S_1'(\log^2 \eta r_\eta - 2\log \eta r_\eta \log \delta + \log^2 \delta) + S_2(\log \eta r_\eta - \log \delta) + g_2 \right) \Longleftrightarrow$$

Intermediate Expansion:

$$\delta^2 \log \delta S_1 + \delta^2 (S_1 \log \eta r_\eta + g_1) + \delta^3 \left(\frac{D_1 \cos \vartheta}{\eta r_\eta} \right)$$

$$. + \delta^4 \log^2 \delta (-S_1 S_1' + \cdots) + \delta^4 \log \delta (g_{21} + S_2 + \cdots)$$

$$+ \delta^4 \left(\frac{E_1 \cos 2\theta}{(\eta r_\eta)^2} + S_1 S_1' \log^2 \eta r_\eta + S_2 \log \eta r_\eta + g_2 + \cdots \right).$$

$$(3.9.30)$$

It is clear that the terms as shown (3.9.30) match. Terms smaller than $O(\delta^4)$ are not matched. Next the outer expansion which exhibits the essential transonic structure is constructed. The outer limit has $\delta \to 0$, $(x, \tilde{y}, \tilde{z}; K, A)$ fixed where $\tilde{y} = \delta y$, $\tilde{z} = \delta z$, $K = \frac{1-M_\infty^2}{\delta^2}$, $A = \frac{\alpha}{\delta}$. The form is

Outer Expansion:

$$\Phi = U \left\{ x + \delta^2 \phi_1(x, \tilde{y}, \tilde{z}; K, A) + \delta^4 \log \delta \phi_{21} + \delta^4 \phi_2 + \cdots \right\}, \qquad (3.9.31)$$

Thus

$$\frac{\Phi_x}{U} = 1 + \delta^2 \phi_{1x} + \delta^4 \log \delta \phi_{21x} + \delta^4 \phi_{2x} + \cdots ,$$

$$\frac{\Phi_{xx}}{U} = \delta^2 \phi_{1xx} + \delta^4 \log \delta \phi_{21x} + \delta^4 \phi_{2x} + \cdots ,$$

$$\frac{\Phi_y}{U} = \delta^3 \phi_{1\tilde{y}} + \delta^5 \log \delta \phi_{21\tilde{y}} + \delta^5 \phi_2 + \cdots ,$$

$$\frac{\Phi_{yy}}{U} = \delta^4 \phi_{1\tilde{y}\tilde{y}} + \delta^6 \log \delta \phi_{21\tilde{y}} + \delta^6 \phi_{2\tilde{y}\tilde{y}} + \cdots ,$$

$$\frac{\Phi_{x\tilde{y}}}{U} = \delta^3 \phi_{1x\tilde{y}} + \cdots ,$$

$$\frac{\Phi_x^2 + \Phi_y^2 + \Phi_z^2}{U^2} = \frac{\Phi_x^2}{U^2} = 1 + \delta^2 2\phi_{1x} + \delta^4 \log \delta 2\phi_{21x} + \delta^4 (2\phi_{2x} + \phi_{1x}^2) + \cdots ,$$

$$\frac{a^2}{U^2} = 1 + K\delta^2 + K^2\delta^4 + \cdots$$
$$+ \frac{\gamma - 1}{2}\left\{1 - \left(1 + \delta^2 2\phi_{1x} + \delta^4 \log \delta 2\phi_{21x} + \delta^4 (2\phi_{2x} + \phi_{1x}^2)\right)\right\},$$
$$= 1 + \delta^2 (K - (\gamma - 1)\phi_{1x})$$
$$- \delta^4 \log \delta (\gamma - 1)\phi_{21x} + \delta^4 \left(K^2 - (\gamma - 1)\phi_{2x} - \frac{\gamma - 1}{2}\phi_{1x}^2\right) + \cdots ,$$

$$\frac{a^2 - \Phi_x^2}{U^2} = \delta^2 (K - (\gamma + 1)\phi_{1x})$$
$$- \delta^4 \log \delta (\gamma + 1)\phi_{21x} + \delta^4 \left(K^2 - (\gamma + 1)\phi_{2x} - \frac{\gamma + 1}{2}\phi_{1x}^2\right) + \cdots .$$

The basic full potential equation (3.9.31) thus becomes

$$\left\{\delta^2 (K - (\gamma + 1)\phi_{1x}) - \delta^4 \log \delta (\gamma + 1)\phi_{21x}\right.$$
$$\left. + \delta^4 \left(K^2 - (\gamma + 1)\phi_{2x} - \frac{\gamma + 1}{2}\phi_{1x}^2\right) + \cdots\right\}$$
$$\times \left\{\delta^2 \phi_{1xx} + \delta^4 \log \delta \phi_{21xx} + \delta^4 \phi_{2xx} + \cdots\right\}$$
$$+ \left\{1 + \delta^2 (K - (\gamma - 1)\phi_{1x}) + \cdots\right\}$$
$$\times \left\{\delta^4 \tilde{\nabla}^2 \phi_1 + \delta^6 \log \delta \tilde{\nabla}^2 \phi_{21} + \delta^6 \tilde{\nabla}^2 \phi_2 + \cdots\right\} + \cdots$$
$$= 2\{1 + \cdots\}\{\phi_{1\tilde{y}}\phi_{1x\tilde{y}} + \phi_{1\tilde{z}}\phi_{1\tilde{y}\tilde{z}}\}\delta^6 + \cdots , \quad \tilde{\nabla}^2 = \frac{\partial^2}{\partial \tilde{y}^2} + \frac{\partial^2}{\partial \tilde{z}^2} . \quad (3.9.32)$$

This procedure yields the following sequence of approximating equations.

$$O(\delta^4) \quad (K - (\gamma + 1)\phi_{1x})\phi_{1xx} + \tilde{\nabla}^2 \phi_1 = 0 , \quad (3.9.33)$$

$$O(\delta^6 \log \delta) \quad (K - (\gamma + 1)\phi_{1x}\phi_{21xx} - (\gamma + 1)\phi_{21x}\phi_{1xx} + \tilde{\nabla}^2 \phi_{21} = 0 , \quad (3.9.34)$$

$$O(\delta^6) \quad (K - (\gamma + 1)\phi_{1x})\phi_{2xx} - (\gamma + 1)\phi_{2x}\phi_{1xx} + \tilde{\nabla}^2 \phi_2$$
$$= -\left(K^2 - \frac{\gamma + 1}{2}\phi_{1x}^2\right)\phi_{1xx} - (K - (\gamma - 1)\phi_{1x})\tilde{\nabla}^2 \phi_1 + \frac{\partial}{\partial x}(\phi_{1\tilde{r}}^2) .$$

This last equation can be put in the form

$$(K - (\gamma + 1)\phi_{1x})\phi_{2xx} - (\gamma + 1)\phi_{2x}\phi_{1xx} + \tilde{\nabla}^2 \phi_2$$
$$= \left(\frac{(2\gamma - 1)(\gamma + 1)}{2}\phi_{1x} - 2\gamma K\right)\phi_{1x}\phi_{1xx} + \frac{\partial}{\partial x}(\phi_{1\tilde{r}}^2) \quad (3.9.35)$$

by using (3.9.33). (3.9.33) is the usual small disturbance transonic equation,
(3.9.34) is its variational equation and (3.9.35) is a forced variational equation.
Matching to the intermediate expansions ($\tilde{r} \to 0$, $r \to \infty$) provides the necessary
boundary conditions for the solutions of (3.9.33-35) in the form of a characterization of the singularity along the axis. The asymptotic behavior of the solutions
to (3.9.33-35) as $\tilde{r} \to 0$ can be obtained by considering them in the form $\tilde{\nabla}^2 =$
RHS, substituting the dominant term in the RHS and integrating. From ([3.9.1],
[3.9.2]) or directly we have

$$\phi_1 = S_1(x) \log \tilde{r} + g_1(x) + \tilde{r}^2 \log^2 \tilde{r} \left(\frac{\gamma+1}{4} S_1' S_1'' \right) + \tilde{r}^2 \log \tilde{r} T(x) + \tilde{r}^2 V(x) + \cdots ,$$

$$(3.9.36)$$

$$\phi_{21} = -2S_1 S_1' \log \tilde{r} + g_{21} + \cdots .   (3.9.37)$$

For ϕ_2 we note that ϕ_2 starts with $\dfrac{D_1(x) \cos \vartheta}{r}$ but

$$\tilde{\nabla}^2 \phi_2 = 2 \frac{S_1 S_1'}{\tilde{r}^2} + \cdots$$

so that, as before

$$\phi_2 = \frac{D_1(x) \cos \vartheta}{\tilde{r}} + S_1 S_1' \log^2 \tilde{r} + S_2(x) \log \tilde{r} + g_2 + \cdots .   (3.9.38)$$

The matching limit between intermediate and outer has r_η fixed where $r_\eta =$
$\eta(\delta)r$, $\delta \ll \eta(\delta) \ll 1$, so that $r = \frac{r_\eta}{\eta} \to \infty$, $\tilde{r} = \frac{\delta r_\eta}{\eta} \to 0$. Note that

$$\log \tilde{r} \to \log \frac{\delta r_\eta}{\eta} = \log \delta + \log \frac{r_\eta}{\eta} ,$$

$$\log^2 \tilde{r} \to \log^2 \delta + 2 \log \delta \log \frac{r_\eta}{\eta} + \log^2 \frac{r_\eta}{\eta} .$$

To check the matching we can write out the intermediate and outer expansions
as follows. (cf. 3.9.24-29,3.9.35-38)

Intermediate:

$$\delta^2 \log \delta S_1 + \delta^2 \left(S_1 \log \frac{r_\eta}{\eta} + g_1 \right) + \delta^3 \left(\frac{D_1(x) \cos \vartheta}{\frac{r_\eta}{\eta}} \right)$$

$$+ \delta^4 \log^2 \delta \left(-S_1 S_1' + \frac{\gamma+1}{4} S_1 S_1'' \frac{r_\eta^2}{\eta^2} \right)$$

$$+ \delta^4 \log \delta \left(g_{21} + S_2 + \frac{r_\eta^2}{\eta^2} \log \frac{r_\eta}{\eta} \frac{\gamma+1}{4} (S_1'^2)' \right.$$

$$\left. + \frac{r_\eta^2}{4} \left((\gamma+1)(g_1' S_1') - K S_1'' - (\gamma+1)(S_1'^2)') \right) \right)$$

$$+ \delta^4 \left(\frac{\gamma+1}{4} S_1' S_1'' \frac{r_\eta^2}{\eta^2} \log^2 \frac{r_\eta}{\eta} + T(x) \frac{r_\eta^2}{\eta^2} \log \frac{r_\eta}{\eta} \right.$$

$$\left. + V(x) \frac{r_\eta^2}{\eta^2} + S_1 S_1' \log^2 \frac{r_\eta}{\eta} + S_2 \log \frac{\eta r_\eta}{\eta} + g_2 + \cdots \right),$$

$$\Longleftrightarrow \qquad\qquad (3.9.39)$$

Outer:

$$\delta^2 \left(S_1 \left(\log \delta + \log \frac{r_\eta}{\eta} \right) + g_1 + \frac{\delta^2 r_\eta^2}{\eta^2} \left(\log^2 \delta + 2 \log \delta \log \frac{r_\eta}{\eta} + \log^2 \frac{r_\eta}{\eta} \right) \frac{\gamma+1}{4} S_1' S_1'' \right.$$

$$\left. + \frac{\delta^2 r_\eta^2}{\eta^2} \left(\log \delta + \log \frac{r_\eta}{\eta} \right) T(x) + \frac{\delta^2 r_\eta^2}{\eta^2} V(x) \right)$$

$$+ \delta^4 \log \delta \left(-2 S_1 S_1' \left(\log \delta + \log \frac{r_\eta}{\eta} \right) + g_{21} + \cdots \right)$$

$$+ \delta^4 \left(\frac{D_1(x) \cos \vartheta}{\dfrac{\delta r_\eta}{\eta}} + S_1 S_1' \left(\log^2 \delta + 2 \log \delta \log \frac{r_\eta}{\eta} + \log^2 \frac{r_\eta}{\eta} \right) \right.$$

$$\left. + S_2 \left(\log \delta + \log \frac{r_\eta}{\eta} \right) + g_2 + \cdots \right)$$

It is clear from a comparison of these expansions that they also are matched. Thus the entire flow field is covered by the three expansions, inner, intermediate, and outer.

The general theoretical procedure for solving this system would be to determine $S_1(x), D_1(x), S_2(x)$ from the solutions to the inner boundary value problems. However it turns out that S_1 and S_2 can be found from the boundary conditions alone (see below), by use of an integral theorem. These functions characterize the singular behavior of the outer potentials $\phi_1, \phi_{21}, \phi_2$ uniquely. When the boundary value problems for $\phi_1, \phi_{21}, \phi_2$ are solved then the previously unknown functions of integration $g_1(x), g_{21}(x), g_2(x)$ are found. This then enables the pressure distribution, drag, and lift of the slender configurations to be found. Some details are now presented.

For the first inner potential $\varphi_1(x, y^*, z^*)$ we have (3.9.6)

$$\frac{\partial^2 \varphi_1}{\partial y^{*2}} + \frac{\partial^2 \varphi_1}{\partial z^{*2}} = 0 \tag{3.9.40}$$

with the boundary condition (3.9.11) at the body surface in a cross-section plane (cf. Fig. 3.9.2)

$$\frac{\partial \varphi_1}{\partial y^*}(x, Ax - F_{u,\ell}, z^*) - \frac{\partial F_{u,\ell}}{\partial z^*}(x, z^*)\frac{\partial \phi_1}{\partial z^*}(x, Ax - F_{u,\ell}, z^*) = \frac{\partial F_{u,\ell}}{\partial x}(x, z^*) - A \ . \tag{3.9.41}$$

The far field is given by (3.9.15) repeated here

$$\varphi_1 = S_1(x)\log r^* + g_1(x) + \frac{D_1(x)\cos\vartheta}{r^*} + \frac{E_1(x)\cos 2\theta}{r^{*2}} + \cdots \ .$$

Expressions for the lift (and side force) up to a station x are easily worked out using a momentum theorem in the inner coordinate system (x, r^*, θ). We note first some useful expressions for various flow quantities to be used in the integral theorems. From the equations following (3.9.5) we have

$$\frac{a^2}{a_\infty^2} = \frac{a^2}{U^2}M_\infty^2 = \frac{a^2}{U^2}(1 - K\delta^2)$$

$$= 1 - \delta^2 \log \delta 2(\gamma - 1)S_1' - \delta^2(\gamma - 1)\left(u_1 + \frac{v_1^2 + w_1^2}{2}\right) + \cdots \ ,$$

$$\text{where}\quad u_1 = \varphi_{1x}, \quad v_1 = \varphi_{1r^*}, \quad w_1 = \frac{1}{r^*}\varphi_{1\theta} \ ,$$

so that

$$\frac{p}{p_\infty} = \left(\frac{a^2}{a_\infty^2}\right)^{\frac{\gamma}{\gamma-1}} = 1 - \delta^2 \log \delta 2\gamma S_1' - \delta^2\gamma\left(u_1 + \frac{v_1^2 + w_1^2}{2}\right) + \cdots \ ,$$

$$\left(\frac{\rho}{\rho_\infty}\right) = \left(\frac{p}{p_\infty}\right)^{\frac{1}{\gamma}} = 1 - \delta^2 \log \delta 2S_1' - \delta^2\left(u_1 + \frac{V_1^2 + w_1^2}{2}\right) + \cdots \ ,$$

and

$$\frac{p - p_\infty}{\rho_\infty U^2} = \left(\frac{p}{p_\infty} - 1\right)\frac{1}{\gamma M_\infty^2} = -\delta^2 \log \delta 2S_1' - \delta^2\left(u_1 + \frac{v_1^2 + w_1^2}{2}\right) + \cdots \ . \tag{3.9.42}$$

The various components of the mass flux vector are

$$
\left\{
\begin{aligned}
\frac{\rho q_x}{\rho_\infty U} &= 1 - \delta^2 \left(\frac{v_1^2 + w_1^2}{2} \right) + \cdots \\
\frac{\rho q_r}{\rho_\infty U} &= \delta v_1 + \delta^3 \log \delta (v_{21} - 2v_1 S_1') + \delta^3 \left(v_2 - u_1 v_1 - v_1 \frac{v_1^2 + w_1^2}{2} \right) + \cdots \\
\frac{\rho q_\theta}{\rho_\infty U} &= \delta w_1 + \delta^3 \log \delta (w_{21} - 2w_1 S_1') + \delta^3 \left(w_2 - u_1 w_1 - w_1 \frac{v_1^2 + w_1^2}{2} \right) + \cdots
\end{aligned}
\right\}
$$

$$(3.9.43)$$

Choosing a cylindrical control surface about the pointed body as shown in Figure 3.9.2, we can express the lift $\ell(x)$ up to the station x in terms of the pressure force and vertical momentum flux on the cylindrical part of the contour and the vertical momentum flux out of the base area $A_c - A(x)$. The control surface will be considered to be in the inner region $r_c = \delta r_c^*$ so that there is no disturbance in the plane $x = 0$.

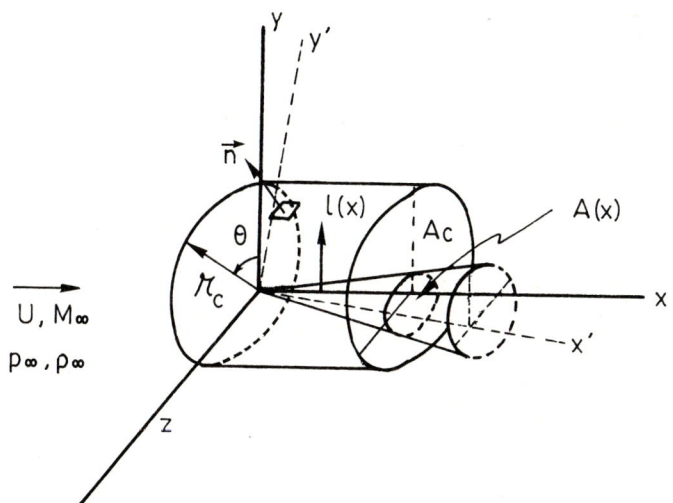

Figure 3.9.2

Control Surface

Applying the momentum theorem in the y-direction we find

$$
\ell(x) = \int_0^x dx' \int_0^{2\pi} r_c d\theta \left(-p(x_1, r_c, \theta) \cos \theta \right)
$$

$$- \int_0^x dx' \int_0^{2\pi} r_c d\theta \, (\rho q_r) q_y - \iint_{A_c - A(x)} (\rho q_x) q_y dA \,, \quad (3.9.44)$$

where $\quad q_y = q_r \cos \theta - q_\theta \sin \theta = \delta \dfrac{\partial \varphi_1}{\partial y^*} + \cdots$

Now expressing the control surface in inner coordinates and using the expressions (3.9.43) we have

$$\frac{\ell(x)}{\rho_\infty U^2} = - \int_0^x dx' \int_0^{2\pi} \delta r_c^* \left(-\delta^2 \log \delta 2 S_1' - \delta^2 \left(u_1 + \frac{v_1^2 + w_1^2}{2} \right) \right) \cos \vartheta \, d\theta$$

$$- \int_0^x dx' \int_0^{2\pi} \delta r_c^* (\delta v_1 + \cdots)(\delta v_1 \cos \vartheta - \delta w_1 \sin \theta) \, d\theta$$

$$- \delta^2 \int \int_{A_c^* - A^*(x)} \left(1 - \delta^2 \left(\frac{v_1^2 + w_1^2}{2} \right) \right) \left(\delta \frac{\partial \varphi_1}{\partial y^*} \right) d\theta \,.$$

Now using (3.9.15)

$$u_1 = \frac{\partial \varphi_1}{\partial x} = S_1' \log r^* + g_1' + \frac{D_1' \cos \vartheta}{r^*} + O\left(\frac{1}{r^*} \right) \,, \quad v_1 = w_1 = O\left(\frac{1}{r^*} \right) \,,$$

we can let $r_c^* \to \infty$. Only the doublet D_1 in the first integral contributes and there remains

$$\frac{\ell(x)}{\rho_\infty U^2 \delta^3} = \pi \int_0^x D_1'(x') dx' - \lim_{r_c^* \to \infty} \int \int_{A_c^* - A(x)} \frac{\partial \varphi_1}{\partial y^*} dy^* dz^*$$

$$= \pi D_1(x) - \lim_{r_c^* \to \infty} \oint \varphi_1(r^*) dz^* + \int_{body} \varphi_1 \, dz^* \,,$$

$$\oint \varphi_1(r_c^*) dz^* = \int_0^{2\pi} \varphi_1(r_c^*) r_c^* \cos \vartheta \, d\theta = D_1'(x) \int_0^{2\pi} \cos^2 \theta \, d\theta = \pi D_1'(x) \,,$$

therefore

$$\frac{\ell(x)}{\rho_\infty U^2 \delta^3} = \oint_{body} \varphi_1(x, F, z^*) dz^* \,. \quad (3.9.45)$$

The line integral of φ_1 against dz^* around the body surface at station x gives the slender body lift. It follows that the side force (in $+z^*$ direction) is

$$\frac{S(x)}{\rho_\infty U^2 \delta^3} = \oint_{body} \varphi_1(x, y^*, z_p^*) dy^* \,. \quad (3.9.46)$$

In the analysis the angle of attack is shown explicitly but the body shape itself can also produce lift. For most cases the body is not designed to produce a side-force except that due to a yaw angle (not written out here). In the special case that the camber of the body does not depend on z, a simple decomposition of the flow results. To see this let

$$\left\{ \begin{array}{l} F_u(x, z^*) = t(x, z^*) + c(x) , \\ F_\ell(x, z^*) = -t(x, z^*) + c(x) , \end{array} \right\} \tag{3.9.47}$$

where $t(x, z^*) = $ thickness distribution,

$$c(x) = \text{camber distribution.}$$

Thus the boundary condition (3.9.11) reads

$$\varphi_{1y^*}(x, t + c - Ax, z^*) - t_{z^*}\varphi_{1z^*}(x, t + c - Ax, z^*) = t_x + c'(x) - A ,$$

and

$$\varphi_{1y^*}(x, -t + c - Ax, z^*) + t_{z^*}\varphi_{1z^*}(x, t + c - Ax, z^*) = -t_x + c'(x) - A .$$

A slight change of coordinates simplifies the boundary conditions; vertical distance is measured from the surface defined by the angle of attack and camber

$$\left\{ \begin{array}{l} y^+ = y^* + Ax - c(x) , \\ z^+ = z^* . \end{array} \right\} \tag{3.9.48}$$

Then the velocity components transform

$$\varphi_{1y^*} = \varphi_{1y^+} , \qquad \varphi_{1z^*} = \varphi_{1z^+} ,$$

and the Laplace equation is invariant. The boundary conditions alone can be written

$$\left\{ \begin{array}{l} \varphi_{1y^+}(x, t(x, z^+, z^+) - t_{z^+}\varphi_{1z^+}(x, t, z^+) = t_x + c' - A , \\ \varphi_{1y^+}(x, -t, z^+) + t_{z^+}\varphi_{1z^+}(x, -t, z^+) = -t_x + c' - A . \end{array} \right\} \tag{3.9.49}$$

In the (y^+, z^+) plane the boundary condition is satisfied on the symmetric upper and lower surfaces $\pm t$, as in Figure 3.9.3.

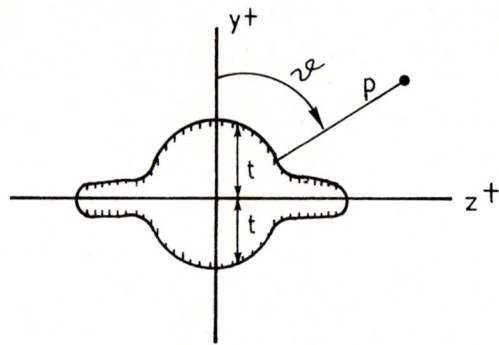

Figure 3.9.3

Symmetric Thickness Distribution

Thus φ_1 can be split into symmetric and anti-symmetric parts with respect to $y^+ = 0$,

$$\varphi_1(y^+, z^+) = \varphi_1^A + \varphi_1^S , \qquad (3.9.50)$$

where

$$\varphi_1^A(y^+, z^+) = -\varphi_1^A(-y^+, z^+) , \qquad \varphi_{1y^+}^A(y^+, z^+) = \varphi_{1y^+}^A(-y^+, z^+) ,$$
$$\varphi_1^S(y^+, z^+) = \varphi_1^S(-y^+, z^+) , \qquad \varphi_{1y^+}^S(y^+, z^+) = -\varphi_{1y^+}^S(-y^+, z^+) .$$

(3.9.49) thus splits into

$$\varphi_{1y^+}^A(x, \pm t, z^+) \mp t_{z^+}\varphi_{1z^+}^A(x, \pm t, z^+) = c' - A , \qquad (3.9.51)$$
$$\varphi_{1y^+}^S(x, \pm t, z^+) \mp t_{z^+}\varphi_{1z^+}^S(x, \pm t, z^+) = \pm t_x . \qquad (3.9.52)$$

The symmetric and anti-symmetric parts can thus be calculated separately. The entire lift is then associated with the anti-symmetric part. The lift integral (3.9.45) becomes

$$\frac{\ell(x)}{\rho_\infty U^2 \delta^2} = \oint_{\text{body}} \varphi_1^A(x, \pm t, z^+) \, dz^+ . \qquad (3.9.53)$$

These results are classical slender-body results for linearized subsonic and supersonic flow which now apply in the transonic range even if shock waves occur. Shock waves in this theory affect $g_1(x)$ which does not enter the lift formula. The notation will be illustrated with a simple example. For a body of revolution at angle of attack α and with meridian shape $r_B^*(x)$

$$t(x, z^+) = \sqrt{r_B^{*2}(x) - z^{+^2}}, \quad c(x) = 0.$$

(3.9.54)

The solution is easily expressed in polar coordinates

$$\left\{ \begin{array}{l} y^+ = \rho \cos \vartheta, \\ z^+ = \rho \sin \vartheta. \end{array} \right\}$$

(3.9.55)

Note that

$$t_{z^+} \cdot = -\frac{z^+}{\sqrt{r_B^{*2} - z^{+^2}}} = -\tan \vartheta \qquad \text{on the surface} \quad \rho = r_B^*(x),$$

since

$$\varphi_{1y^+} = \varphi_{1\rho} \cos \vartheta - \frac{1}{\rho} \varphi_{1u} \sin \vartheta,$$

$$\varphi_{1z^+} = \varphi_{1\rho} \sin \theta + \frac{1}{\rho} \varphi_{1u} \cos \vartheta.$$

The boundary condition for the anti-symmetric part (3.9.51) becomes

$$\varphi_{1\rho}^A(x, r_B^*, \theta) = -A \cos \vartheta.$$

(3.9.56)

The flow is that due to a doublet in these coordinates,

$$\varphi_1^A = A \frac{r_B^{*2}(x)}{\rho} \cos \vartheta.$$

(3.9.57)

On the surface $\varphi_1^A(x, r_B^*, \vartheta) = A r_B^*(x) \cos \vartheta$. Since $dz^+ = r_B^*(x) \cos \vartheta \, d\vartheta$ on the surface the lift integral becomes

$$\frac{\ell(x)}{\rho_\infty U^2 \delta^3} = \int_0^{2\pi} \left(A r_B^*(x) \cos \vartheta \right) r_B^*(x) \cos \vartheta \, d\vartheta = \pi A r_B^{*2}(x).$$

(3.9.58)

If the body does have camber then it is clear from (3.9.51) that only the local effective angle of attack matters. The lift is obtained by replacing A in (3.9.58) with $A - c'(x)$. The lift coefficient c_L based on the base area at any station is

$$c_L = \frac{\ell}{\dfrac{\rho_\infty U^2}{2} (\pi \delta^2 r_B^{*2})} = 2 \delta A = 2\alpha \qquad \text{since} \quad A = \frac{\alpha}{\delta}.$$

(3.9.59)

Theoretically the closed body of revolution has no lift; lift on such bodies is connected with flow separation. This lift is that of a flat wing at angle α with the same planform $z^*_{LE} = r^*_B(x)$. The lift coefficient based on wing area W is

$$c_L = \frac{\ell(x)}{\frac{\rho_\infty U^2}{2}(W)} = 2\delta^3 \pi A r^{*2}_B = \frac{\pi}{2}\alpha(AR)$$

$$\text{where} \quad AR = \text{aspect ratio} = \frac{\left(2\delta r^*_B(x)\right)^2}{W}.$$

The calculations of drag can be carried out in a similar way. The occurence of shock waves in the inner control volume does not affect the momentum considerations for drag since both mass and momentum are conserved across the shocks. First we discuss the solution a little.

The source strength in the far field (cf. (3.9.15)) can be evaluated by using the expression for the potential φ_1 in terms of a surface distribution of sources and doublets.

$$\varphi_1 = \oint \left(G\frac{\partial \varphi_1}{\partial n^*} - \varphi_1 \frac{\partial G}{\partial n^*}\right)_B d\ell, \qquad (3.9.60)$$

where $G = \text{source solution} \quad = \frac{1}{2\pi}\log\{(y^* - y^*_B)^2 + (z^* - z^*_B)^2\}^{\frac{1}{2}}$,

$(y^*_B, z^*_B) = \text{coordinates of a point on the body,}$

$\mathbf{n}^* = \text{normal to surface in a cross plane (cf. Figure 3.9.4).}$

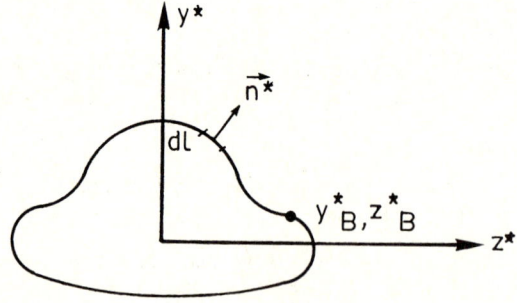

Figure 3.9.4
Cross-Plane

Thus, as $r^* \to \infty$

$$\varphi_1(r^*, \theta) = \frac{1}{2\pi} \log r^* \oint \left(\frac{\partial \varphi_1}{\partial n^*}\right)_B d\ell + g_1(x) + O\left(\frac{1}{r^*}\right), \tag{3.9.61}$$

$$\left(\frac{\partial \varphi_1}{\partial n^*}\right)_B = \mathbf{n}^* \cdot \nabla \varphi_1 ; \quad \text{and} \quad \mathbf{n}^* = \frac{\nabla^* B^*}{|\nabla^* B^*|} = \frac{\mathbf{j} - \dfrac{\partial F_{u,\ell}}{\partial z^*} \mathbf{k}}{\sqrt{1 + \left(\dfrac{\partial F_{u,\ell}}{\partial z^*}\right)^2}} ;$$

$$\left(\frac{\partial \varphi_1}{\partial n^*}\right)_B = \left(\frac{\partial \varphi_1}{\partial y^*} - \frac{\partial F_{u,\ell}}{\partial z^*} \frac{\partial \varphi_1}{\partial z^*}\right) \frac{1}{\sqrt{1 + \left(\dfrac{\partial F_{u,\ell}}{\partial z^*}\right)^2}}, \tag{3.9.62}$$

$$d\ell = \sqrt{dy^{*2} + dz^{*2}} = dz^* \sqrt{1 + \left(\frac{\partial F_{u,\ell}}{\partial z^*}\right)^2}.$$

Thus

$$\varphi_1(r^*, \theta) = \frac{1}{2\pi} \log r^* \oint \left(\frac{\partial \varphi_1}{\partial y^*} - \frac{\partial F_{u,\ell}}{\partial z^*} \frac{\partial \varphi_1}{\partial z^*}\right) dz^* + g_1(x) + O\left(\frac{1}{r^*}\right).$$

We see that the boundary condition (3.9.11) prescribes the normal derivative and using this

$$\varphi_1 = \frac{1}{2\pi} \log r^* \oint \left(\frac{\partial F}{\partial x} - A\right) dz^* + g_1 + O\left(\frac{1}{r^*}\right)$$

$$= \frac{1}{2\pi} \log r^* \frac{\partial}{\partial x} \oint F(x, z^*) dz^* + g_1 + O\left(\frac{1}{r^*}\right),$$

$$\oint F \, dz^* = \int (F_u - F_\ell) \, dz^* = A^*(x) = \text{normalized cross-section area.}$$

Then

$$\varphi_1 = \frac{\dfrac{\partial A^*}{\partial x}}{2\pi} \log r^* + g_1 + O\left(\frac{1}{r^*}\right). \tag{3.9.63}$$

The source strength $2\pi S_1 = \frac{\partial A^*}{\partial x} = $ rate of change of cross-section area.

Next we can write down the components of the x-momentum flux vector appropriate for the control surface by using the mass-flux vector (3.9.43) and the basic expansion (3.9.5).

We find

$$\frac{(\rho q_x) q_x}{\rho_\infty U^2} = 1 + \delta^2 \log \delta (2S_1') + \delta^2 \left(u_1 - \frac{v_1^2 + w_1^2}{2}\right) + \cdots \tag{3.9.64}$$

$$\frac{(\rho q_x) q_r}{\rho_\infty U^2} = \delta v_1 + \delta^3 \log \delta v_{21} + \delta^3 \left(v_{21} - \frac{v_1}{2}(v_1^2 + w_1^2)\right) + \cdots . \tag{3.9.65}$$

Applying the x-momentum integral to the cylindrical contour of Figure 3.9.2 and extending the contour to $x = 1$ where the body is assumed to end, in general, in a base we also find

$$D_N = \text{nose drag} =$$
$$\iint_{A_c} (p + \rho q_x^2)_{x=0}\, dA - \int \int_{A_c - A_b} (p + \rho q_x^2)_{x=1}\, dA - \iint_{r=r_c} \rho q_x q_r\, dA\,, \quad (3.9.66)$$
    where $A_b = \text{base area}.$

From the expressions above (3.9.66) becomes $\left(S_1'(0) = 0 \right)$

$$\frac{D_N}{\rho_\infty U^2} = \frac{p_\infty A_b}{\rho_\infty U^2} + A_b$$

$$+ \delta^2 \iint_{A_c^* - A_b^*} r^*\, dr^*\, d\theta \left\{ \delta^2 \log \delta 2 S_1'(1) + \delta^2 \left(u_1 + \frac{v_1^2 + w_1^2}{2} \right) \right.$$

$$\left. - \delta^2 \log \delta 2 S_1'(1) - \delta^2 \left(u_1 - \frac{v_1^2 + w_1^2}{2} \right) \right\}$$

$$- \delta \int_0^1 dx' \int_0^{2\pi} x_c^*\, d\theta \left\{ \delta v_1 + \delta^3 \log \delta v_{21} + \delta^3 \left(v_2 - v_1 \left(\frac{v_1^2 + w_1^2}{2} \right) \right) \right\}\,.$$

With the pressure on the base arbitrarily put equal to p_b the base drag is $D_b = -p_b A_b / \rho_\infty U^2$ so that the actual drag is

$$\frac{D}{\rho_\infty U^2} = \delta^2 A_b^* + \delta^4 \iint_{A_c^* - A_b^*} (v_1^2 + w_1^2) r^*\, dr^*\, d\theta$$

$$- \delta^2 r_c^* \int_0^1 dx' \int_0^{2\pi} v_1 d\theta - \delta^4 \log \delta r_c^* \int_0^1 dx' \int_0^{2\pi} v_{21} d\theta$$

$$- \delta^4 r_c^* \int_0^1 dx' \int_0^{2\pi} d\theta \left(v_2 - v_1 \frac{v_1^2 + w_1^2}{2} \right)\,, \quad (3.9.67)$$
    where $A_b^* = \delta^2 A_b\,.$

By using the mass flux integral it can be shown that this expression is actually independent of the location r_c^* of the control surface, as it must be. Mass conservation tells us that

$$\iint_{A_c} \rho q_x\, dA = \iint_{A_c - A_b} \rho q_x\, dA + \iint_{r_c} \rho q_x dA\,,$$

or

$$\delta^2 A_c^* = \delta^2 (A_c^* - A_b^*) - \delta^4 \iint_{A_c^* - A_b^*} \left(\frac{v_1^2 + w_1^2}{2} \right) r^* \, dr^* \, d\theta$$

$$+ \delta r_c^* \int_0^1 dx' \int_0^{2\pi} d\theta \left\{ \delta v_1 + \delta^3 \log \delta (v_{21} - 2 v_1 S_1') \right.$$

$$\left. + \delta^3 \left(v_2 - u_1 v_1 - v_1 \frac{v_1^2 + w_1^2}{2} \right) \right\} .$$

Thus

$$O(\delta^2) \qquad A_b^* = r_c^* \int_0^1 dx' \int_0^{2\pi} d\theta - v_1 , \qquad (3.9.68)$$

$$O(\delta^4 \log \delta) \qquad 0 = \int_0^1 dx' \int_0^{2\pi} d\theta \{ v_{21} - 2 v_1 S_1' \} , \qquad (3.9.69)$$

$$O(\delta^4) \qquad \iint_{A_c^* - A_b^*} \left(\frac{v_1^2 + w_1^2}{2} \right) r^* \, dr^* \, d\theta$$

$$= r_c^* \int_0^1 dx' \int_0^{2\pi} d\theta \left\{ v_2 - u_1 v_1 - v_1 \frac{v_1^2 + w_1^2}{2} \right\} . \qquad (3.9.70)$$

Using these expressions to eliminate v_2, v_{21} from the drag integral we have the first order expression for the drag,

$$\frac{D}{\rho_\infty U^2} = \delta^4 \log \delta r_c^* \int_0^1 dx' \int_0^{2\pi} d\theta \, 2 v_1 S_1'(x)$$

$$+ \delta^4 \iint_{A_c^* - A_b^*} \left(\frac{v_1^2 + w_1^2}{2} \right) r^* \, dr^* \, d\theta - \delta^4 r_c^* \int_0^1 dx' \int_0^{2\pi} d\theta \, u_1 v_1 .$$

Returning to (3.9.15) we see that

$$v_1 = \frac{\partial \varphi_1}{\partial r^*} = \frac{S_1(x)}{r^*} - \frac{D_1(x) \cos \vartheta}{r^{*2}} - 2 \frac{E_1(x) \cos 2\theta}{r^{*3}} + \cdots ,$$

$$\int_0^{2\pi} v_1 S_1' \, d\theta = \frac{2\pi}{r^*} S_1 S_1' ; \qquad \text{so that on} \quad r^* = r_c^*$$

$$\int_0^1 dx' \int_0^{2\pi} d\theta \, 2 v_1 S_1' = 4\pi \int_0^1 S_1 S_1'(x') \, dx' = \frac{2\pi}{r_c^*} S_1^2(1) .$$

Also $\iint_{A_c^* + A_b^*} \left(\dfrac{v_1^2 + w_1^2}{2} \right) r^* \, dr^* \, d\theta$ is an expression for the perturbation kinetic energy of the transverse flow in the base cross-section plane. It can be expressed in terms of boundary values by use of Green's theorem

$$\iint \psi \nabla^2 \psi \, dA = \int \psi \frac{\partial \psi}{\partial n} \, d\ell - \iint (\nabla \psi)^2 \, dA . \qquad (3.9.71)$$

Applied to this case we have

$$\iint_{A_c^* - A_b^*} (v^2 + w^2)_{x=1} r^* \, dr^* \, d\theta = r_c^* \int_0^{2\pi} (\varphi_1 v_1)_{x=1} d\theta - \int_0^{2\pi} \left(\varphi_1 \frac{\partial \varphi_1}{\partial n^*} \right)_{\substack{B^* \\ x=1}} d\theta \, . \tag{3.9.72}$$

Further

$$\int_0^{2\pi} (\varphi_1 v_1)_{\substack{x=1 \\ r_c^*}} d\theta$$

$$= \int_0^{2\pi} d\theta \left\{ S_1(1) \log r_c^* + g_1(1) + \frac{D_1(1) \cos \vartheta}{r_c^*} + \frac{E_1(1) \cos 2\theta}{r^{*2}} + \cdots \right\}$$

$$\times \left\{ \frac{S_1(1)}{r_c^*} - \frac{D_1(1) \cos \theta}{r^{*2}} - 2 \frac{E_1(1) \cos 2\theta}{r^{*3}} + \cdots \right\}$$

$$= \frac{2\pi S_1^2(1) \log r^*}{r_c^*} + \frac{2\pi S_1(1) g_1(1)}{r_c^*} - \frac{\pi D_1^2(1)}{r_c^{*3}} - \frac{\pi E_1^2(1)}{r_c^{*4}} - \text{etc.} \,, \tag{3.9.73}$$

and

$$\int_0^1 dx' \int_0^{2\pi} d\theta \phi_{1x} \phi_{1r^*}$$

$$= \int_0^1 dx' \int_0^{2\pi} d\theta \left\{ S_1'(x') \log r_c^* + g_1'(x') + \frac{D_1'(x') \cos \vartheta}{r_c^*} + \frac{E_1'(x') \cos 2\theta}{r_c^{*2}} + \cdots \right\}$$

$$\times \left\{ \frac{S_1(x')}{r_c^*} - \frac{D_1(x') \cos \vartheta}{r_c^{*2}} - \frac{E_1(x') \cos 2\vartheta}{r^{*2}} + \cdots \right\}$$

$$= \int_0^1 dx' \left\{ \frac{2\pi \log r_c^*}{r_c^*} S_1 S_1' + \frac{2\pi S_1 g_1'}{r_c^*} - \frac{\pi D_1 D_1'}{r_c^{*2}} - \frac{\pi E_1 E_1'}{r_c^{*3}} + \text{etc.} \right\} \,,$$

$$\int_0^1 dx' \int_0^{2\pi} d\theta \, \phi_{1x} \phi_{1r^*} = \frac{\pi S_1^2(1) \log r_c^*}{r_c^*} + \frac{2\pi}{r_c^*} \int_0^1 S_1(x) g_1'(x) \, dx$$

$$- \frac{\pi}{2} \frac{D_1^2(1)}{r_c^{*2}} - \frac{\pi}{2} \frac{E_1^2(1)}{r_c^{*3}} - \cdots \, . \tag{3.9.74}$$

The term-on the body surface in (3.9.72) can also be simplified by using the boundary conditions of tangent flow. Thus the drag expression becomes

$$\frac{D}{\rho_\infty U^2} = \delta^4 \log \delta 2\pi S_1^2(1) + \delta^4 \pi S_1(1) g_1(1) - 2\pi \int_0^1 S_1(x) g_1'(x) \, dx$$

$$- \oint \varphi_1 \left(\frac{\partial \varphi_1}{\partial y^*} - \frac{\partial F_{u,\ell}}{\partial z^*} \frac{\partial \varphi_1}{\partial z^*} \right) dz^* \, .$$

By partial integration

$$\int_0^1 S_1(x) g_1'(x) \, dx = S_1(1) g_1(1) - \int_0^1 S_1'(x) g_1(x; K) dx \, ,$$

and from the boundary condition (3.9.11) this expression becomes

$$\frac{D}{\rho_\infty U^2} = \delta^4 \log \delta 2\pi S_1^2(1) + \delta^4 \left(2\pi \int_0^1 S_1'(x) g_1(x; K)\, dx - \pi S_1(1) g_1(1) \right)$$
$$- \delta^4 \frac{1}{2} \int_{body} \varphi_1 \left(\frac{\partial F_{u,\ell}}{\partial x} - A \right) dz^* .$$

Using the expression (3.9.45) for the lift $L = \ell(1)$ we can write the drag as

$$\frac{D}{\rho_\infty U^2} = \delta^4 \log \delta 2\pi S_1^2(1) + \delta^4 \left(2\pi \int_0^1 S_1'(x) g_1(x; K)\, dx - \pi S_1(1) g_1(1) \right)$$
$$+ \frac{A\delta \frac{1}{2} L}{\rho_\infty U^2} - \delta^4 \frac{1}{2} \int_{-z_{LE}^*(1)}^{z_{LE}^*(1)} \varphi_1 \left(\frac{\partial F_u}{\partial x} - \frac{\partial F_\ell}{\partial x} \right) dz^* , \qquad (3.9.75)$$

where $A = \dfrac{\alpha}{\delta}$.

Note $S_1(x) = \dfrac{1}{2\pi} \dfrac{\partial(A^*)}{\partial x}$ where $A^* = $ normalized cross-section area

$$= \oint F_{u,\ell}\, dz^* = \int_{-z^*_{LE}(x)}^{z^*_{LE}(x)} (F_u - F_\ell)\, dz^* .$$

Equation (3.9.75) is the principal result of the analysis and gives an explicit formula for the drag once $g_1(x; K)$ is known. Details of the shape other than cross-section area distribution appear only in the integral in the base plane in (3.9.75) and in the lift. Remember that $g_1(x; K)$ also only depends on the cross-section area distribution since the singularity at the axis for the first outer potential ϕ_1 depends only on $S_1(x)$. (cf. (3.9.36)). If this integral vanishes then we have an area rule. For a given lift the drag depends only on the cross-section area distribution, and not the details of the body shape.

For example

(i) $\dfrac{\partial(F_u)}{\partial x} = \dfrac{\partial(F_\ell)}{\partial x}$; the rate of change of shape is zero at the tail. In general $F_{ux} - F_{\ell x} = 2t_x = 0$ (cf. 3.9.47, but here $c = c(x, z^*)$ is all right). Then $S_1(1) = 0$ so that

$$\frac{D}{\rho_\infty U^2} = \delta^4 2\pi \int_0^1 S_1'(x) g_1(x; K)\, dx + \frac{1}{2} \frac{\alpha L}{\rho_\infty U^2} . \qquad (3.9.76)$$

The drag due to lift in (3.9.76) can be considered to be the induced drag.

(ii) $\dfrac{\partial F_u}{\partial x}(1, z^*) = \text{constant}$, $\dfrac{\partial F_\ell}{\partial x}(1, z^*) = \text{constant}$ - as occurs for a wing with finite trailing edge angle. If the upper and lower trailing edge angles do not depend on z near the trailing edge we have

$$F_u = a_u(1 - x) + \cdots , \qquad F_\ell = -a_\ell(1 - x) + \cdots , \qquad a_{u,\ell} > 0 . \qquad (3.9.77)$$

Then $\displaystyle\int_{-z_{\text{LE}}^*(1)}^{z_{\text{LE}}^*(1)} \varphi_1 \left(\frac{\partial F_u}{\partial x} - \frac{\partial F_\ell}{\partial x} \right) dz^* = (a_\ell - a_u) \oint_{\substack{\text{body} \\ x=1}} \varphi_1 \, dz^*$

$$= \frac{(a_\ell - a_u)L}{\rho_\infty U^2 \delta^3} \quad \text{as above. The cross-}$$

section area near the tail is $A^* = 2z_{\text{LE}}^*(1)(a_u - a_\ell)(1 - x) + \cdots$, $S_1(1) = 2z_{\text{LE}}^*(1)(a_u - a_\ell) = \text{constant}$.

(iii) If the body is pointed at the tail the lift goes to zero, $S_1(1) = 0$ and an area rule holds,

$$\frac{D}{\rho_\infty U^2} = \delta^4 2\pi \int_0^1 S_1'(x)g_1(x; K)dx . \qquad (3.9.78)$$

(iv) If the body is axisymmetric at the tail

$$\int \varphi_1 \left(\frac{\partial F_u}{\partial x} - \frac{\partial F_\ell}{\partial x} \right) dz^* = r_B^* \frac{dr_B^*}{dx} \int_0^{2\pi} \varphi_1(x, r_B^*)d\theta$$

$$= r_B^* \frac{dr_B^*}{dx} \int_0^{2\pi} \left(S_1(1) \log r_B^* + g_1(x) \right) ,$$

or

$$\int \varphi_1 \left(\frac{\partial F_u}{\partial x} - \frac{\partial F_\ell}{\partial x} \right) dz^* = 2\pi S_1^2(1) \log r_b^*(1) + 2\pi S_1(1)g_1(1) . \qquad (3.9.79)$$

Again there is an area rule for given lift,

$$\frac{D}{\rho_\infty U^2} = \frac{\frac{1}{2}\alpha L}{\rho_\infty U^2} + \delta^4 \log \frac{\delta}{\sqrt{r_B^*(1)}} \left(\pi S_1^2(1) \right)$$

$$+ \delta^4 \left(2\pi \int_0^1 S_1'(x)g_1(x; K)dx - 2\pi S_1(1)g_1(1) \right) \qquad (3.9.80)$$

References

[3.9.1] Cole, J. D. and Messiter, A. F., Expansion Procedures and Similarity Laws for Transonic Flow, *Z. Ang. Math. Physic.* **8**, 1957, pp 1-25.

[3.9.2] Cole, J. D., Studies in Transonic Flow I: Transonic Area Rule - Slender Bodies, *UCLA Eng Rpt 7257*, Aug. 1972.

3.10 Lift and Drag Integrals

Lift and drag integrals are worked out in this section for the problems of two and three dimensional lifting wings. The following recapitulation of formulas is useful.

For the two-dimensional case:

$$\Phi = U\left\{x + \delta^{\frac{2}{3}}\phi(x, \tilde{y}; K) + \cdots\right\}, \quad K = \frac{1 - M_\infty^2}{\delta^{\frac{2}{3}}}, \quad \tilde{y} = \delta^{\frac{1}{3}}y$$

$$\left(K - (\gamma + 1)\phi_x\right)\phi_{xx} + \phi_{\tilde{y}\tilde{y}} = 0 \ \ (\text{K-G}), \quad \phi_{\tilde{y}}(x, 0\pm) = \frac{\partial F_{u,\ell}}{\partial x}$$

$$u = \phi_x, \quad v = \phi_{\tilde{y}}$$

$$\left\{ \begin{array}{ll} \left(K - (\gamma + 1)u\right)u_x + v_{\tilde{y}} = 0 & \text{continuity} \\[2mm] u_{\tilde{y}} - v_x = 0 & \text{irrotationality} \end{array} \right\} \ \text{TSD system}$$

Shock Jumps   (3.10)-BE

$$\left[Ku - \frac{\gamma + 1}{2}u^2\right]_s d\tilde{y}_s - [v]_s\, dx_s = 0$$

$$[u]_s\, dx_s + [v]_s\, d\tilde{y}_s = 0$$

Pressure Increment $\quad P - P_\infty = -\rho_\infty U^2 \delta^{\frac{2}{3}}\phi_x$

The boundary value problem for the two-dimensional airfoil is as shown in Figure 3.10.1.

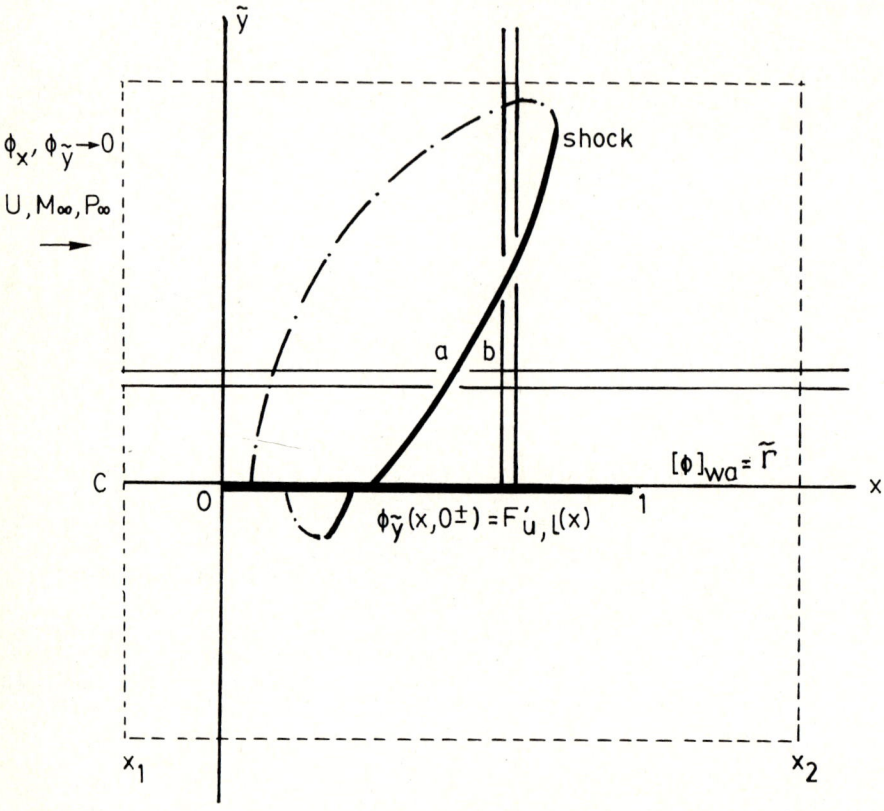

Figure 3.10.1

Two Dimensional Airfoil - Boundary Value Problem

The lift force L on the airfoil is given by (chord $= 1$)

$$L = \int_0^1 \left(P(x,\,0-) - P(x,\,0+)\right)\,dx \qquad (3.10.1)$$

or

$$\tilde{L} = \frac{L}{\rho_\infty U^2 \delta^{\frac{2}{3}}} = \int_0^1 \left(\phi_x(x,\,0+) - \phi_x(x,\,0-)\right)\,dx\;. \qquad (3.10.2)$$

This can be expressed as a line integral $\oint u\,ds$ on the airfoil which is the circulation $\oint u\,dx + v\,d\tilde{y}$ for the special contour on the airfoil. The circulation is independent of the path as is easily seen below (existence of a potential). Along $\tilde{y} = 0$ there can be no jump in ϕ_x or pressure except across the lifting airfoil surface $(0 < x < 1)$. Integration of ϕ_x along the upper and lower surfaces produces a jump in ϕ at the tail

$$[\phi]_{\text{TE}} = \phi(1, 0+) - \phi(1, 0-) . \tag{3.10.3}$$

The K-J condition says that $[\phi_x]_{\text{TE}} = 0$. Further, for $x > 1$, $[\phi_x]_{\tilde{y}=0} = 0$. Therefore $[\phi]_{\text{wake}} = \phi(x, 0+) - \phi(x, 0-)$, $x > 1 = [\phi]_{\text{TE}} = \tilde{\Gamma}$ where $\tilde{L} = \tilde{\Gamma} = \oint \phi_x \, ds$, $\tilde{y} = 0$, ϕ jumps across $\tilde{y} = 0$, $x = 1$, for a lifting airfoil. But for any path enclosing the airfoil

$$\Gamma = \oint u \, dx + v \, d\tilde{y}$$

$$= \oint d\phi = [\phi]_{\text{wake}} .$$

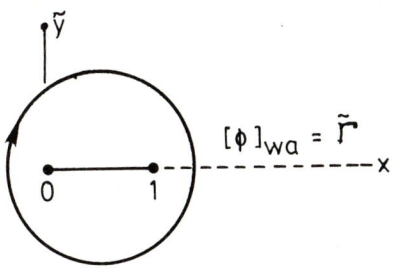

Figure 3.10.2

The drag on this airfoil must also be connected with an integral theorem. A convenient starting point is pressure drag integrated on the airfoil surface.

$$D = \int_0^1 (P - P_\infty) \delta F_u'(x) \, dx - \int_0^1 (P - P_\infty) \delta F_\ell'(x) \, dx , \quad F(0) = F(1) = 0 \tag{3.10.4}$$

Using (3.10)-BE

$$D = -\rho_\infty U^2 \delta^{\frac{5}{3}} \int_0^1 (uv)_{0+} \, dx + \rho_\infty U^2 \delta^{\frac{5}{3}} \int_0^1 (uv)_{0-} dx$$

or

$$\tilde{D} = \frac{D}{\rho_\infty U^2 \delta^{\frac{5}{3}}} = -\int_0^1 [uv]_w \, dx \tag{3.10.5}$$

where w is the wing and $[\ \ ]_w = (\ )_{\tilde{y}=0+} - (\ )_{\tilde{y}=0+}$.

If a conservative (divergence) form can be found involving uv as a boundary term an integral theorem for drag can be found. This is done directly by multiplying the continuity equation by u and irrotationality by v in the TSD system:

$$\left\{ \begin{array}{l} (K - (\gamma + 1)u)\, uu_x + uv_{\tilde{y}} = 0 \\ vu_{\tilde{y}} - vv_x = 0 \end{array} \right\}.$$

Adding these together we have

$$\left(K\frac{u^2}{2} - \frac{v^2}{2} - \frac{\gamma + 1}{3} u^3 \right)_x + (uv)_{\tilde{y}} = 0 \quad \text{divergence form} \tag{3.10.6}$$

This does not correspond to a physical conservation principle and we cannot expect the form to be conserved across shock waves. Consider at first flow subsonic at infinity. Integrate (3.10.6) in the control contour C in Figure 3.10.1 letting the upper and lower boundaries tend to infinity. In addition to boundary terms we have to account for jumps across the shocks. E.g.,

$$\int_{x_1}^{x_a} + \int_{x_b}^{x_2} \quad \text{implies} \quad \left[K\frac{u^2}{2} - \frac{v^2}{2} + \frac{\gamma + 1}{3} u^3 \right]_s .$$

Thus

$$\begin{aligned}
0 &= \iint_c \left(K\frac{u^2}{2} - \frac{v^2}{2} - \frac{\gamma + 1}{3} u^3 \right)_x dx\, dy + \iint (uv)_{\tilde{y}}\, d\tilde{y}\, dx \\
&= -\int_{-\infty}^{\infty} \left(K\frac{u^2}{2} - \frac{v^2}{2} - \frac{\gamma + 1}{3} u^3 \right)_{x=x_1} d\tilde{y} \\
&\quad + \int_{-\infty}^{\infty} \left(K\frac{u^2}{2} - \frac{v^2}{2} - \frac{\gamma + 1}{3} u^3 \right)_{x=x_2} d\tilde{y} \\
&= \int_{\text{shocks}} \left[K\frac{u^2}{2} - \frac{v^2}{2} - \frac{\gamma + 1}{3} u^3 \right]_s d\tilde{y} - \underbrace{\int_0^1 [uv]_w\, dx}_{\tilde{D}} + \int_{\text{shocks}} [uv]_s\, dx . \tag{3.10.7}
\end{aligned}$$

Now, for flow subsonic at infinity we can expect the solution of the (K-G) equation to approach the solution of

$$K\phi_{xx} + \phi_{\tilde{y}\tilde{y}} = 0 . \tag{3.10.8}$$

Details of far fields are given later, but here we note that dominant far-field should be the same as that of linearized theory

$$\phi = -\frac{\tilde{\Gamma}}{2\pi}\theta + \cdots , \quad \theta = \tan^{-1}\frac{\sqrt{K}\tilde{y}}{x} , \quad r = \sqrt{x^2 + K\tilde{y}^2} \to \infty . \quad (3.10.9)$$

This gives estimates for u, v

$$u = \frac{\tilde{\Gamma}}{2\pi}\sqrt{K}\frac{\tilde{y}}{x^2 + K\tilde{y}^2} + \cdots \quad = \frac{\tilde{\Gamma}}{2\pi}\frac{\sin\theta}{\tilde{r}} + \cdots$$

$$v = \frac{\tilde{\Gamma}}{2\pi}\sqrt{K}\frac{x}{x^2 + K\tilde{y}^2} + \cdots \quad = -\frac{\tilde{\Gamma}}{2\pi}\sqrt{K}\frac{\cos\theta}{\tilde{r}} + \cdots . \quad (3.10.10)$$

Thus we can show that the boundary terms $x = x_{1,2}$ drop out as $x_1 \to -\infty$, $x_2 \to +\infty$. The integral theorem (3.10.7) becomes

$$\tilde{D} = -\int_{\text{shocks}} [uv]_s \, dx_s - \left[K\frac{u^2}{2} - \frac{v^2}{2} - \frac{\gamma+1}{3}u^3\right]_s dy_s . \quad (3.10.11)$$

Now this expression can be simplified using the shock relations. Note

$$[uv]_s = [u]_s\langle v\rangle_s + \langle u\rangle_s[v]_s , \quad \left[\frac{v^2}{2}\right]_s = \langle v\rangle_s[u]_s ,$$

$$\text{where} \quad \langle \; \rangle_s = \text{average} = \frac{1}{2}((\;)_a + (\;)_b) .$$

Thus

$$\neg[uv]_s \, dx_s + \frac{v^2}{2}\, dy_s = [u]_s\langle v\rangle_s \, dx_s + \langle u\rangle_s[v]_s \, dx_s + \langle v\rangle_s[u]_s \, dy_s$$

$$= \langle u\rangle_s[v]_s \, dx_s$$

(3.10.11) becomes

$$\tilde{D} = -\int_{\text{shocks}} \langle u\rangle_s[v]_s \, dx_s - \left[K\frac{u^2}{2} - \frac{\gamma+1}{3}u^3\right]_s dy_s . \quad (3.10.12)$$

Using the shock jumps to replace $[v]\,dx$ we find

$$\tilde{D} = -\int_{\text{shocks}} \left(\langle u\rangle_s\left[Ku - \frac{\gamma+1}{2}u^2\right]_s - \left[K\frac{u^2}{2} - \frac{\gamma+1}{3}u^3\right]_s\right) dy_s$$

or

$$\tilde{D} = (\gamma + 1) \int_{\text{shocks}} \left(\langle u \rangle_s \left[\frac{u^2}{2} \right]_s - \left[\frac{u^3}{3} \right]_s \right) d\tilde{y}_s . \tag{3.10.13}$$

Note that

$$[u^3] = [u^2]\langle u \rangle + [u]\langle u^2 \rangle = 2\langle u \rangle^2 [u] + \langle u^2 \rangle [u] .$$

But

$$\langle u \rangle^2 = \frac{1}{4}\left\{ u_a^2 + 2u_a u_b + u_b^2 \right\} = \frac{1}{2}\langle u^2 \rangle + \frac{1}{2} u_a u_b$$

$$[u]^2 = u_a^2 - 2u_a u_b + u_b^2$$

therefore　　$\langle u \rangle^2 + \frac{1}{4}[u]^2 = \frac{1}{2}(u_a^2 + u_b^2) = \langle u^2 \rangle .$

(3.10.13) is thus, since $[u^3] = 3\langle u \rangle^2 [u] + \frac{1}{4}[u]^3$

$$\tilde{D} = -(\gamma + 1) \int_{\text{shocks}} \left(\langle u \rangle_s^2 [u]_s + \frac{1}{12}[u]_s^3 - \langle u \rangle_s^2 [u]_s \right) d\tilde{y}_s$$

or

$$\boxed{\tilde{D} = -\frac{(\gamma + 1)}{12} \int_{\text{shocks}} [u]_s^3 \, d\tilde{y}_s .} \tag{3.10.14}$$

Since $[u]_s < 0$ always, $\tilde{D} > 0$. Thus, for flow subsonic at infinity the drag is totally connected to the occurrence of shock waves. Shock waves necessitate drag and vice versa. Thus there are two different ways to calculate drag, by a pressure integral on the surface and by an integral along the shocks. (3.10.14) has a simple physical interpretation. Referring to the expression for the jump of specific entropy across a weak shock

$$\frac{[S]}{c_v} = \frac{\gamma(\gamma - 1)}{12}\varepsilon^3 + \cdots .$$

Since $\varepsilon = 1 - \dfrac{\rho_a}{\rho_b} = \dfrac{[\rho]_s}{\rho_\infty} + \cdots$, and from (cf. 3.1.3)

$$\frac{\rho}{\rho_\infty} = 1 - \delta^{\frac{2}{3}} u , \qquad \frac{[\rho]}{\rho_\infty} = -\delta^{\frac{2}{3}}[u]$$

we have

$$\tilde{D} = \frac{D}{\rho_\infty U^2 \delta^{\frac{5}{3}}} = -\frac{\gamma + 1}{12} \int_{\text{shocks}} \frac{[S]}{c_v \gamma(\gamma - 1)} \frac{d\tilde{y}}{\delta^2}$$

or

$$D = \int\limits_{\text{shocks}} \frac{\rho_\infty U^2}{c_v \gamma (c_p - c_v)} \frac{\delta^{\frac{5}{3}} \delta^{\frac{1}{3}}}{\delta^2} \, dy \, [S] \ .$$

Using $\dfrac{U^2}{\gamma R T_\infty} = 1 + \cdots$, we find

$$\boxed{D = \rho_\infty T_\infty \int\limits_{\text{shocks}} [S] \, dy \ .}$$

(3.10.15)

The drag is related to the total entropy production in the shock system (3.10.15).

This result can be extended to the case of three dimensional wings as described in Section (3.1.1). See Figure 3.10.3.

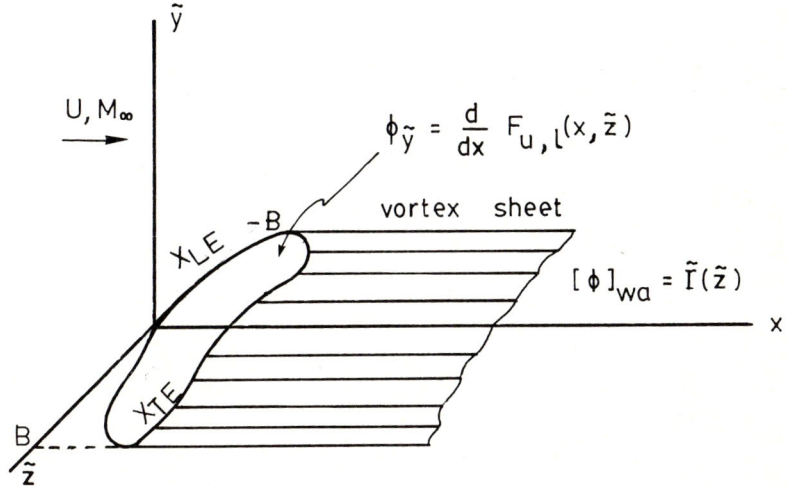

Figure 3.10.3

BVP for Three-Dimensional Lifting Surfaces

The expression for the lift is now

$$L = \int_{-b}^{b} dz \int_{x_{\text{LE}}}^{x_{\text{TE}}} \left(P(x, 0-) - P(x, 0+) \right) dx \ , \qquad \tilde{z} = \delta^{\frac{1}{3}} z$$

$$= \rho_\infty U^2 \delta^{\frac{1}{3}} \int_{-B}^{B} d\tilde{y} \int_{x_{\text{LE}}(\tilde{z})}^{x_{\text{TE}}(\tilde{z})} [\phi_x]_w \, dz$$

$$\tilde{L} = \frac{L}{\rho_\infty U^2 \delta^{\frac{1}{3}}} = \int_{-B}^{B} \tilde{\Gamma}(\tilde{z}) \, dz \ ; \qquad \tilde{\Gamma} = \oint \phi_x \, dx = [\phi]_{\text{TE}}(\tilde{z}) \ . \quad (3.10.16)$$

The lift is given by the spanwise integral of the circulation.

The system of equations which generalizes the transonic system of (3.10)-BE is

$$\left\{\begin{array}{l} (K - (\gamma + 1)u)\,u_x + v_{\tilde{y}} + w_{\tilde{y}} = 0 \\[2mm] u_{\tilde{y}} - v_x = 0 , \qquad u_{\tilde{z}} - w_x = 0 . \end{array}\right\} , \qquad w = \phi_{\tilde{z}} \qquad (3.10.17)$$

The corresponding divergence form for drag is

$$\left(K\frac{u^2}{2} - \frac{v^2 + w^2}{2} - \frac{\gamma - 1}{3} u^3 \right)_x + (uv)_{\tilde{y}} + (uw)_{\tilde{z}} = 0 . \qquad (3.10.18)$$

Now the integration can be carried out over a volume, bounded by x_1, x_2. However disturbances do not die out as $x_2 \to \infty$, due to the vortex sheet. This can be seen from the three-dimensional K-G equation (cf. 3.1.38)

$$(K - (\gamma + 1)\phi_x)\,\phi_{xx} + \phi_{\tilde{y}\tilde{y}} + \phi_{\tilde{z}\tilde{z}} = 0 . \qquad (3.10.19)$$

As $x \to \infty$ ($\tilde{y}$, $\tilde{z}$ fixed) the flow approaches a potential flow in a cross-plane $(\frac{\partial}{\partial x} \to 0)$

$$\phi_{\tilde{y}\tilde{y}} + \phi_{\tilde{z}\tilde{z}} = 0 . \qquad (3.10.20)$$

There is a vortex-sheet producing a jump in sidewash w

$$[w]_{\text{wake}} = [\phi_{\tilde{z}}]_{\text{wake}} = \frac{d\tilde{\Gamma}}{d\tilde{z}} .$$

See Figure 3.10.4.

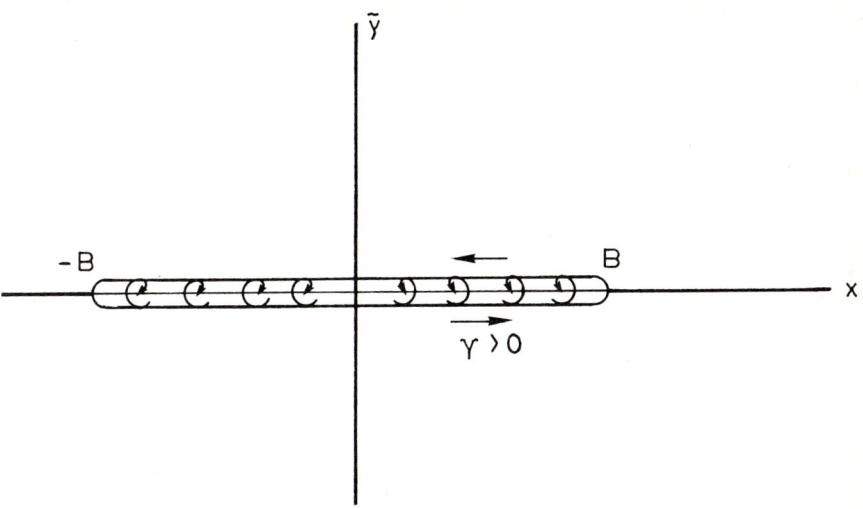

Figure 3.10.4

Potential Flow at $x_2 = \infty$

The calculation of this flow is a classical problem in potential theory. The flow in this plane can be thought of as produced by two-dimensional vortices

$$\phi(\tilde{y}, \tilde{z}) = -\frac{1}{2\pi} \int_{-b}^{b} \gamma(\varsigma) \tan^{-1} \left(\frac{\tilde{y}}{\tilde{z} - \varsigma} \right) d\varsigma \qquad (3.10.21)$$

$$w = \phi_{\tilde{z}} = \frac{1}{2\pi} \tilde{y} \int_{-b}^{b} \gamma(\varsigma) \frac{1}{\tilde{y}^2 + (\tilde{z} - \varsigma)^2} d\varsigma \quad \rightarrow \pm \frac{\gamma(\tilde{z})}{2}, \quad \tilde{y} \rightarrow 0\pm$$

thus

$$\gamma(\tilde{z}) = 2w(0+, \tilde{z}) = [w]_{\text{wake}} = \frac{d\Gamma}{d\tilde{z}} . \qquad (3.10.22)$$

The vortex strength is related to the spanwise lift distribution.

The drag is still related to the $[uv]_w$ integrated over the wing surface. Corresponding to (3.10.5) we have

$$\tilde{D} = \frac{D}{\rho_\infty U^2 \delta^{\frac{4}{3}}} = -\int_{-B}^{B} d\tilde{z} \int_{x_{\text{LE}}(\tilde{z})}^{x_{\text{TE}}(\tilde{z})} [uv]_w \, dx . \qquad (3.10.23)$$

When $x_1 \to -\infty$, $x_2 \to +\infty$ we are left with a term representing the kinetic energy of the flow produced by the vortex sheet at $x_2 = \infty$ in addition to the shock jump terms, which work out as before. The final result is

$$\tilde{D} = \lim_{x_2 \to \infty} \iint_{-\infty}^{\infty} \left(\frac{v^2 + w^2}{2} \right) d\tilde{y} \, d\tilde{z} - \frac{\gamma + 1}{12} \iint_{\text{shocks}} [u]_s^3 \, d\tilde{y} \, d\tilde{z} \, . \qquad (3.10.24)$$

The first term represents so-called induced drag which is present even if there are no shocks. This expression for the induced drag is identical with that of linearized theory. A well-known result is that for a given lift $\tilde{L}$, the induced drag is minimized if the spanwise lift distribution is elliptical. Further, a larger aspect ratio B reduces the minimum induced drag for a given lift. If this could be accomplished without increasing the second term, the shock drag, then it would be a good design policy.

3.11 Unsteady Transonic Flow

In this section unsteady transonic flow at relatively low frequencies is considered, as in [3.11.1]. The analysis is carried out for vibrating or plunging motions of the airfoil. The equation derived in the transonic limit reduces, under steady flow conditions, to the steady transonic equation (3.1.19). Two new similarity parameters arise with the introduction of time dependence. (The original derivation of the basic equation (3.1.13) is due to Timman [3.11.2])

Shock waves continue to be weak in this unsteady flow so that the flow remains isentropic to the order that we are considering (see 2.4.22). So, as for steady flow,

$$\frac{P}{\rho^\gamma} = \text{constant} = \frac{P_\infty}{\rho_\infty^\gamma} \tag{3.11.1}$$

and

$$a^2 = \frac{\gamma P}{\rho} = \gamma \frac{P_\infty}{\rho_\infty^\gamma} \rho^{\gamma-1} . \tag{3.11.2}$$

Also, since shock waves move slowly the flow has negligible vorticity. Hence, a potential Φ, $\mathbf{q} = \nabla\Phi$, exists.

We first derive the full unsteady potential equation, then consider the transonic limit. Using the potential Φ, the momentum equation

$$\mathbf{q}_T + \nabla_{X,Y} \frac{\mathbf{q}^2}{2} + \frac{1}{\rho} \nabla_{X,Y} P = 0$$

can be written

$$\nabla \left(\Phi_T + \frac{(\nabla_{X,Y}\Phi)^2}{2} + \frac{a^2}{\gamma-1} \right) = 0 .$$

We can integrate this once and, since the flow is assumed sready at upstream infinity, we obtain the Bernoulli equation

$$\Phi_T + \frac{(\nabla_{X,Y}\Phi)^2}{2} + \frac{a^2}{\gamma-1} = \frac{U^2}{2} + \frac{a_\infty^2}{\gamma-1} , \tag{3.11.3}$$

where U is the velocity at upstream infinity.

Using the potential Φ and condition (3.11.2) the continuity equation

$$\frac{1}{\rho}(\rho_T + \mathbf{q} \cdot \nabla_{X,Y}\rho) + \nabla \cdot \mathbf{q} = 0$$

becomes

$$\frac{1}{\gamma-1} \left((a^2)_T + \nabla_{X,Y}\Phi \cdot \nabla_{X,Y} a^2 \right) + a^2 \nabla_{X,Y}^2 \Phi = 0 .$$

Finally, eliminating derivatives of a^2 by using the Bernoulli equation (3.11.3) we obtain the unsteady potential equation

$$(a^2 - \Phi_X^2)\Phi_{XX} - 2\Phi_X\Phi_Y\Phi_{XY} + (a^2 - \Phi_Y^2)\Phi_{YY}$$
$$- 2\Phi_X\Phi_{XT} - 2\Phi_Y\Phi_{YT} - \Phi_{TT} = 0 \tag{3.11.4}$$

where a^2 is given by

$$\Phi_T + \frac{\Phi_X^2 + \Phi_y^2}{2} + \frac{a^2}{\gamma - 1} = \frac{U^2}{2} + \frac{a_\infty^2}{\gamma - 1}. \tag{3.11.5}$$

The last three terms of equation (3.11.14) are new terms which arise because the flow is unsteady as is the first term in equation (3.11.15). The local speed of sound, a^2, is itself a function of time.

Now consider the specific problem of a plunging airfoil as shown in Figure (3.11.1).

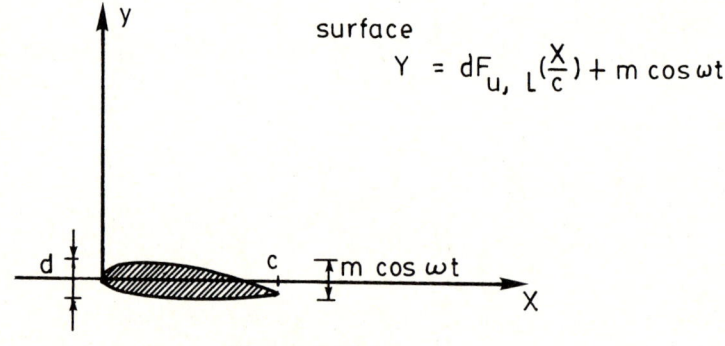

Figure 3.11.1

Plunging Airfoil

The thickness is d, the chord c, and the airfoil surface is given by $y = dF_{u,\ell}\left(\frac{x}{c}\right) + m\cos\omega t$, where $(F_u - F_\ell)_{max} = 1$. The boundary condition of tangent flow on the airfoil surface is

$$W_T + q\cdot\nabla W = 0 \quad \text{on} \quad W = 0 \tag{3.11.6}$$

where

$$W = Y - dF_{u,\ell}\left(\frac{X}{c}\right) - m\cos\omega t. \tag{3.11.7}$$

The other boundary conditions are the Kutta condition, that the flow leave the trailing edge smoothly, uniform flow at upstream infinity, and the shock conditions.

The transonic limit is a generalization of that used for steady flow. The small parameter is the airfoil thickness ratio

$$\delta = \frac{d}{c}.$$

The spatial coordinates are

$$x = \frac{X}{c}, \quad \tilde{y} = \delta^{\frac{1}{3}} \frac{Y}{c}. \tag{3.11.8}$$

The time coordinate is

$$\tilde{t} = \beta(\delta) \frac{UT}{c}, \tag{3.11.9}$$

where $\beta \ll 1$ is to be determined. The potential expansion is

$$\Phi = U\left\{x + \delta^{\frac{2}{3}} c\phi(x, \tilde{y}, \tilde{t}) + \cdots\right\}, \tag{3.11.10}$$

since we expect that steady flow should be a limiting case of this unsteady flow.

Now

$$\frac{\Phi_X}{U} \sim 1 + \delta^{\frac{2}{3}} \phi_x, \quad \frac{\Phi_Y}{U} \sim \delta\phi_{\tilde{y}}, \quad \frac{\Phi_T}{U} \sim \delta^{\frac{2}{3}} \beta(\delta)\phi_{\tilde{t}}, \tag{3.11.11}$$

so that in this transonic limit the Bernoulli equation (3.11.5) becomes

$$\frac{a^2}{U^2} = 1 + \delta^{\frac{2}{3}} \left(K - (\gamma + 1)\phi_x\right) + \cdots. \tag{3.11.12}$$

Note that the explicit unsteady terms have dropped out of the Bernoulli equation. The first three terms in the continuity equation (3.11.4) thus become $\delta^{\frac{4}{3}} \left\{\left(K - (\gamma + 1)\phi_x\right)\phi_{xx} - \phi_{\tilde{y}\tilde{y}}\right\} + \cdots$ as in the steady case. Since

$$\frac{\Phi_{TT}}{U^2} \sim \delta^{\frac{2}{3}} \beta^2(\delta)\phi_{\tilde{t}\tilde{t}}, \quad \frac{\Phi_X \Phi_{XT}}{U^2} \sim \delta^{\frac{2}{3}} \beta(\delta)\phi_x \phi_{x\tilde{t}},$$

$$\frac{\Phi_Y \Phi_{YT}}{U^2} \sim \delta^2 \beta(\delta)\phi_{\tilde{y}} \phi_{\tilde{y}\tilde{t}},$$

the distinguished limit of the continuity equation (3.11.5) occurs with

$$\beta = \delta^{\frac{2}{3}},$$

$$\left(K - (\gamma + 1)\phi_x\right)\phi_{xx} + \phi_{\tilde{y}\tilde{y}} - 2\phi_{x\tilde{t}} = 0 \,. \tag{3.11.13}$$

It is clear that this approximation will not be valid for high frequencies since we have assumed $\Phi_{TT} \ll \Phi_T$.

In this limit the boundary condition (3.11.6) becomes

$$\phi_{\tilde{y}}(x, 0\pm, \tilde{t}) = F'_{u,\ell}(x) - N \sin \Omega\tilde{t} \tag{3.11.14}$$

where the similarity parameters

$$N = \frac{m\omega}{\delta U}, \quad \Omega = \frac{\omega\ell}{m\delta^{\frac{2}{3}}} \tag{3.11.15}$$

must be kept fixed as $\delta \to 0$ in order to preserve the time variation in the problem.

The pressure coefficient is still

$$c_p = -2\delta^{\frac{2}{3}}\phi_x \,,$$

that is the only time dependence is through ϕ. The Kutta condition is then

$$\left[\phi_x(1, 0, \tilde{t})\right] = 0 \,.$$

Thus the unsteady wake looks, at any instant of time, steady. This is a consequence of the fact that we have assumed that the stream-wise flow, the shedding of vorticity, is much faster than the plunging motion.

Finally, consider the shocks. Note that the unsteady transonic equation (3.11.13) is, in conservation form,

$$\frac{\partial}{\partial\tilde{t}}(-2\phi_x) + \frac{\partial}{\partial x}\left(K\phi_x - \frac{\gamma+1}{2}\phi_x^2\right) + \frac{\partial}{\partial\tilde{y}}(\phi_{\tilde{y}}) = 0 \,. \tag{3.11.16}$$

In order to see that this is the physical conservation law note that the density ratio is

$$\frac{\rho}{\rho_\infty} = 1 - \delta^{\frac{2}{3}}\phi_x + \delta^{\frac{4}{3}}\left\{K\phi_x - \frac{\gamma-1}{2}\phi_x^2 - \phi_{\tilde{t}}\right\} + \cdots$$

and the mass flux is

$$\frac{\rho\mathbf{q}}{\rho_\infty U} = \left(1 + \delta^{\frac{4}{3}}\left(K\phi_x - \frac{\gamma+1}{2}\phi_x^2 - \phi_{\tilde{t}}\right), \ \delta\phi_{\tilde{y}}\right) + \cdots \,,$$

Thus with

$$\nabla = \left(\frac{\partial}{\partial x}, \ \delta^{\frac{1}{3}}\frac{\partial}{\partial\tilde{y}}\right) \,,$$

and the time changes measured by

$$\delta^{\frac{2}{3}} \frac{\partial}{\partial \tilde{t}} \, ,$$

we see that equation (3.11.16) states that the divergence of the mass flux vector is equal to the time change of mass. The other shock condition is that there is no jump in tangential velocity across the shock,

$$[\phi_x] dx_s + [\phi_{\tilde{y}}] d\tilde{y}_s = 0 \, .$$

The shock geometry is shown in Figure (3.11.2).

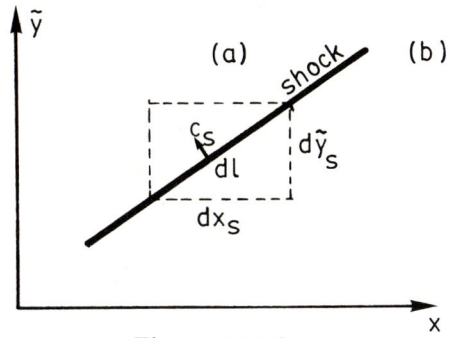

Figure 3.11.2

Shock Element

If c_s is the velocity of the shock then the conditions are

$$2c_s[\phi_x] = \left[K\phi_x - \frac{\gamma+1}{2}\phi_x^2 \right] \left(\frac{d\tilde{y}}{d\ell} \right)_s - [\phi_{\tilde{y}}] \left(\frac{dx}{d\ell} \right)_s \, , \qquad (3.11.17)$$

$$0 = [\phi_x] \left(\frac{dx}{d\ell} \right)_s + [\phi_{\tilde{y}}] \left(\frac{d\tilde{y}}{d\ell} \right)_s \, , \qquad (3.11.18)$$

where

$$\left(\frac{dx}{d\ell} \right)_s = \sin\theta \, , \qquad \left(\frac{d\tilde{y}}{d\ell} \right)_s = \cos\theta \, . \qquad (3.11.19)$$

The unsteady transonic equation (3.11.16) is always of hyperbolic type. However, the structure of the characteristic surfaces changes as the flow is locally subsonic, sonic, or supersonic.

To examine the characteristic surfaces locally, let

$$K^* = K - \overline{(\gamma + 1)}\phi_x$$

in the neighborhood of a given point. Then the characteristic surface $\xi(x, \tilde{y}, \tilde{t})$ for the equation

$$K^*\phi_{xx} + \phi_{\tilde{y}\tilde{y}} - 2\phi_{x\tilde{t}} = 0$$

is given by the solution to

$$K^*\xi_x^2 + \xi_{\tilde{y}}^2 - 2\xi_x\xi_{\tilde{t}} = 0 \ .$$

First, looking for plane surfaces,

$$\xi(x, y, \tilde{t}) = \tilde{t} + \alpha x + \beta\tilde{y} \ ,$$

a one parameter family is found where the condition is

$$K^*\alpha^2 + \beta^2 - 2\alpha = 0 \ .$$

For $K^* > 0$, locally subsonic flow this defines an ellipse, $K^*\left(\alpha - \frac{1}{K^*}\right)^2 + \beta^2 = \frac{1}{K^*}$. This ellipse defines the locus of all possible wave fronts. For $K^* < 0$, locally supersonic flow, the locus is a hyperbola, $-\beta^2 + |K^*|\left\{\alpha + \frac{1}{|K^*|}\right\}^2 = \frac{1}{|K^*|}$. For $K^* = 0$, locally sonic flow, the locus is a parabola (Figure (3.11.3)).

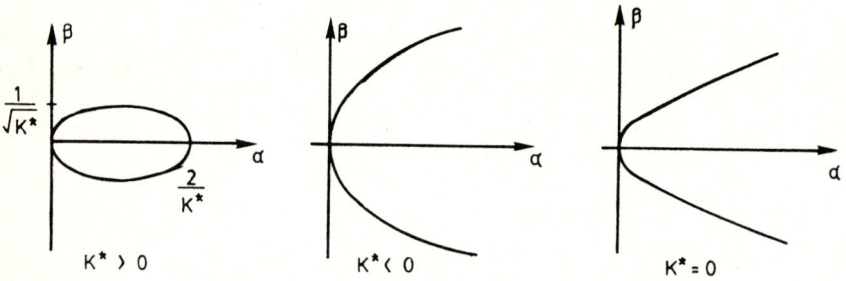

Figure 3.11.3

Parameter values for plane wave characteristic surfaces

The envelope of these plane waves describes the characteristic cone. The equations determining the envelope are

$$\xi = \tilde{t} + \alpha x \pm \sqrt{2\alpha - K^*\alpha^2}\,\tilde{y} = 0 \,, \tag{3.11.20}$$

and

$$\xi_\alpha = 0 \,. \tag{3.11.21}$$

Evaluating ξ_α from (3.11.20) gives

$$\xi_\alpha = x \pm \frac{1 - K^*\alpha}{\sqrt{2\alpha - K^*\alpha^2}}\,\tilde{y} \,.$$

Thus equation (3.11.21) is

$$x \pm \frac{1 - K^*\alpha}{\sqrt{2\alpha - K^*\alpha^2}}\,\tilde{y} = 0 \,.$$

Solving for $\alpha(x, y)$ gives

$$\alpha(x, y) = \frac{1}{K^*}\left(1 - \frac{x}{\sqrt{K^*\tilde{y}^2 + x^2}}\right) \,.$$

Here the minus sign was chosen in order that the waves propagate out from the origin. Then

$$-\sqrt{2\alpha - K^*\alpha^2} = \beta = \frac{-\tilde{y}}{\sqrt{K^{*2}(K^*\tilde{y}^2 + x^2)}} \,.$$

Substituting into (3.11.20) and simplifying we find the equation of the characteristic surface

$$\tilde{y}^2 = K^*\tilde{t}^2 + 2x\tilde{t} \,. \tag{3.11.22}$$

For $K^* > 0$, locally subsonic flow, for fixed $\tilde{t}$ the surface is a parabola. As $\tilde{t}$ increases the parabola moves left, resulting in a conoid in $x, \tilde{y}, \tilde{t}$ space (Figure 3.11.4a). For $K^* < 0$, locally supersonic flow, the constant $\tilde{t}$ surface is still a parabola but now as $\tilde{t}$ varies the envelope of these parabolas traces out the Mach line $\tilde{y} = \sqrt{-K^*}\,x$. (Figure (3.11.4b)). For $K^* = 0$, locally sonic flow the parabola widens as $\tilde{t}$ increases, but always passes through $x = 0$ (Figure 3.11.4c).

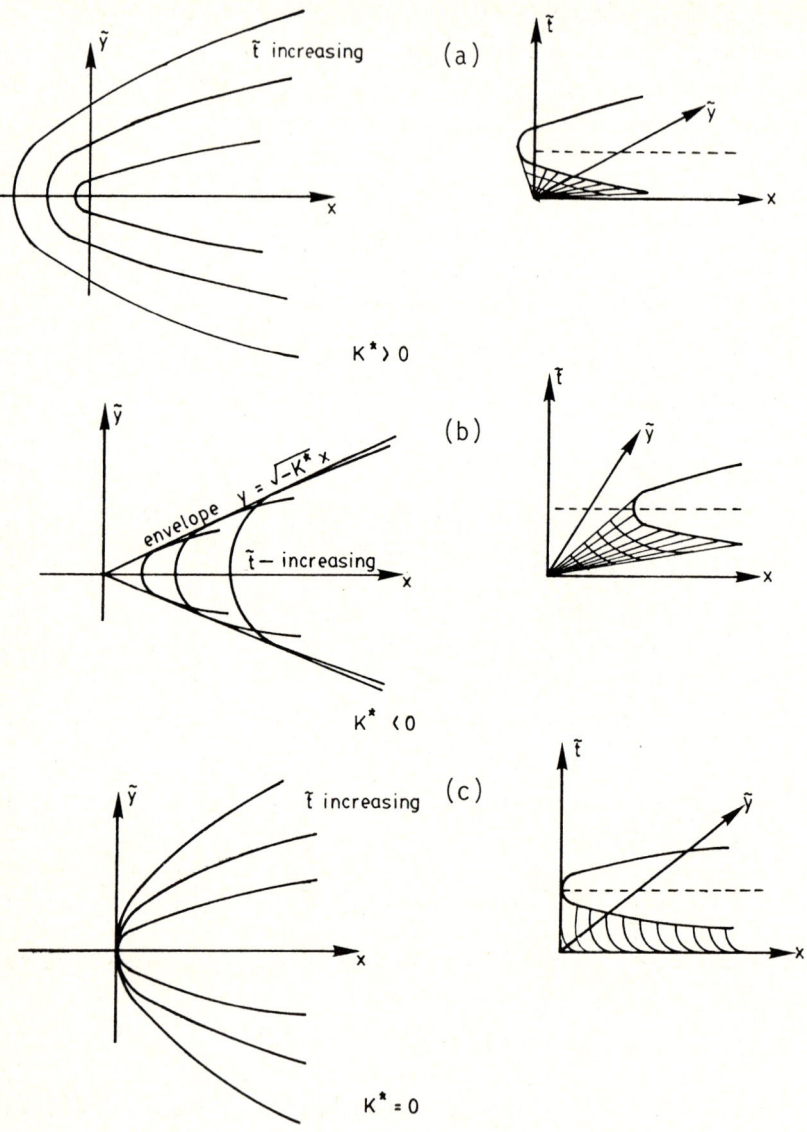

Figure 3.11.4a,b,c
Characteristic Surfaces

Note that all this behavior is local. The surfaces are open for large x since waves propagate infinitely fast in the downstream direction due to the transonic approximation. The regions of influence are the shaded areas in figures (3.11.4a-c) and must be accounted for in any numerical scheme.

References

[3.11.1] Krupp, J. A. and Cole, J. D., Unsteady Transonic Flow, Studies in Transonic Flow IV, *UCLA School of Engineering and Applied Science Report*, UCLA-Eng-76104, Oct 1976.

[3.11.2] Timman, R., Unsteady Motion in Transonic Flow, *Symposium Transsonicum*, Springer Verlag, 1964, K. Oswatitsch, ed.

4. Transonic Far Fields

In this chapter some details will be worked out about the far fields of transonic flow. This serves several purposes. These flow fields are examples which show the qualitative difference of subsonic, transonic, and supersonic fields. They are essential for understanding wind tunnel wall corrections or interference. They can be important elements in sophisticated computation schemes. In these computation schemes the flow past objects is calculated in the near field by finite difference or finite element methods and the far field serves as a fairly accurate outer boundary condition. The special problem of "transonic boom" is connected with the pressure signature on the ground due to slightly supersonic flight. This pressure signature can be expressed in terms of the supersonic far field. Finally the similarity methods introduced in this section are applied again later in connection with local solutions which represent the local singularities and in obtaining aspect ratio corrections due to finite span at $M_\infty = 1$, and in analyzing the stabilization law.

In the following subsections we take up first the flow at $M_\infty = 1$ ($K = 0$) which has the features most typical of transonic flows. Then the slightly supersonic case $M_\infty > 1$ ($K < 0$) which resembles the sonic case in some features, is considered. Lastly, we consider the subsonic freestream which seems to be completely different from the sonic case.

4.1 Far Field at $M_\infty = 1$ ($K = 0$).

The nature of the far field at $M_\infty = 1$ is completely different from the far field at subsonic speeds or supersonic speeds. The qualitative features appear as in Figure 4.1.1, which holds equally for planar or axi-symmetric flow. The sonic line runs to infinity. Only a small portion of the supersonic flow over the body can influence the flow upstream, namely that portion whose Mach lines reach the sonic line. A limit Mach wave exists which is asymptotic to the sonic line at infinity. The solution downstream of the limit Mach line cannot influence the solution upstream, which is thus calculated separately. Afterwards, the flow downstream can be found. This description is valid for the entire flow field, in particular for the flow near infinity. As $M_\infty \to 1-$, the rear half of the airfoil or body has a weaker and weaker effect on the flow upstream so that the $M_\infty = 1$ case is a natural dividing case. At supersonic speeds this feature of division of the flow field into front and rear persists.

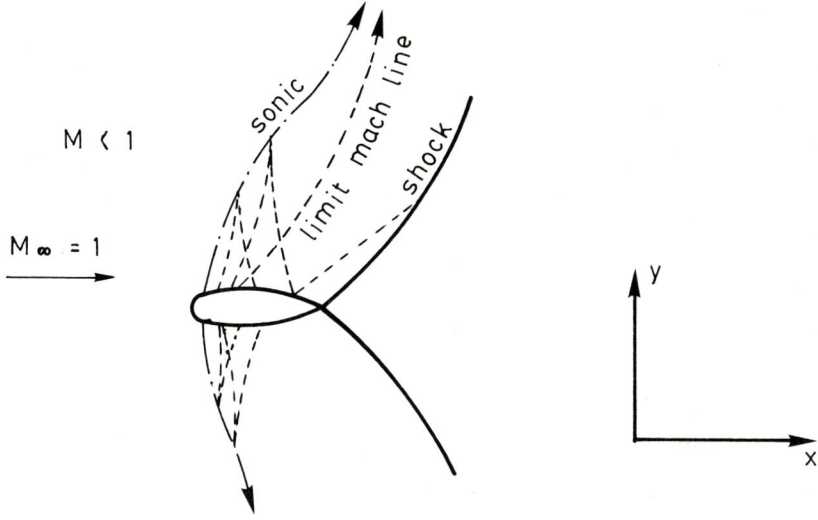

Figure 4.1.1
Flowfield at $M_\infty = 1$

The nature of the flow at a large distance upstream can not be characterized as that due to a source or singular force. The location of the sonic line on the airfoil and of the limit Mach wave vary in some way unknown a priori depending on details of the body shape. Also it turns out that the effect of lift dies out more rapidly at infinity than the effect of thickness. This is in direct contrast to the subsonic case discussed later. Thus recourse must be had to general similarity arguments.

4.1.1 Two Dimensional Flow

When $K = 0$ we must solve the transonic equation (3.1.16)

$$(\gamma + 1)\phi_x\phi_{xx} - \phi_{\tilde{y}\tilde{y}} = 0 \tag{4.1.1}$$

with boundary conditions on $0 \le x \le 1$

$$\phi_{\tilde{y}}(x, 0\pm) = F'_{u,\ell}(x) . \tag{4.1.2}$$

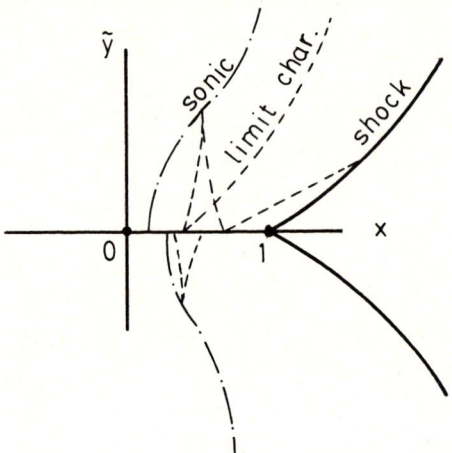

Figure 4.1.2

Transonic Plane

We are interested in the far field as $(x, \tilde{y} \to \infty)$ ahead of the airfoil. The length scale used in (4.1.1) was (arbitrarily) set as (chord $= 1$). But near infinity this scale is irrelevant. As an observer tends to infinity the airfoil appears to shrink to a point. In order to describe the approach to infinity let

$$x - x_0 = \frac{X}{\lambda} \,, \quad \tilde{y} = \frac{Y}{\mu} \,, \quad \text{where} \quad 0 \le x_0 \le 1 \,. \tag{4.1.3}$$

Then as the parameters $\lambda, \mu \to 0$ for fixed coordinates (X, Y) $x \to \infty$, $\tilde{y} \to \infty$. x_0 is an arbitrary origin. This may be regarded as rewriting the basic equation in an arbitrary set of new length scales, larger than before so that the airfoil shrinks to a point. Now the asymptotic potential $\varphi(X, Y)$ can be expected to be of the form

$$\varphi(X, Y) = \lim_{\lambda, \mu \to 0} \nu \phi \left(x_0 + \frac{X}{\lambda}, \frac{Y}{\mu} \right) \tag{4.1.4}$$

The parameter ν scales the size of the potential. Now if we make the following assumptions, we can deduce the form of the far field.

(i) $\varphi(X, Y)$ must represent a flow (in a new scale) and thus must satisfy

$$(\gamma + 1)\varphi_X \varphi_{XX} - \varphi_{YY} = 0 \,.$$

(ii) $\varphi(X, Y)$ must have a form which is independent of the parameters (ν, λ, μ) in the limit.

To satisfy (i) note that

$$\varphi_X = \frac{\nu}{\lambda}\phi_x, \ \varphi_Y = \frac{\nu}{\mu}\phi_{\tilde{y}} \quad \text{etc}.$$

Thus (4.1.1) becomes

$$(\gamma + 1)\frac{\lambda^3}{\nu^2}\varphi_X\varphi_{XX} - \frac{\mu^2}{\nu}\varphi_{YY} = 0$$

and φ becomes a small disturbance flow if

$$\frac{\mu^2\nu}{\lambda^3} = 1.$$

This provides one relationship and leaves a two parameter family of scalings. The following argument is typical of group theory [4.1.1]. Since one length scale (say μ) is arbitrary, we consider

$$\lambda = \lambda(\mu)$$

and study the changes in (4.1.4) as μ varies. First

$$\varphi(X, Y) = \lim_{\mu \to 0} \frac{\lambda^3(\mu)}{\mu^2}\phi\left(x_0 + \frac{X}{\lambda(\mu)}, \frac{Y}{\mu}\right) \tag{4.1.5}$$

We want $\varphi(X, Y)$ independent of μ. Thus

$$\frac{\partial\varphi}{\partial\mu} = 0 = \left(\frac{3\lambda^2}{\mu^2}\frac{d\lambda}{d\mu} - 2\frac{\lambda^3}{\mu^3}\right)\phi + \frac{\lambda^3}{\mu^2}\left(-\frac{X}{\lambda^2}\phi_x\frac{d\lambda}{d\mu} - \frac{Y^2}{\mu^2}\phi_{\tilde{y}}\right)$$

$$= \left(\frac{3\lambda^2}{\mu^2}\frac{d\lambda}{d\mu} - 2\frac{\lambda^2}{\mu^3}\right)\phi(x, \tilde{y}) - \frac{\lambda^2}{\mu^2}\frac{d\lambda}{d\mu}x\phi_x - \frac{\lambda^3}{\mu^3}\tilde{y}\phi_{\tilde{y}}.$$

If now

$$\frac{d\lambda}{d\mu} = \kappa\frac{\lambda}{\mu}; \ \kappa = \text{constant as } \lambda, \mu \to 0,$$

and

$$(3\kappa - 2)\phi(x, \tilde{y}) - \kappa x\phi_x(x, \tilde{y}) - \tilde{y}\phi_{\tilde{y}}(x, \tilde{y}) = 0, \tag{4.1.6}$$

then, for any κ, as $\mu \to 0$ the resulting form is independent of μ. (4.1.6) is a first-order partial differential equation which can be integrated to provide the asymptotic form of $\phi(x, y)$. If

$$\kappa x\phi_x + \tilde{y}\phi_{\tilde{y}} = (3\kappa - 2)\phi,$$

then

$$\frac{dx}{\kappa x} = \frac{d\tilde{y}}{\tilde{y}} = \frac{d\phi}{(3\kappa - 2)\phi} \qquad (4.1.7)$$

are the characteristic equations. Integration of the first two

$$\log x - \kappa \log \tilde{y} = \log \xi$$

defines the similarity curves

$$\xi = \frac{x}{\tilde{y}^{\kappa}} \qquad (4.1.8)$$

Integrating the last two of (4.1.7) along $\xi = $ constant

$$\log \phi = (3\kappa - 2) \log \tilde{y} + \log \frac{f(\xi)}{(\gamma + 1)} , \qquad (4.1.9)$$

or the resulting form is

$$(\gamma + 1)\phi(x, \tilde{y}) = \tilde{y}^{3\kappa - 2} f(\xi), \, \xi = \frac{x}{\tilde{y}^{\kappa}} . \qquad (4.1.10)$$

The factor $(\gamma + 1)$ is inserted for convenience. It follows from (4.1.1) that $f(\xi)$ will satisfy an ordinary differential equation. (4.1.10) can be regarded as the first term of an asymptotic expansion of the flow near ∞ based on $\tilde{y} \to \infty$ along lines $\xi = $ constant. The "exact" potential is thus represented by

$$\Phi = U\left\{x + \frac{\delta^{\frac{2}{3}}}{\gamma + 1}\left(\tilde{y}^{3\kappa - 2}f(\xi) + \tilde{y}^{\sigma}f_1(\xi) + \cdots\right) + \cdots\right\} \qquad (4.1.11)$$

where $\sigma < 3\kappa - 2$. Since $x = \xi\tilde{y}^{\kappa}$ we can expect the asymptotic form to be valid near infinity if

$$\kappa > 3\kappa - 2 \quad \text{or} \quad \kappa < 1 .$$

This guarantees $\phi_x, \phi_{\tilde{y}} \to 0$. The form of the similarity curves thus appears in Figure 4.1.3 for $\tilde{y} > 0$:

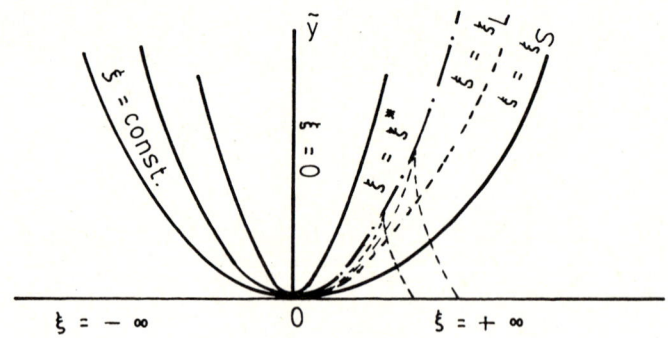

Figure 4.1.3

Similarity Coordinates

Velocity perturbations are thus, from (4.1.10),

$$w = (\gamma + 1)\phi_x = \tilde{y}^{2\kappa - 2} f'(\xi) \left.\right\}$$
$$\vartheta = (\gamma + 1)\phi_{\tilde{y}} = \tilde{y}^{3\kappa - 3}\left\{(3\kappa - 2)f - \kappa\xi f'\right\} \left.\right\} \tag{4.1.12}$$

Thus, the sonic line occurs when $f'(\xi) = 0$ which we label $\xi = \xi^*$. Of course $f'(\xi) > 0$ in the supersonic region and $f'(\xi) < 0$ in the subsonic region. Also the slope of the Mach lines is

$$\frac{dx}{d\tilde{y}} = \pm\sqrt{(\gamma + 1)\phi_x} = \pm\sqrt{w} \ .$$

For the limit Mach wave we have the $+$ family and

$$\frac{dx}{d\tilde{y}}_L = \sqrt{w_L} = \tilde{y}^{\kappa - 1}\sqrt{f'(\xi_L)} \quad \text{on} \quad x = \xi_L\tilde{y}^\kappa \ ,$$

$$\text{or} \quad dx_L = \kappa\xi_L\tilde{y}^{\kappa - 1}d\tilde{y}_L \ .$$

It follows that the limit Mach wave occurs on $\xi = \xi_L$ where

$$f'(\xi_L) = \xi_L^2\kappa^2 \tag{4.1.13}$$

We can also expect the tail shock to occur on $\xi = \xi_s > \xi_L$.

Now the problem is to discover the suitable value of κ so that the flow proceeds from $-\infty$ with $f'(\xi) < 0$, then accelerates through the sonic $f'(\xi) > 0$ and passes *smoothly* to the limit Mach wave. Continuation of the flow past the limit Mach wave through a tail shock and subsequent acceleration to sonic would complete the picture. This program can be carried out by studying the properties of the differential equation for $f(\xi)$. Note

$$(\gamma + 1)\phi_{xx} = \tilde{y}^{\kappa - 2}f''(\xi) \ ,$$

$$(\gamma + 1)\phi_{\tilde{y}\tilde{y}} = \tilde{y}^{3\kappa - 3}\left\{(3\kappa - 3)f' - \kappa\xi f''\right\}\left\{-\kappa\frac{\xi}{\tilde{y}}\right\}$$
$$+ (3\kappa - 3)\tilde{y}^{3\kappa - 4}\left\{(3\kappa - 2)f - \kappa\xi f'\right\} \ .$$

Thus the transonic equation (4.1.1) becomes

$$(f' - \kappa^2\xi^2)f'' - \kappa(5 - 5\kappa)\xi f' + (3 - 3\kappa)(3\kappa - 2)f = 0 \ . \tag{4.1.14}$$

Note that $f''(\xi_L)$ is in general infinite (cf 4.1.13). This equation has the following group property,

$$\left\{ \begin{array}{l} f^+ = A^3 f \\ \xi^+ = A\xi \end{array} \right\} \tag{4.1.15}$$

which leaves the equation invariant for any choice of the scale factor A. That is, the equation for $f^+(\xi^+)$ is the same as (4.1.14). If $f = F(\xi)$ is a solution, then $f^+ = F(\xi^+)$ is also, or

$$f = \frac{1}{A^3} F(A\xi) , \quad A = \text{constant} , \tag{4.1.16}$$

is also a solution. Exploration of this group property allows (4.1.1) to be reduced to a first-order differential equation. Since A is arbitrary, we could look for a solution where $\xi_L = 1$ (or $\xi^* = 1$). To match this far field up with the flow around a given airfoil a suitable value of A must then be chosen.

Now note that the flow in which the dominant term at infinity is a circulation independent of the path would have $\kappa = 2/3$, $(\gamma + 1)\phi = f(\xi)$. For this form at some value of ξ it is possible to have a jump in ϕ and this would be independent of the path as $\tilde{y} \to \infty$. This is the form of the circulating flow which is analogous to that for subsonic speeds. It can be shown that for this flow the velocity perturbation is normal to the lines $\xi = $ constant. Also, as shown below, it is not possible to find a solution, if $\kappa = 2/3$, which connects $-\infty$ smoothly to the limit Mach wave. It is not suprising then that the far field is symmetric with respect to $\tilde{y}$,

$$\phi(x, \tilde{y}) = \phi(x, -\tilde{y}) ,$$

even if the airfoil is lifting, and that

$$\boxed{\frac{2}{3} < \kappa < 1 \, .}$$

The symmetric field dies out more slowly toward infinity than the circulating one ahead of the airfoil. Further at sonic both increase and decrease of speed produces a widening of stream tubes (cf. Figure 3.1.2), so that a symmetric component of the far field is always produced.

First, we study the behavior as $\xi \to \infty$ and then the expansion near the limit Mach line and sonic line so that a complete qualitative picture emerges. Near $\xi = -\infty$ the basic equation (4.1.1) becomes $\phi_{\tilde{y}\tilde{y}} = 0$ so that

$$(\gamma + 1)\phi = A_0(x) + A_1(x)\tilde{y} + \cdots = \tilde{y}^{3\kappa - 2} f(\xi) , \quad \xi \to -\infty .$$

Thus f must have the form

$$A_0(x) + A_1(x)\tilde{y} + \cdots = \tilde{y}^{3\kappa-2}\left\{a_0(-\xi)^{3-\frac{2}{\kappa}} + b_0(-\xi)^{3-\frac{3}{\kappa}} + \cdots\right\}$$

and

$$A_0(x) = a_0(-x)^{3-\frac{2}{\kappa}}, \quad A_1(x) = b_0(-x)^{3-\frac{3}{\kappa}},$$

$$(\gamma+1)\phi = a_0(-x)^{3-\frac{2}{\kappa}} + b_0(-x)^{3-\frac{3}{\kappa}}\tilde{y} + \cdots, \quad \xi \to -\infty.$$

The same form can be deduced from a study of the indicial equation of (4.1.14) near $\xi = -\infty$. If we start integration at $\xi = -\infty$, (a_0, b_0) are the two arbitrary constants. If we seek the solution symmetric in $\tilde{y}$, then $b_0 \equiv 0$. In addition to the previous argument hodograph considerations due originally to Frankl and Guderley indicate also that $b_0 = 0$, that is strict symmetry for the fundamental solution. Details appear later when the closed form hodograph solution is derived. Then, we obtain the following expansion of the far field as $\xi \to -\infty$

$$f(\xi) = a_0(-\xi)^{3-\frac{2}{\kappa}} + a_1(-\xi)^{3-\frac{4}{\kappa}} + \cdots,$$

$$f'(\xi) = -\left(3 - \frac{2}{\kappa}\right)a_0(-\xi)^{2-\frac{2}{\kappa}} - \left(3 - \frac{4}{\kappa}\right)a_1(-\xi)^{2-\frac{4}{\kappa}} + \cdots. \tag{4.1.17}$$

Substitution in (4.1.14) shows

$$a_1(a_0) = \frac{1}{\kappa^3}(3\kappa - 2)^2(1 - \kappa)a_0^2 > 0. \tag{4.1.18}$$

For the velocity components, we have (cf. 4.1.12)

$$(\gamma+1) = w = \tilde{y}^{2\kappa-2}\left\{-\left(3 - \frac{2}{\kappa}\right)a_0(-\xi)^{2-\frac{2}{\kappa}} - \left(3 - \frac{4}{\kappa}\right)a_1(-\xi)^{2-\frac{4}{\kappa}}\right\}$$

$$= -\left(3 - \frac{2}{\kappa}\right)a_0(-x)^{2-\frac{2}{\kappa}} - \left(3 - \frac{4}{\kappa}\right)a_1(-x)^{2-\frac{4}{\kappa}}\tilde{y}^2 + \cdots, \tag{4.1.19}$$

$$(\gamma+1)\phi_{\tilde{y}} = \vartheta = \tilde{y}^{3\kappa-3}\left\{(3\kappa - 2)a_1(-\xi)^{3-\frac{4}{\kappa}} + \kappa(-1)\left(3 - \frac{4}{\kappa}\right)a_1(-\xi)^{3-\frac{4}{\kappa}} + \cdots\right\}$$

$$= \tilde{y}^{3\kappa-3}\left\{(2a_1(-\xi)^{3-\frac{4}{\kappa}} + \cdots\right\}$$

$$= 2a_1(-x)^{3-\frac{4}{\kappa}}\tilde{y} + \cdots \tag{4.1.20}$$

Near the sonic line $\xi = \xi^*$, where $f' = 0$, (4.1.14) shows that

$$f'' > 0 \quad \text{if} \quad f > 0.$$

At the limit Mach wave $\xi = \xi_L$, we can seek a regular expansion (cf. (4.1.13)).

$$f(\xi) = m_0 + \kappa^2 \xi_L^2 (\xi - \xi_L) + \frac{1}{2} m_2 (\xi - \xi_L)^2 + \cdots \qquad (4.1.21)$$

Upon substituting into the basic equation (4.1.14) m_0 takes a given value and m_2 must satisfy a quadratic equation. One root corresponds to acceleration through the limit Mach wave. We obtain:

$$f(\xi) = \frac{5\kappa^3}{3(3\kappa - 2)} \xi_L^3 + \kappa^2 \xi_L^3 \left(\frac{\xi}{\xi_L} - 1 \right)$$
$$+ \kappa(3 - 2\kappa)\xi_L^3 \left(\frac{\xi}{\xi_L} - 1 \right)^2 + \cdots . \qquad (4.1.22)$$

From these general considerations, the course of the solution must be as shown below.

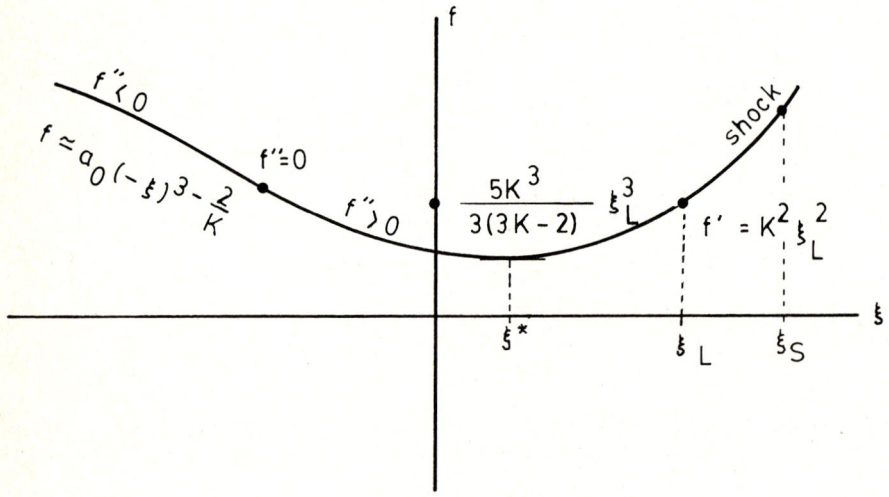

Figure 4.1.4

Far Field 2-D Case

Now the nature of the problem is clear. Two conditions must be satisfied at $\xi = \xi_L$, * $f(\xi_L) = \frac{5\kappa^3 \xi_L^3}{3(3\kappa - 2)}$, $f'(\xi_L) = \kappa^2 \xi_L^2$ and further the solution must be

---

* ξ_L is considered to be chosen arbitrarily (e.g. $\xi_L = 1$). The solution for any other value of ξ_L can be found by using the group property (4.1.16).

in the accelerating branch. Only one constant of integration a_0 is still arbitrary. Therefore, one cannot expect to find the solution in general but only for a special value of the parameter κ.

The existence of the solution was demonstrated first by Frankl [4.1.2] and Guderley [4.1.3] and the unique value of κ was found. Later workers [4.1.4], [4.1.5], [4.1.6], [4.1.7] were able to find a closed form by further direct analysis of (4.1.14). We follow this latter method because the results can be achieved this way for the case of axial symmetry also, where a linear hodograph theory is not available. A sketch of the hodograph result is given later. The details here follow Müller and Matschat's elegant paper [4.1.4]. Guderley introduced the idea of studying (4.1.14) in a suitable phase plane by use of the group property. Any set of coordinates invariant under the group allow the equation to be reduced to a first order equation which can be studied in a "phase" plane. The full solution is then obtained by quadrature. One set of invariant coordinates is

$$t = \xi^{-2}\frac{df}{d\xi}\,, \quad s = \xi^{-3}f(\xi)\,,$$

but it seems a little simpler to work directly with velocity components *

$$w \doteq (\gamma+1)\phi_x = \tilde{y}^{2\kappa-2}f'(\xi) = \tilde{y}^{2\kappa-2}\xi^2 t(\xi)$$
$$= \frac{x^2}{\tilde{y}^2}t(\xi)\,,$$
$$\vartheta = (\gamma+1)\phi_{\tilde{y}} = \tilde{y}^{3\kappa-3}\big\{(3\kappa-2)f - \kappa\xi f'\big\} \qquad (4.1.23)$$
$$= \tilde{y}^{3\kappa-3}\big\{(3\kappa-2)s - \kappa t\big\}\xi^3$$
$$= \frac{x^3}{\tilde{y}^3}\sigma(\xi)\,.$$

The basic system (cf. 4.1.1) is

$$\left.\begin{array}{c} ww_x - \vartheta_{\tilde{y}} = 0 \\[2mm] w_{\tilde{y}} - \vartheta_x = 0 \end{array}\right\} \qquad (4.1.24)$$

where for $w > 0$ the flow is supersonic, for $w < 0$ subsonic. We have

$$w_x = \frac{x}{\tilde{y}^2}\left\{\xi\frac{dt}{d\xi} + 2t\right\}\,, \quad w_{\tilde{y}} = \frac{-x^2}{\tilde{y}^3}\left\{\kappa\xi\frac{dt}{d\xi} + 2t\right\}\,,$$
$$\vartheta_x = \frac{x^2}{\tilde{y}^3}\left\{\xi\frac{d\sigma}{d\xi} + 3\sigma\right\}\,, \quad \vartheta_{\tilde{y}} = -\frac{x^3}{\tilde{y}^4}\left\{\kappa\xi\frac{d\sigma}{d\xi} + 3\sigma\right\}\,,$$

---

* See also [4.1.8] where a systematic study is made of similarity solutions in general.

so that (4.1.24) is

$$\begin{cases} t\xi\dfrac{dt}{d\xi} + 2t^2 + \kappa\xi\dfrac{d\sigma}{d\xi} + 3\sigma = 0 \,, \\[2mm] \kappa\xi\dfrac{dt}{d\xi} + 2t + \xi\dfrac{d\sigma}{d\xi} + 3\sigma = 0 \,, \end{cases}$$

or

$$\begin{cases} (t - \kappa^2)\xi\dfrac{dt}{d\xi} = -2t^2 + 2t\kappa - 3\sigma(1 - \kappa) \\[2mm] (t - \kappa^2)\xi\dfrac{d\sigma}{d\xi} = -2(1 - \kappa)t^2 - 3\sigma(t - \kappa) \end{cases} \tag{4.1.25}$$

Thus, dividing the second equation by the first, we have

$$\boxed{\dfrac{d\sigma}{dt} = \dfrac{2(1 - \kappa)t^2 + 3\sigma(t - \kappa)}{2t^2 - 2\kappa t + 3\sigma(1 - \kappa)}} \,. \tag{4.1.26}$$

The solution along any path of (4.1.26) is mapped back onto ξ by integration by any of (4.1.25), e.g.

$$\frac{d\xi}{\xi} = -\frac{(t - \kappa^2)\,dt}{2t^2 - 2\kappa t + 3\sigma(1 - \kappa)} \quad \text{along a path of (4.1.26)} \,. \tag{4.1.27}$$

As will be seen below the origin ($\sigma = t = 0$) corresponds to $\xi = -\infty$, the region of infinity in (σ, t) to $\xi = 0$, and a saddle point singularity on $t = \kappa^2 = \xi_L^{-2}\frac{df}{d\xi}(\xi_L)$ coresponds to the limiting Mach wave. The problem is thus to discover that value of κ such that the path which leaves the origin in a definite way runs through infinity and enters the saddle point (cf. Figure 4.1.5).

The hodograph methods used by Guderley and Frankl use analogous conditions and determine κ. For the two-dimensional case, Müller and Matschat introduce a transformation which makes (4.1.26) linear and are able to develop a whole class of special solutions in closed form one of which has the desired properties. For the axially-symmetric case a closed form is written down, esentially by inspection, knowing the desired properties (next section) [4.1.4], [4.1.6]. Near the origin $\sigma = 0$, $t = 0$, ($\xi = -\infty$) we have (4.1.18), (4.1.19), (4.1.20),

$$w = -\left(3 - \frac{2}{\kappa}\right)a_0(-x)^{2-\frac{2}{\kappa}} = \frac{x^2}{\tilde{y}^2}t \,,$$

$$\vartheta = \frac{2}{\kappa}\left(3 - \frac{2}{\kappa}\right)^2(1 - \kappa)a_0^2(-x)^{3-\frac{4}{\kappa}}\tilde{y} = \frac{x^3}{\tilde{y}^3}\sigma \,,$$

or

$$t \to -\left(3 - \frac{2}{\kappa}\right) a_0 \tilde{y}^2 (-x)^{-\frac{2}{\kappa}},$$

$$\sigma \to -\frac{2}{\kappa}(1 - \kappa)\left(3 - \frac{2}{\kappa}\right)^2 a_0^2 \tilde{y}^4 (-x)^{-\frac{4}{\kappa}},$$

or the path near the origin is

$$\sigma = -\frac{2}{\kappa}(1 - \kappa)t^2 + \cdots . \tag{4.1.28}$$

This is an exceptional path of the node at the origin and it can be checked in (4.1.26). Near $t = 0$, according to (4.1.27)

$$\frac{d\xi}{\xi} = -\frac{\kappa}{2}\frac{dt}{t} \quad \text{or} \quad (-\xi) = C_{-\infty}(-t)^{-\frac{\kappa}{2}} + \cdots$$

but $C_{-\infty}$ is given by the formulas above in terms of a_0 (arbitrary). cf. Figure (4.1.5). Now this path runs toward ($\sigma = -\infty$, $t = -\infty$), where $x = 0$ or $\xi = 0$. Since the solution is completely regular at $\xi = 0$ ($w < 0$, $\vartheta > 0$) (4.1.23) shows

$$t \to \frac{\tilde{y}^{2\kappa}}{x^2} f'(0) \to \frac{f'(0)}{\xi^2}$$

$$\sigma \to \frac{\tilde{y}^{3\kappa}}{x^3}(3\kappa - 2) f(0)$$

$$\to \frac{f(0)(3\kappa - 2)}{\xi^3},$$

so that as $x \to 0$,

$$\sigma \to \pm C_0(-t)^{\frac{3}{2}}, \quad \begin{cases} + & \text{if} \quad x \to 0+ \\ - & \text{if} \quad x \to 0- \end{cases}. \tag{4.1.29}$$

The path reappears at $\sigma = +\infty$, $t = -\infty$ as x crosses 0 and then runs towards the saddle point SP. The saddle point SP is located at

$$t = \kappa^2, \quad \sigma = \frac{2}{3}\kappa^3. \tag{4.1.30}$$

We want behavior according to (4.1.22) on the path approaching the saddle. It follows from (4.1.23) that

$$w = \tilde{y}^{2\kappa-2} f'(\xi) = \tilde{y}^{2\kappa-2}\left\{\kappa^2 \xi_L^2 + 2\kappa\xi_L(3 - 2\kappa)(\xi - \xi_L) + \cdots\right\}$$

$$= \frac{x^2}{\tilde{y}^2}t,$$

$$\vartheta = \tilde{y}^{3\kappa-3}\left\{(3\kappa - 2)f - \kappa\xi f'\right\}$$

$$= \tilde{y}^{3\kappa-3}\left\{\frac{2}{3}\kappa^3 \xi_L^3 + (6\kappa - 8)\kappa^2 \xi_L^2(\xi - \xi_L) + \cdots\right\} = \frac{x^3}{\tilde{y}^3}\sigma.$$

Thus

$$t \to \kappa^2 + 6\kappa(1 - \kappa)\frac{\xi - \xi_L}{\xi_L} + \cdots \quad \text{as} \quad \xi \to \xi_L \,,$$

$$\sigma \to \frac{2}{3}\kappa^3 - 4\kappa^2(2 - \kappa)\frac{\xi - \xi_L}{\xi_L} + \cdots \quad \text{as} \quad \xi \to \xi_L \,, \tag{4.1.31}$$

or the path into the saddle SP runs along

$$\sigma - \frac{2}{3}\kappa^3 = -\frac{2}{3}\frac{\kappa(2 - \kappa)}{1 - \kappa}(t - \kappa^2) \,, \tag{4.1.32}$$

and this can be verified again in (4.1.26). The solution to (4.1.26) which satisfies all the conditions can be found only for

$$\boxed{\kappa = \frac{4}{5}\,,}$$

and can be written

$$\sigma = \begin{cases} \sigma_- = \dfrac{4}{3} - 2t - \dfrac{4}{3}(1 - t)^{\frac{3}{2}} \,, & \sigma < 0\,, \quad -\infty < \xi < 0\,, \\[2mm] \sigma_+ = \dfrac{4}{3} - 2t + \dfrac{4}{3}(1 - t)^{\frac{3}{2}} \,, & \sigma > 0\,, \quad 0 < \xi < \xi_L\,. \end{cases} \tag{4.1.33}$$

The path crosses the sonic line at $t = 0$, $\sigma = \frac{8}{3}$. By integration of (4.1.27) using the equation of the path (4.1.33) the distribution $\xi(t)$ along the path is easily found.

$$\left(\frac{\xi^*}{\xi}\right)^5 = \begin{cases} -\dfrac{1}{4}(1 - t)^{\frac{3}{2}}\left\{1 - (1 - t)^{\frac{1}{2}}\right\}^2 & \begin{aligned} &-\infty < \xi < 0\,, \\ &0 > t > -\infty\,, \end{aligned} \\[3mm] \dfrac{1}{4}(1 - t)^{\frac{3}{2}}\left\{1 + (1 - t)^{\frac{1}{2}}\right\}^2 & \begin{aligned} &0 < \xi < \xi_L\,, \\ &-\infty < t < \kappa^2 = \dfrac{16}{25}\,. \end{aligned} \end{cases} \tag{4.1.34}$$

(4.1.33) and (4.1.34) provide, in implicit form, the desired solution. The velocity components are obtained from (4.1.23).

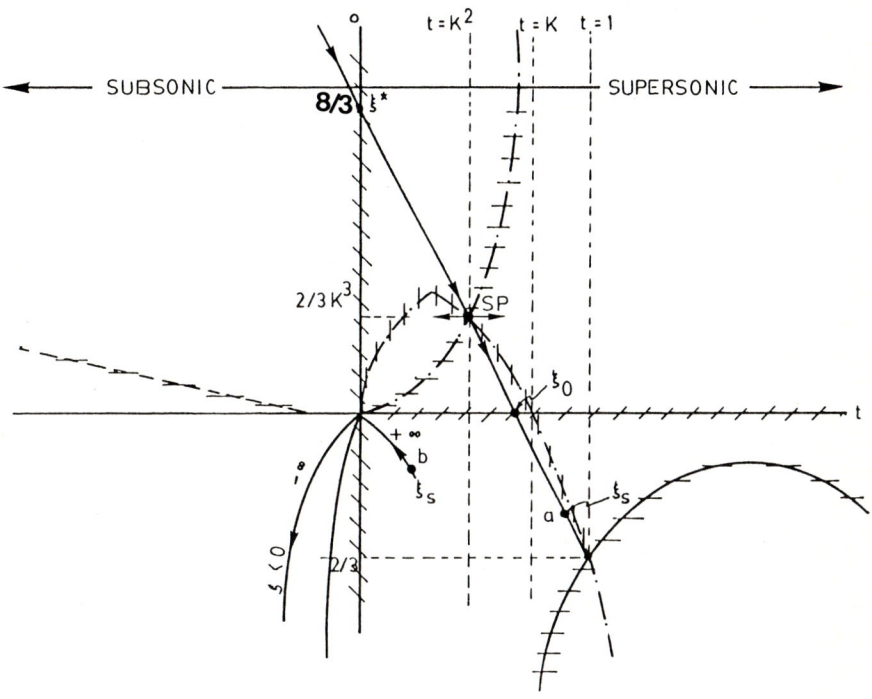

Figure 4.1.5

Phase Plane

Since the solution passes smoothly through the singular point $t = \kappa^2 = \frac{16}{25}$ corresponding to the limit Mach wave the $+$ branch of $(4.1.33)$ can be continued. It crosses $\sigma = 0$ for a value of $\xi_0 > \xi_L$ and on ξ_0, $\vartheta = 0$, the streamlines are horizontal. As σ increases further and t increases toward 1 this path $\sigma_+(t)$ heads for the singular point at $t = 1$, $\sigma = -\frac{2}{3}$.

This singular point is a node so that $\xi \to +\infty$ (downstream $\tilde{y} = 0$ axis) if the solution is continued along this curve right into the singular point. However, since (t, σ) are finite this would imply infinite velocity. Hence a shock wave must occur at some $\xi_s > \xi_L$ so that the solution can get onto the exceptional path running into the origin ($\sigma = t = 0$). Since the local considerations which lead to $(4.1.19)$ $(4.1.20)$ for the upstream axis are equally valid near the downstream axis (with $(-x)$ replaced by x, a_0 by a_0') the $\sigma(t)$ may be continued to $t > 0$.

This represents the flow near $\xi = \infty$. Now, a shock wave is a jump in (w, ϑ) and induces a corresponding jump in (t, σ). The jump must carry from $\sigma_+(t)$ to $\sigma_-(t)$ and satisfy the shock jump conditions, which will now be worked out in (t, σ).

Following the method of section (3.6), we may write, from (4.1.24)

$$\left\{ \begin{array}{c} \frac{1}{2}[w^2]d\tilde{y}_s + [\vartheta]\,dx_s = 0 \\ [w]\,dx_s + [\vartheta]\,d\tilde{y}_s = 0 \end{array} \right\},$$

or in general

$$\left\{ \begin{array}{c} \langle w \rangle [w] + [\vartheta]\left(\dfrac{dx}{d\tilde{y}}\right)_s = 0 \\ [w]\left(\dfrac{dx}{d\tilde{y}}\right)_s + [\vartheta] = 0 \end{array} \right\}. \tag{4.1.35}$$

However, the shock must occur on a similarity curve ξ_s where

$$x = \xi_s \tilde{y}^\kappa\,,$$

$$\left(\frac{dx}{d\tilde{y}}\right)_s = \kappa\xi_s\tilde{y}^{\kappa-1} = \kappa\frac{x}{\tilde{y}}\,, \quad \left(\kappa = \frac{4}{5}\right)\,,$$

so

$$[w] = \frac{x^2}{\tilde{y}^2}[t]\,, \quad \langle w \rangle = \frac{x^2}{\tilde{y}^2}\langle t \rangle\,, \quad [\vartheta] = \frac{x^3}{\tilde{y}^3}[\sigma]\,,$$

so that (4.1.35) becomes

$$\left\{ \begin{array}{c} \langle t \rangle [t] + \kappa[\sigma] = 0 \\ \kappa[t] + [\sigma] = 0 \end{array} \right\}. \tag{4.1.36}$$

It is clear that there is no solution to this linear system for jumps unless the determinant

$$\langle t \rangle - \kappa^2 = 0\,. \tag{4.1.37}$$

Let $(\ )_a$ = state ahead of shock, $(\ )_b$ = state behind shock. Thus, we have directly

$$t_a + t_b = 2\kappa^2 = 2\left(\frac{4}{5}\right)^2\,,$$

$$\sigma_b - \sigma_a = \kappa(t_b - t_a) = -\frac{4}{5}(t_b - t_a)\,. \tag{4.1.38}$$

Further, the jumps must end on the integral curves (4.1.34)

$$\left(\frac{\xi^*}{\xi_s}\right)^5 = \frac{1}{4}(1 - t_a)^{\frac{3}{2}}\left\{1 + (1 - t_a)^{\frac{1}{2}}\right\}^2$$

$$= \frac{1}{4}(1 - t_b)^{\frac{3}{2}}\left\{1 - (1 - t_b)^{\frac{1}{2}}\right\}^2\,. \tag{4.1.39}$$

The solution satisfying these relations is

$$t_a = \left(\frac{4}{5}\right)^2 + \frac{\sqrt{3}}{2}\left(\frac{3}{5}\right)^2 , \qquad t_b = \left(\frac{4}{5}\right)^2 - \frac{\sqrt{3}}{2}\left(\frac{3}{5}\right)^2 ,$$

$$\sigma_a = -\frac{23}{75} - \frac{2}{\sqrt{3}}\left(\frac{3}{5}\right)^3 , \qquad \sigma_b = -\frac{23}{75} + \frac{2}{\sqrt{3}}\left(\frac{3}{5}\right)^3 . \qquad (4.1.40)$$

The shock thus has supersonic velocity behind and the flow recompresses smoothly to sonic as $\xi \to +\infty$. These values must also satisfy (4.1.33), $\sigma_+(t_a) = \sigma_a$, $\sigma_-(t_b) = \sigma_b$.

These results appear in a paper of Barish and Guderley [4.1.9], where they were obtained numerically. The relation between sonic line, limit Mach line, and shock location is

$$\xi_L = \frac{\xi^*}{a_0} = \frac{\xi_s}{b_0} , \qquad (4.1.41)$$

where $a_0 = 2^{\frac{4}{5}}3^{\frac{3}{5}}5^{-1} \approx .67$, $b_0 = 2^{\frac{8}{5}}11^{-\frac{2}{5}}(9\sqrt{3})^{\frac{1}{5}} \approx 2.03$.

Wind tunnel experiments verify that $x \sim \tilde{y}^{\frac{4}{5}}$ along the asymptotic shock. [4.1.10]

Hodograph Representation

The hodograph plane equations derived in Section 3.5 for two-dimensional flow are

$$\left.\begin{matrix} w\dfrac{\partial \tilde{y}}{\partial \vartheta} - \dfrac{\partial x}{\partial w} = 0 \\[2mm] \dfrac{\partial x}{\partial \vartheta} - \dfrac{\partial \tilde{y}}{\partial w} = 0 \end{matrix}\right\} \qquad (4.1.42)$$

where $x = x(w,\vartheta)$, $\tilde{y} = \tilde{y}(w,\vartheta)$. For the case of sonic free stream

$$w = (\gamma + 1)\phi_x$$
$$\vartheta = (\gamma + 1)\phi_{\tilde{y}}$$

where $w > 0$ corresponds to supersonic flow, $w < 0$ to subsonic flow. The far-field of the physical plane is represented as a singular point at the origin in the hodograph. Disturbances vanish far away from any object in $(x, \tilde{y})$ so that all streamlines with different values $\tilde{y} = $ constant run into the origin ($w = \vartheta = 0$). The equation satisfied by $\tilde{y}(w,\vartheta)$ is the Tricomi equation

$$w\tilde{y}_{\vartheta\vartheta} - \tilde{y}_{ww} = 0 . \qquad (4.1.43)$$

A solution of (4.1.43) must be found with a singularity at $(w,\vartheta) = 0$, many values of $\tilde{y}$, and $x \to -\infty$, but with no singularity on the characteristic leading out of the origin. These characteristics are the images of the limiting characteristics in the physical plane. A sketch of the streamlines and characteristics appears in Figure 4.1.6.

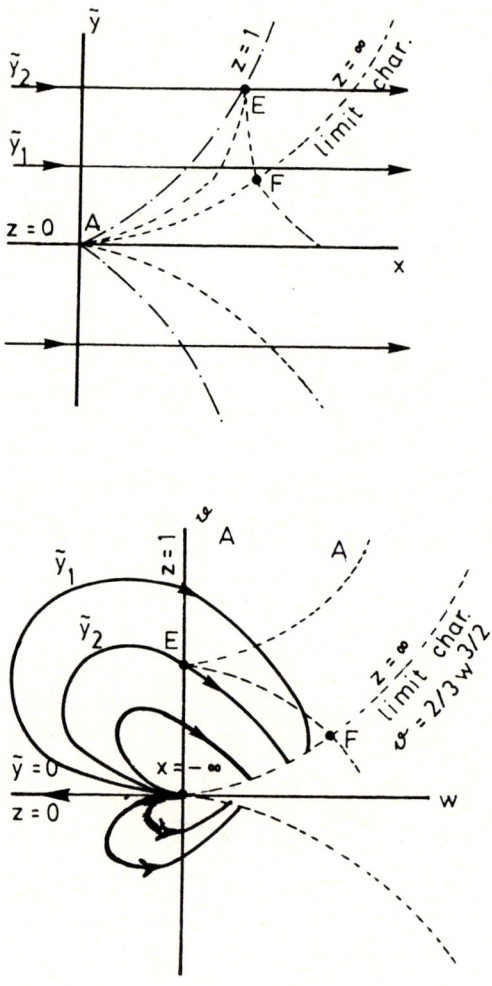

Figure 4.1.6

Sketch of Free Stream Singularity

The calculation of this free stream singularity which yields similarity [4.1.2], [4.1.3] curves $\xi = \dfrac{x}{\tilde{y}^{\frac{4}{5}}}$ was actually discussed first in the hodograph by Frankl and Guderley. Here the results are reproduced using solutions of the Tricomi equation in its canonical elliptical form. A fairly detailed treatment is given here since this solution will also be used later. Introducing

$$\tau = \frac{2}{3}(-w)^{\frac{3}{2}} \qquad (4.1.44)$$

one can write the system (4.1.42)

$$\left\{ \begin{array}{l} \left(\dfrac{3\tau}{2}\right)^{\frac{1}{3}} \tilde{y}_{\vartheta} - x_{\tau} = 0 \\[2mm] \left(\dfrac{3\tau}{2}\right)^{\frac{1}{3}} \tilde{y}_{\tau} - x_{\vartheta} = 0 \end{array} \right\} . \qquad (4.1.45)$$

Then the Tricomi equation reads

$$\tilde{y}_{\tau\tau} + \frac{1}{3\tau}\tilde{y}_{\tau} + \tilde{y}_{\vartheta\vartheta} = 0 \qquad (4.1.46)$$

the canonical elliptical form. Now introducing polar coordinates in the τ, ϑ plane (cf. Figure 4.1.7)

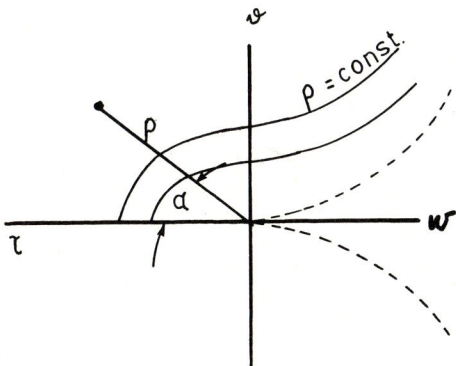

Figure 4.1.7

Polar Coordinates in Hodograph

the system (4.1.45) can be replaced by

$$\left\{\begin{array}{l} \dfrac{\partial x}{\partial \rho} = \left(\dfrac{3\rho}{2}\right)^{\frac{1}{3}} \left(\cos^{\frac{1}{3}}\alpha\right) \dfrac{1}{\rho}\dfrac{\partial \tilde{y}}{\partial \alpha} \\[4mm] \dfrac{1}{\rho}\dfrac{\partial x}{\partial \alpha} = -\left(\dfrac{3\rho}{2}\right)^{\frac{1}{3}} \left(\cos^{\frac{1}{3}}\alpha\right)\dfrac{\partial \tilde{y}}{\partial \rho} \end{array}\right\} \tag{4.1.47}$$

The Tricomi equation (4.1.46) is

$$\frac{\partial^2 \tilde{y}}{\partial \rho^2} + \frac{4}{3\rho}\frac{\partial \tilde{y}}{\partial \rho} + \frac{1}{\rho^2}\frac{\partial^2 \tilde{y}}{\partial \alpha^2} - \frac{1}{3\rho^2}\tan\alpha\frac{\partial \tilde{y}}{\partial \alpha} = 0 . \tag{4.1.48}$$

The singularity is now at $\rho = 0$. Solutions can be found by separation of variables in the form

$$\tilde{y} = R(\rho) A(\alpha)$$

so that

$$\rho^2 \frac{R''}{R} + \frac{4}{3}\rho\frac{R'}{R} = -\frac{A''}{A} + \frac{\tan\alpha}{3}\frac{A'}{A} = \lambda^2 = \text{constant} . \tag{4.1.49}$$

The equation for R is homogeneous in ρ

$$R'' + \frac{4}{3}\frac{R'}{\rho} - \frac{\lambda^2}{\rho^2}R = 0 . \tag{4.1.50}$$

We use the solutions

$$R = \rho^{-\frac{1}{6}-\beta} , \quad \text{where} \quad \beta = \sqrt{\frac{1}{36} + \lambda^2} . \tag{4.1.51}$$

For $A(\alpha)$ we have

$$\frac{d^2 A}{d\alpha^2} - \frac{1}{3}\tan\alpha\frac{dA}{d\alpha} + \lambda^2 A = 0 . \tag{4.1.52}$$

This can also be written in self adjoint form $\dfrac{d}{d\alpha}\left(\cos^{1/3}\alpha\dfrac{dA}{d\alpha}\right) + (\cos^{1/3}\alpha)\lambda^2 A = 0$
Introducing

$$z = \sin^2\alpha = \frac{\vartheta^2}{\rho^2} = \frac{\vartheta^2}{\vartheta^2 - \frac{4}{9}w^3} , \tag{4.1.53}$$

equation (4.1.52) takes the standard hypergeometric form

$$z(1-z)\frac{d^2 A}{dz^2} + \left(\frac{1}{2} - \frac{7}{6}z\right)\frac{dA}{dz} + \frac{\lambda^2}{4}A = 0 . \tag{4.1.54}$$

With the notation $F(a, b; c; z) = 1 + \dfrac{ab}{c}z + \dfrac{a(a+1)b(b+1)}{c(c+1)2!}z^2 + \cdots$, (4.1.54) has

$c = \dfrac{1}{2}$, $a + b + 1 = \dfrac{7}{6}$, $ab = -\dfrac{\lambda^2}{4}$ when compared with standard form of the hypergeometric equation. Thus

$$a, b = \frac{1}{12} \pm \sqrt{\frac{1}{144} + \frac{\lambda^2}{4}} = \frac{1}{12} \pm \frac{\beta}{2} \quad \text{say} . \tag{4.1.55}$$

Thus the general family of linearly independent singular solutions is

$$\tilde{y}_I(\rho, \alpha) = \rho^{-\frac{1}{6} - \beta} F\left(\frac{1}{12} + \frac{\beta}{2}, \frac{1}{12} - \frac{\beta}{2}; \frac{1}{2}; \sin^2 \alpha\right) \qquad \text{symmetric in } \alpha$$

$$\tag{4.1.56}$$

$$\tilde{y}_{II}(\rho, \alpha) = \rho^{-\frac{1}{6} - \beta} \sin \alpha \, F\left(\frac{7}{12} + \frac{\beta}{2}, \frac{7}{12} - \frac{\beta}{2}; \frac{3}{2}; \sin^2 \alpha\right) \quad \text{anti-symmetric in } \alpha$$

$$\tag{4.1.57}$$

with $\beta > -\frac{1}{6}$. This last solution is the second linearly independent hypergeometric solution near $z = 0$ of the form $z^{1-c}F(a + 1 - c, b + 1 - c; 2 - c; z).$* It's not necessary to consider $\rho^{-\frac{1}{6} + \beta}$ separately since the solutions are exactly the same as (4.1.56, 57). It is clear that the symmetric solution in the hodograph, $\tilde{y}_I$, can not represent the free stream since both positive and negative values of $\tilde{y}$ are needed as infinity is approached along the sonic line.

It can be seen from (4.1.53) that $z = 0$ corresponds to the x-axis upstream of the origin (body), $z = 1$ corresponds to the sonic line and $z = \infty$ to the limit characteristic $\vartheta = \pm\frac{2}{3}w^{\frac{3}{2}}$. The hypergeometric functions in (4.1.56),(4.1.57) are regular expansions about $(z = \alpha = 0)$ as written. The behavior on the limiting characteristic must be found by analytic continuation of these functions past the sonic line $(z = 1, \alpha = \frac{\pi}{2})$ to the limiting characteristic $z = \infty$. Only certain values of β allow the solutions to be regular ($\tilde{y}$ finite) as $z \to \infty$.

The following formula for analytic continuation of the hypergeometric function is useful [4.1.11]

$$F(a, b; c; z) = A_1 z^{-a} F\left(a, a + 1 - c; a + b + 1 - c; 1 - \frac{1}{z}\right)$$

$$+ A_2 z^{a-c}(1 - z)^{c-a-b} F\left(c - a, 1 - a; c + 1 - a - b; 1 - \frac{1}{z}\right)$$

---

* For a collection of formulas involving hypergeometric functions see [4.1.11] or any standard work.

where

$$A_1 = \frac{\Gamma(c)\Gamma(c-a-b)}{\Gamma(c-a)\Gamma(c-b)}, \quad A_2 = \frac{\Gamma(c)\Gamma(a+b-c)}{\Gamma(a)\Gamma(b)} \quad \text{for } |\arg z| < \pi \quad (4.1.58)$$

From this we can get the continuation of $\tilde{y}_{II}$ past $z = 1$ on the real axis

$$\tilde{y}_{II} = \vartheta\rho^{-\frac{7}{6}-\beta} F\left(\frac{7}{12}+\frac{\beta}{2}, \frac{7}{12}-\frac{\beta}{2}; \frac{3}{2}; \frac{\vartheta^2}{\rho^2}\right) \quad (4.1.59)$$

$$= \vartheta\rho^{-\frac{7}{6}-\beta}\left\{ \frac{\Gamma\left(\frac{3}{2}\right)\Gamma\left(\frac{1}{3}\right)}{\Gamma\left(\frac{11}{12}-\frac{\beta}{2}\right)\Gamma\left(\frac{11}{12}+\frac{\beta}{2}\right)} \left(\frac{\vartheta^2}{\rho^2}\right)^{-\frac{7}{12}-\frac{\beta}{2}} \right.$$

$$\times F\left(\frac{7}{12}+\frac{\beta}{2}, \frac{1}{12}+\frac{\beta}{2}; \frac{2}{3}; 1-\frac{\rho^2}{\vartheta^2}\right)$$

$$+ \frac{\Gamma\left(\frac{3}{2}\right)\Gamma\left(-\frac{1}{3}\right)}{\Gamma\left(\frac{7}{12}+\frac{\beta}{2}\right)\Gamma\left(\frac{7}{12}-\frac{\beta}{2}\right)} \left(\frac{\vartheta^2}{\rho^2}\right)^{-\frac{11}{12}+\frac{\beta}{2}} \left(1-\frac{\vartheta^2}{\rho^2}\right)^{\frac{1}{3}}$$

$$\left. \times F\left(\frac{11}{12}-\frac{\beta}{2}, \frac{5}{12}-\frac{\beta}{2}; \frac{4}{3}; 1-\frac{\rho^2}{\vartheta^2}\right) \right\} \quad (4.1.60)$$

Now to approach the limit characteristic $\vartheta = \pm\frac{2}{3}w^{\frac{3}{2}}$ we let $\rho \to 0$, ϑ fixed. Using

$$F(a,b;\,c;\,1) = \frac{\Gamma(c)\,\Gamma(c-a-b)}{\Gamma(c-a)\,\Gamma(c-b)} \quad \text{for} \quad \begin{array}{l} c-a-b > 0 \\ c \ne 0,\,-1,\dots, \end{array} \quad (4.1.61)$$

and the identity

$$F(a,b;\,c;\,z) = (1-z)^{c-a-b}F(c-a,\,c-b;\,c;\,z) \quad (4.1.62)$$

we find

$$F(a,b,c;\,z) \sim (1-z)^{c-a-b}\frac{\Gamma(c)\Gamma(a+b-c)}{\Gamma(a)\Gamma(b)} \quad \text{as } z \sim 1 \text{ if } c-a-b < 0 .$$

To show the singularity as $z \to 1$, we write (4.1.60) as $\rho \to 0$ using (4.1.61) and (4.1.62)

$$\tilde{y}_{II} \to \vartheta^{-\frac{1}{6}-\beta}\rho^{-2\beta}\frac{\Gamma\left(\frac{3}{2}\right)\Gamma(\beta)}{\Gamma\left(\frac{11}{2}+\frac{\beta}{2}\right)\Gamma\left(\frac{7}{12}+\frac{\beta}{2}\right)}$$

$$\times \left\{ \frac{\Gamma\left(\frac{2}{3}\right)\Gamma\left(\frac{1}{3}\right)}{\Gamma\left(\frac{11}{12}-\frac{\beta}{2}\right)\Gamma\left(\frac{1}{12}+\frac{\beta}{2}\right)} - \frac{\Gamma\left(-\frac{1}{3}\right)\Gamma\left(\frac{4}{3}\right)}{\Gamma\left(\frac{7}{12}-\frac{\beta}{2}\right)\Gamma\left(\frac{5}{12}+\frac{\beta}{2}\right)} \right\} + \text{regular terms} .$$

$$(4.1.63)$$

Using

$$\Gamma(z)\Gamma(1 - z) = \frac{\pi}{\sin \pi z}$$

the bracket in (4.1.63) becomes

$$\frac{1}{\sin \frac{\pi}{3}} \left\{ \sin \pi \left(\frac{1}{12} + \frac{\beta}{2} \right) + \sin \pi \left(\frac{5}{12} + \frac{\beta}{2} \right) \right\} .$$

Thus for no singularity on the limit characteristic ($\rho \to 0$, ϑ fixed say)

$$\frac{\pi}{12} + \frac{\beta \pi}{2} = - \left(\frac{5\pi}{12} + \frac{\beta \pi}{2} \right) + 2k\pi \quad k = 0, 1, 2 \cdots$$

,or

$$\boxed{\beta = -\frac{1}{2} + 2k} , \quad k = 1, 2, \ldots . \tag{4.1.64}$$

The dominant singular terms in (4.1.63) vanish for this set of β and the remaining terms, which are worked out explicitly below, contribute a finite value of $\tilde{y}_{II}$ on the limit characteristic. However not all the values of β are admissible if $\tilde{y}_{II}$ is to represent the far field.

If $\tilde{y}_{II}$ represents the dominant term in the far field, then necessarily $\tilde{y}_{II} > 0$ for $\vartheta > 0$. We now show that only $\beta = -\frac{1}{2}$ guarantees this condition. Now $\tilde{y}_{II} = \rho^{-\frac{1}{6} - \beta} A(z)$ from (4.1.57) where A satisfies (4.1.54) or equivalently

$$\frac{d}{dz} \left(\sqrt{z}(1 - z)^{\frac{2}{3}} \frac{dA}{dz} \right) + \frac{\lambda^2}{4} \frac{1}{\sqrt{z}(1 - z)^{\frac{1}{3}}} A = 0 \tag{4.1.65}$$

with $\lambda^2 = \beta^2 - \frac{1}{36}$ and A, $\frac{dA}{dz}$ continuous on $(0, 1]$, $A \underset{z \to 0}{\sim} \sqrt{z}$. Let A_1, A_2 be two solutions of (4.1.65) of this form with $\lambda = \lambda_1, \lambda_2$ respectively, $\lambda_2 > \lambda_1$. Then, if α, β are successive zeros of A_1, A_2 has a zero on (α, β). To show this, suppose not. That is, suppose A_1, $A_2 > 0$ on (α, β). Then

$$\int_\alpha^\beta \left(A_2 \left\{ \frac{d}{dz} \sqrt{z}(1 - z)^{\frac{2}{3}} \frac{dA_1}{dz} + \frac{\lambda_1^2}{4} \frac{1}{\sqrt{z}(1 - z)^{\frac{1}{3}}} A_1 \right\} \right.$$
$$\left. - A_1 \left\{ \frac{d}{dz} \sqrt{z}(1 - z)^{\frac{2}{3}} \frac{dA_2}{dz} + \frac{\lambda_2^2}{4} \frac{1}{\sqrt{z}(1 - z)^{\frac{1}{3}}} A_2 \right\} \right) dz = 0$$

or, after integration by parts once and using the fact that $A_1(\alpha) = A_1(\beta) = 0$,

$$\sqrt{z}(1 - z)^{\frac{2}{3}} A_2 \frac{dA_1}{dz} \Big|_\alpha^\beta = \frac{\alpha_2^2 - \alpha_1^2}{4} \int_\alpha^\beta \frac{A_1 A_2}{\sqrt{z}(1 - z)^{\frac{1}{3}}} dz .$$

Now $A_1(\alpha) = A_1(\beta) = 0$, $A_1(z) > 0$ in (α, β), so that $\frac{dA_1}{dz}(\alpha) \geq 0$, $\frac{dA_1}{dz}(\beta) \leq 0$. Hence the left hand side is ≤ 0, the right hand side is > 0 which is a contradiction. The result remains true even if $\alpha = 0$ despite the fact that $\frac{dA}{dz}$ is undefined since one takes the limit of the above equations as $\alpha \to 0$. Since $A_{1,2} \underset{z \to 0}{\sim} \sqrt{z}$, $A'_{1,2} \underset{z \to 0}{\sim} \frac{1}{2\sqrt{z}}$ one obtains

$$0 = \frac{\lambda_2^2 - \lambda_1^2}{4} \int_0^\beta \frac{A_1 A_2}{\sqrt{z}(1-z)^{\frac{1}{3}}} dz .$$

Now, consider A_k, the solution of (4.1.65) with $\beta = -\frac{1}{2} + 2k$. Along the sonic line, $z = 1$,

$$A_k = \frac{\Gamma\left(\frac{3}{2}\right) \Gamma\left(\frac{1}{3}\right)}{\Gamma\left(\frac{11}{2} - \frac{\beta}{2}\right) \Gamma\left(\frac{11}{2} + \frac{\beta}{2}\right)}$$

with $\beta = -\frac{1}{2} + 2k$, from (4.1.60). But,

$$\Gamma\left(\frac{11}{2} - \frac{\beta}{2}\right) = \Gamma\left(\frac{7}{6} - k\right) > 0 \quad \text{if} \quad k = 1, 3, 5, \ldots$$
$$< 0 \quad \text{if} \quad k = 2, 4, 6, \ldots .$$

Thus A_{2k} becomes negative within $[0, 1]$ and hence $\tilde{y}_{II}$ can not be the dominant far field term if k is even. Since $A_{2k}(0) = 0$, $A_{2k}(\beta) = 0$ for some $\beta < 1$, then $A_{2k+1}(\alpha) = 0$ for some α, $0 < \alpha < \beta < 1$. Hence A_{2k+1} is not allowable. We are left with only

$$\tilde{y}_{II} = \rho^{-\frac{1}{6} - \beta} A_1 ,$$

with $\beta = -\frac{1}{2}$. This is always positive in $(0, 1)$ as can be seen from the closed form of the solution in (4.1.71).

Therefore the proper singular solution can only occur for the case $k = 1$,

$$\boxed{\beta = \frac{3}{2}} , \tag{4.1.66}$$

$$\tilde{y}_{IIs} = \vartheta^{-\frac{5}{3}} \sin^{\frac{8}{3}} \alpha F\left(\frac{4}{3}, -\frac{1}{6}; \frac{3}{2}; \sin^2 \alpha\right) . \tag{4.1.67}$$

Since $\alpha = $ constant is a similarity curve in the hodograph, $\tilde{y}_{IIs} \sim \vartheta^{-\frac{5}{3}}$ along the similarity curves and from (4.1.47) $x_{IIs} \sim \vartheta^{\frac{1}{3}} \tilde{y}_{IIs} \sim \vartheta^{-\frac{4}{3}}$. Thus, similarity curves in the hodograph for the singular solution map to similarity curves

$$\xi = \frac{x}{\tilde{y}^{\frac{4}{5}}}$$

in the physical plane, as was used earlier in this section.

Now as shown by Germain [4.1.8] the singular solutions can be expressed in a simple closed form by using transformations of the hypergeometric functions. This can be accomplished by writing (4.1.67) as

$$\tilde{y}_{IIs} = \rho^{-\frac{5}{3}} \sin \alpha \, F\left(\frac{4}{3}, -\frac{1}{6}; \frac{3}{2}; \sin^2 \alpha\right) \tag{4.1.68}$$

and using Goursat's quadratic transformation. [4.1.11].

$$2\frac{\Gamma\left(-\frac{1}{2}\right)\Gamma\left(a+b-\frac{1}{2}\right)}{\Gamma\left(a-\frac{1}{2}\right)\Gamma\left(b-\frac{1}{2}\right)} z^{\frac{1}{2}} F\left(a, b; \frac{3}{2}; z\right)$$

$$= F\left(2a-1, 2b-1; a+b-\frac{1}{2}; \frac{1-z^{\frac{1}{2}}}{2}\right)$$

$$- F\left(2a-1, 2b-1; a+b-\frac{1}{2}; \frac{1+z^{\frac{1}{2}}}{2}\right). \tag{4.1.69}$$

So,

$$\tilde{y}_{IIs} = \rho^{-\frac{5}{3}} \frac{\Gamma\left(\frac{5}{6}\right)\Gamma\left(-\frac{2}{3}\right)}{\Gamma\left(-\frac{1}{2}\right)\Gamma\left(\frac{2}{3}\right)}$$

$$\times \left\{ F\left(\frac{5}{3}, -\frac{4}{3}; \frac{2}{3}; \frac{1-\sin\alpha}{2}\right) - F\left(\frac{5}{3}, -\frac{4}{3}; \frac{2}{3}; \frac{1+\sin\alpha}{2}\right) \right\}. \tag{4.1.70}$$

Since there is an integer difference between parameters c, a in (4.1.70) the solution can be brought to closed form using the identity (4.1.62).

$$F\left(\frac{5}{3}, -\frac{4}{3}; \frac{2}{3}; \frac{1-\sin\alpha}{2}\right) = \left(1 - \frac{1-\sin\alpha}{2}\right)^{\frac{1}{3}} F\left(-1, 2; \frac{2}{3}; \frac{1-\sin\alpha}{2}\right)$$

$$= \frac{\left(1+\frac{\vartheta}{\rho}\right)^{\frac{2}{3}}}{2^{\frac{1}{3}}} \left(1 - 3\frac{1-\frac{\vartheta}{\rho}}{2}\right)$$

$$= -(2\rho)^{-\frac{4}{3}} (\vartheta + \rho)^{\frac{1}{3}} (3\vartheta - \rho).$$

Thus;

$$\tilde{y}_{IIs} = \frac{\Gamma\left(\frac{5}{6}\right)\Gamma\left(-\frac{2}{3}\right)}{2^{\frac{5}{3}}\Gamma\left(-\frac{1}{2}\right)\Gamma\left(\frac{2}{3}\right)} \rho^{-3} \left\{(\rho+\vartheta)^{\frac{1}{3}}(3\vartheta-\rho) + (\rho-\vartheta)^{\frac{1}{3}}(3\vartheta-\rho)\right\}.$$

Since the normalization is arbitrary we can choose as the basic singular solution

$$\tilde{y}_s = \rho^{-3} \left\{ (\rho + \vartheta)^{\frac{1}{3}} (3\vartheta - \rho) + (\rho - \vartheta)^{\frac{1}{3}} (3\vartheta + \rho) \right\} ,$$

where

$$\rho^2 = \vartheta^2 - \frac{4}{3} w^3 .$$

(4.1.71)

In application the singular solution which represents the far field can be multiplied by a "source" strength Q. Now we can see that as the limit characteristic is approached ($\rho \to 0$, ϑ fixed) $\tilde{y}_s$ approaches a finite value.

$$(\rho + \vartheta)^{\frac{1}{3}} = \vartheta^{\frac{1}{3}} \left(1 + \frac{\rho}{\vartheta} \right)^{\frac{1}{3}}$$

$$= \vartheta^{\frac{1}{3}} \left\{ 1 + \frac{1}{3} \frac{\rho}{\vartheta} + \frac{\frac{1}{3} \left(-\frac{2}{3} \right)}{2} \left(\frac{\rho}{\vartheta} \right)^2 + \frac{\frac{1}{3} \left(-\frac{3}{3} \right) \left(-\frac{5}{3} \right)}{3!} \left(\frac{\rho}{\vartheta} \right)^3 + \cdots \right\}$$

Thus, from (4.1.71)

$$\tilde{y}_s = \vartheta^{\frac{1}{3}} \rho^{-3} \left\{ 3\vartheta \left(\frac{2}{3} \frac{\rho}{\vartheta} + \frac{10}{81} \frac{\rho^3}{\vartheta^3} \right) - 2\rho \left(1 - \frac{1}{9} \frac{\rho^2}{\vartheta^2} + \cdots \right) \right\}$$

$$\to \frac{16}{27} \vartheta^{-\frac{5}{3}} \quad \text{as} \quad \rho \to 0 .$$

Finally using the equations (4.1.47) it is possible to find x_s

$$\tilde{y}_s(\rho, \alpha) = \rho^{-\frac{5}{3}} \left\{ (1 + \sin \alpha)^{\frac{1}{3}} (3 \sin \alpha - 1) + (1 - \sin \alpha)^{\frac{1}{3}} (3 \sin \alpha + 1) \right\}$$

$$\frac{\partial \tilde{y}_s}{\partial \alpha} = \rho^{-\frac{5}{3}} \cos \alpha \left\{ \begin{array}{l} \frac{1}{3} (1 + \sin \alpha)^{-\frac{2}{3}} (3 \sin \alpha - 1) + 3(1 + \sin \alpha)^{\frac{1}{3}} \\[2mm] -\frac{1}{3} (1 - \sin \alpha)^{-\frac{2}{3}} (3 \sin \alpha - 1) + 3(1 - \sin \alpha)^{\frac{1}{3}} \end{array} \right\}$$

$$= \frac{4}{3} \rho^{-\frac{5}{3}} \cos \alpha \left\{ (2 + 3 \sin \alpha)(1 - \sin \alpha)^{-\frac{2}{3}} + (2 - 3 \sin \alpha)(1 - \sin \alpha)^{-\frac{2}{3}} \right\} ,$$

$$\frac{\partial x_s}{\partial \rho} = \left(\frac{3}{2} \right)^{\frac{1}{3}} \frac{4}{3} \rho^{-\frac{7}{3}} \cos^{\frac{4}{3}} \alpha$$

$$\times \left\{ (2 + 3 \sin \alpha)(1 + \sin \alpha)^{-\frac{2}{3}} + (2 - 3 \sin \alpha)(1 - \sin \alpha)^{-\frac{2}{3}} \right\} ,$$

$$x_s = -\left(\frac{3}{2}\right)^{\frac{1}{3}} \rho^{-\frac{4}{3}} \cos^{\frac{4}{3}} \alpha$$

$$\times \left\{ (2 + 3\sin\alpha)(1 + \sin\alpha)^{-\frac{2}{3}} + (2 - 3\sin\alpha)(1 - \sin\alpha)^{-\frac{2}{3}} \right\}.$$

Finally, in terms of $\rho^2 = \vartheta^2 - \frac{4}{9}w^3$, and ϑ

$$\boxed{x_s = -\frac{2}{3}w^2\rho^{-3}\left\{ (2\rho + 3\vartheta)(\rho + \vartheta)^{-\frac{2}{3}} + (2\rho - 3\vartheta)(\rho - \vartheta)^{-\frac{2}{3}} \right\}.}\qquad (4.1.72)$$

For example along $\vartheta = 0$, $\tilde{y} = 0$, $x_s = -\dfrac{8}{3}\dfrac{1}{w^2}\left(\dfrac{9}{4}\right)^{\frac{5}{4}} \to -\infty$ as $w \to 0$. The decay of the disturbance ahead of the object is rather slow $w \sim \dfrac{1}{\sqrt{-x}}$. Expanding x_s in a similar way as $\tilde{y}_s$ shows that on the limit characteristic

$$x_s \to A_s^{\frac{1}{5}}\frac{5 \cdot 2^{\frac{5}{3}}}{3^{\frac{8}{3}}}\vartheta^{-\frac{4}{3}}.\qquad (4.1.77)$$

Thus, for the limit characteristic,

$$\xi_L = \frac{x_s}{\tilde{y}^{\frac{4}{5}}} = A_s^{\frac{1}{5}}\frac{5 \cdot 2^{\frac{5}{3}}\, 3^{\frac{12}{5}}}{3^{\frac{8}{3}}\, 2^{\frac{16}{5}}}.\qquad (4.1.78)$$

For a standard solution

$$\xi_L = 1, \qquad A_s = 5^{-5}2^{\frac{23}{3}}3^{\frac{4}{3}}.\qquad (4.1.78)$$

Parametric Representation.

It is also possible to find a parametric form for the velocity components in another way based on the hodograph solutions and some remarkable properties uncovered by Frankl [4.1.2]. From the velocity components the potential expressed by $f(\xi)$ can also easily be given a parametric representation. These representations have analogues in the axisymmetric case and are useful in further developments.

The method relies on the fact that the nozzle flow of Section 3.7 and the far field solutions are both members of a family of hypergeometric functions

expressible in algebraic form and related by differentiations. The components of
the nozzle flow (3.7.23, 24) can be rewritten

$$w(x, \tilde{y}) = x + \frac{1}{2}\tilde{y}^2 \tag{4.1.79}$$

$$\vartheta(x, \tilde{y}) = x\tilde{y} + \frac{1}{6}\tilde{y}^3 \tag{4.1.80}$$

where the substitution $(\gamma + 1)^{1/2}x^*$, $(\gamma + 1)^{1/2}y = \tilde{y}$ has been made. Because of
the homogeneity of the hodograph system (4.1.24) in $(x, \tilde{y})$ this is also a solution.
Replacing x in (4.1.80) we have the cubic relation mentioned earlier

$$\tilde{y}^3 - 3w\tilde{y} + 3\vartheta = 0 \quad \text{for} \quad \tilde{y} = \tilde{y}_N(w, \vartheta) \tag{4.1.81}$$

which we now regard as applying for the hodograph solution $\tilde{y} = \tilde{y}_N(w, \vartheta)$, the
nozzle flow. For this flow $\tilde{y}$ is evidently anti-symmetric in ϑ. We can see from
(4.1.79, 80) or from (4.1.81) that this solution is a hodograph similarity solution
of the type expressed by $\tilde{y}_I$, $\tilde{y}_{II}$ in (4.1.56, 57). In fact $\tilde{y}_N = \vartheta^{1/3} fn\left(\dfrac{w}{\vartheta^{2/3}}\right)$.
Thus because of the symmetry $\tilde{y}_N$ is represented by $\tilde{y}_{II}$ with $\beta = -\frac{1}{2}$

$$\tilde{y}_N = \rho^{\frac{1}{3}} \sin \alpha\, F\left(\frac{1}{3}, \frac{5}{6}; \frac{3}{2}; \sin^2 \alpha\right). \tag{4.1.82}$$

Since $\tilde{y}(w, \vartheta)$ is the real root of (4.1.81) for $w < 0$, F is expressible in finite
algebraic terms. Since $\tilde{y}(w, \vartheta)$ is a solution of the Tricomi equation (4.1.43), $\dfrac{\partial \tilde{y}}{\partial \vartheta}$
is also a solution. Thus, solutions are

$$\tilde{y}_N = \vartheta^{\frac{1}{3}} fn\left(\frac{w}{\vartheta^{\frac{2}{3}}}\right) \qquad \text{anti-symmetric in} \quad \vartheta\,,$$

$$\frac{\partial \tilde{y}_N}{\partial \vartheta} = \vartheta^{-\frac{2}{3}} fn\left(\frac{w}{\vartheta^{\frac{2}{3}}}\right) \qquad \text{symmetric in} \quad \vartheta\,,$$

$$\frac{\partial^2 \tilde{y}_N}{\partial \vartheta^2} = \vartheta^{-\frac{5}{3}} fn\left(\frac{w}{\vartheta^{\frac{2}{3}}}\right) \qquad \text{anti-symmetric in} \quad \vartheta\,.$$

This last expression has the behavior $\vartheta^{-5/3}$ along the similarity curves of the far
field singular solution $\tilde{y}_{II_s}$ (4.1.67) and thus must be proportional to it. This
implies (as we have already seen) that $\tilde{y}_{II_s}$ can be expressed in finite algebraic
terms. A parametric representation of the singular solution $\tilde{y}_s$ can be found for
$\xi = \dfrac{x}{\tilde{y}^{4/5}}$, $w\tilde{y}^{2/5}$, $\vartheta\tilde{y}^{3/5}$ based on the cubic (4.1.81). Using B_s as a scale factor
let

$$\tilde{y}_s = B_s \frac{\partial^2 \tilde{y}_N}{\partial \vartheta^2}\,. \tag{4.1.83}$$

Note

$$3\tilde{y}_N^2 \frac{\partial \tilde{y}_N}{\partial \vartheta} - 3w \frac{\partial \tilde{y}_N}{\partial \vartheta} + 3 = 0$$

or

$$\frac{\partial \tilde{y}_N}{\partial \vartheta} = \frac{-1}{\tilde{y}_N^2 - w}$$

$$\frac{\partial^2 \tilde{y}_N}{\partial \vartheta^2} = \frac{2\tilde{y}_N \frac{\partial \tilde{y}_N}{\partial \vartheta}}{(\tilde{y}_N^2 - w)^2} = -\frac{2\tilde{y}_N}{(\tilde{y}_N^2 - w)^3} \tag{4.1.84}$$

$$\tilde{y}_s = -2B_s \frac{\tilde{y}_N}{(\tilde{y}_N^2 - w)^3} \ . \tag{4.1.85}$$

We also need x_N and $\dfrac{\partial^2 x_N}{\partial \vartheta^2}$, and these are also solutions. It follows from (4.1.79) that

$$x_N(w, \vartheta) = w - \frac{1}{2}\tilde{y}_N^2(w, \vartheta) \ .$$

Then

$$\frac{\partial x_N}{\partial \vartheta} = -\tilde{y}_N \frac{\partial \tilde{y}_N}{\partial \vartheta} \ ,$$

$$\frac{\partial^2 x_N}{\partial \vartheta^2} = -\tilde{y}_N \frac{\partial^2 \tilde{y}_N}{\partial \vartheta^2} - \left(\frac{\partial \tilde{y}_N}{\partial \vartheta}\right)^2 = \frac{2\tilde{y}_N^2}{(\tilde{y}_N^2 - w)^3} - \frac{1}{(\tilde{y}_N^2 - w)^2} = \frac{\tilde{y}_N^2 + w}{(\tilde{y}_N^2 - w)^3} \ .$$

Thus

$$x_s = B_s \frac{\tilde{y}_N^2 + w}{(\tilde{y}_N^2 - w)^3} \ . \tag{4.1.86}$$

Now a parameter s, constant on similarity curves is introduced by

$$s = \frac{\tilde{y}_N^2}{\tilde{y}_N^2 - w} \tag{4.1.87}$$

or

$$w = \tilde{y}_N^2 \left(\frac{s-1}{s}\right) \ ,$$

and

$$\tilde{y}_N^2 - w = \tilde{y}_N^2 \left(1 - \frac{s-1}{s}\right) = \frac{\tilde{y}_N^2}{s} \ , \quad \tilde{y}_N^2 + w = \tilde{y}_N^2 \left(\frac{2s-1}{s}\right) \ .$$

For $\xi = \dfrac{x_s}{\tilde{y}_s^{4/5}}$ we have

$$\xi = \frac{B_s^{\frac{1}{4}}}{2^{\frac{4}{5}}} \frac{\tilde{y}_N^2 + w}{(\tilde{y}_N^2 - w)^3} \frac{(\tilde{y}_N^2 - w)^{\frac{12}{5}}}{\tilde{y}_N^{\frac{4}{5}}} = \frac{B_s^{\frac{1}{5}}}{2^{\frac{4}{5}}} \left(\frac{2s-1}{s}\right) s^{\frac{3}{5}}$$

or

$$\boxed{\frac{x_s}{\tilde{y}_s^{\frac{4}{5}}} = \xi = \frac{1}{2}(2B_s)^{\frac{1}{5}}(2s-1)s^{-\frac{2}{5}} .}$$

(4.1.88)

$s = 0$ corresponds to $\xi = -\infty$, $s = \frac{1}{2}$ to $\xi = 0$. Corresponding velocity components can be found

$$w\tilde{y}_s^{\frac{2}{5}} = \hat{y}_N^2\left(\frac{s-1}{s}\right)B_s^{\frac{2}{5}}2^{\frac{2}{5}}\frac{\tilde{y}_N^{\frac{2}{5}}}{(\tilde{y}_N^2-w)^{\frac{6}{5}}} = 2^{\frac{2}{5}}B_s^{\frac{2}{5}}\left(\frac{s-1}{s}\right)s^{\frac{6}{5}}$$

or

$$\boxed{w\tilde{y}_s^{\frac{2}{5}} = (2B_s)^{\frac{2}{5}}(s-1)s^{\frac{1}{5}} .}$$

(4.1.89)

$s = 1$ thus corresponds to the sonic line $\xi = \xi^*$.

The limit characteristic follows from the condition

$$\left(\frac{dx}{dy}\right)_L = \sqrt{w} = k\xi_L\tilde{y}^{k-1} = \frac{4}{5}\tilde{y}^{-\frac{1}{5}}\xi_L \quad \text{for} \quad k = \frac{4}{5}$$

or

$$w\tilde{y}_s^{\frac{2}{5}} = \frac{16}{25}\xi_L^2 .$$

(4.1.90)

Using (4.1.89, 88)

$$(s-1)s^{\frac{1}{5}} = \frac{16}{25} \times \frac{1}{4}(2s-1)^2 s^{-\frac{4}{5}} ,$$

or s is a root of

$$s^2 - s - \frac{4}{9} = \left(s - \frac{4}{3}\right)\left(1 + \frac{1}{2}\right) = 0 .$$

Thus

$$\boxed{s_L = \frac{4}{3} .}$$

The value of ξ_L depends on B_s. For a standard solution with $\xi_L = 1$

$$2B_s = 2^9 3^3 5^{-5} .$$

$\vartheta\tilde{y}_s^{3/5}$ follows from (4.1.81)

$$\vartheta\tilde{y}_s^{\frac{3}{5}} = \left(\frac{-\tilde{y}_N^3 + 3w\tilde{y}_N}{3}\right)\left(\frac{(-2B_s)^{\frac{3}{5}}\tilde{y}_N^{\frac{6}{5}}}{(\tilde{y}_N^2 - w)^{\frac{9}{5}}}\right)$$

$$= \tilde{y}_N^{\frac{8}{5}}\left(\frac{\tilde{y}_N^2 - 3\tilde{y}_N^2\frac{s-1}{s}}{3}\right)\left(\frac{(2B_s)^{\frac{3}{5}}s^{\frac{9}{5}}}{\tilde{y}_N^{\frac{18}{5}}}\right)$$

$$\boxed{\vartheta \tilde{y}_s^{\frac{3}{5}} = +\frac{(2B_s)^{\frac{3}{5}}}{3}(3 - 2s)s^{\frac{4}{5}} \; .}$$

(4.1.91)

Since the velocity components are known $f(\xi)$ which gives the potential found in (4.1.14) and (4.1.16) is also easily found. It follows from (4.1.14) since $\kappa = \frac{4}{5}$ that

$$\frac{2}{5}f = \vartheta \tilde{y}_s^{\frac{3}{5}} + \frac{4}{5}\xi f' = \vartheta \tilde{y}_s^{\frac{3}{5}} + \frac{4}{5}\xi w \tilde{y}_s^{\frac{2}{5}} \; ,$$

$$f = \frac{5}{2}\vartheta \tilde{y}_s^{\frac{3}{5}} + 2\xi w \tilde{y}_s^{\frac{2}{5}} \; .$$

(4.1.92)

Using (4.1.88, 89, 91) we find

$$f = \frac{5}{2}\frac{(2B_s)^{\frac{3}{5}}}{3}(3 - 2s)s^{\frac{4}{5}} + (2B_s)^{\frac{1}{5}}(2s - 1)s^{-\frac{2}{5}}(2B_s)^{\frac{2}{5}}(s - 1)s^{\frac{1}{5}}$$

$$= (2B_s)^{\frac{3}{5}}s^{-\frac{1}{5}}\left\{ \left(\frac{5}{2} - \frac{5}{3}s\right)s + (2s - 1)(s - 1)\right\} \; ,$$

$$\boxed{f = (2B_s)^{\frac{3}{5}}\left\{\frac{s^2}{3} - \frac{s}{2} + 1\right\}s^{-\frac{1}{5}} \; .}$$

(4.1.93)

This parametric representation can be continued only up to the tail shock. Applying the shock conditions as before and using

$$s = \frac{\tilde{y}_N^2}{w - \tilde{y}_N^2} > 0 \quad \text{downstream of the shock}$$

there is a jump in s at the shock. The result is (cf. [4.1.2])

downstream $\quad s_b = \frac{1}{6}(5\sqrt{3} - 8) = -11 \; , \quad$ upstream $s_a = \frac{1}{6}(5\sqrt{3} + 8) = 2.78$

(4.1.94)

and the parametric representation downstream of the shock is

$$\xi = x\tilde{y}_s^{-\frac{4}{5}} = \frac{1}{2}(2C_s)^{\frac{1}{5}}(2C_s + 1)s^{-\frac{2}{5}} \; ,$$

$$w\tilde{y}_s^{\frac{2}{5}} = (2C_s)^{\frac{2}{5}}(2C_s + 1)s^{\frac{1}{5}} \; ,$$

$$\vartheta \tilde{y}_s^{\frac{3}{5}} = -(2C_s)^{\frac{3}{5}}(3 + 2s)s^{\frac{4}{5}} \; ,$$

$$f = (2C_s)^{\frac{3}{5}}(2s^2 + 3s + 6)s^{-\frac{1}{5}} \; ,$$

(4.1.95)

and

$$\frac{C_s}{B_s} = \frac{9\sqrt{3} + 1}{9\sqrt{3} - 1} = 1.14 \; .$$

The course of the parameter s is then summarized in the following table.

ξ	$-\infty$	0	ξ^*	ξ_L	ξ_s	$+\infty$
s	0	$\frac{1}{2}$	1	$\frac{4}{3}$	$s_a\|s_b$	0

These sets of relations provide complete information about the far field solution for the planar case for $M_\infty = 1$.

Higher Order Terms

The orders of the subsequent terms in ϕ are found by assuming that as $\tilde{y} \to \infty$ for ξ fixed,

$$(\gamma + 1)\phi_0 = \tilde{y}^{\frac{2}{5}} f(\xi) + \tilde{y}^{\sigma_0} c_0 f_0(\xi) + \tilde{y}^{\sigma_1} c_1 f_1(\xi) + \cdots , \qquad (4.1.96)$$

where $\sigma_1 < \sigma_0 < 2/5$. Since these terms will be needed in Section 6.3 as well as in Section 5.6, we proceed to find them here.

Substitution into (4.1.1) shows that if $\alpha = \sigma_{0,1}$, $g = f_{0,1}$, then for $\sigma_1 > 2\sigma_0 - 2/5$, g satisfies the equation

$$\left(f' - \frac{16}{25}\xi^2 \right) g'' + \left(f'' + \frac{4}{5}\left(2\alpha - \frac{9}{5} \right)\xi \right) g' - \alpha(\alpha - 1)g = 0 , \qquad (4.1.97)$$

which has a regular singular point at the limit characteristic $\left(f'(\xi_L) = \frac{16}{25}\xi_L^2 \right)$, and a singular point along the entire negative x-axis ($\xi = -\infty$). This then is a linear eigenvalue problem, we must find those values of α for which there are solutions g such that $\tilde{y}^\alpha g$ is smooth from the negative x-axis ($\xi \to -\infty$) through the limit Mach line ($\xi_L = 1$).

Equation (4.1.97) is reducible to hypergeometric form using several changes of variable including just the parametric representation (4.1.88), (4.1.93).

Under one more change of variables,

$$g = s^{-\frac{\alpha}{2}} \ell(s) , \quad t = \frac{3}{4}s , \quad h(t) = \ell(s) , \qquad (4.1.98)$$

the equation that h must satisfy is

$$(1 - t)th'' + \left(\frac{1}{2} - \frac{1}{3}(5\alpha + 4)t \right)h' + \frac{5}{12}\alpha(5\alpha - 2)h = 0 , \qquad (4.1.99)$$

for $0 \le t \le 1$. This is a hypergeometric equation with

$$a = \frac{5}{2}\alpha \,, \quad b = \frac{1}{6}(2 - 5\alpha) \,, \quad c = \frac{1}{2} \,.$$

Two linearly independent solutions of (4.1.99) are

$$h^S(t) = F\left(\frac{5}{2}\alpha, \ \frac{2 - 5\alpha}{6} \ ; \ \frac{1}{2} \ ; \ t\right) \,,$$

$$h^A(t) = t^{\frac{1}{2}} F\left(\frac{5\alpha - 1}{2}, \ \frac{5}{6}(1 - \alpha); \ \frac{3}{2} \ ; \ t\right) \,,$$
(4.1.100)

where

$$F(a, b, c, t) = \sum_{r=0}^{\infty} \frac{\Gamma(a + n)\,\Gamma(b + n)}{\Gamma(a)\,\Gamma(b)\,\Gamma(c + n)} \Gamma(c) t^n \,.$$

Thus, if either a or b is a negative integer the seriese terminates after a finite number of terms and $F(a, b, c; t)$ is a polynomial in t and hence smooth from $t = 0$ to $t = 1$. In fact, $F(a, b, c; t)$ is smooth through $t = 1$ if and only if a or b is a negative integer. This can be seen from the analytic continuation case since, if $c - a - b$ is not an integer,

$$F(a, b; c; t) = \frac{\Gamma(c)\,\Gamma(c - a - b)}{\Gamma(c - a)\,\Gamma(c - b)} F(a, b; 1 + a + b - c; 1 - t)$$

$$+ (1 - t)^{c-a-b} \frac{\Gamma(c)\,\Gamma(a + b - c)}{\Gamma(a)\,\Gamma(b)} F(c - a, c - b; 1 - a - b + c; 1 - t) \,,$$

if $c - a - b$ is an integer,

$$F(a, b; a + b + n; t)$$

$$= \frac{\Gamma(n)\,\Gamma(a + b + n)}{\Gamma(a + n)\,\Gamma(b + n)} \sum_{k=0}^{n-1} \frac{(a)_k (b)_k}{k!\,(1 - n)_k} (1 - t)^k$$

$$+ (-1)^n \frac{\Gamma(a + b + n)}{\Gamma(a)\,\Gamma(b)} \sum_{k=0}^{\infty} \frac{(a + n)_k (b + n)_k}{k!\,(k + n)!} \{h_{abnk} - \log(1 - t)\}(1 - t)^{n+k} \,,$$

where h_{abnk} is a number, $\sum_{k=0}^{n-1}$ is zero if $n - 1 \le -1$. Thus, $h^s(t)$ (respectively $h^A(t)$) is regular through $t = 1$ if and only if,

$$\alpha = -\frac{2}{5}n \,, \quad \alpha = \frac{2 + 6n}{5} \quad (\text{respectively } \alpha = \frac{-2n - 1}{5} \,, \quad \alpha = 1 + \frac{6}{5}n) \,,$$
(4.1.101)

$$n = 0, 1, 2, \ldots .$$

This takes care of the singular behavior at the limiting characteristic. We must now check which of these solutions satisfy $\tilde{y}^\alpha g$ is smooth along the negative x-axis.

Now, as $\xi \to -\infty$ ($x < 0$, $\tilde{y} \sim 0$)

$$t = \frac{3}{4}s \sim \frac{3}{4}a_1^{\frac{1}{2}} 2^{-\frac{5}{2}} \xi^{-\frac{5}{2}}$$

$$\sim \frac{3}{4}a_1^{\frac{1}{2}} 2^{-\frac{5}{2}} x^{-\frac{5}{2}} \tilde{y}^2 \equiv bx^{-\frac{5}{2}} \tilde{y}^2 .$$

So, we have that as we approach the negative x axis,

$$\tilde{y}^\alpha g(s) = \tilde{y}^\alpha s^{-\frac{\alpha}{2}} h(t)$$

$$\sim x^{-\frac{5\alpha}{4}} bh(t) .$$

Since $h^S(t)$ is a polynomial in t, hence is smooth in $\tilde{y}^2$ as $\tilde{y} \to 0$, and since $h^A(t)$ is a polynomial in $t^{1/2}$, hence is smooth in $\tilde{y}$ as $\tilde{y} \to 0$, then both solutions $\tilde{y}^\alpha h^S(t)$ and $\tilde{y}^\alpha h^A(t)$ are smooth from $\tilde{y} = 0$, $x < 0$ through the limit characteristic as long as (4.1.101) is satisfied.

Therefore in (4.1.96), using (4.1.101), we have

$$\sigma_0 = 0, \quad \sigma_1 = -\frac{1}{5}, \quad f_0 = 1, \quad h_1 = t \Rightarrow f_1 = \frac{\sqrt{3}}{2}s^{\frac{3}{5}} .$$

Returning now to the group property for f (4.1.16) we see that the general far field is given by

$$(\gamma + 1)\phi = \tilde{y}^{\frac{2}{5}} \frac{f(a\xi)}{a^3} + c_0 + c_1 \tilde{y}^{-\frac{1}{5}} \frac{f_1(a\xi)}{a^3} + \cdots \qquad (4.1.102a)$$

where

$$a\xi = \frac{1}{2}a_1^{\frac{1}{5}} s^{-\frac{2}{5}} (2s - 1) ,$$

$$f = \frac{1}{6}a_1^{\frac{3}{5}} s^{-\frac{1}{5}} (2s^2 - 3s + 6) , \qquad (4.1.102b)$$

$$f_1 = \frac{\sqrt{3}}{2}s^{\frac{3}{5}} ,$$

and

$$a_1 = 2^9 \cdot 3^3 \cdot 5^{-5} .$$

The representation of the higher order terms in the hodograph plane is quite simple to find. Since the governing equation (4.1.43), Tricomi's equation, is linear corrections satisfy the same equation as the leading terms. Hence, with

$$\tilde{y} = \tilde{y}_s(\rho, \alpha) + \tilde{y}_c(\rho, \alpha) + \cdots ,$$

for $\vartheta \to 0$, $\dfrac{\vartheta}{\rho}$ fixed, where $\tilde{y}_s$ is given in (4.1.68), $\tilde{y}_c$ is given by one of the solutions (4.1.56) or (4.1.57). Continuation of y_{II} through the limit characteristic showed that there is no singularity on the limit characteristic if

$$\beta = -\frac{1}{2} + 2k , \quad \text{or} \quad \beta = -\frac{1}{6} - \frac{2k}{3} , \quad k = 1, 2, \ldots . \tag{4.1.103a}$$

Note the second case was not discussed in (4.1.64) since there only singular solutions were considered. Similar analysis on y_I shows there is no singularity on the limit characteristic if

$$\beta = -\frac{3}{2} + 2k , \quad \text{or} \quad \beta = \frac{1}{2} - \frac{2k}{3} , \quad k = 1, 2, \ldots . \tag{4.1.103b}$$

Thus the next most singular solution after $\tilde{y}_s$ is that with $\beta = \frac{1}{2}$, so that as $\vartheta \to 0$, $\dfrac{\vartheta}{\rho}$ fixed,

$$\tilde{y} = K\rho^{-\frac{5}{3}} \sin \alpha F\left(\frac{4}{3}, -\frac{1}{6}; \frac{3}{2}; \sin^2 \alpha\right) + b_1 \rho^{-\frac{2}{3}} F\left(\frac{1}{3}, -\frac{1}{6}; \frac{1}{2}; \sin^2 \alpha\right) + \cdots . \tag{4.1.104}$$

The constants a, c_1 in the far field expansion (4.1.102) can be found by using various conservation laws. The problem is to connect the far field behavior with the actual shape of the airfoil. We will use some hodograph representation so let $w = \phi_{0x}$, $\vartheta = \phi_{0\tilde{y}}$ be the usual hodograph coordinates so that in the far field $w \sim \tilde{y}^{-2/5} f'(a\xi)/a^2$, and $\vartheta \sim \tilde{y}^{-3/5}\{\frac{2}{5} f(a\xi) - \frac{4}{5} a\xi f'(a\xi)\}/a^3$.

Now, we want to look for pairs of functions $(-X, Y)$, where $X(w, \vartheta)$, $Y(w, \vartheta)$, which are divergence free i.e.,

$$\frac{\partial}{\partial x}(-X) + \frac{\partial}{\partial \tilde{y}}(Y) = 0 . \tag{4.1.105}$$

Once these functions are found, then of course

$$0 = \oint (X \, d\tilde{y} + Y \, dx) \tag{4.1.106}$$

around any closed contour which does not encircle the body or the wake.

It can be shown [4.1.8] that any two functions $X(w, \vartheta)$, $Y(w, \vartheta)$, which satisfy the Tricomi system

$$\left\{ \begin{aligned} (-X)_w + wY_\vartheta = 0 \\ (-X)_\vartheta + Y_w = 0 \end{aligned} \right\} \tag{4.1.107}$$

generate a conservation law of the form (4.1.105). In fact in [4.1.8] Germain generates a whole family of conservation laws which have the basic form

$$Y = \rho^{-\frac{5}{3}} \left\{ \rho^{\frac{1}{6} + \beta} \sin \alpha \, F\left(\frac{1}{12} + \frac{\beta}{12}, \frac{7}{12} - \frac{\beta}{2}; \frac{3}{2}; \sin^2 \alpha \right) \right\}, \tag{4.1.108}$$

where F is the hypergeometric function, and the appropriate associated x.

To compute a we consider (4.1.108) with $\beta = 3/2$. In particular the pair,

$$Y = (\rho + \vartheta)^{\frac{1}{3}} (3\vartheta - \rho) + (\rho - \vartheta)^{-\frac{1}{3}} (3\vartheta + \rho),$$

$$X = -\frac{2}{3} w^2 \left\{ (\rho + \vartheta)^{-\frac{2}{3}} (3\vartheta + 2\rho) + (\rho - \vartheta)^{-\frac{2}{3}} (2 - 3\vartheta) \right\},$$

is related to the hodograph form of the free stream singularity for two dimensional sonic flow (4.1.71). It is clear then that $(-X, Y)$ satisfies the Tricomi system (4.1.107) and furthermore on the limit characteristic, $\vartheta = \frac{2}{3} w^{3/2}$, $\rho = 0$, hence $X = Y = 0$.

Now, choose the path of integration in (4.1.106) $C = I_B + I_L + I_1 + I_2 + I_3$ as shown in Figure (4.1.8).

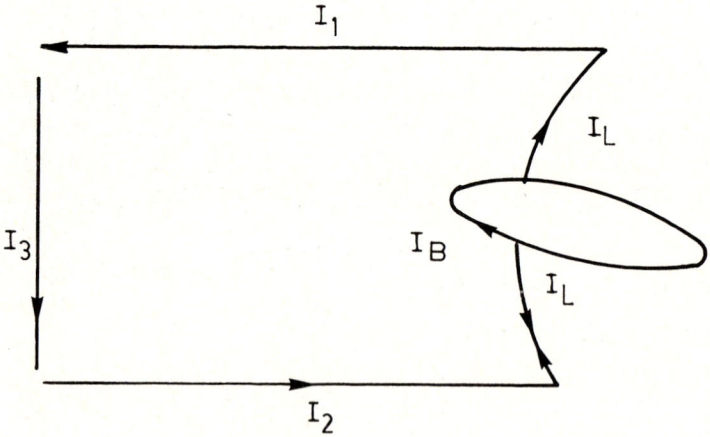

Figure 4.1.8
Control Surface

In particular I_L is the integral along the limit characteristic, I_B along the body, I_1, I_2 along the horizontal sides, I_3 along the vertical side. Then from (4.1.106)

$$0 = \oint (X\,d\tilde{y} + Y\,dx) = \int_{I_B} (X\,d\tilde{y} + Y\,dx) + \int_{I_1 \cup I_2} Y\,dx + \int_{I_3} X\,d\tilde{y}, \quad (4.1.109)$$

since $X \equiv Y \equiv 0$ along I_L. It is easily checked that $X = 0(\tilde{y}^{-1})$ as $\tilde{y} \to -\infty$, hence as I_3 moves off to negative infinity, $\int_{I_3} X\,d\tilde{y} \to 0$. Hence, (4.1.109) gives

$$\int_{x_L^+}^{-\infty} Y(x, y_1)\,dx + \int_{-\infty}^{x_L^-} Y(x, -y_1)\,dx = -\int_{\text{Body}} (X\,d\tilde{y} + Y\,dx), \quad (4.1.110)$$

where x_L^+, x_L^- are the coordinates of the upper, lower limit characteristics with $\tilde{y} = \tilde{y}_1$, $-\tilde{y}_1$ respectively. Now as $|y_1| \to 0$, $x_L^+ \to x_L^-$, that is the far field is symmetric to first order. Furthermore, this means that $Y(x, y_1) = -Y(x, -y_1)$. Hence (4.1.110) reads

$$-2 \int_{-\infty}^{x_L} Y(x, y_1)\,dx = \int_{\text{Body}} (X\,d\tilde{y} + Y\,dx). \quad (4.1.111)$$

We can compute the left hand side of (4.1.111) exactly. In fact a (or rather K, the hodograph analogue of a) was computed by Germain. We proceed with a slightly different method of evaluation in order that we can also use the simplified technique to calculate the integrals involved in computing c_1.

In terms of the variable s (4.1.102b) we can rewrite the left hand side of (4.1.111) as

$$-2 \int_{-\infty}^{x_L} Y(x, y_1)\,dx = \frac{-2 \cdot 3^{-\frac{4}{3}} a_1}{5 a^5(z^*)} \int_0^{\frac{4}{3}} m(s)\,ds,$$

where

$$m(s) = \frac{3s+1}{s} \left(\left\{ (-3s+4)^{\frac{1}{2}} + s^{\frac{1}{2}}(-2s+3) \right\}^{\frac{1}{3}} \left\{ -(-3s+4)^{\frac{1}{2}} + 3s^{\frac{1}{3}}(2s+3) \right\} \right.$$

$$\left. + \left\{ (-3s+4)^{\frac{1}{2}} - s^{\frac{1}{2}}(-2s+3) \right\}^{\frac{1}{3}} \left\{ (-3s+4)^{\frac{1}{2}}(-2s+3) \right\} \right).$$

$$(4.1.112)$$

The integral can be further simplified by introducing yet one more change of variable

$$s = \frac{4}{3} \cos^2 \theta. \quad (4.1.113)$$

Then

$$\int_0^{\frac{4}{3}} m(s)\, ds = 2^{\frac{19}{3}} 3^{-\frac{1}{2}} \int_0^{\frac{\pi}{2}} \sin^4\theta(1 + 4\cos^2\theta)\, d\theta = \frac{2^{\frac{7}{3}} 5\pi}{3^{\frac{1}{2}}} ,$$

so that

$$\int_{I_1 \cup I_2} Y\, dx = \frac{2^{\frac{37}{3}} 3^{\frac{7}{6}} \pi}{5^5 a^5 (z^*)} = -2 \int_{-\infty}^{x_L} Y(x, y_1)\, dx . \tag{4.1.114}$$

Thus, substituting into (4.1.111) we are able to derive an expression for a,

$$a^{-5} = \frac{5^5}{2^{\frac{37}{3}} 3^{\frac{7}{6}} \pi} \int_{\text{Body}} \left\{ (\rho + \vartheta)^{\frac{1}{3}} (3\vartheta - \rho) + (\rho - \vartheta)^{\frac{1}{3}} (3\vartheta + \rho) \right\} dx , \tag{4.1.115}$$

where the integral along the body extends from the lower limit characteristic around the nose to the upper limit characteristic. Note that ϑ is known on the body, it is the body shape, but w (hence ρ), as well as the location of the limit characteristics on the body are not known in advance. The constant a can not be determined until after the flow field has been calculated. In this respect it is analogous to the circulation Γ about a subsonic airfoil.

The situation for c_1 is similar. In this case we use the conservation law of the form (4.1.108) with $\beta = -\frac{1}{6}$. Hence

$$\begin{aligned} Y &= (\rho + \vartheta)^{\frac{1}{3}} + (\rho - \vartheta)^{\frac{1}{3}} , \\ X &= 3^{\frac{1}{2}} 2^{-\frac{4}{3}} \left\{ (\rho - \vartheta)^{\frac{2}{3}} - (\rho + \vartheta)^{\frac{2}{3}} \right\} . \end{aligned} \tag{4.1.116}$$

Again we integrate around the contour C as in Figure (4.1.8). As before $\int_{I_3} = 0$ as I_3 moves off to infinity, and $\int_{I_L} = 0$. Thus,

$$\int_{\text{Body}} Y\, dx + X\, d\tilde{y} = - \int_{I_1 \cup I_2} Y\, dx = 2 \int_{-\infty}^{x_L} Y(x, y^*)\, dx . \tag{4.1.117}$$

In terms of the s variables the right hand side is

$$\int_{I_1 \cup I_2} Y\, dx = \frac{-3^{\frac{1}{6}} c_1(z^*)}{a^2 5 a_1^{-\frac{1}{5}}} \int_0^{\frac{4}{3}} g(s)\, ds , \tag{4.1.118}$$

where

$$\begin{aligned} g(s) = \frac{1}{s} \Big(&\left\{ s^{\frac{1}{2}} (s - 3)(-3s + 4)^{-\frac{1}{2}} - (3s - 1) \right\} \left\{ (-3s + 4)^{\frac{1}{2}} + s^{\frac{1}{2}} (-2s + 3) \right\}^{-\frac{2}{3}} \\ &+ \left\{ s^{\frac{1}{2}} (s - 3)(-3s + 4)^{-\frac{1}{2}} + (3s - 1) \right\} \left\{ (-3s + 4)^{\frac{1}{2}} - s^{\frac{1}{2}} (2s + 3) \right\}^{-\frac{2}{3}} \Big) . \end{aligned}$$

Using the change of variables (4.1.113) we finally find

$$\int_0^{\frac{3}{4}} g(s)\, ds = \int_{\frac{\pi}{2}}^0 \frac{-3^{\frac{1}{2}}5}{2^{\frac{5}{3}}\sin\theta\cos\theta}\left(-\frac{\theta}{3}\sin\theta\cos\theta\right)d\theta = \frac{-2^{\frac{1}{3}}5\pi}{3^{\frac{1}{2}}}\ . \tag{4.1.119}$$

Thus, substituting this result into (4.1.117) and (4.1.118) we get

$$c_1 = \frac{-2^{\frac{22}{15}}3^{\frac{14}{15}}a^2}{5\pi} \int\limits_{\text{Body}} \left\{(\rho+\vartheta)^{\frac{1}{3}} + (\rho-\vartheta)^{\frac{1}{3}}\right\} dx\ , \tag{4.1.120}$$

where a is determined from (4.1.115) and the integral along the body passes from the lower limit characteristic around the nose to the upper limit characteristic.

Note that this analysis applies only to that portion of the flow in front of the shock surface. It was derived for the region in front of the limit characteristic, but can be continued back to the shock. It is difficult at this point to see how the lift enters, even in the ϕ_0 problem since, for example, there is no lift evident in the far field of ϕ_0. Hence, we continue to find the behavior behind the shock.

Now in front of the shock we had (4.1.102a, b) that as $\tilde{y} \to \infty$

$$\phi \sim \frac{\tilde{y}^{\frac{2}{5}} f(a\xi)}{a^3} + c_0 + c_1 \frac{\tilde{y}^{-\frac{1}{5}} f_1(a\xi)}{a^3} + O(\tilde{y}^{-\frac{2}{5}})\ , \tag{4.1.121}$$

where $\xi = \dfrac{x}{\tilde{y}^{\frac{4}{5}}}$. This expansion holds up to the shock. Behind the shock we expect an expansion of the form

$$\tilde{\phi} \sim \tilde{y}^{\frac{2}{5}}\frac{f(a\xi)}{a^3} + \tilde{y}^{n_1}\frac{\tilde{f}_1(a\xi)}{a^3} + \tilde{y}^{n_2}\frac{\tilde{f}_2(a\xi)}{a^3} + \cdots\ , \tag{4.1.122}$$

where, since $\tilde{\phi}$ and ϕ satisfy the same differential equation, (4.1.1), the $\tilde{f}_1$ satisfy equations similar to that satisfied by f_1, namely (4.1.97). Hence $\tilde{f}_1$ and $\tilde{f}_2$ satisfy

$$\left(f' - \frac{16}{25}a^2\xi^2\right)\tilde{f}_1'' + \left(f'' + \frac{4}{5}\left(2n_1 - \frac{9}{5}\right)a\xi\right)\tilde{f}_1' - n_1(n_1 - 1)\tilde{f}_1 = 0\ , \tag{4.1.123a}$$

$$\left(f' - \frac{16}{25}a^2\xi^2\right)\tilde{f}_2'' + \left(f'' + \frac{4}{5}\left(n_2 - \frac{9}{5}\right)a\xi\right)\tilde{f}_2' - n_2(n_2 - 1)\tilde{f}_2$$

$$= \begin{cases} 0 & \text{if } n_2 > 2n_1 - \dfrac{2}{5}\ , \\[2mm] -\tilde{f}_1'\tilde{f}_1'' & \text{if } n_2 = 2n_1 - \dfrac{2}{5}\ . \end{cases} \tag{4.1.123b}$$

The solution functions are not the same as those found for (4.1.102) since the boundary conditions are quite different. The limit characteristic (the singular point $f'(a\xi) = \frac{16}{25}a^2\xi_L^2$) is not in the domain of the $\tilde{f}_i$. Instead, the $\tilde{f}_i$ must satisfy appropriate jump conditions at the shock, as well as guarantee that $\tilde{\phi}_x$, $\tilde{\phi}_{\tilde{y}}$ are smooth through the positive x-axis, $(\xi = +\infty)$.

The shock conditions are (3.1.25), (3.1.26),

$$-\frac{1}{2}[\phi_x^2][\phi_x]_s + [\phi_{\tilde{y}}]_s^2 = 0 , \qquad (4.1.124)$$

$$[\phi]_s = 0 . \qquad (4.1.125)$$

The zeroth order shock locus can itself be expanded out in the far field. To first order it is the similarity curve ξ_0 (4.1.41) so that the actual location, ξ_S, for $\tilde{y} > 0$, is

$$\xi_S^+ = a\xi_S = \xi_0^+ + \xi_1^+ \tilde{y}^{m_1} + \xi_2^+ \tilde{y}^{m_2} + \cdots , \qquad (4.1.126)$$

where $m_2 < m_1 < 0$. Thus along the shock we have

$$ax_S = \xi_0^+ \tilde{y}^{\frac{4}{5}} + \xi_1^+ \tilde{y}^{\frac{4}{5}+m_1} + \xi_2^+ \tilde{y}^{\frac{4}{5}+m_2} + \cdots , \qquad (4.1.127)$$

or

$$a\,dx_S^+ = \left\{ \frac{4}{5}\xi_0^+ \tilde{y}^{-\frac{1}{5}} + \left(\frac{4}{5} + m_1\right) \xi_1^+ \tilde{y}^{-\frac{1}{5}+m_1} + \cdots \right\} d\tilde{y}_S . \qquad (4.1.128)$$

For any function $f(\xi)$ we have

$$f|_S = f|_{s_0} + \xi_1^+ \tilde{y}^{m_1} f'|_{s_0} + \xi_2^+ \tilde{y}^{m_2} f'|_{s_0} + \xi_1^+ \tilde{y}^{2m_1} \left.\frac{f''}{2}\right|_{s_0} + \cdots \qquad (4.1.129)$$

where

$$s_0 : a\xi = \xi_0^+ . \qquad (4.1.130)$$

Hence substituting the far field expansions (4.1.121), (4.1.122) into the shock jump conditions (4.1.124), (4.1.125) and using (4.1.129) we have that condition (4.1.125) is equivalent to

$$\tilde{y}^{\frac{2}{5}}[f] - \tilde{y}^{n_1}\tilde{f}_1 + \tilde{y}^{m_1+\frac{2}{5}}\xi_1^+[f'] + \xi_2^+\tilde{y}^{m_2+\frac{2}{5}}[f']$$
$$+ \xi^{+^2}\tilde{y}^{2m_1+\frac{2}{5}}[\frac{f''}{2}] - \tilde{y}^{n_2}\tilde{f}_2 - \tilde{y}^{m_1+n_1}\xi_1^+\tilde{f}_1' + \cdots = 0 , \qquad (4.1.131)$$

and condition (4.1.124) is equivalent to

$$
\tilde{y}^{-\frac{6}{5}}\left(-\frac{1}{2}\right)\left\{[f'^2] + 2\xi_1^+ \tilde{y}^{m_1}[f'f''] + 2\tilde{y}^{n_1-\frac{2}{5}}f_1'\tilde{f}_1\right\}
$$
$$
\times\left\{[f'] + \tilde{y}^{n_1-\frac{2}{5}}\tilde{f}_1 + \tilde{y}^{n_2-\frac{2}{5}}f_2' + \tilde{y}^{m_1}\xi_1^+[f'']\right\}
$$
$$
+ \tilde{y}^{-\frac{2}{5}}\left\{[\tfrac{2}{5}f] - \tfrac{4}{5}\xi_0^+[f']\right\}^2 2\tilde{y}^{m_1}\xi_1'\left\{\tfrac{2}{5}[f] - \tfrac{4}{5}\xi_0^+[f']\right\}\left\{-\tfrac{2}{5}[f'] - \tfrac{4}{5}\xi_0^+[f'']\right\}
$$
$$
+ \tilde{y}^{n_1-\frac{2}{5}}2\left\{\left(-n_1\tilde{f}_1 + \tfrac{4}{5}\xi_0^+\tilde{f}_1\right)\left(\tfrac{2}{5}[f] - \tfrac{4}{5}\xi_0^+[f']\right)\right\} + \cdots = 0 , \qquad (4.1.132)
$$

where all functions and all jumps are evaluated at the zeroth order shock location $\xi^+ = \xi_0^+$. Thus, the first order terms give

$$
[f]_{s_0} = 0 , \qquad (4.1.133)
$$

and

$$
-\frac{1}{2}[f'^2]_{s_0}[f']_{s_0} + \left\{\frac{2}{5}[f]_{s_0} - \frac{4}{5}\xi_0[f']_{s_0}\right\}^2 = 0 .
$$

Using (4.1.133) this last expression can be written as

$$
\langle f'\rangle_{s_0} = \left(\frac{4}{5}\xi_0^+\right)^2 . \qquad (4.1.134)
$$

Using (4.1.133) in the equation governing f, (4.1.97), one finds also

$$
\langle f''\rangle_{s_0} = \frac{4}{5}\xi_0^+ . \qquad (4.1.135)
$$

Before proceeding to the next order, the relative sizes of $m_1, n_1, 0$, must be fixed. Note that if $n_1, m_1 + \frac{2}{5} > 0$, and if:

(i) $m_1 + \frac{2}{5} < n_1$, then since $[f']_{s_0} \neq 0$ from (4.1.131), $\xi_1 = 0$, hence $\tilde{f}_1|_{s_0} = 0$. Then from (4.1.132) $\tilde{f}_1'|_{s_0} = 0$. Since $\tilde{f}_1$ satisfies a second order differential equation (4.1.123), with zero initial conditions, $\tilde{f}_1 \equiv 0$;

(ii) $m_1 + \frac{2}{5} < n_1$, then it can be shown, in a similar fashion to (i), that $\tilde{f}_1 \equiv 0$:

Thus, the first possibly interesting case is

$$
m_1 + \frac{2}{5} = n_1 > 0 . \qquad (4.1.136)
$$

With this condition, to order $\tilde{y}^{n_1}$, (4.1.131) gives

$$\tilde{f}_1|_{s_0} = -\xi_1^+[f']_{s_0} \,, \tag{4.1.137}$$

and to order $\tilde{y}^{n_1-8/5}$, (4.1.132) gives (using(4.1.133))

$$- \xi_1[f'f''] + \tilde{f}_1'f'[f'] + \frac{1}{2}[f'^2](+\tilde{f}_1' + [f''])$$

$$+ \frac{32}{25}\xi_0^+\xi_1^+[f']^2 + \frac{8}{5}\xi_0^+[f']\left(\frac{2}{5}[f'] + \frac{4}{5}\xi_0^+[f'']\right) \tag{4.1.138}$$

$$- 2\left(n_1\tilde{f}_1 - \frac{4}{5}\xi_0^+\tilde{f}_1'\right)\left(-\frac{4}{5}\xi_0^+[f']\right) = 0 \,.$$

Again all jumps are evaluated at s_0, all other quantities are evaluated at ξ_0^+. In the case of $f'|_{\xi_0^+}$ the value is clearly that just behind the shock. Simplifying this last condition, using (4.1.134), (4.1.135), (4.1.137) gives

$$\tilde{f}_1'|_{s_0} = \frac{8}{25}\xi_1^+\xi_0^+(10n_1 - 1) \,. \tag{4.1.139}$$

The analysis up to this point has been for $\tilde{y} > 0$. For $\tilde{y} < 0$ the expansion (equivalent to (4.1.122)) is

$$\tilde{\phi}_0 = \tilde{y}^{\frac{2}{5}}\frac{f_\ell(\xi^+)}{a^3} + (-\tilde{y})^{n_1}\frac{\tilde{f}_{1\ell}(\xi^+)}{a^3} + \cdots \,, \tag{4.1.140}$$

and we have

$$\xi_s^+ = \xi_0^+ + \xi_{1\ell}^+(-\tilde{y})^{m_1} + \cdots \,. \tag{4.1.141}$$

The relationships obtained across the shock are precisely the same as those calculated for $\tilde{y} > 0$ with ξ_1^+ replaced by $\xi_{1\ell}^+$ etc., namely (4.1.133), (4.1.134), (4.1.137), (4.1.139). In fact $f_\ell = f$ can be clearly seen.

Equation (4.1.123a) can be rewritten as a hypergeometric function, similar to the case in front of the shock, by using a parametrization of the solution f. Behind the shock the s variables are given by (4.1.95).

The variable s now ranges from its value of $(5\sqrt{3}-8)/6$ just behind the shock to zero along the positive x-axis. In terms of these variables (4.1.123) becomes

$$(3s + 4)s^2\frac{d^2\tilde{f}_1}{ds^2} + 2(1 + 2n_1)(1 + 2s)s\frac{d\tilde{f}_1}{ds} + n_1(n_1 - 1)(1 - 3s)\tilde{f}_1 = 0 \,. \tag{4.1.142}$$

Finally, in terms of

$$\tau = -t = \frac{3}{4}s , \quad \tilde{f}_1(s) = s^{-\frac{n_1}{2}} \tilde{h}(t) , \tag{4.1.143}$$

the equation is

$$t(t-1)\tilde{h}'' + \{c - (a+b+1)t\}\tilde{h}' - ab\tilde{h} = 0 , \tag{4.1.144}$$

where

$$c = \frac{1}{2} , \quad a = \frac{5}{2}n_1 , \quad b = \frac{1}{6}(2 - 5n_1) .$$

Two linearly independent solutions are

$$\tilde{h}^S = F\left(\frac{5n_1}{2} , \frac{1}{6}(2 - 5n_1); \frac{1}{2}; -\tau\right)$$

$$\tilde{h}^A = -\tau^{\frac{1}{2}} F\left(\frac{5n_1 + 1}{2} , \frac{5(1 - n_1)}{6}; \frac{3}{2}; -\tau\right) .$$

Thus

$$\tilde{f}_{1u} = (\tilde{c}_1 \tilde{h}^S + \tilde{c}_2 \tilde{h}^A)s^{-\frac{n_1}{2}} ,$$
$$\tilde{f}_{1\ell} = (\tilde{d}_1 \tilde{h}^S + \tilde{d}_2 \tilde{h}^A)s^{-\frac{n_1}{2}} .$$

So,

$$\tilde{\phi} = \frac{\tilde{y}^{\frac{2}{5}} f}{a^3} + \frac{|\tilde{y}|^{n_1}}{a^3}\{\tilde{c}_1 \tilde{h}^S + \tilde{c}_2 \tilde{h}^A\}s^{-\frac{n_1}{2}} + \cdots , \text{if} \quad \tilde{y} > 0, \tag{4.1.145}$$

$$\tilde{\phi} = \tilde{y}^{\frac{2}{5}}\frac{f}{a^3} + \frac{|\tilde{y}|^{n_1}}{a^3}\{\tilde{d}_1 \tilde{h}^S + \tilde{d}_2 \tilde{h}^A\}s^{-\frac{n_1}{2}} + \cdots \text{ if } \quad \tilde{y} < 0. \tag{4.1.146}$$

Now, $\tilde{\phi}_x$, $\tilde{\phi}_{\tilde{y}}$ must be continuous across the wake. As $\tilde{y} \to 0$, $\xi \to \infty$, $s, t \to 0$, $f \sim -a_0 \xi^{+1/2}$ and $\tilde{f}_1 \sim \tilde{c}_1 s^{-\frac{n_1}{2}} + \bar{c}_1 s^{(-n_1+2)/2} + \tilde{c}_2 \frac{\sqrt{3}}{4} s^{(-n_1+1)/2} + \bar{c}_2 \frac{\sqrt{3}}{4} s^{(-n_1-1)/2}$, etc. Also $s \sim \ell\xi^{+-\frac{5}{2}} + \bar{\ell}\xi^{+-5} + \cdots$.

$$a^2 \tilde{\phi}_x \sim -\frac{1}{2}a_0' x^{-\frac{1}{2}} + \tilde{c}_1 x^{(5n_1/4)-1} + \bar{c}_1 \tilde{y} + \cdots , \quad \tilde{y} > 0 ,$$

$$a^2 \tilde{\phi}_x \sim -\frac{1}{2}a_0' x^{-\frac{1}{2}} + \tilde{d}_1 x^{(5n_1/4)-1} + \tilde{d}_2 |\tilde{y}| + \cdots , \quad \tilde{y} < 0 .$$

Therefore continuity of $\tilde{\phi}_x$ across $\tilde{y} = 0$ requires

$$\tilde{c}_1 = \tilde{d}_1 . \tag{4.1.147}$$

Then

$$a^3 \tilde{\phi}_{\tilde{y}} \sim \tilde{y}^{-\frac{3}{5}} \left\{ \frac{2}{5} a_1' (\xi^+)^{-2} \right\}$$

$$+ \tilde{y}^{n_1 - 1} \left\{ \tilde{c}_1 \left(n_1 \xi^{+\frac{5n_1}{4}} \right) + c_1 \xi^{+\frac{5}{4}(n_1 - 2)} \right.$$

$$\left. - n_1 \tilde{c}_1 \xi^{+\frac{5n_1}{4}} - \frac{5}{4}(n_1 - 2) c_2 \xi^{+\frac{5}{4}(n_1 - 2)} + \cdots \right\}$$

$$\sim 2a_1' x^{-2} \tilde{y} + 2\bar{c}_1 \tilde{y} x^{\frac{5}{4}(n_1 - 2)} + \tilde{c}_2 x^{\frac{5}{4}(n_1 - 1)} + \cdots , \quad \tilde{y} > 0 .$$

Similarly

$$a^3 \tilde{\phi}_{\tilde{y}} \sim 2a_1' x^{-2} \tilde{y} + 2\bar{d}_1 \tilde{y} x^{\frac{5}{4}(n_1 - 2)} - d_2 x^{\frac{5}{4}(n_1 - 1)} + \cdots , \quad \tilde{y} < 0 .$$

Hence, for $\tilde{\phi}_{\tilde{y}}$ to be continuous across the wake we must have

$$\tilde{d}_2 = -\tilde{c}_2 . \tag{4.1.148}$$

Now, returning to the shock conditions (4.1.137), (4.1.139) in the three unknowns $f_1|_{s_0}$, $f_1'|_{s_0}$, ξ_1^+, we can eliminate ξ_1^+ so that the condition is

$$\frac{\tilde{f}_1'}{\tilde{f}_1} = \frac{8}{25} \frac{(10n_1 - 1)}{[f']} \xi_0^+ . \tag{4.1.149}$$

Substituting our solution (4.1.145), (4.1.146), (4.1.147), (4.1.148) the conditions are

$$\frac{\tilde{c}_1 \tilde{f}_1'^s + \tilde{c}_2 \tilde{f}_1'^A}{\tilde{c}_1 \tilde{f}_1^s + \tilde{c}_2 \tilde{f}_1^A} = \frac{8}{25} \frac{(10n_1 - 1)\xi_0^+}{[f']} , \quad \tilde{y} > 0 , \tag{4.1.150}$$

and

$$\frac{\tilde{c}_1 \tilde{f}_1'^s - \tilde{c}_2 \tilde{f}_1'^A}{\tilde{c}_1 \tilde{f}_1^s - \tilde{c}_2 \tilde{f}_1^A} = \frac{8(10n_1 - 1)\xi_0^+}{25[f]} , \quad \text{where} \quad \tilde{y} < 0 . \tag{4.1.151}$$

Here

$$\tilde{f}_1^\ell(\xi^+) = \tilde{h}^\ell(t) s^{-\frac{n_1}{2}} .$$

For $\bar{c}_1^2 + \tilde{c}_2^2 \neq 0$, (4.1.150) and (4.1.151) can not simultaneously be true, since the two solutions are linearly independent, hence their Wronskian can not be zero. Thus if there is a nonzero solution $\tilde{f}_1$ it must be purely even in $\tilde{y}$ or purely odd. It can not be a nontrivial linear combination of the two.

In the τ variables condition (4.1.149) on the solution $\tilde{h}^s$ or h^A respectively is

$$S^s = 30\tau(1+\tau)\frac{dh^s}{d\tau} - \{(8-65n_1)\tau + 3(1-5n_1)\}h^s \,, \qquad (4.1.152)$$

or

$$S^A = 30\tau(1+\tau)\frac{d}{d\tau}(\tau^{-\frac{1}{2}}h^A) + \{3(4+5n_1)+(65n_1+7)\tau\}\tau^{-\frac{1}{2}}h^A \,, \qquad (4.1.153)$$

must be zero at $\tau = (5\sqrt{3}-8)/8$. Thus the question is for what (if any) n_1, $0 < n_1 < 2/5$, can these conditions be satisfied.

Euvrard [4.1.12] has shown that (4.1.152) has only one solution for $n_1 \in (0, 2/5)$ which occurs at $n_1 = 1/5$. Similarly he has shown that (4.1.153) has no solutions in the range, $n_1 \in (0, 2/5)$. He shows this by explicitly calculating the coefficients in the power series expansions of S^S, S^A.

For $n_1 = 1/5$, equation (4.1.123) is an exact derivative, namely

$$\left(\left(f' - \frac{16}{25}\xi^{+2}\right)\tilde{f}_1' + \frac{4}{25}\xi^+\tilde{f}_1\right)' = 0 \,. \qquad (4.1.154)$$

The symmetric solution is

$$\tilde{f}_1 = \tilde{C}_1 s^{-\frac{1}{10}}(3s+4)^{-\frac{1}{6}} \,. \qquad (4.1.155)$$

Thus the values of ξ_1^+, $\tilde{f}_1|_{s_0}$, $\tilde{f}_1'|_{s_0}$ can be evaluated exactly. Euvrard finds them to be

$$\xi_1^+ \approx .1852427\,\tilde{C}_1\xi_0^+ \,.$$

We now want to proceed one term further – our goal is to find whether or not there is a lift term, corresponding to a jump in ϕ_0 across the wake. Thus we are looking for terms of order y^0.

So far we have

$$\phi_0 = y^{\frac{2}{5}}\frac{f(a\xi)}{a^3} + |y|^{\frac{1}{5}}\frac{\tilde{f}_1(a\xi)}{a^3} + y^{n_2}\frac{\tilde{f}_2^{u,\ell}(a\xi)}{a^3} + \cdots \qquad (4.1.156)$$

and

$$\xi_s^+ = \xi_0^+ + \xi_1^+|y|^{-\frac{1}{5}} + \xi_2^{u,\ell}|y|^{m_2} + \cdots \,. \qquad (4.1.157)$$

Also the equations and shock conditions governing $\tilde{f}_2$ are the same as those governing $\tilde{f}_1$ if $n_2 > 2n_1 - \frac{2}{5} = 0$. Thus the first possible solution is for

$$n_2 = 0 \,, \quad m_2 = -\frac{2}{5} \,. \qquad (4.1.158)$$

In this case the equation governing $\tilde{f}_2$ is a forced equation, (4.1.123b). The shock conditions are

$$\tilde{f}_2|_{s_0} = -\frac{8}{25}\xi_0^+ {\xi_1^+}^2 - \frac{{\xi_1^+}^2}{2}[f''] - \xi_2^+[f'] + a^3 c_0 \tag{4.1.159}$$

$$\tilde{f}'_2|_{s_0} = \frac{-2}{[f']}\left\{\frac{2}{25}\xi_0^+ {\xi_1^+}^2[f''] + \frac{64}{625}\xi_0^+ {\xi_1^+}^2\right\} - \frac{8}{25}\xi_0^+ \xi_2^+ + \frac{28}{25}{\xi_1^+}^2 - \frac{8}{5}\xi_1^+ \tag{4.1.160}$$

where as usual all jumps are evaluated at s_0, and f', f'' are the values just behind the shock.

In the s variables the governing equation (4.1.123b) with $n_2 = 0$ is

$$s(3s+4)\frac{d^2\tilde{f}_2}{ds^2} + 2(1+2s)\frac{d\tilde{f}_2}{ds} = -4a_2^{-\frac{3}{5}}(-3s+1)^{-3}(3s+4)^{-\frac{10}{3}}(2s+1)$$

$$\times (78s^3 + 52s^2 + 81s + 6)\tilde{C}_1^2 . \tag{4.1.161}$$

Euvrard found a similar but more complex equation since he was not doing small disturbance theory. The general solution is

$$\tilde{f}_2(s) = \tilde{e}_0^{u,\ell} + \tilde{e}_1^{u,\ell}\int_0^S \sigma^{-\frac{1}{2}}(3\sigma+4)^{-\frac{5}{3}}d\sigma$$

$$+ \frac{1}{2}a_2^{-\frac{3}{5}}\tilde{C}_1^2(1-3s)^{-1}(3s+4)^{-\frac{7}{3}}(2s^3 + 5s^2 - s - 4) , \tag{4.1.162}$$

where u, ℓ refers to $\tilde{y} > 0$ ($\tilde{y} < 0$) respectively. The coefficients in front of the homogeneous solution, $\tilde{e}_0^{u,\ell}$, $\tilde{e}_1^{u,\ell}$ are determined by the shock conditions, as well as the condition of continuity of ϕ_{0x}, $\phi_{0\tilde{y}}$ across the wake.

Since as $s \to 0$,

$$\tilde{f}_2(s) = \tilde{e}_0^{u,\ell} + \tilde{e}_1^{u,\ell}\sqrt{s} - a_2^{-\frac{3}{5}}\tilde{C}_1^2 2^{-\frac{4}{3}} + \cdots ,$$

$$(\tilde{f}_2)_x = {}_{\cdot}\tilde{y}^{-\frac{4}{5}}s^{\frac{7}{5}}f_{2s} = \tilde{y}^{-\frac{4}{5}}s^{\frac{7}{5}}\frac{\tilde{e}_1^{u,\ell}}{2\sqrt{s}} + \cdots$$

$$\sim \tilde{y}x^{-\frac{9}{4}}\tilde{e}_1^{u,\ell}\frac{\left(\frac{2a}{a_2^{\frac{1}{5}}}\right)^{-\frac{9}{5}}}{2} .$$

and

$$a^2(\tilde{f}_2)_{\tilde{y}} \sim -\frac{4}{5}\tilde{y}^{-1}\xi^+ f_{2\xi^+} \sim -\frac{4}{5}\frac{x}{y^{\frac{9}{5}}}s^{\frac{7}{5}}\left\{\frac{\tilde{e}_1^{u,\ell}}{2\sqrt{s}} + \cdots\right\}$$

$$\sim -\frac{2}{5}x^{-\frac{5}{4}}bar{e}_1^{u,\ell}\left(\frac{2a}{a_2^{\frac{1}{5}}}\right)^{-\frac{9}{5}} .$$

Hence $\tilde{e}_0^{u,\ell}$ is still arbitrary, but

$$\tilde{e}_1^u = -\tilde{e}_1^\ell \equiv \tilde{e}_1 \ ,$$

in order that ϕ_{0y} be continuous across the wake.

Now, eliminating ξ_2^+ from the two shock conditions (4.1.159), (4.1.160) we obtain a relationship between $\tilde{f}_2|_{s_0}$ and $\tilde{f}_2'|_{s_0}$ of the form

$$\tilde{f}_2'|_{s_0} = A + B\tilde{f}_2 \ . \tag{4.1.163}$$

Here

$$B = \frac{8}{25}\frac{\xi_0^+}{[f']} \ ,$$

and

$$A = -\frac{2}{[f']}\left\{ \frac{2}{25}\xi_0^+ {\xi_1^+}^2 [f''] + \frac{64}{625}{\xi_0^+}^2 {\xi_1^+}^2 + \frac{4}{25}a^3 c_0 + \frac{32}{625}{\xi_0^+}^2 {\xi_1^+}^2 \right.$$
$$\left. + \frac{2}{25}\xi_0^+ {\xi_1^+}^2 \frac{[f'']}{[f']^2} \right\} \ .$$

Also, we have (from 4.1.62)

$$\tilde{f}_2|_{s_0} = \tilde{e}_0^{u,\ell} + \hat{e}_1(\operatorname{sgn} y) + \hat{e}_2 \tilde{C}_1^2 \ ,$$

where

$$\hat{e}_1 = \tilde{e}_1 \int_0^{\frac{5\sqrt{3}-8}{6}} \sigma^{-\frac{1}{2}}(3\sigma + 4)^{-\frac{5}{3}}\, d\sigma$$

$$\hat{e}_2 = \frac{1}{2}a_2^{-\frac{5}{3}}\left(-3 - \frac{5}{2}\sqrt{3}\right)^{-1}(5\sqrt{3})^{-\frac{7}{3}}$$
$$\times \left(\left(\frac{5\sqrt{3}-8}{2}\right)^3 + \left(\frac{5}{6}(5(\sqrt{3}-8))^2 - \frac{5\sqrt{3}}{6} - \frac{8}{3}\right)\right) \ ,$$

and

$$\tilde{f}_2'|_{s_0} = L_1(\operatorname{sgn} y) + L_2 \tilde{C}_1^2 \ ,$$

where

$$L_1 = \tilde{e}_1 \tilde{\mathcal{L}}_1$$
$$= \tilde{e}_1 \left\{ 5a_2^{-\frac{1}{5}}\left(\frac{5\sqrt{3}-8}{6}\right)^{\frac{7}{5}}\left(\frac{5\sqrt{3}}{2}-5\right)^{-1} \right\} \left\{ \left(\frac{5\sqrt{3}-8}{6}\right)^{-\frac{1}{2}}\left(\frac{5\sqrt{3}}{2}\right)^{-\frac{5}{3}} \right\} \ ,$$

$$L_2 = 5a_2^{-\frac{1}{5}}\left(\frac{5\sqrt{3}-8}{6}\right)^{\frac{7}{5}}\left(\frac{5\sqrt{3}}{2}-5\right)^{-1}$$
$$\times \left\{ \frac{d}{ds}(1-3s)^{-1}(3s+4)^{-\frac{7}{3}}(2s_2 + 5s^2 - s - 4) \right\}\Bigg|_{s=\frac{5\sqrt{3}-8}{6}}$$

Hence, the shock conditions (4.1.163) state that

$$L_1 + L_2\tilde{c}_1^2 = A + B(\tilde{e}^u + \hat{e}_1 + \hat{e}_2\tilde{C}_1^2) \quad \text{for} \quad \tilde{y} > 0 \,,$$

$$-L_1 + L_2\tilde{c}_1^2 = A + B(\tilde{e}^\ell - \hat{e}_1 + \hat{e}_2\tilde{C}_1^2) \quad \text{for} \quad \tilde{y} < 0 \,.$$

Hence,

$$2L_1 = B(\tilde{e}_0^u - \tilde{e}_0^\ell) + 2B\hat{e}_1 \,,$$

$$2L_2\tilde{C}_1^2 = 2A + B(\tilde{e}_0^u + \tilde{e}_0^\ell) + 2B\hat{e}_2\tilde{C}_1^2 \,.$$

$$\langle \tilde{e}_0 \rangle_W = \frac{1}{B}\left(L_2\tilde{C}_1^2 - A - B\bar{e}_2\tilde{C}_1\right) \,,$$

$$[\tilde{e}_0]_W = \frac{1}{B}\left(2\tilde{L}_1 - 2B\bar{e}_1\right) \,.$$

Here $\tilde{C}_1$ was the constant in front of $\tilde{f}_1$ (4.1.55), so that now we know $\tilde{e}_0(\tilde{e}_1)$.

Finally, returning to (4.1.159), the value of the shock shift can be found,

$$\xi_2^{+u,\ell} = -\frac{1}{[f']}\left\{ \tilde{f}_2^{u,\ell} + \frac{8}{25}\xi_0^+ {\xi_1^+}^2 + \frac{{\xi_1^+}^2}{2}[f''] + a^3 c_0 \right\} \,, \tag{4.1.164}$$

where all the quantities are evaluated at the shock.

At this stage the lift term is evident

$$c_{0L} \propto \oint \phi_{0x}\, d\ell = \tilde{e}_0^\ell - \tilde{e}_0^u \,.$$

The as yet undetermined constants $\tilde{c}_1$, $\tilde{e}_1$ can presumably be determined by conservation laws as were a, c_1.

4.1.2 Axially Symmetric Flow

When $K = 0$, the transonic equation for axisymmetric flow is (3.9.33)

$$-(\gamma + 1)\phi_x\phi_{xx} + \phi_{\tilde{r}\tilde{r}} + \frac{1}{\tilde{r}}\phi_{\tilde{r}} = 0 , \tag{4.1.166}$$

and the boundary condition on the body surface $0 \le x \le 1$ is, (3.9.63), as $\tilde{r} \to 0$

$$\phi \sim \frac{\left(F^2(x)\right)'}{2} \log \tilde{r} . \tag{4.1.167}$$

Note that the governing equation (4.1.166) is the same as that governing plane flow (4.1.1) except for the addition of the last (lower order) term. Thus, as can be expected, the procedure we follow to find the far field for axisymmetric flow is the same as that of Section 4.1.1 for planar flow except that the simplicity of the hodograph plane is no longer available. Unfortunately, this does limit the results. We begin by finding the flow upsteam of the limit characteristic, this flow is then continued on downstream to the shock and then behind the shock.

The similarity form for ϕ is

$$(\gamma + 1)\phi(x, \tilde{r}) = \tilde{r}^{3k-2} f(\xi) \tag{4.1.168}$$

where

$$\xi = \frac{x}{\tilde{r}^k} , \tag{4.1.169}$$

and k must be determined. Velocity perturbations are, therefore,

$$\left.\begin{array}{l} w = (\gamma + 1)\phi_x = \tilde{r}^{2k-2} f'(\xi) , \\ \vartheta = (\gamma + 1)\phi_{\tilde{r}} = \tilde{r}^{3k-2}\{(3k - 2)f - k\xi f'\} \end{array}\right\} \tag{4.1.170}$$

and also we have

$$(\gamma + 1)\phi_{xx} = \tilde{r}^{3k-2} f''(\xi) . \tag{4.1.171}$$

Substituting these into the governing equation (4.1.166) we find the equation for f,

$$(f' - k^2\xi^2)f'' + (5k - 5 + \lambda)k\xi f' - (3k - 2)(3k - 3 + \lambda)f = 0 , \tag{4.1.172}$$

with $\lambda = 1$, or

$$(f' - k^2\xi^2)f'' + (5k - 4)k\xi f' - (3k - 2)^2 f = 0 . \tag{4.1.173}$$

Note that (4.1.172) with $\lambda = 0$ is the equation governing the far field behavior of two dimensional flow (4.1.14).

From (4.1.170) we see that the sonic line coresponds to $\xi = \xi^*$ such that $f'(\xi^*) = 0$, the limit characteristic corresponds to $\xi = \xi_L$ where $f'(\xi_L) = k^2 \xi_L^2$, and the tail shock occurs for some ξ_S such that $\xi_S > \xi_L$.

Equation (4.1.173) has singular points at $\xi = -\infty$ (corresponding ot the negative x axis) and at the limit characteristic ξ_L. Thus, again, we have a nonlinear eigenvalue problem to solve; find k for which we can find f, a solution of (4.1.173) such that f joins $\xi = -\infty$ smoothly through $\xi = \xi_L$. We expect as ξ increases from $-\infty$ that f first decreases ($f' < 0$, subsonic flow) then passes through a minimum ($f' = 0$, sonic), then increases ($f' > 0$, supersonic flow). (See Figure 4.1.4). We also require that $k < 2/3$ in order that the velocity die out along similarity curves as $\tilde{r} \to \infty$.

Originally, the determination of k was done numerically; today an exact solution is available. It is useful to understand the properties of this solution hence we proceed to analyze the solutions of (4.1.173).

Equation (4.1.173) has the group property that if $F(\xi)$ is a solution, so is

$$\frac{1}{A^3} F(A\xi) , \qquad (4.1.174)$$

hence the equation can be reduced to a first order equation. Using the invariance (4.1.174) we look for a solution with $\xi_L = 1$, then find the appropriate scaling A for a given body. One can use the coordinates

$$s = \xi^{-3} f(\xi) , \quad t = \xi^{-2} f'(\xi) ,$$

or in line with the two-dimensional analysis, the velocity components

$$\left. \begin{aligned} w &= (\gamma + 1)\phi_x = \tilde{r}^{2k-2} f'(\xi) = \frac{x^2}{\tilde{r}^2} t(\xi) \\ \vartheta &= (\gamma + 1)\phi_{\tilde{r}} = \tilde{r}^{3k-3} \{(3k - 2) f(\xi) - k\xi f'(\xi)\} \\ &= \tilde{r}^{3k-3} \{(3k - 2)s - kt\} \xi^3 = \frac{x^3}{\tilde{r}^3} \sigma(\xi) . \end{aligned} \right\} \qquad (4.1.175)$$

Then, we find,

$$w_x = \frac{x}{\tilde{r}^2} \{\xi t_\xi + 2t\} \qquad w_{\tilde{r}} = -\frac{x^2}{\tilde{r}^2} \{k\xi t_\xi + 2t\}$$

$$\vartheta_x = \frac{x^2}{\tilde{r}^2} \{\xi \sigma_\xi + 3\sigma\} \qquad \vartheta_{\tilde{r}} = -\frac{x^3}{\tilde{r}^3} \{k\xi \sigma_\xi + 3\sigma\} ,$$

Thus, on substitution into the basic system (4.1.66)

$$-ww_x + \vartheta_{\tilde{r}} + \frac{1}{\tilde{r}}\vartheta = 0 \,,$$
$$w_{\tilde{r}} - \vartheta_x = 0 \,,$$

we get,

$$\left\{ \begin{array}{l} t\xi t_\xi + 2t^2 + k\xi\sigma_\xi + 2\sigma = 0 \,, \\ k\xi t_\xi + 2t + \xi\sigma_\xi + 3\sigma = 0 \,, \end{array} \right\} \tag{4.1.176}$$

or,

$$\left\{ \begin{array}{l} (t - k^2)\xi t_\xi + 2t(t - k) + (2 - 3k)\sigma = 0 \,, \\ (t - k^2)\xi\sigma_\xi + 2t^2(1 - k) + (3t - 2k)\sigma = 0 \,. \end{array} \right\} \tag{4.1.177}$$

Thus, eliminating ξ we have the phase plane equation

$$\frac{d\sigma}{dt} = \frac{2t^2(1 - k) + \sigma(3t - 2k)}{2t(t - k) + \sigma(2 - 3k)} \,. \tag{4.1.178}$$

The solution in terms of ξ can be obtained from that of (4.1.178), $\sigma(t)$, by integration of one of (4.1.177) e.g.

$$\frac{d\xi}{\xi} = \frac{-(t - k^2)\,dt}{2t^2 - 2kt + 2\sigma - 3k\sigma} \,. \tag{4.1.179}$$

The singular points of (4.1.178) are $(t, \sigma) = (0, 0)$, $(n^2, \dfrac{2n^3(n - 1)}{3n - 2})$ and $(2/3, -4/9)$. They are, respectively a node, saddle point, and a node (if $k < 2/3$). Closer investigation of these points allows us to clarify the solution path.

Near $(t, \sigma) = (0, 0)$ the integral curves solve

$$\frac{d\sigma}{dt} = \frac{-2k\sigma}{-2kt + (2 - 3k)\sigma} \,,$$

so that

$$\sigma = 0 \quad \text{or} \quad t = \frac{2 - 3k}{2}\sigma \ln C\sigma \,.$$

In the second case, using (4.1.179), we see that

$$\sigma \sim \xi^{-\frac{2}{k}} \,.$$

Thus, $\phi_{\tilde{r}} \sim \frac{1}{\tilde{r}}x^{3-2/k}$, $\phi_x \sim x^{2-3/k}(2 - 3k)\ln\xi$, or for $0 < k < 1$, the solution is singular on the x-axis. Therefore, our solution, which must not have a singularity on the x-axis, corresponds to the other path of approach to $(t,\sigma) = (0,0)$ namely $\sigma = 0$. From (4.1.179) then,

$$\frac{d\xi}{\xi} = -\frac{k}{2t}\,dt\,,\tag{4.1.180}$$

or

$$(-\xi) = C_{-\infty}(-t)^{-\frac{k}{2}}\,,$$

precisely as in the two-dimensional analysis. This particular path runs from the origin in the (t,σ) plane to $(-\infty,\infty)$ which must correspond (4.1.175) to $x = 0$. Near $x = 0$ the solution is regular $(\xi = 0)$ and

$$t \sim \frac{\tilde{y}^{2k}}{x^2}f'(0) = \frac{f'(0)}{\xi^2}\,,$$

$$\sigma \sim \frac{\tilde{y}^{3k}}{x^3}(3k-2)f(0) = f(0)\frac{(3k-2)}{\xi^3}\,,$$

so that the path must follow

$$\sigma = \pm C_0(-t)^{\frac{3}{2}} \quad \text{as} \quad x \to 0^{\pm}\,.\tag{4.1.181}$$

This path reappears at $(t,\sigma) = (-\infty,\infty)$ as x crosses zero and must now run towards the saddle point

$$t = k^2\,,\quad \sigma = \frac{2k^3(k-1)}{3k-2}\,.$$

In order to find the path of approach to the saddle point, one can work with the properties of f as in section (4.1.1) or use the phase plane equation directly. The path of approach to the saddle point is

$$\sigma - \frac{2k^3(k-1)}{3k-2} = m(t-k^2)$$

where

$$m = \frac{-k - k\sqrt{25k^2 - 56k + 32}}{2(2-3k)}\,.$$

In Figure 4.1.9, two figures are drawn. In a, the path which passes through $(t,\sigma) = 0$ is sketched for $1 > k > 1/2$, and in b, the desired path is sketched. In particular in a use is made of the exact solution for $k = 2/3$. If $k = 2/3$ (as if $k = 1/2$ or $k = 1$) a first integral of (4.1.173) can be found. The path which passes through $(t,\sigma) = (0,0)$ is given by $3\sigma^2 + 4t\sigma + 2t^3 = 0$.

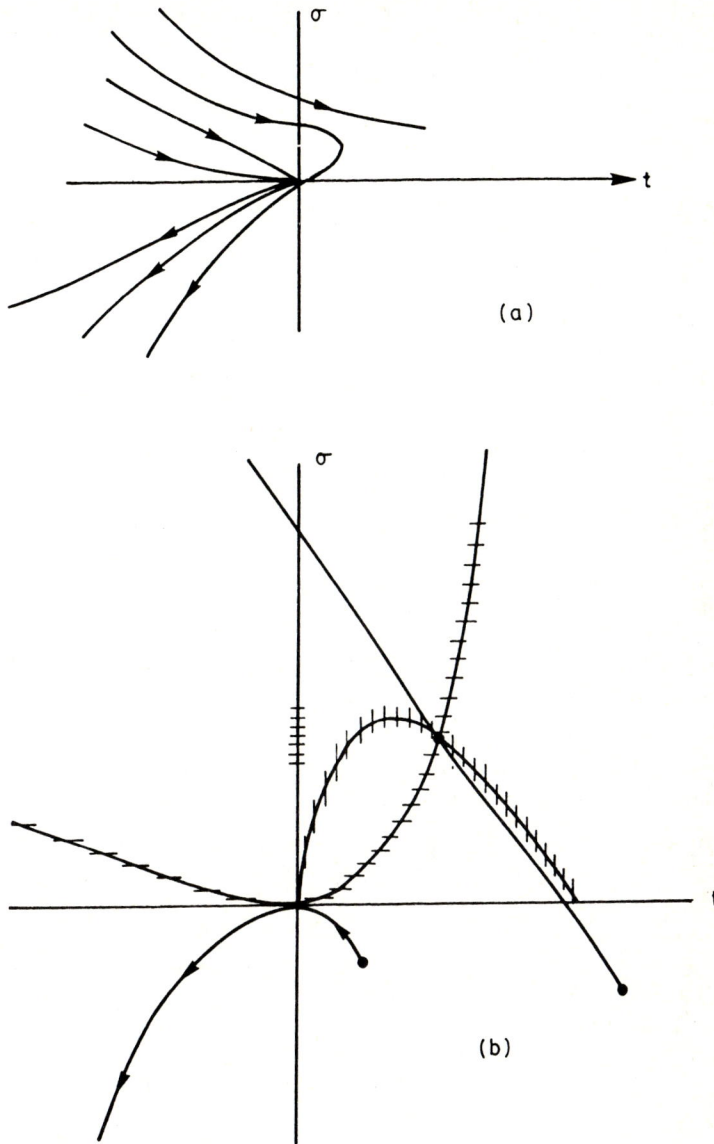

Figure 4.1.9

Phase Plane, Axisymmetric Flow

The solution to (4.1.178) which satisfies these conditions can only be found if

$$k = \frac{4}{7},$$

and can be written as

$$\sigma = \begin{cases} \dfrac{8}{9} - 2t - \dfrac{8}{9}\left(1 - \dfrac{3}{2}t\right)^{\frac{5}{2}}, & \sigma < 0, \quad -\infty < \xi < \infty, \\[3mm] \dfrac{8}{9} - 2t + \dfrac{8}{9}\left(1 + \dfrac{3}{2}t\right)^{\frac{5}{2}}, & \sigma > 0, \quad 0 < \xi < \xi_L. \end{cases} \tag{4.1.182}$$

Integration of (4.1.179) using this solution gives $\xi(t)$,

$$\left(\frac{\xi^*}{\xi}\right)^7 = \begin{cases} -\dfrac{1}{4}\left(1 - \dfrac{3}{2}t\right)^{\frac{5}{2}}\left\{1 - \left(1 - \dfrac{3}{2}t\right)^{\frac{1}{2}}\right\}^2 & \begin{array}{l} -\infty < \xi < \infty, \\ 0 > t > -\infty, \end{array} \\[4mm] \dfrac{1}{4}\left(1 - \dfrac{3}{2}t\right)^{\frac{5}{2}}\left\{1 + \left(1 - \dfrac{3}{2}t\right)^{\frac{1}{2}}\right\}^2 & \begin{array}{l} 0 < \xi < \xi_L, \\ k^2 > t > -\infty. \end{array} \end{cases} \tag{4.1.183}$$

(4.1.182) and (4.1.183) give the desired solution .

Now, to continue behind the limit characteristic, note the solution we want passes smoothly through the limit characteristic so the plus branch of (4.1.182) can be continued past $t = k^2$. That is past the limit characteristic, $t = k^2$, the solution is given by the second line of (4.1.182), (4.1.183). The path passes through $\sigma = 0$ ($\vartheta = 0$, no radial flow) to $\sigma < 0$ ($\vartheta > 0$), and heads directly for the node $(t, \sigma) = (2/3, -4/9)$. It can not be allowed to approach the node since then $\xi \to \infty$, t, σ are bounded, hence (4.1.175) the velocities are infinite in a region corresponding to the downstream x axis. So, a shock wave occurs at some $\xi_S > \xi_L$ so that the solution can get onto the exceptional path running into the origin. This path is given from (4.1.180) by $\xi = C_{(-\infty)}t^{-\frac{k}{2}}$. That is as $\xi \to \infty$ we must have $(t, \sigma) \to (0, 0)$ so that velocities remain bounded. The shock conditions are (see Section 3.6)

$$\langle w \rangle [w] - [\vartheta]\left(\frac{dx}{d\tilde{r}}\right)_S = 0,$$

$$[w]\left(\frac{dx}{d\tilde{r}}\right)_S + [\vartheta] = 0. \tag{4.1.184}$$

Since the shock occurs on a similarity curve, $x/\tilde{r}^k = \xi_s$, we have that

$$\left(\frac{dx}{d\tilde{r}}\right)_S = \frac{4}{7}\frac{x}{\tilde{r}}.$$

From (4.1.75),

$$[w] = \frac{x^2}{\tilde{r}^2}[t] , \quad \langle w \rangle = \frac{x^2}{\tilde{r}^2}\langle t \rangle , \quad [\vartheta] = \frac{x^3}{\tilde{r}^3}[\sigma] ,$$

so that (4.1.184) is equivalent to the system

$$\begin{cases} \langle t \rangle [t] + \dfrac{4}{7}[\sigma] = 0 , \\[2mm] \dfrac{4}{7}[t] + [\sigma] = 0 . \end{cases} \tag{4.1.185}$$

A nontrivial solution to this linear system for $[t]$, $[\sigma]$ can exist iff

$$\langle t \rangle - \left(\frac{4}{7}\right)^2 = 0 ,$$

or

$$t_b = 2\left(\frac{4}{7}\right)^2 - t_a . \tag{4.1.186}$$

Then,

$$\sigma_b = -\frac{4}{7}[t] + \sigma_a . \tag{4.1.187}$$

In addition to these conditions ξ must be continuous across the shock so that

$$\left(\frac{\xi^*}{\xi_s}\right)^7 = \frac{1}{4}\left(1 - \frac{3}{2}t_a\right)^{\frac{5}{2}}\left\{1 + \left(1 - \frac{3}{2}t_a\right)^{\frac{1}{2}}\right\}^2$$
$$= \frac{1}{4}\left(1 - \frac{3}{2}t_b\right)^{\frac{5}{2}}\left\{1 - \left(1 - \frac{3}{2}t_b\right)^{\frac{1}{2}}\right\}^2 . \tag{4.1.188}$$

From (4.1.188) and (4.1.186),

$$t_a = \left(\frac{4}{7}\right)^2 + \frac{1}{\sqrt{3}}\left(\frac{5}{7}\right)^2 , \quad t_b = \left(\frac{4}{7}\right)^2 - \frac{1}{\sqrt{3}}\left(\frac{5}{7}\right)^2 , \tag{4.1.189}$$

and then from (4.1.182) and (4.1.187),

$$\sigma_a = -\frac{58}{7^3} - \frac{1}{\sqrt{3}}\frac{4}{7}\left(\frac{5}{7}\right)^2 , \quad \sigma_b = -\frac{58}{7^3} + \frac{1}{\sqrt{3}}\left(\frac{5}{7}\right)^2 . \tag{4.1.190}$$

These results were obtained numerically by Barish and Guderley [4.1.9] and explicitly by Fal'kovich and Chernov [4.1.6], Müller and Matschat [4.1.4], and

Randall [4.1.13]. They represent the basic solution to (4.1.173) implicitly. The relationship between ξ_L, ξ^* and ξ_S is given by

$$\xi_L = \frac{\xi^*}{a_0} = \frac{\xi_S}{b_0}$$

where

$$a_0 = 6^{\frac{2}{7}} 5^{\frac{5}{7}} 7^{-1} \approx .75 \quad \text{and} \quad b_0 = 2 \cdot 3^{\frac{1}{7}} (\sqrt{3} - 1)^{\frac{1}{7}} \approx 2.23 .$$

A useful parametrization of the solution in terms of a parameter s is given by

$$\xi = \frac{12s - 5}{7s^{\frac{2}{7}}} , \quad f = \frac{2^5}{7^3} s^{\frac{1}{7}} (12s^2 - 15s - 25)$$

in front of the shock, and

$$\xi = C \frac{12s + 5}{7s^{\frac{2}{7}}} , \quad f = \frac{2^5}{7^3} C^3 s^{\frac{1}{7}} (12s^2 + 15s - 25)$$

behind the shock, where

$$C = \left(2 - \sqrt{3}\right)^{\frac{1}{7}} ,$$

and at the shock

$$\xi_s = 2 \cdot 3^{\frac{1}{7}} (\sqrt{3} - 1)^{\frac{1}{7}} , \quad s_a = \frac{7\sqrt{3} + 12}{12} , \quad s_b = \frac{7\sqrt{3} - 12}{12} .$$

The course of s is summarized below.

ξ	$-\infty$	0	ξ^*	ξ_L	ξ_s	$+\infty$
s	0	$\frac{5}{12}$	$\frac{5}{6}$	1	$s_a \vert s_b$	0

Returning now to the group property (4.1.174) for f we see that

$$\phi \sim \tilde{r}^{-\frac{2}{7}} \frac{f(a\xi)}{a^3}$$

where in front of the shock

$$a\xi = \frac{12s - 5}{7s^{\frac{2}{7}}} , \quad f = \frac{2^5}{7^3} s^{\frac{1}{7}} (12s^2 - 15s - 25) , \tag{4.1.191}$$

and behind the shock

$$a\xi = C \frac{12s + 5}{7s^{\frac{2}{7}}} , \quad f = \frac{2^5}{7^3} C^3 s^{\frac{1}{7}} (12s^2 + 5s - 25) \tag{4.1.192}$$

where C is $(2 - \sqrt{3})^{-1/7}$ and at the shock

$$a\xi_S = 2 \cdot 3^{\frac{1}{7}}(\sqrt{3} - 1)^{\frac{1}{7}} .$$

Higher Order Terms

The orders of the subsequent terms in the far field are found by assuming the expansion

$$\phi_0 = \frac{\tilde{r}^{-\frac{2}{7}} f(a\xi)}{a^3} + \tilde{r}^{\sigma_0} \frac{c_0}{a^3} f_0(a\xi) + \frac{\tilde{r}^{\sigma_1} c_1 f_1(a\xi)}{a^3} + \cdots , \qquad (4.1.193)$$

as $\tilde{r} \to \infty$ with ξ fixed, where $\sigma_1 < \sigma_0 < -2/7$. Letting $\alpha = \sigma_{0,1}$, $g = f_{0,1}$, and substituting into the governing equation (4.1.166) we find that g satisfies the equation

$$\left(f' - \left(\frac{4}{7} a\xi \right)^2 \right) g'' + \left(f'' + \frac{4}{7} \left(2\alpha - \frac{4}{7} \right) a\xi \right) g' - \alpha^2 g = 0 \qquad (4.1.194)$$

if $\sigma_1 > 2\sigma_0 + 2/7$.

Using the parametrization (4.1.191) for f, (4.1.194) becomes

$$\frac{4s^2(1-s)}{1+6s} \frac{d^2g}{ds^2} + \frac{4(1+\alpha)s(5-12s)}{s(1+6s)} \frac{dg}{ds} + \alpha^2 g = 0 . \qquad (4.1.195)$$

Now let

$$g(s) = s^{-\frac{\alpha}{2}} h(s) , \qquad (4.1.196)$$

then h satisfies the hypergeometric equation

$$s(1-s)h'' + \left(1 - \frac{1}{5}s(12 + 7\alpha) \right) h' + \frac{7}{20}\alpha(7\alpha + 2)h = 0 , \qquad (4.1.197)$$

where

$$a = \frac{7}{10} \left\{ 1 + \alpha + \sqrt{6\alpha^2 + \frac{24}{7}\alpha + 1} \right\} , \quad b = \frac{7}{10} \left\{ 1 + \alpha - \sqrt{6\alpha^2 + \frac{24}{7}\alpha + 1} \right\} ,$$

$$c = 1 .$$

The values for α must be chosen so that there is a solution h joining $\xi = -\infty$ $(s = 0)$ smoothly to ξ_L $(s = 1)$.

Two linearly independent solutions of (4.1.197) are

$$h^i(s) = F(a, b, 1; s)$$

$$h^{ii}(s) = F(a, b, 1; s) \ln s + \sum_{k=1}^{\infty} \frac{(a)_k (b)_k}{(k!)^2} s^k g_{a,b,k} ,$$

where $g_{a,b,k}$ are numbers.

Now as $s \to 0$ ($\xi \to -\infty$) the second solution behaves like $\ln s$, hence $g(s) = O(s^{-\alpha/2} \ln s)$. But $\xi = O(s^{-2/7})$, so that $g(s) \underset{\xi \to \infty}{=} O(\xi^{7\alpha/4} \ln \xi)$. The contribution of this term to the velocity in the x direction would be $O(x^{+\frac{7\alpha}{4} - 1} \log \xi)$, and to the radial velocity would be $O(x^{\frac{7\alpha}{4}} \tilde{r}^{-1})$, as $\xi \to -\infty$, that is, near the negative x axis. Thus, this solution is singular near the x axis, hence must be rejected. Similar analysis of the first solution shows that its contribution to the velocity is smooth as $\tilde{r} \to 0$. Thus

$$h(s) = F(a, b, 1; s) . \tag{4.1.198}$$

As was shown in Section 4.1.1, this solution is smooth through the limit characteristic $s = 1$ iff a or b is a negative integer or zero. Thus

$$\frac{7}{10} \left\{ 1 + \alpha \pm \sqrt{6\alpha^2 + \frac{24}{7}\alpha + 1} \right\} = -k , \quad k = 0, 1, 2, \dots ,$$

or

$$\alpha = \frac{1}{7} \left\{ 2k - 1 \pm \sqrt{24k^2 + 24k + 1} \right\} , \quad k = 0, 1, 2, \dots . \tag{4.1.199}$$

Since we expect $\alpha < -\frac{2}{7}$,

$$\alpha = \frac{1}{7} \left\{ 2k - 1 - \sqrt{24k^2 + 24k + 1} \right\} , \quad k = 0, 1, 2, \dots . \tag{4.1.200}$$

Therefore, in (4.1.193) using (4.1.199) we have the desired values

$$\sigma_0 = -\frac{6}{7} , \quad \sigma_1 = \frac{3 - \sqrt{145}}{7} \approx -\frac{9.04}{7} . \tag{4.1.201}$$

and functions

$$f_0 = 1 - \frac{6}{5}s ,$$

$$f_1 = 1 - 2\lambda s + \frac{\lambda(\lambda + 1)}{2} s^2 , \quad \text{with} \quad \lambda = 4 - \frac{\sqrt{145}}{5} . \tag{4.1.202}$$

The constants a, c_0, c_1 in the far field expansion must still be determined. It is the values of these constants which will link the far field behavior to the profile shape. Unfortunately, the conservation laws as in Section 4.1.1 are no longer useful since this flow is not governed by Tricomi's equation. The determination of a, c_0, c_1 is still open.

Note that the representation of the far field obtained so far,

$$\phi = \frac{\tilde{r}^{-\frac{2}{7}} f(a\xi)}{a^3} + \tilde{r}^{-\frac{6}{7}} \frac{c_0 f_0(a\xi)}{a^3} + \tilde{r}^{\frac{(3-\sqrt{145})}{7}} c_1 \frac{f_1(a\xi)}{a^3} + \cdots ,$$

where f_0, f are given in (4.1.202) and $s(\xi)$ is given by (4.1.191) is valid only in front of the shock. In order to find the representation behind the shock we must use the shock conditions. We expect that behind the shock, the far field has an expansion of the form

$$\phi = \frac{\tilde{r}^{-\frac{2}{7}} f(a\xi)}{a^3} + \frac{\tilde{r}^{n_0} \tilde{f}_0(a\xi)}{a^3} + \frac{\tilde{r}^{n_1} \tilde{f}_1(a\xi)}{a^3} + \cdots \qquad (4.1.203)$$

Letting $\alpha = n_0$, n_1, and $\tilde{g} = \tilde{f}_0$, $\tilde{f}_1$, and substituting (4.1.203) into the governing equation (4.1.166) we find the equation governing $\tilde{g}$,

$$\left\{ f' - \left(\frac{4}{7}a\xi\right)^2 \right\}\tilde{g}'' + \left\{ f'' + \frac{4}{7}\left(2\alpha - \frac{4}{7}\right)a\xi \right\}\tilde{g}'' - \alpha^2 \tilde{g}^2$$

$$= \begin{cases} 0 \quad \text{if} \quad \begin{aligned} &\alpha = n_0, \quad \text{or if} \\ &\alpha = n_1 > 2n_0 + \frac{2}{7} \end{aligned} \\ -\tilde{f}_0' \tilde{f}_0'' \quad \text{if} \quad \alpha = n_1 = 2n_0 + \frac{2}{7} \end{cases} \qquad (4.1.204)$$

Note that this is the same equation as that governing the terms in front of the shock. The solutions, however, are different because we now have different boundary conditions. The solution g must now take on prescribed values at the shock, ξ_s, and carry itself smoothly through $\xi = +\infty$.

In terms of the s parametrization, which is given for behind the shock by (4.1.192) the homogeneous equation for $\tilde{h}$, where

$$\tilde{g}(s) = s^{-\frac{\alpha}{2}} \tilde{h}(s) \qquad (4.1.205)$$

is

$$s(s+1)\tilde{h}'' + \left(1 + \frac{7\alpha + 12}{5}s\right)\tilde{h}' + \frac{7}{20}\alpha(7\alpha + 2)\tilde{h} = 0 . \qquad (4.1.206)$$

This is a hypergeometric equation with

$$a = \frac{7}{10}\left\{\alpha + 1 - \sqrt{6\alpha^2 + \frac{24}{7}\alpha + 1}\right\}, \quad b = \frac{7}{10}\left\{\alpha + 1 + \sqrt{6\alpha^2 + \frac{24}{7}\alpha + 1}\right\},$$

$$c = 1 . \tag{4.1.207}$$

One solution is $F(a, b, 1; -s)$, the other linearly independent solution contains a $\log s$ term. Since we want a solution which is smooth as $\xi \to \infty$, hence as $s \to 0$, the second solution must be rejected. Hence,

$$\tilde{h}_0(s) = cF(a, b, 1; -s) . \tag{4.1.208}$$

Note that the contribution of this correction term to the x velocity is $O(x^{(\frac{7\alpha}{4})-1})$ as $\tilde{r} \to 0$, and to the radial velocity is $O(\tilde{r}x^{(\frac{7\alpha}{4})-\frac{7}{2}})$ as $\tilde{r} \to 0$.

In order to determine the parameter value α, hence a and b, we must introduce the shock conditions (see section 3.9)

$$-\frac{1}{2}[\phi_x^2]_s[\phi_x]_s + [\phi_{\tilde{r}}]_s^2 = 0 ,$$

$$[\phi]_s = 0 . \tag{4.1.209}$$

The shock location is given by

$$\xi_s^+ = a\xi_s = \xi_0^+ + \xi_1^+ \tilde{r}^{m_1} + \xi_2^+ \tilde{r}^{m_2} + \cdots , \tag{4.1.210}$$

where ξ_0^+ is the zeroth order shock location and $m_2 < m_1 < 0$. Thus

$$ax_s = \xi_0^+ \tilde{r}^{\frac{4}{7}} + \xi_1^+ \tilde{r}^{m_1+\frac{4}{7}} + \xi_2^+ \tilde{r}^{m_2+\frac{4}{7}} + \cdots , \tag{4.1.211}$$

and for a given function f, using its Taylor series expansion about the zeroth order shock location,

$$f|_s = f|_{s_0} + \xi_1^+ \tilde{r}^{m_1} f'|_{s_0} + \left(\xi_2^+ \tilde{r}^{m_2} \frac{f''}{2}\right)\Big|_{s_0} + \cdots .$$

Expanding the conditions (4.1.209), we find

$$-\frac{1}{2}\tilde{r}^{-\frac{18}{7}}[f'][f'^2] - \frac{\xi_1^+}{2}\tilde{r}^{m_1-\frac{18}{7}}[f''][f'^2] - \frac{\xi_1^+}{2}\tilde{r}^{m_1-\frac{18}{7}}[2f'f''][f']$$

$$+ \tilde{r}^{n_0-\frac{16}{7}}f'\tilde{f}_0'[f'] + \frac{1}{2}\tilde{r}^{n_0-\frac{16}{7}}[f'^2]\tilde{f}_0'$$

$$+ r^{-\frac{18}{7}}\left[-\frac{2}{7}f - \frac{4}{7}\xi_0 f'\right]^2 + 2\xi_1\tilde{r}^{m_1-\frac{18}{7}}\left[-\frac{2}{7}f - \frac{4}{7}\xi_0 f'\right]\left[-\frac{2}{7}f' - \frac{4}{7}(\xi f')'\right]$$

$$- \tilde{r}^{n_0-\frac{16}{7}}\left\{n_0\tilde{f}_0 - \frac{4}{7}\xi\tilde{f}_0'\right\}\left[-\frac{2}{7}f - \frac{4}{7}\xi f\right] + \cdots = 0 \tag{4.1.212}$$

and

$$\tilde{r}^{-\frac{2}{7}}[f] + \xi_1 \tilde{r}^{n_1 - \frac{2}{7}}[f'] - \tilde{r}^{n_0} \tilde{f}_0 + \cdots = 0 \qquad (4.1.213)$$

where $[\;\;]$ are all evaluated at the zeroth order shock location ξ_0^+. Thus, to lowest order, $O(\tilde{r}^{-18/7})$,

$$[f] = 0, \quad \langle f' \rangle = \left(\frac{4}{7}\right)^2 \xi_1 .$$

Here we've used the fact that $[(\;\;)^2] = 2\langle(\;\;)\rangle[(\;\;)]$ where $\langle(\;\;)\rangle$ is the average value. To next order we must have $n_0 = m_1 - \frac{2}{7}$. The argument follows precisely the line of that for the two-dimensional analogue namely that if $n_0 > m_1 - \frac{2}{7}$, since $[f']_{s_0} \neq 0$, from (4.1.216) we'd need $\xi_1 = 0$ and then $\tilde{f}_0|_{s_0} = 0$. Then, from (4.1.215) $\tilde{f}_0'|_{s_0} = 0$. Then $f_0 \equiv 0$ since it is the solution of a second order ordinary differential equation with zero initial conditions.

With $m_1 - \frac{2}{7} = n_0$, from (4.1.213), we have

$$\tilde{f}_0 = \xi_1 [f'] , \qquad (4.1.214)$$

and after simplification from (4.1.212), we have

$$\tilde{f}_0' = \frac{16}{7} \xi_0 \xi_1 \left(n_0 + \frac{2}{7}\right) . \qquad (4.1.215)$$

Thus the compatibility condition for $\tilde{f}_0$ at the shock is,

$$\frac{\tilde{f}_0'}{\tilde{f}_0} = \frac{16}{7} \xi_0^+ \frac{\left(n_0 + \frac{2}{7}\right)}{[f']} .$$

In terms of the s parametrization and $\tilde{h}_0$ (4.1.205), this becomes

$$\left\{ 10t(t+1)\frac{d\tilde{h}_0}{dt} + \left\{ \left(19\left(n_0 + \frac{2}{7}\right) + \frac{10}{7}\right)t + 5\left(n_0 + \frac{4}{7}\right) \right\}\tilde{h}_0 \right\}\Big|_{t_b} = 0 .$$

$$(4.1.216)$$

Tournemine [4.1.13] has shown by substitution of (4.1.208) into (4.1.216) that the first n_0, $n_0 < -\frac{2}{7}$, is $n_0 = -\frac{4}{7}$. For this value of α the solution of the governing equation (4.1.207) corresponding to (4.1.208) is

$$\tilde{f}_0 = c_0 s^{\frac{2}{7}} (s+1)^{\frac{2}{5}} ,$$

where $c_0(\xi_1)$ can be determined from (4.1.215).

Carrying these computations to higher orders Tournemine has shown that behind the shock,

$$\phi \sim \tilde{r}^{-\frac{2}{7}} f + \tilde{r}^{-\frac{4}{7}} f_0 + \tilde{r}^{-\frac{6}{7}} \tilde{f}_1 + O(1) \,,$$

and

$$\xi_s \sim \xi_0 + \xi_1 \tilde{r}^{-\frac{2}{7}} + \xi_2 \tilde{r}^{-\frac{4}{7}} + \cdots .$$

Note that $\tilde{f}_1$ is then the solution of the inhomogeneous equation (4.1.204). Note that, as expected for axisymmetric flow, there is no lift.

The results obtained here for the higher order terms in the far field were obtained in front of the shock by Euvrard [4.1.12] and behind the shock by Tournemine [4.1.14]. Euvrard actually included dependence on the angle Θ in his work so that he was calculating the far field of a three-dimensional airfoil.

References

[4.1.1] G. W. Bluman and J. D. Cole, *Similarity Methods for Differential Equations*, Springer Verlag, 1974.

[4.1.2] Frankl, F. I., *Über eine Klasse von Lösungen der gasdynamischen Gleichungen von S.A.. Tschaplygin: Uchebnia Zapiskü*. Also: *Dokl. Akad. Nauk SSR* **57** (1947) 7, 5161–164. And: Landau, L. and Lifschitz, E. M., *Fluid Mechanics*, Pergamon 1969.

[4.1.3] Guderley, G., *Theorei Schallnaher Stromungen*, Springer 1957, and earlier references in that book.

[4.1.4] Muller, E. A. and Matschat, K. Ahnlichkeitslosungen der Transonischen Gleichungen bei der Anstrom-Machzahl 1., *Proceedings of the Eleventh International Congress of Applied Mechanics*, Munich (1964), pp. 1061–1068, Springer.

[4.1.5] Falkovich, S. V., Plane Transonic Gas Flow with Singularities on the Sonic Line, *PMM* **25** (1961), pp. 324–338. (English Translation)

[4.1.6] Falkovich, S. V. and Chernov, I. A., Flow of a Sonic Gas Stream Past a Body of Revolution, *PMM* Jan. 1965, pp. 342–347. (English Translation)

[4.1.7] Szaniawski, A., Two Parametrical Forms of the Self-Similar Transonic

Guderley-Frankl solutions, *Kleine Mittenlungen Z.A.M.M.*, v. **47**, No. 5, 1967, p. 342.

[4.1.8] Germain. P., Ècoulements Transsoniques Homogeénes, *Progress in Aeronautical Sciences*, v. 5, 1964. Also: *O.N.E.R.A. Tire A* Part 242, 1965.

[4.1.9] Barish, D. T. and Guderley, G., Asymptotic Forms of Shock Waves in Flows over Symmetrical Bodies at Mach One, *J. Aero. Sci.*, v. **20**, p. 491, 1963.

[4.1.10] Van-Guy, Nguyen, Verification Experimentale de la Solution Homogene pour un Ecoulement Plan a $M = 1$, These Lille U., 1962.

[4.1.11] Goursat, M. E., Sur l'équation différentielle linéaire qui admit pour intégrale la série hypergéométrique, *Annales Scientifiques de l'école normale Supérieure*, Supplément au Tome X - Année 1881 Deuxième série, pps. 1–142.

[4.1.12] Euvrard, D., Etude asymptotique de l'écoulement à grande distance d'un obstacle se déplacent à la vitesse du son, Première Partie, *J. de Méchanique*, **6**, #4, 1967, pp. 547–592, Deuxième Partie, *J. de Méchanique*, **7**, #1, 1968, pp. 7–139, Troisième Partie, *J. de Méchanique*, **7**, 1968, pp. 291–307.

[4.1.13] Randall, D. G., Some results on the theory of almost axisymmetric flow at transonic speed, *AIAAJ*, **3**, 12, 1965, pp. 2339–2341.

[4.1.14] Tournemine, Georges, Compartement asymptotique de l'écoulement sonique autour d'un corps de révolution finies en avail de l'onde de choc, *J. de Méchanique*, **7**, 3, pp. 309–333.

4.2 Far Field at $M_\infty > 1$ ($K < 0$)

For two-dimensional and axisymmetric flow at $M_\infty > 1$ the flow field is divided into three regions. In front of the head shock there is no flow perturbation, the body has no upstream influence. In this part the flow differs from the sonic ($M_\infty = 1$) flow. The flow between the head and tail shock is independent of the flow behind the tail shock. This is similar to the Mach one flow. (see Figure 4.2.1)

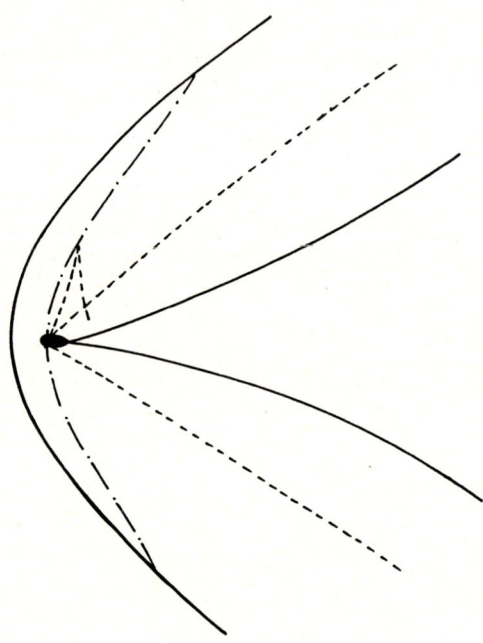

Figure 4.2.1

Flow at $M_\infty > 1$

In the far field the flow approaches that of linearized theory namely the disturbance is carried by the linearized streamwise characteristic surface. To systematize this idea, that one expects changes along the characteristic surface to be much smaller than changes across characteristics, coordinates based on the linearized characteristics are used.

The result is that the far field between the shocks is, as in the sonic case, dominated by symmetric flow. No lift appears in this portion of the flow. In order to find the lift one is forced, as in the sonic case, to examine the flow behind the tail shock.

4.2.1. Two Dimensional Flow

For $K < 0$ the equation to be solved,

$$\left(-|K| - (\gamma + 1)\phi_x\right)\phi_{xx} + \phi_{\tilde{y}\tilde{y}} = 0 , \tag{4.2.1}$$

is hyperbolic in the far field ($\phi_x \rightarrow 0$). In fact, as $\phi_x \rightarrow 0$ equation (4.2.1) approaches the linear equation

$$|K|\hat{\phi}_{xx} - \hat{\phi}_{\tilde{y}\tilde{y}} = 0 . \tag{4.2.2}$$

The flow field for the linear equation consists of no flow ($\hat{\phi}$ = constant) in front of and behind the wave zone which is defined by the characteristics $x \pm \sqrt{|K|}\tilde{y} = 0$, $x \pm \sqrt{|K|}\tilde{y} = 1$, for $\tilde{y}$ less than, or greater than, zero respectively. (Here it is assumed that the airfoil is located along $\tilde{y} = 0$, $0 \leq x \leq 1$). In the wave zone the disturbance is carried along the characteristics $x \pm \sqrt{|K|}\tilde{y}$ = constant.

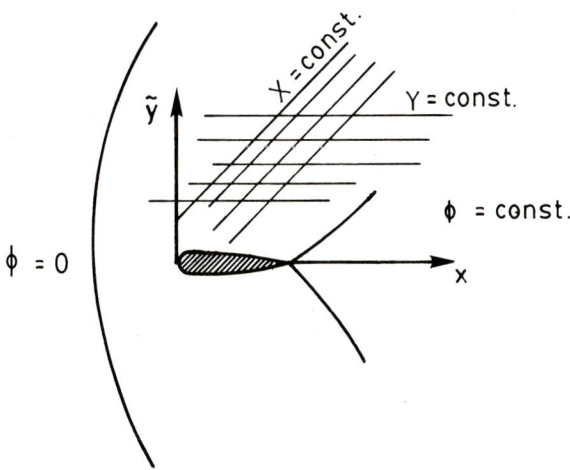

Figure 4.2.2

Transonic Plane, $M_\infty > 1$

The situation is somewhat similar for the nonlinear equation (4.2.1). That is, there is a head shock and a tail shock. There is no flow in front of the head shock, $\phi = 0$. The shocks are curved lines, not straight, due to the nonlinearity in (4.2.1). (See Figure 4.2.2.)

In order to analyze the flow in the wave zone between the head and tail shocks it is convenient to use coordinates based on the characteristics of the linearized equation (4.2.2),

$$X = x \mp \sqrt{|K|}\,\tilde{y} - x_0\,, Y = \tilde{y}\,, \qquad (4.2.3)$$

so that for a fixed range of X as Y varies one remains in the wave zone. Instead of these oblique coordinates orthogonal coordinates based on the linearized characteristics could be used but no advantage is obtained.

In the new coordinates equation (4.2.1) becomes

$$(\gamma + 1)\phi_X\phi_{XX} \pm 2\sqrt{|K|}\,\phi_{XY} - \phi_{YY} = 0\,. \qquad (4.2.4)$$

Here X roughly measures the horizontal distance from a center location in the wave zone, x_0 is undetermined so far. Changes in flow across characteristics (in the direction of flow) are expected to be much larger than changes along characteristics. Thus we expect $\frac{\partial}{\partial X} \gg \frac{\partial}{\partial Y}$, so that to first order equation (4.2.4) should be approximated by

$$(\gamma + 1)\tilde{\phi}_X\tilde{\phi}_{XX} \pm 2\sqrt{|K|}\,\tilde{\phi}_{XY} = 0$$

or, in terms of $\tilde{u} = (\gamma + 1)\tilde{\phi}_X$,

$$\tilde{u}\tilde{u}_X \pm 2\sqrt{|K|}\,\tilde{u}_Y = 0\,.$$

We now proceed to systematize this idea.

We look for a solution in similarity form, that is a far field solution of the form

$$\phi = |Y|^\alpha f(\xi) + |Y|^\delta g(\xi) + \cdots\,, \qquad (4.2.5)$$

where $\delta < a$, $\xi = X/|Y|^\beta$ and α, β must be determined. Also we expect $\beta < 1$, so that $\frac{\partial}{\partial X} \gg \frac{\partial}{\partial Y}$ as $|Y| \to \infty$.

To find the equation governing $f(\xi)$, and the restrictions on α, β, substitute the expansion (4.2.5) into equation (4.2.4) and balance the terms of highest order in Y. Since

$$\phi_Y = \pm |Y|^{\alpha-1}(\alpha f - \beta \xi f') \pm |Y|^{\delta-1}(\delta g - \beta \xi g')$$

$$\phi_{YY} = |Y|^{\alpha-2}\{(\alpha-1)(\alpha f - \beta \xi f') - \beta \xi (\alpha f - \beta \xi f')'\} + O(|Y|^{\delta-2})$$

$$\phi_X = |Y|^{\alpha-\beta} f' + |Y|^{\delta-\beta} g' \tag{4.2.6}$$

$$\phi_{XX} = |Y|^{\alpha-2\beta} f'' + |Y|^{\delta-2\beta} g''$$

$$\phi_{XY} = \pm |Y|^{\alpha-\beta-1}(\alpha f - \beta \xi f') \pm |Y|^{\delta-\beta-1}(\delta g - \beta \xi g')'$$

equation (4.2.4) becomes, to highest order

$$(\gamma+1)|Y|^{2\alpha-3\beta} f' f'' + 2\sqrt{|K|}|Y|^{\alpha-\beta-1}\{\alpha f - \beta \xi f'\}' + O(|Y|^{\alpha-2}) = 0 . \tag{4.2.7}$$

Thus, to balance the highest order terms

$$2\alpha - 3\beta = \alpha - \beta - 1 ,$$

or

$$\alpha = 2\beta - 1 .$$

Thus we have a one parameter family of solutions

$$\phi \sim |Y|^{(2\beta-1)} f(X/|Y|^{\beta}) , \tag{4.2.8}$$

where f satisfies the equation

$$(\gamma+1) f' f'' + 2\sqrt{|K|}\{(2\beta-1)f - \beta \xi f'\}' = 0 , \tag{4.2.9}$$

or, after one integration,

$$\frac{(\gamma+1)}{2}(f')^2 + 2\sqrt{|K|}\{(2\beta-1)f - \beta \xi f'\} = A \tag{4.2.10}$$

where A is a constant to be determined.

The shock polar is (3.6.22)

$$\left[K\phi_x - \frac{(\gamma+1)}{2}\phi_x^2 \right]_s [\phi_x]_s + [\phi_{\hat{y}}]_s^2 = 0 .$$

For the head shock this can be written quite simply since $u_a = v_a = 0$, so that

$$|K|u_b^2 + \frac{\gamma+1}{2}u_b^3 = v_b^2 . \tag{4.2.11}$$

In terms of the X, Y coordinates we have

$$u = \phi_x = \phi_X = |Y|^{\beta-1} f'(\xi)$$
$$v = \phi_{\bar{y}} = \phi_Y \mp \sqrt{|K|} \phi_X = \mp |Y|^{\beta-1} \sqrt{|K|} f'(\xi)$$
$$\pm |Y|^{2\beta-2} \left((2\beta - 1) f(\xi) - \beta \xi f'(\xi) \right) .$$

Substituting these expressions into the shock polar (4.2.11) gives

$$\frac{\gamma+1}{2} |Y|^{3\beta-3} (f')^3 = -2\sqrt{|K|} |Y|^{3\beta-3} f' \left((2\beta - 1) f - \beta \xi f' \right)$$
$$+ |Y|^{4\beta-4} \left((2\beta - 1) f - \beta \xi f' \right)^2$$

or, to leading order (since $\beta < 1$), $f' = 0$ $(u = v = 0)$, or

$$\frac{\gamma+1}{2} (f')^2 = -2\sqrt{|K|} \left((2\beta - 1) f - \beta \xi f' \right) .$$

Thus, the constant in (4.2.10) is determined, $A = 0$, and

$$\frac{\gamma+1}{2} (f')^2 + 2\sqrt{|K|} \{ (2\beta - 1) f - \beta \xi f' \} = 0 \qquad (4.2.12)$$

in the wave zone.

Equation (4.2.12) is a first order ordinary differential equation with a parameter β. The boundary condition is that $f = 0$ at the head shock (ϕ does not jump across the shock) and furthermore the flow must first compress ($f' < 0$), then expand ($f' > 0$) as the wave zone is traversed. (Figure 4.2.3) We must determine β so that such a solution exists, then determine the corresponding f.

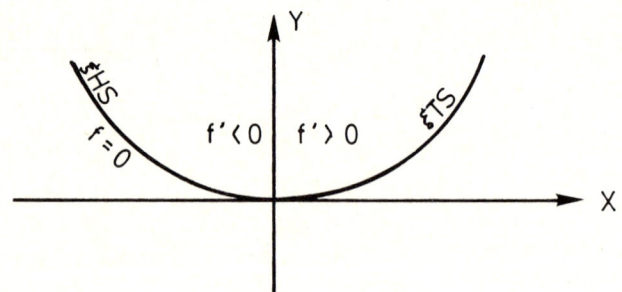

Figure 4.2.3

The Similarity Coordinate ξ and the behavior of f

Consider equation (4.2.12) in the new coordinates

$$t = \frac{f'}{\xi}, \quad s = \frac{f}{\xi^2}$$

namely

$$\frac{\gamma+1}{4\sqrt{|K|}}t^2 - \beta t + (2\beta - 1)s = 0, \qquad (4.2.13)$$

then the solution can be found by integrating

$$\frac{d\xi}{\xi} = \frac{ds}{t - 2s}$$

along the curve, i.e., a solution of (4.2.13). We are looking for a solution which starts at $s = 0$, $t > 0$ ($f = 0$, $\xi < 0$, $f' < 0$), and proceeds through $s < 0$, $t > 0$ ($f < 0$, $\xi < 0$, $f' < o$).

For $\beta \neq 1/2$, from equation (4.2.9), we see that when ξ passes through zero $f' \neq 0$, $f \neq 0$. (f' passes through zero only once and f must still be negative there.) Thus the curve must go through infinity ($\xi = 0$), and must pass through $t = 0$ before passing through $s = 0$. The curve passes from $t > 0$, $s < 0$ to $t < 0$, $s < 0$ ($f' < 0$, $\xi > 0$, $f < 0$ or $f' > 0$, $\xi < 0$) and then to $t > 0$, ($f' > 0$, $\xi > 0$). Thus the curve must look like Figure 4.2.4a or 4.2.4b if $\beta \neq 1/2$.

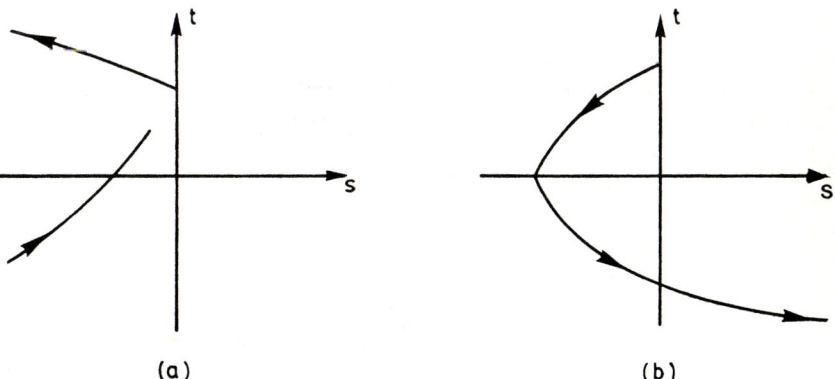

(a)   (b)

Figure 4.2.4

Desired Behavior of f

However (4.2.10) has the form of Figure 4.2.4a,b for $0 < \beta < 1/2$, $1 > \beta > 1/2$ respectively. Thus these two cases are ruled out.

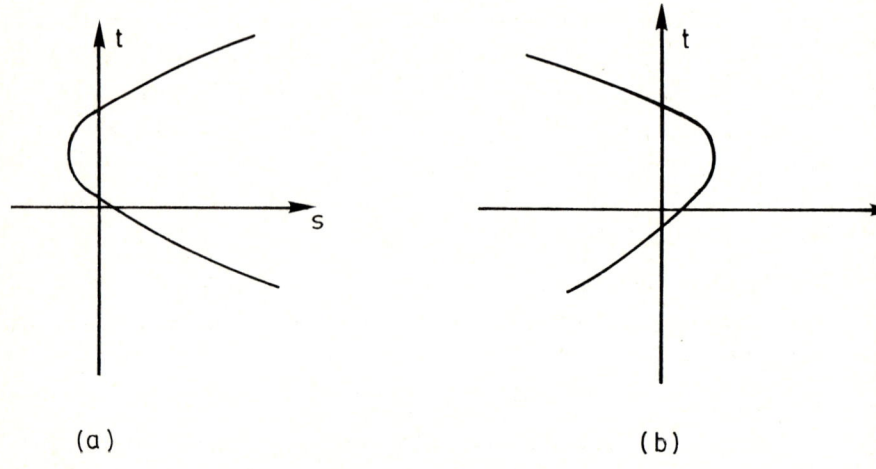

(a)   (b)

Figure 4.2.5

Behavior of (4.2.10) for $\beta \pm 1/2$.

For $\beta = 1/2$ we have

$$\frac{\gamma + 1}{4\sqrt{|K|}} f'^2 - \frac{1}{2}\xi f' = 0 ,$$

hence $f' = 0$, $(u = v = 0)$, or

$$f' = +\frac{2\sqrt{|K|}\xi}{\gamma + 1}$$

(where the positive root is taken since the flow must first compress, then expand, Figure 4.2.3). Integrating once,

$$f(\xi) = +\frac{\sqrt{|K|}}{\gamma + 1}(\xi^2 - \lambda^2) , \qquad (4.2.14)$$

where λ is as yet unknown. Then

$$\phi = \frac{\sqrt{|K|}}{\gamma + 1}\left(\frac{X^2}{Y} - \lambda^2\right) , \qquad (4.2.15)$$

and

$$(\gamma + 1)u = (\gamma + 1)\phi_x = 2\sqrt{|K|}\frac{X}{|Y|} = 2\sqrt{|K|}\,|Y|^{-\frac{1}{2}}\xi. \tag{4.2.16}$$

The head shock and tail shock are at $\xi = \pm\lambda$ respectively and the pressure perturbation is linear in ξ. The constant λ cannot be determined from far field conditions only but must be connected back to the body. The flow is that of an N-wave far from the profile (Figure 4.2.6) (for example on the ground) giving the sonic boom. The shock wave is a parabola to this approximation and ϕ is constant on the similarity curves $\xi = X/\sqrt{|Y|} = $ constant. The pressure perturbation dies out $O(|Y|^{-1/2})$ on similarity curves.

In order to find the first correction to this N-wave flow, namely $|Y|^\delta g$, we must examine the $O(|Y|^{-2})$ terms in equation (4.2.7). The expansions (4.2.6), with values of α, β, f substituted, are;

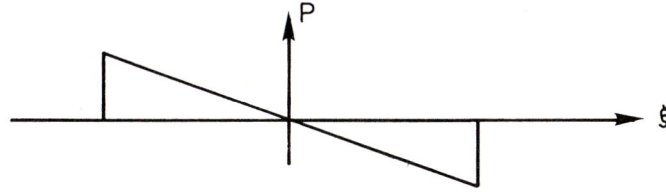

Figure 4.2.6a

Pressure Distribution in the Far Field

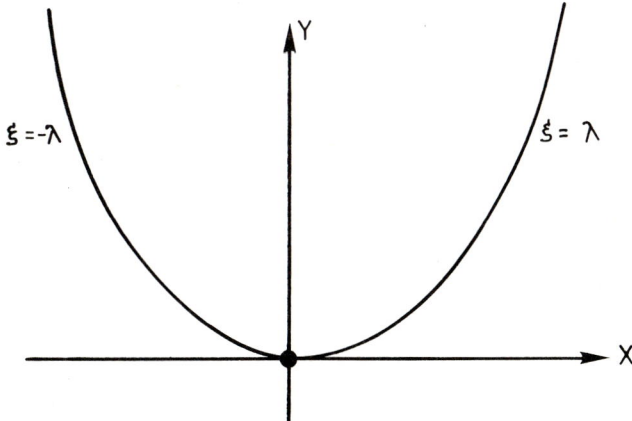

Figure 4.2.6b

Shocks in the Far Field

$$\phi_x = |Y|^{-\frac{1}{2}}\frac{2\sqrt{|K|}}{\gamma+1}\xi + |Y|^{\delta-\frac{1}{2}}g' + \cdots ,$$

$$\phi_{XX} = \frac{2\sqrt{|K|}}{\gamma+1}|Y|^{-1} + |Y|^{\delta-1}g'' + \cdots ,$$

$$\phi_Y = \pm\frac{\sqrt{|K|}}{\gamma+1}\xi^2|Y|^{-1} \pm |Y|^{\delta-1}\left(\delta g - \frac{1}{2}\xi g'\right) + \cdots ,$$

$$\phi_{YY} = \frac{2\sqrt{|K|}}{\gamma+1}\xi^2|Y|^{-2} + \cdots ,$$

$$\phi_{XY} = \pm\frac{-\sqrt{|K|}}{\gamma+1}\xi|Y|^{-\frac{3}{2}} \pm |Y|^{\delta-\frac{3}{2}}\left(\delta g' - \frac{1}{2}(\xi g')'\right) + \cdots .$$

Substituting into the governing equation (4.2.4) we find

$$|Y|^{-2}\frac{2\sqrt{|K|}}{\gamma+1}\xi^2 - 2\sqrt{|K|}\,|Y|^{\delta-\frac{3}{2}}\left(\delta g' - \frac{1}{2}(\xi g')'\right)$$

$$= (\gamma+1)|Y|^{\delta-\frac{3}{2}}\frac{2\sqrt{|K|}}{\gamma+1}(\xi g'' + g') + \cdots .$$

Thus for a nontrivial solution,

$$\delta = -\frac{1}{2}$$

and, collecting terms,

$$\frac{2}{\gamma+1}\xi^2 = \xi g'' .$$

Integation gives

$$g = \frac{1}{3(\gamma+1)}\xi^3 + C\xi + D .$$

Finally we must consider the boundary conditions at the shock. The shock is located at

$$\xi_s = -\lambda + \lambda_1|Y|^{-\frac{1}{2}} ,$$

and the shock polar (4.2.11) in terms of U, V, where $u = U$, $v = V - \sqrt{|K|}\,U$ is

$$\frac{\gamma+1}{2}U^3 = V^2 \mp 2\sqrt{|K|}\,UV . \qquad (4.2.18)$$

Now

$$U = |Y|^{-\frac{1}{2}}\frac{2\sqrt{|K|}}{\gamma+1}\xi + |Y|^{-1}g' + \cdots ,$$

$$V = \pm\frac{\sqrt{|K|}}{\gamma+1}\xi^2|Y|^{-1} \mp |Y|^{-\frac{3}{2}}\frac{1}{2}(g + \xi g') + \cdots .$$

In order to find the appropriate shock jump conditions we expand about the zeroth order shock locations, $\xi = -\lambda$. Thus,

$$
\begin{aligned}
U|_s &= \frac{2\sqrt{|K|}}{\gamma+1}|Y|^{-\frac{1}{2}}\left(-\lambda + \lambda_1|Y|^{-\frac{1}{2}}\right) + |Y|^{-1}g'|_{\xi=-\lambda} + \cdots , \\
&= -\frac{2\sqrt{|K|}}{\gamma+1}\lambda|Y|^{-\frac{1}{2}} + |Y|^{-1}\left(\frac{2\sqrt{|K|}}{\gamma+1}\lambda_1 + g'(-\lambda)\right) + \cdots ,
\end{aligned}
$$

(4.2.19)

$$
\begin{aligned}
V|_s &= \frac{-\sqrt{|K|}}{\gamma+1}|Y|^{-1}\left(\lambda^2 - 2\lambda\lambda_1|Y|^{-\frac{1}{2}}\right) - |Y|^{-\frac{3}{2}}\frac{1}{2}\left(g(-\lambda) - \lambda g'(-\lambda)\right) + \cdots , \\
&= \frac{-\sqrt{|K|}}{\gamma+1}\lambda^2 Y^{-1} + Y^{-\frac{3}{2}}\left(2\lambda\lambda_1\frac{\sqrt{|K|}}{\gamma+1} - \frac{1}{2}\left(g(-\lambda) - \lambda g'(-\lambda)\right)\right) + \cdots .
\end{aligned}
$$

(4.2.20)

Since

$$
[\phi_s] = 0 ,
$$

$$
g|_s = \frac{2\sqrt{|K|}}{\gamma+1}\lambda\lambda_1 .
$$

(4.2.21)

Substituting the expansions (4.2.19), (4.2.20) for U, V at the shock into the shock polar (4.2.18) we find to leading order, $O(Y^{-2})$,

$$
g'|_s = \frac{1}{2(\gamma+1)}\left\{-4\sqrt{|K|}\lambda_1 + \lambda^2\right\} .
$$

(4.2.22)

Here we have substituted for $g|_s$ using (4.2.21). Equations (4.2.21), (4.2.22) give us two relations among the three unknowns C, D and the shock shift λ_1.

4.2.2. Axisymmetric Flow

The axisymmetric small disturbance equation to be solved for $K < 0$ is (3.9.33)

$$
(-|K| - (\gamma+1)\phi_x)\phi_{xx} + \phi_{\tilde{r}\tilde{r}} + \frac{1}{\tilde{r}}\phi_{\tilde{r}} = 0 .
$$

(4.2.23)

The analysis of the far field is similar to that for two-dimensional flow (Section 4.2.1), as might be expected since (4.2.23) differs from the two-dimensional equation only in the addition of the last term. Thus the equation is hyperbolic in the far field, ($\phi_x \to 0$ and there is no flow ($\phi = $ constant) in front of the wave zone which is bounded by the head and tail shocks.

To analyze the far field of equation (4.2.23) it is convenient (as in the 2-D case) to use coordinates based on the linearized characteristics $x - \sqrt{|K|}\tilde{r} =$ constant. Thus we use coordinates

$$X = x - \sqrt{|K|}\,\tilde{r} - x_0\,, \quad R = \tilde{r}\,, \tag{4.2.24}$$

and look for a solution of similarity form so that

$$\phi(x,\tilde{r}) = R^\alpha f(\xi) + R^\delta g(\xi) + \cdots\,, \tag{4.2.25}$$

where

$$\xi = \frac{X}{R^\beta} \tag{4.2.26}$$

and $\delta < \alpha$, $\beta > 0$.

Differentiating expansion (4.2.25), we have that

$$\phi_X = R^{\alpha-\beta} f' + R^{\delta-\beta} g' + \cdots \tag{4.2.27}$$

$$\phi_{XX} = R^{\alpha-2\beta} f'' + R^{\delta-2\beta} g'' + \cdots \tag{4.2.28}$$

$$\phi_R = R^{\alpha-1}\{\alpha f - \beta\xi f'\} + R^{\delta-1}\{\delta g - \beta\xi g'\} + \cdots \tag{4.2.29}$$

$$\phi_{RR} = R^{\alpha-2}\{\alpha(\alpha-1)f - \beta(\alpha-1)\xi f' - \beta\alpha\xi f' + \beta^2\xi(\xi f')'\}$$
$$+ R^{\delta-2}\{\delta(\delta-1)g - \beta(\delta-1)\xi g' - \beta\xi\delta g' + \beta^2\xi(\xi g')'\}$$
$$+ \cdots\,, \tag{4.2.30}$$

$$\phi_{XR} = R^{\alpha-\beta-1}\{(\alpha-\beta)f' - \beta\xi f''\} + R^{\delta-\beta-1}\{(\delta-\beta)g' - \beta\xi g''\}$$
$$+ \cdots\,. \tag{4.2.31}$$

Since $\frac{\partial}{\partial x} = \frac{\partial}{\partial X}$, $\frac{\partial}{\partial r} = \frac{\partial}{\partial R} - \sqrt{|K|}\frac{\partial}{\partial X}$, we must have that equation (4.2.23) in X, R coordinates is

$$-(\gamma+1)\phi_X\phi_{XX} + \phi_{RR} - 2\sqrt{|K|}\,\phi_{XR} + \frac{1}{R}\left(\phi_R - \sqrt{|K|}\,\phi_X\right) = 0\,. \tag{4.2.32}$$

Substituting the expansions for the derivatives of ϕ into the equation (4.2.32) gives, to leading order (if $\beta < 1$)

$$-(\gamma+1)R^{2\alpha-3\beta} f'f'' - 2\sqrt{|K|}\,R^{\alpha-\beta-1}\left\{\left(\alpha-\beta+\frac{1}{2}\right)f' - \beta\xi f''\right\} = 0\,. \tag{4.2.33}$$

Thus for a balance

$$2\alpha - 3\beta = \alpha - \beta - 1\,,$$

or, as in the two-dimensional case,

$$\alpha = 2\beta - 1 , \tag{4.2.34}$$

and the equation to be solved for f is,

$$-(\gamma + 1)f'f'' + 2\beta\xi\sqrt{|K|}f'' - \sqrt{|K|}(2\beta - 1)f' = 0 . \tag{4.2.35}$$

Integrating once gives

$$\frac{-(\gamma + 1)f'^2}{2} + 2\beta\sqrt{|K|}\xi f' - (4\beta - 1)\sqrt{|K|}f = A , \tag{4.2.36}$$

where A is to be determined by the shock conditions.

The shock polar is

$$\left[K\tilde{r}\phi_x - \frac{\gamma + 1}{2}\tilde{r}\phi_x^2 \right] [\phi_x] + [\tilde{r}\phi_{\tilde{r}}][\phi_{\tilde{r}}] = 0 ,$$

hence with ϕ ahead constant, for ϕ behind the shock,

$$-|K|\phi_x^2 - \frac{\gamma + 1}{2}\phi_x^3 + \phi_{\tilde{r}}^2 = 0 , \tag{4.2.37}$$

or in X, R coordinates,

$$-\frac{\gamma + 1}{2}\phi_X^3 + \phi_R^2 - 2\sqrt{|K|}\phi_X\phi_R = 0 . \tag{4.2.38}$$

Substituting from (4.2.27), (4.2.29) we have to highest order

$$R^{3\beta - 3}\left(-\frac{\gamma + 1}{2}f'^3 - 2\sqrt{|K|}(2\beta - 1)ff' + 2\sqrt{|K|}\beta\xi f'^2 \right) = 0 , \tag{4.2.39}$$

hence $f' = 0$, or

$$-\frac{\gamma + 1}{2}f'^2 - 2\sqrt{|K|}(2\beta - 1)f + 2\sqrt{|K|}\beta\xi f' = 0 , \tag{4.2.40}$$

behind the shock. Since f at the head shock is zero (ϕ does not jump across the shock) we see that (4.2.40) agrees with equation (4.2.36) at the shock if $A = 0$.

So, we must solve the nonlinear eigenvalue problem

$$-\frac{\gamma + 1}{2}f'^2 + 2\beta\sqrt{|K|}\xi f' - (\beta - 1)\sqrt{|K|}f = 0 , \tag{4.2.41}$$

where, as in (4.2.1) we require a solution f so that f is zero, then negative ($f' < 0$, the flow is slowed by the shock), then $f' > 0$ as the flow speeds up again. With $t = f'/\xi$, $s = f/\xi^2$, the equation is

$$-\frac{(\gamma + 1)}{2}t^2 + 2\beta\sqrt{|K|}\,t - (4\beta - 1)\sqrt{|K|}\,s = 0 .$$

This equation is the same as (4.2.13) except for the $(4\beta - 1)$ as coefficient instead of $2\beta - 1$, and the requirements on f, f' are the same. Thus that argument holds once again so that the desired solution exists only if

$$\beta = \frac{1}{4} . \qquad (4.2.42)$$

Then we have

$$\left(-\frac{\gamma + 1}{2}f' + \frac{1}{2}\sqrt{|K|}\,\xi\right)f' = 0 ,$$

hence $f' = 0$ or

$$f'(\xi) = \frac{\sqrt{|K|}\,\xi}{(\gamma + 1)} , \qquad (4.2.43)$$

$$f(\xi) = \frac{\sqrt{|K|}}{2(\gamma + 1)}(\xi^2 - \lambda^2) . \qquad (4.2.44)$$

This is the same function (up to a factor of two) as obtained for the two-dimensional case, the potential and velocities however, are

$$\phi = R^{-\frac{1}{2}}f(\xi) = \tilde{r}^{-\frac{1}{2}}f(\xi)$$

$$u = \phi_x = R^{-\frac{3}{4}}f'(\xi) = \tilde{r}^{-\frac{3}{4}}\frac{\sqrt{|K|}\,\xi}{(\gamma + 1)} ,$$

$$v = \phi_{\tilde{r}} = R^{-\frac{3}{2}}\left\{-\frac{1}{2}f - \frac{1}{4}\xi f'\right\} - R^{-\frac{3}{4}}\sqrt{|K|}f' ,$$

or

$$v = \frac{-\sqrt{|K|}\,\xi}{(\gamma + 1)\tilde{r}^{\frac{3}{4}}} - \frac{\sqrt{|K|}\,\xi^2}{4(\gamma + 1)\tilde{r}^{\frac{3}{2}}} + \frac{\sqrt{|K|}\,\lambda\xi}{4(\gamma + 1)\tilde{r}^{\frac{3}{2}}} .$$

So, once again the head and tail shocks are located asymptotically at $\xi = \pm\lambda$, and the presure perturbation is linear in ξ. The presure now dies out $O(\tilde{r}^{-3/4})$ on the similarity curves $\xi = $ constant. To first order we have the shock polar.

$$v = -\sqrt{|K|}\,u - \frac{\gamma + 1}{\sqrt{|K|}}\frac{u^2}{4} .$$

In order to determine the correction to the first order flow, we redo the analysis including the next order terms. Substituting the derivative expansions (4.2.27-4.2.31) into the governing equation (4.2.32), and using values of α, β, f, we have to next order,

$$R^{\delta - \frac{5}{4}} \left\{ -(\gamma + 1)(f''g' + f'g'') - 2\sqrt{|K|}\left(\delta - \frac{1}{4}\right)g' + \frac{\sqrt{|K|}}{2}\xi g'' - \sqrt{|K|}g' \right\}$$
$$+ R^{-\frac{5}{2}}\left\{ \frac{1}{4}f + \frac{5}{16}\xi f' + \frac{1}{16}\xi^2 f'' \right\} = 0 .$$

Thus

$$\delta = -\frac{5}{4} , \tag{4.2.45}$$

and the equation to be solved for g is

$$-\frac{\xi}{2}g'' + g' = -\frac{1}{2(\gamma + 1)}\left(\xi^2 - \frac{\lambda^2}{4}\right) .$$

So

$$\frac{g''}{\xi^2} - \frac{2}{\xi^3}g' = \frac{1}{(\gamma + 1)}\left(\frac{1}{\xi} - \frac{\lambda^2}{4\xi^3}\right) ,$$

$$g' = \frac{\xi^2}{(\gamma + 1)}\left(\ln|\xi| + \frac{\lambda^2}{4\xi^2} + \bar{B}\right) ,$$

$$g = \frac{\xi^3}{3(\gamma + 1)}\ln|\xi| + \frac{\lambda^2}{8(\gamma + 1)}\xi + B\xi^3 + C . \tag{4.2.46}$$

There are also the shock conditions to be satisfied. The shock is located at

$$\xi_s = -\lambda + \lambda_1 R^{-\frac{3}{4}} + \cdots , \tag{4.2.47}$$

$$\phi_X|_s = -R^{-\frac{3}{4}}\frac{\sqrt{|K|}}{\gamma + 1}\lambda + R^{-\frac{3}{2}}\left\{ \frac{\sqrt{|K|}}{\gamma + 1}\lambda_1 + g'|_s \right\} + \cdots , \tag{4.2.48}$$

$$\phi_R|_s = -R^{-\frac{3}{2}}\frac{\lambda^2\sqrt{|K|}}{4(\gamma + 1)} + R^{-\frac{9}{4}}\left\{ \frac{\sqrt{|K|}\lambda\lambda_1}{(\gamma + 1)} - \frac{5}{4}g + \frac{\lambda}{4}g' \right\} + \cdots . \tag{4.2.49}$$

Since the potential can't jump across a shock, and since

$$\phi|_s = R^{-\frac{5}{4}}\left\{ -\frac{\sqrt{|K|}}{\gamma + 1}\lambda\lambda_1 + g|_s \right\} + \cdots ,$$

we have

$$g|_s = \frac{\sqrt{|K|}}{\gamma + 1} \lambda \lambda_1 . \tag{4.2.50}$$

Substituting the expansions for $\phi_x|_s$, $\phi_R|_s$, (4.2.48) and (4.2.49) into the shock polar (4.2.38) we have to leading order, $O(R^{-3})$,

$$g'|_s = \frac{\lambda^2}{40(\gamma + 1)} \left\{ 8\lambda_1 \sqrt{|K|} + \lambda^2 \right\} . \tag{4.2.51}$$

Here the value for $g|_s$, (4.2.50), has been used. Equations (4.2.50) and (4.2.51) are the two relations to use to determine the constants λ_1, B, and C.

4.3 Far Fields at $M_\infty < 1$ ($K > 0$)

The flow at $M_\infty < 1$ differs strongly from that for sonic ($M_\infty = 1$) or supersonic ($M_\infty > 1$) Mach numbers. For $M_\infty < 1$ there is no division of the flow into independent regions. The supersonic zone is confined to a bubble near the airfoil surface. The flow outside this bubble is subsonic, hence all points influence all other points in the field.

In the far field the flow approaches that of linearized theory - namely governed by a modified Laplace equation. The far field behavior can be found by the development of an expansion assuming to leading order the known solution of linearized theory. The solution in the far field can also be found by analyzing the asymptotic behavior of the integral equation governing the flow. This last method has the advantage that it gives a representation for the coefficients in the expansion.

The result is that unlike $M_\infty > 1$ and $M_\infty = 1$, for lifting flow at $M_\infty < 1$ the far field expansion is dominated by the effect of lift. The far field can be characterized as due to a vortex, source, then various doublets - in fact singularities located at the origin.

4.3.1 Two-Dimensional Flow

For $M_\infty < 1$, ($K > 0$) the far field flow is subsonic. The governing equation (3.1.16)

$$\left(K - (\gamma + 1)\phi_x\right)\phi_{xx} + \phi_{\hat{y}\hat{y}} = 0 \,,$$

under the change of variables

$$\hat{y} = \sqrt{K}\tilde{y} = \sqrt{1 - M_\infty^2}\, y \tag{4.3.1}$$

becomes

$$\phi_{xx} + \phi_{\hat{y}\hat{y}} = \frac{(\gamma + 1)}{K}\phi_x\phi_{xx} \,. \tag{4.3.2}$$

Thus in the far field, $\phi_x \to 0$, the basic flow in the $x, \tilde{y}$ coordinates is governed by Laplace's equation. The far field behavior can be obtained either by the development of an expansion starting from the known behavior of linearized flow, or by expanding an integral equation. We begin by following the first procedure since it is simpler. The second procedure will be discussed after since, although more complicated, it does give more information on the unknown constants.

For a lifting airfoil the lift is connected to the circulation,

$$\frac{L}{\gamma P_\infty \delta^{\frac{1}{3}}} = \oint u\, d\ell = \int_0^1 [u]_{\hat{y}=0}\, dx = [\phi]_{\substack{\hat{y}=0 \\ x>1}} = \Gamma .$$ (4.3.3)

For linearized theory the dominant term in the far field is that of circulation,

$$\phi \sim \frac{-\Gamma}{2\pi}\theta ,$$ (4.3.4)

where

$$\theta = \tan^{-1}\frac{\sqrt{|K|}\hat{y}}{x} = \tan^{-1}\frac{\hat{y}}{x} ,$$ (4.3.5)

and we will use

$$r = \sqrt{x^2 + K\hat{y}^2} = \sqrt{x^2 + \hat{y}^2} .$$ (4.3.6)

The possible $\log r$ term is missing from the expansion since the airfoil is closed. The next terms in linearized theory would be doublets $O(1/r)$, but due to the nonlinear terms in equation (4.3.2) other corrections enter first.

Successive terms in the asymptotic expansion for ϕ as $r \to \infty$ can be found by solving equation (4.3.2), where at each stage the right hand side (nonlinear terms) are known. In terms of the r, θ coordinates equation (4.3.2) is

$$\phi_{rr} + \frac{1}{r}\phi_r + \frac{1}{r^2}\phi_{\theta\theta} = \frac{\gamma+1}{K}\phi_x\phi_{xx} .$$ (4.3.7)

With the asymptotic form of ϕ assumed to be

$$\phi = \frac{-\Gamma\theta}{2\pi} + \cdots ,$$

we have

$$\phi_x = \frac{\Gamma}{2\pi}\frac{\sin\theta}{r} + \cdots ,$$

$$\phi_{xx} = \frac{-\Gamma}{\pi}\frac{\sin\theta\cos\theta}{r^2} + \cdots ,$$

so that

$$\phi_x\phi_{xx} = -\frac{\Gamma^2}{2\pi^2}\frac{\sin^2\theta\cos\theta}{r^3}$$

$$= -\frac{\Gamma^2}{8\pi^2 r^3}\{\cos\theta - \cos 3\theta\} + \cdots .$$

Then equation (4.3.7) is, to first order,

$$\phi_{rr} + \frac{1}{r}\phi_r + \frac{1}{r^2}\phi_{\theta\theta} = -\frac{(\gamma+1)\Gamma^2}{8\pi^2 K}\left(\frac{\cos\theta}{r^3} - \frac{\cos 3\theta}{r^3}\right) + \cdots .$$

Particular solutions of this equation are easily found of the form $\cos\theta(\frac{\log r}{r})$, $\frac{\cos 3\theta}{r}$, so that

$$\phi = -\frac{\Gamma\theta}{2\pi} + \frac{\Gamma^2(\gamma+1)\log r}{16\pi^2 Kr}\cos\theta - \frac{\Gamma^2(\gamma+1)\cos 3\theta}{64\pi^2 Kr} + \frac{D\cos\theta}{2\pi r} + \frac{E\sin\theta}{2\pi r} + \cdots .$$
(4.3.8)

The first term in (4.3.8) is the (linearized) circulation with which we started. The next two terms are nonlinear corrections. The $\frac{\log r}{r}$ term is necessary since $\frac{\cos\theta}{r}$ is a solution of the homogeneous (Laplace) equation. The fourth and fifth terms are doublet solutions of the homogeneous Laplace equation. The values of the unknown constants can not be determined at this stage since the details of the body have been ignored in the computation. The form of this expansion (4.3.8) agrees with that found by Finn and Gilbarg [4.3.1] and by Ludford [4.3.2].

Note that the expansion for ϕ, (4.3.8), is not uniform as $K \to 0$ unless, at a similar rate, $r \to \infty$.

The drag is given by (3.10.7),

$$\frac{D}{\delta^{\frac{2}{3}}\gamma P_\infty} = \oint \left(\left\{ \frac{v^2}{2} + \frac{\gamma+1}{3}u^3 - \frac{Ku^2}{2} \right\} d\tilde{y} + u\,dx \right) + \int [uv]_s\,dx .$$

From the asymptotic form of ϕ we see that $u = O(1/r)$, $v = O(1/r)$, so that as $r \to \infty$ the contribution from the contour integral vanishes. Thus the drag arises only from the shock, $\tilde{D} = -\frac{\gamma+1}{12}\int_{\text{Shock}} [u]_s^3\,dx$. (See section 3.10)

The hodograph form of this far field can also be found and sketched. The far field of the physical plane corresponds to the point $w = -K$, $\vartheta = 0$ (where $w = (\gamma+1)\phi_x - K$, $\vartheta = (\gamma+1)\phi_{\hat{y}}$) in the hodograph so that the hodograph form of the far field is a singularity located at $w = -K$, $\vartheta = 0$.

Since $\phi_x \sim \frac{\Gamma\hat{y}}{2\pi r^2}$, $\phi_{\hat{y}} = \frac{-\Gamma x}{2\pi r^2}$, we have $(w+K)^2 + \vartheta^2 = (\gamma+1)\frac{\Gamma^2}{4\pi^2 r^2}$. So,

$$\hat{y} \sim \frac{(\gamma+1)(w+K)\Gamma}{2\pi\left((w+K)^2 + \vartheta^2\right)} .$$
(4.3.9)

The lines $\hat{y} = \text{constant}$ correspond to the circles

$$\left(w + K - \frac{1}{2C}\right)^2 + \vartheta^2 = \frac{1}{4C^2} ,$$
(4.3.10)

so that the singularity in the hodograph looks like a doublet (Figure 4.3.1a). This was for a lifting airfoil, $\Gamma \neq 0$. For a non-lifting airfoil, $\Gamma = 0$, ϕ is dominated by the doublet terms of (4.3.8). Then the far field singularity in the hodograph is a branch point. (Figure (4.3.1b)).

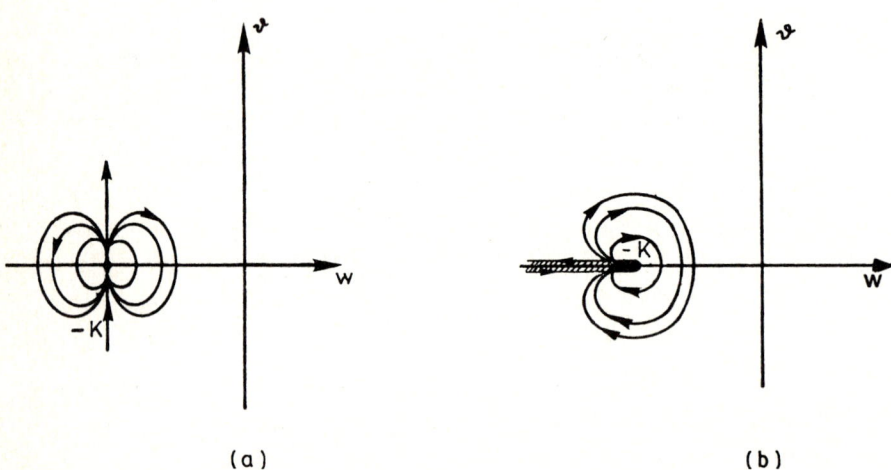

Figure 4.3.1

Far field of a (a) lifting (b) nonlifting airfoil in the hodograph plane

In order to find the far field representation more rigorously the exact integral equation, equivalent to the full boundary problem, equation (4.3.2) with the shock conditions (3.1.25), (3.1.26), boundary conditions (3.1.16, BC1, BC2) and the Kutta condition (3.11b, KJ), can be formulated. Then the asymptotic properties of the integral equation are examined as $r \to \infty$.

In formulating this integral equation we consider the linear portion of the equation as the basic equation, the nonlinear portion as the driving function. So, the source solution of Laplace's equation $S(x, \hat{y})$, satisfying

$$S_{xx} + S_{\hat{y}\hat{y}} = \delta(x - \xi)\delta(\hat{y} - \eta) \tag{4.3.11}$$

with bounded derivatives at infinity is used,

$$S(x - \xi,\, \hat{y} - \eta) = \frac{1}{2\pi} \log \sqrt{(x - \xi)^2 + (\hat{y} - \eta)^2}\,.$$

Then,

$$\iint\limits_{D} \left(S\nabla^2\phi - \phi\nabla^2 S\right) d\xi\, d\eta = -\int_0^1 S(\xi,0)[\phi_n]_w\, d\xi + \int_0^\infty S_\eta(\xi,0)[\phi]_{\hat{y}=0}\, d\xi$$

$$- \int\limits_{\text{shocks}} S(\xi, \eta_s(\xi))\left\{[\phi_\xi]\, d\eta_s - [\phi_\eta]\, d\xi\right\}. \quad (4.3.12)$$

Substituting for $\nabla^2\phi$ from (4.3.2)

$$\nabla^2\phi = \frac{(\gamma+1)}{K}\phi_\xi\phi_{\xi\xi} = \frac{\gamma+1}{2K}\frac{\partial}{\partial\xi}\phi_\xi^2,$$

and for $\nabla^2 S$ from (4.3.11), equation (4.3.12) becomes,

$$\phi(x,y) = \frac{\gamma+1}{2K}\iint\limits_{D} S\frac{\partial}{\partial\xi}(\phi_\xi)^2\, d\xi\, d\eta + \int_0^1 S(\xi,0)[\phi_n]_W\, d\xi$$

$$- \int_0^\infty S_\eta(\xi,0)[\phi]_{\hat{y}=0}\, d\xi + \int\limits_{\text{shocks}} S\left\{[\phi_n]\, d\eta_s - [\phi_\eta]\, d\xi\right\}.$$

Integrating by parts once on the first integral and using (4.3.3), we get,

$$\phi(x,y) = -\frac{\gamma+1}{2K}\iint\limits_{D} S_\xi\phi_\xi^2\, d\xi\, d\eta + \int_0^1 S(\xi,0)[\phi_n]_W\, d\xi$$

$$- \int_0^1 S_\eta(\xi,0)[\phi]_{\hat{y}=0}\, d\xi - \Gamma\int_1^\infty S_\eta(\xi,0)\, d\xi$$

$$+ \int\limits_{\text{shocks}} S\left\{\left([\phi_\xi] - \frac{\gamma+1}{2K}[\phi_\xi^2]\right) d\eta_s - [\phi_\eta]\, d\xi\right\}$$

$$= I_1 + I_2 + I_3 + I_4 + I_5.$$

The integral over the shock, I_5, is zero since the shock condition is contained in the equation, that is the quantity in the parenthesis is zero. The difficulty here is in analyzing the first integral, I_1; I_2, I_3, I_4 are much simpler.

I_2 is, using the boundary condition (3.1.16, BC2)

$$\frac{1}{2\pi}\int_0^1 (F_u' - F_\ell')\log\sqrt{r^2 - 2r\xi\cos\theta + \xi^2}\, d\xi,$$

where r, θ are as in (4.3.5), (4.3.6), so that as $r \to \infty$ we have

$$\frac{1}{2\pi} \log r \int_0^1 (F_u' - F_\ell') \, d\xi - \frac{1}{2\pi} \frac{\cos \theta}{r} \int_0^1 \xi (F_u' - F_\ell') \, d\xi + \cdots ,$$

or, assuming a closed airfoil so that $F_u\binom{0}{1} = F_\ell\binom{0}{1}$,

$$I_2 = \frac{1}{2\pi} \frac{\cos \theta}{r} \int_0^1 (F_u - F_\ell) \, d\xi + \cdots = A_w \frac{\cos \theta}{2\pi r} + \cdots ,$$

where A_w is the cross-sectional area of the airfoil.

Now as $r \to \infty$

$$S_\eta(\xi, 0) = -\frac{1}{2\pi} \frac{r \sin \theta}{r^2 - 2r\xi \cos \theta + \xi^2} = -\frac{\sin \theta}{2\pi r} + \cdots ,$$

so that

$$I_3 = +\frac{\sin \theta}{2\pi r} \int_0^1 [\phi(\xi, 0)] \, d\xi .$$

Now S_η has a singularity in the domain of integration of I_4 so a little more care must be taken.

$$I_4 = \frac{\Gamma \hat{y}}{2\pi} \int_1^\infty \frac{1}{(x - \xi)^2 + \hat{y}^2} \, d\xi ,$$

so that substituting $\xi = x + \hat{y} \tan \theta$ we have as $r \to \infty$,

$$\begin{aligned}
I_4 &= -\frac{\Gamma}{2\pi} \tan^{-1} \frac{\hat{y}}{x - 1} \\
&= \frac{\Gamma}{2\pi} \tan^{-1} \frac{\hat{y}}{x} \left(1 + \frac{1}{x} + \cdots \right) \\
&= -\frac{\Gamma}{2\pi} \tan^{-1} \frac{\hat{y}}{x} - \frac{\Gamma}{2\pi} \frac{\hat{y}}{x^2 + \hat{y}^2} + \cdots \\
&= -\frac{\Gamma}{2\pi} \theta - \frac{\Gamma \sin \theta}{2\pi r} + \cdots .
\end{aligned}$$

So far then,

$$\phi(x, \hat{y}) = I_1 - \frac{\Gamma \theta}{2\pi} + \frac{\sin \theta}{2\pi r} + \left\{ -\Gamma + \int_0^1 [\phi(\xi, 0)] \, d\xi \right\} + A_w \frac{\cos \theta}{2\pi r} + \cdots .$$

Now to evaluate I_1 we break the integral up into two regions, one inside a fixed circle of radius M, the other outside the circle. Outside the circle we add and subtract the circulation term hence,

$$\frac{-2K}{\gamma+1}I_1 = \int_0^M \int_0^{2\pi} S_\xi \phi_\xi^2 \rho \, d\rho \, d\vartheta + \int_M^\infty \int_0^{2\pi} S_\xi \left(\phi_\xi^2 - \left(\frac{\Gamma}{2\pi} \right)^2 \frac{\sin^2 \vartheta}{\rho^2} \right) \rho \, d\rho \, d\vartheta$$

$$+ \int_M^\infty \int_0^{2\pi} S_\xi \left(\frac{\Gamma}{2\pi} \right)^2 \frac{\sin^2 \vartheta}{\rho} \, d\rho \, d\vartheta$$

$$= J_1 + J_2 + J_3 \, ,$$

where

$$\xi^2 + \eta^2 = \rho^2 \, , \quad \phi = \tan^{-1} \frac{\eta}{\xi} \, ,$$

and the first integral has a slit along $\eta = 0$ from $\xi = 0$ to $\xi = 1$.

Now

$$S_\xi = -\frac{1}{2\pi} \frac{r \cos \theta - \rho \cos \vartheta}{r^2 - 2r\rho \cos(\theta - \vartheta) + \rho^2}$$

so that for ρ bounded as $r \to \infty$,

$$S_\xi = -\frac{1}{2\pi} \frac{\cos \theta}{r} + \cdots . \tag{4.3.13}$$

So, as $r \to \infty$,

$$J_1 = -\frac{1}{2\pi} \frac{\cos \theta}{r} \int_0^M \int_0^{2\pi} \phi_\xi^2 \rho \, d\rho \, d\vartheta + \cdots ,$$

where again the integral has a slit along $\eta = 0$ from $\xi = 0$ to $\xi = 1$. Then also

$$J_2 = -\frac{1}{2\pi} \frac{\cos \theta}{r} \int_M^\infty \int_0^{2\pi} \left(\phi_\xi^2 - \left(\frac{\Gamma}{2\pi} \right)^2 \frac{\sin^2 \vartheta}{\rho^2} \right) \rho \, d\rho \, d\vartheta \, .$$

This last is not immediately obvious since the asymptotic form (4.3.13) of S_ξ is not valid throughout the range of integration. However, the remaining portion of the integrand, $\{\phi_\xi^2 - (\Gamma^2/(2\pi)^2)(\sin^2 \vartheta/\rho^2)\} = O(1/\rho^2)$ as $\rho \to \infty$. In fact, as we will see, it is $O((\log \rho)/\rho^3)$. Thus, when the asymptotic form of S_ξ is no longer valid, the remaining portion of the integral is small. Writing

$$J_2 = \frac{\cos \theta}{r} \int_M^\infty \int_0^{2\pi} \phi_\xi^2 - \left(\frac{\Gamma}{2\pi} \right)^2 \frac{\sin^2 \vartheta}{\rho^2} \rho \, d\rho \, d\vartheta$$

$$+ \int_M^\infty \int_0^{2\pi} \left(S_\xi - \frac{\cos \theta}{r} \right) \left(\phi_\xi^2 - \left(\frac{\Gamma}{2\pi} \right)^2 \frac{\sin^2 \vartheta}{\rho^2} \right) \rho \, d\rho \, d\vartheta \, ,$$

it can be shown that the second integral is $O((\log r)/r^2)$ as $r \to \infty$.

Finally we must evaluate J_3,

$$J_3 = \left(\frac{\Gamma}{2\pi}\right)^2 \int_M^\infty \rho\, d\rho \int_0^{2\pi} -\frac{1}{2\pi} \frac{r\cos\theta - \rho\cos\vartheta}{r^2 - 2r\rho\cos(\theta - \vartheta) + \rho^2}\, \frac{\sin^2\vartheta}{\rho^2}\, d\vartheta$$

$$= -\frac{\Gamma^2}{(2\pi)^3} \int_M^\infty \rho^{-1}\, d\rho \int_0^{2\pi} \frac{(r\cos\theta - \rho\cos\vartheta)\sin^2\vartheta}{r^2 - 2r\rho\cos(\theta - \vartheta) + \rho^2}\, d\vartheta\ .$$

Letting $\theta - \vartheta = x$, expanding the integrand, and using

$$\int_0^\pi \frac{\cos nx\, dx}{1 - 2a\cos x + a^2} = \begin{cases} \dfrac{\pi}{1 - a^2}a^n & \text{if } a^2 < 1, \\[2mm] \dfrac{\pi}{a^n(a^2 - 1)} & \text{if } a^2 > 1, \end{cases}$$

the quantity in brackets can be evaluated. It is,

$$\{\quad\} = \begin{cases} 2\pi\left(\dfrac{\cos\theta}{2r} - \dfrac{\cos 3\theta}{4}\dfrac{\rho^2}{r^3}\right) & \text{if } \rho < r, \\[3mm] \dfrac{2\pi r\cos\theta}{4\rho^2} & \text{if } \rho > r. \end{cases}$$

So,

$$J_3 = -\frac{\Gamma^2}{(2\pi)^2} \int_M^r \left(\frac{\cos\theta}{2r\rho} - \frac{\cos 3\theta\rho}{4r^3}\right) d\rho + \int_r^\infty \frac{r\cos\theta}{4\rho^3}\, d\rho$$

$$= -\left(\frac{\Gamma}{2\pi}\right)^2 \left\{\frac{\cos\theta\log r}{2r} - \frac{\cos\theta\log M}{2r} - \frac{\cos 3\theta}{8r} + \frac{\cos\theta}{8r} + \cdots\right\}\ .$$

Thus, putting all the results together we have

$$\phi = -\frac{\Gamma\theta}{2\pi} + \frac{\gamma + 1}{4K}\left(\frac{\Gamma}{2\pi}\right)^2 \frac{\log r}{r}\cos\theta - \frac{\gamma + 1}{16K}\left(\frac{\Gamma}{2\pi}\right)^2 \frac{\cos 3\theta}{r}$$

$$+ \frac{D\cos\theta}{2\pi r} + \frac{E\sin\theta}{2\pi r} + \cdots \tag{4.3.14}$$

where

$$\frac{D}{2\pi} = \frac{A_w}{2\pi} + \frac{\gamma + 1}{16K}\left(\frac{\Gamma}{2\pi}\right)^2 + \frac{\gamma + 1}{4K\pi}\int_0^M \int_0^{2\pi} \rho\phi_\xi^2\, d\rho\, d\vartheta$$

$$+ \frac{\gamma + 1}{4K\pi}\left\{\int_M^\infty \int_0^{2\pi} \left(\phi_\xi^2 - \left(\frac{\Gamma}{2\pi}\right)^2 \frac{\sin^2\vartheta}{\rho^2}\right)\rho\, d\rho d\vartheta + \left(\frac{\Gamma}{2\pi}\right)^2 \pi\log M\right\}, \tag{4.3.15}$$

$$\frac{E}{2\pi} = \frac{1}{2\pi} \int_0^1 \left[\phi(\xi, 0) \right] d\xi - \frac{\Gamma}{2\pi} \ . \tag{4.3.16}$$

Note that the dependence on M is illusory. The $\log M$ term is actually subtracted back out by the second term in the integral in the same brackets when evaluated at the bottom limit. Similarly the first term in that integral has its M dependence cancelled by the first integral in D.

Note that the far field obtained by iteration (4.3.8) agrees with that obtained from the integral equation (4.3.14). The latter also gives the doublet strengths (4.3.15), (4.3.16).

4.3.2 Axisymmetric Flow

For nonlifting axisymmetric flow at $M_\infty < 1$ the far field is dominated by doublet terms, as it is for incompressible flow. In this section this result is found from the integral representation of the potential ϕ and thus a representation for the doublet coefficient is found. The analysis follows that of Section 4.3.1.

With $\tilde{r} = \sqrt{K(\tilde{y}^2 + \tilde{z}^2)}$ the equation governing axisymmetric flow, (3.9.33), is

$$\phi_{xx} + \phi_{\tilde{r}\tilde{r}} + \frac{1}{\tilde{r}} \phi_{\tilde{r}} = \frac{\gamma+1}{2K} (\phi_x^2)_x \ .$$

The source function S for Laplace's equation is $\frac{1}{\tilde{r}}$, so by Green's Theorem

$$\phi = \frac{\gamma+1}{2K} \iiint_D S(\phi_\xi^2)_\xi \, dv + \iint_{\delta D} \left(S \frac{\partial \phi}{\partial n} - \frac{\partial S}{\partial n} \right) da \ , \tag{4.3.17}$$

where D is three-space bounded by the body, which is an infinitesimal cylinder along the x axis, $0 \le x \le 1$, a surface at infinity, S_∞, and the shocks.

We will see that for large R, $\phi = O(x/R^3)$, where $R = (x^2 + \tilde{r}^2)^{\frac{1}{2}}$. So, if S_∞ is a cylinder of radius $\tilde{r}_\infty$, height x_∞, the integral over S_∞ vanishes as $\tilde{r}_\infty$, $x_\infty \to \infty$.

Integrating the first integral of (4.3.17) by parts once we find it is

$$-\frac{\gamma+1}{2K} \iiint_D S_\xi \phi_\xi^2 \, dv + \lim_{\varepsilon \to 0} \iint_{c_\varepsilon} \left(S \frac{\partial \phi}{\partial n} - \phi \frac{\partial s}{\partial n} \right) da \tag{4.3.18}$$

where c_ε is a cylinder of radius ε, $0 \le x \le 1$. Note that the shock term is missing, it was

$$\iint_S \left(S \frac{\partial \phi}{\partial n} - \frac{\gamma+1}{2K} S \phi_\xi^2 \right) da$$

which is zero since the integrand is the shock jump condition (3.9.36).

Now from (3.9.36) we know that

$$\phi \underset{r \to 0}{\sim} \left(\frac{F^2(x)}{2} \right)' \log \tilde{r} ,$$

so that the expression for ϕ, (4.3.18) is

$$\phi = -\frac{\gamma + 1}{2K} \iiint S_\xi \phi_\xi^2 \, dv + \int_0^1 \int_0^{2\pi} S|_{\rho=0} \left(\frac{F^2(x)}{2} \right)' \, d\vartheta \, dx$$

$$= I_1 + I_2 ,$$

where

$$S = \left((x - \xi)^2 + \tilde{r}^2 + \rho^2 - 2\tilde{r}\rho \cos(\theta - \vartheta) \right)^{-\frac{1}{2}} .$$

Thus

$$I_1 = \frac{1}{2} \int_0^1 \int_0^{2\pi} \frac{\left(F^2(\xi) \right)'}{\sqrt{(x - \xi)^2 + \tilde{r}^2}} \, d\vartheta \, d\xi$$

$$\underset{r \to \infty}{\sim} \pi \int_0^1 \frac{x\xi \left(F^2(\xi) \right)'}{R^3} \, d\xi$$

or, after integration by parts once,

$$I_1 \underset{r \to \infty}{\sim} \frac{-\pi x}{R^3} \int_0^1 F^2(\xi) \, d\xi .$$

Here we have assumed a closed body, $F\binom{0}{1} = 0$.

Now for ξ, ρ bounded, as $\tilde{r} \to \infty$

$$S_\xi = \frac{x - \xi}{\left((x - \xi)^2 + \tilde{r}^2 - 2\tilde{r}\rho \cos(\theta - \vartheta) + \rho^2 \right)^{\frac{3}{2}}} \tag{4.3.19}$$

$$= \frac{x}{R^3} + O\left(\frac{1}{R^3} \right) .$$

Then

$$I_2 = -\frac{\gamma + 1}{2K} \iiint_D S_\xi \phi_\xi^2 \, dv$$

$$= -\frac{(\gamma + 1)x}{2K R^3} \int_0^{2\pi} \int_{-\infty}^{\infty} \int_0^{\infty} \phi_\xi^2 \rho \, d\rho \, d\xi \, d\vartheta$$

$$- \frac{\gamma + 1}{2K} \int_0^{2\pi} \int_{-\infty}^{\infty} \int_0^{\infty} \left(S_\xi - \frac{x}{R^3} \right) \phi_\xi^2 \rho \, d\rho \, d\xi \, d\vartheta . \tag{4.3.20}$$

Using (4.3.19), and the fact that $\phi \underset{R\to\infty}{=} O(1/R^2)$, it can be seen that the second integral in (4.3.20) is $O(1/R^3)$ as $R \to \infty$. Thus

$$\phi_1 \underset{R\to\infty}{\sim} \frac{Dx}{R^3}$$

where

$$D = -\pi \left\{ \int_0^1 F^2(\xi)\, d\xi + \frac{\gamma+1}{K} \int_{-\infty}^{\infty} \int_0^{\infty} \phi_\xi^2 \rho\, d\rho\, d\xi \right\} .$$

4.3.3 Lifting Three-Dimensional Wing

In order to find the far field behavior of a lifting three-dimensional wing an integral equation is found for the potential ϕ, and then the behavior of this equation is analyzed asymptotically. The analysis is thus similar to that for the two dimensional wing, Section 4.3.1.

This far field was found by Klunker [4.3.3] and corrections to his work were later made by Murman [4.3.4].

Under the change of variables $(\hat{\ }) = \sqrt{K}(\tilde{\ })$, the three-dimensional transonic equation is

$$\phi_{xx} + \phi_{\hat{y}\hat{y}} + \phi_{\hat{z}\hat{z}} = \frac{\gamma+1}{2K}(\phi_x^2)_x . \tag{4.3.21}$$

The equation holds in a region bounded by the wing surface W, the trailing vortex sheet V, and the shocks S. The surfaces V, W are in the $x, \tilde{z}$ plane, $-B \leq \tilde{z} \leq B$, $(-\sqrt{K}B \leq \hat{z} \leq \sqrt{K}B)$ (Figure 4.3.2). Hence by Green's Theorem, with the source function for the Laplacian

$$S(x, \hat{y}, \hat{z}) = \left(4\pi \sqrt{(x-\xi)^2 + (\hat{y}-\eta)^2 + (\hat{z}-\varsigma)^2} \right)^{-1} ,$$

$$\iiint_D (S\nabla^2\phi - \phi\nabla^2 S)\, dv = \iint_{\delta D} \left(\phi \frac{\partial S}{\partial n} - S \frac{\partial\phi}{\partial n} \right) da ,$$

or, using equation (4.3.21)

$$\phi(x, \hat{y}, \hat{z}) = \iiint_D S \frac{\gamma+1}{2K}(\phi_\xi^2)_\xi\, dv + \iint_{\delta D} \left(S \frac{\partial\phi}{\partial n} - \phi \frac{\partial S}{\partial n} \right) da , \tag{4.3.22}$$

where n is the inward normal to δD and δD consists of $W \cup V \cup S \cup S_\infty$. Here S_∞ is the enclosing surface at infinity.

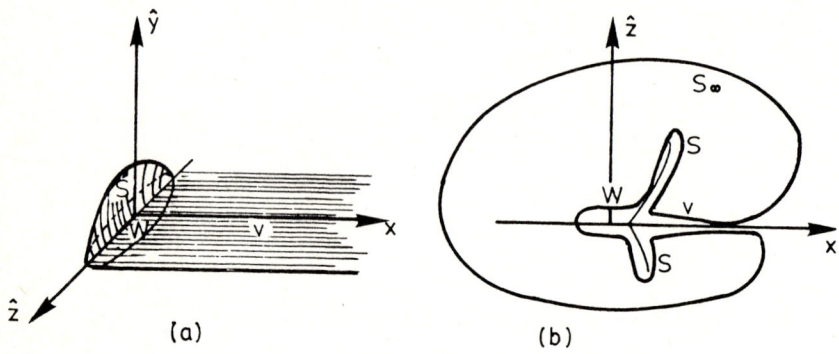

Figure 4.3.2

Three-Dimensional Wing and Cross Section

Equation (4.3.22) for ϕ can be further simplified by several steps. The first integral can be integrated by parts once giving a new integral over D plus an integral over the shock surface S. The second integral is

$$\int_S S\frac{\partial \phi}{\partial n}\, da - \int_W \left(\phi\frac{\partial S}{\partial n} - S\frac{\partial \phi}{\partial n}\right) da - \int_V \frac{\partial S}{\partial n}\, da\ .$$

Here we have used the fact that ϕ does not jump across the shock, and that ϕ_x does not jump across the vortex sheet, so that their contributions cancel. Now ϕ_x is continuous across V hence the jump in ϕ across V is a function only of z. This fact can be used to eliminate any integrals over the vortex sheet. That is, integrating by parts once we have

$$-\iint_{W\cup V} \phi\frac{\partial S}{\partial n}\, da$$

$$= \frac{\hat{y}}{4\pi} \iint_W [\phi_\xi]\left(1 + \frac{x-\xi}{\sqrt{(x-\xi)^2 + \hat{y}^2 + (\hat{z}-\varsigma)^2}}\right) \frac{1}{\hat{y}^2 + (\hat{z}-\varsigma)^2}\, d\xi\, d\varsigma\ ,$$

where now $[\phi_\xi]$ is $\phi_\xi(\xi, 0^+, \varsigma) - \phi_\xi(\xi, 0^-, \varsigma)$, the jump in ϕ across the wing. Finally, collecting all the shock contributions, we have

$$\int_S S\left[-\frac{\gamma+1}{2K}\phi_x^2 + \frac{\partial \phi}{\partial n}\right] da\ .$$

The quantity in brackets is the shock jump condition (3.1.43), so that this last integral is zero. Hence, collecting terms, and using boundary condition (3.1.36)

$$\phi = \frac{\gamma+1}{2K} \iiint_D S_\xi \phi_\xi^2 \, d\xi \, d\eta \, d\varsigma - \iint_W S[F_\xi] \, d\xi d\varsigma$$
$$+ \frac{\hat{y}}{4\pi} \iint_W [\phi_\xi] \frac{1}{\hat{y}^2 + (\hat{z}-\varsigma)^2} \left(1 + \frac{x-\xi}{\sqrt{(x-\xi)^2 + \hat{y}^2 + (\hat{z}-\varsigma)^2}} \right) d\xi \, d\varsigma$$
$$= I_1 + I_2 + I_3 \,, \tag{4.3.23}$$

and $[\;\;] = (\;\;)_u - (\;\;)_\ell$.

Note that if ϕ is independent of z (two-dimensional flow), (4.3.23) can be integrated in the z direction and the result agrees with equation (4.3.12). Also note that the integral over S_∞ has disappeared. This has to be checked after the far field has been obtained. The second integral in equation (4.3.23) is a superposition of sources over the wing surface. The third integral is a superposition of horseshoe vortices along the wing surface.

For $R = \sqrt{x^2 + \hat{y}^2 + \hat{z}^2} \gg 1$, ξ, ς bounded, $\eta = 0$,

$$S(x, \hat{y}, \hat{z}) \sim \frac{1}{4\pi R} \left(1 + \frac{x\xi}{R^2} + \frac{\hat{z}\varsigma}{R^2} + \cdots \right),$$

so that

$$\iint_W S[F_\xi] \, d\xi \, d\varsigma \sim \frac{x}{4\pi R^3} \iint_W \xi[F_\xi] \, d\xi \, d\varsigma \,.$$

For a closed wing $\int_W [F_\xi] \, d\xi = 0$, so that after integration by parts once we have

$$I_2 = - \iint_W S[F_\xi] \, d\xi \, d\varsigma \sim -\frac{x}{4\pi R^3} \iint_W [F] \, d\xi d\varsigma \,.$$

In analyzing the asymptotes of I_3 note that there are two regions to be considered, one near the wake (x large, $r = O(1)$), the other far from the wake (r large).

For $R \gg 1$, $r = \hat{y}^2 + \hat{z}^2 \gg 1$, that is away from the wake, and ξ, ς bounded,

$$\frac{1}{\hat{y}^2 + (\hat{z}-\varsigma)^2} \left(1 + \frac{x-\xi}{\sqrt{(x-\xi)^2 + \hat{y}^2 + (\hat{z}-\varsigma)^2}} \right) \sim \frac{1}{r^2} \left(1 + \frac{2\varsigma\hat{z}}{r^2} \right)$$
$$\times \left(1 + \frac{x}{R} + \frac{x\hat{z}\varsigma - r^2\xi}{R^3} \right),$$

so that

$$I_3 \sim \frac{\hat{y}}{4\pi r^2}\left(1 + \frac{x}{R}\right)\iint_W [\phi_\xi]\,d\xi\,d\varsigma$$

$$+ \frac{\hat{z}\hat{y}}{4\pi r^2}\left(\frac{2}{r^2}\left(1 + \frac{x}{R}\right) + \frac{x}{R^3}\right)\iint_W \varsigma[\phi_\xi]\,d\xi\,d\varsigma - \frac{\hat{y}}{4\pi R^3}\iint_W \xi[\phi_\xi]\,d\xi\,d\varsigma \ .$$

Note that if ϕ_ξ is even in ς, that is the wings are symmetric left and right, then the second term is zero and

$$I_3 \sim \frac{\hat{y}}{4\pi r^2}\left(1 + \frac{x}{R}\right)\iint_W [\phi_\xi]\,d\xi\,d\varsigma - \frac{\hat{y}}{4\pi R^3}\iint_W \xi[\phi_\xi]\,d\xi\,d\varsigma \ .$$

For $R \gg 1$, r small, that is in a neighborhood of the vortex wake, we can no longer expand $\hat{y}^2 + (\hat{z} - \varsigma)^2$. Then we have

$$I_3 \sim \frac{\hat{y}}{2\pi}\iint_W \frac{[\phi_\xi]}{(\hat{z} - \varsigma)^2 + \hat{y}^2}\,d\xi\,d\varsigma - \frac{\hat{y}}{8\pi x^2}\iint_W [\phi_\xi]\,d\xi d\varsigma \ ,$$

$$\sim \frac{\hat{y}}{2\pi}\int_W \frac{\Gamma(\varsigma)}{\hat{y}^2 + (\hat{z} - \varsigma)^2}\,d\varsigma - \frac{\hat{y}}{8\pi x^2}\int_W \Gamma(\varsigma)\,d\varsigma \ .$$

Finally we deal with the asymptotics of I_1. As long as the integration volume is bounded we have

$$S_\xi \sim \frac{x}{4\pi R^3} \ .$$

On the other hand for unbounded regions of integration we have (from I_2, I_3) that

$$\phi = O\!\left(\frac{1}{R^2}\right) \ .$$

Hence

$$I_1 \sim \frac{x}{4\pi R^3}\frac{\gamma + 1}{2K}\iiint \phi_\xi^2\,d\xi\,d\eta\,d\varsigma \ .$$

Finally then we have that

$$\phi \sim \phi_{\text{lift}} + \frac{x}{4\pi R^3}\left(-\iint_W [F]\,d\xi\,d\varsigma + \frac{\gamma + 1}{2K}\iiint_D \phi_\xi^2\,d\xi\,d\eta\,d\varsigma\right) \qquad (4.3.24)$$

where

$$\phi_{\text{lift}} = I_3$$

$$
\underset{R \to \infty}{\sim}
\begin{cases}
\dfrac{\hat{y}}{r^2}\left(1 + \dfrac{x}{R}\right)L - \dfrac{\hat{y}}{R^3}M_y + \dfrac{\hat{z}\hat{y}}{r^2}\left(\dfrac{2}{r^2}\left(1 + \dfrac{x}{R}\right) + \dfrac{x}{R^3}\right)M_x & \text{if } r \gg 1, \\[3mm]
\dfrac{\hat{y}}{2\pi}\displaystyle\int_W \dfrac{\Gamma(\varsigma)}{\hat{y}^2 + (\hat{z} - \varsigma)^2}\, d\varsigma - \dfrac{\hat{y}}{8\pi x^2}\displaystyle\int_W \Gamma(\varsigma)\, d\varsigma & \text{if } r \ll 1,
\end{cases}
$$

$$(4.3.25)$$

where

$$L = \frac{1}{4\pi}\iint_W [\phi_\xi]\, d\xi\, d\varsigma, \qquad M_y = \frac{1}{4\pi}\iint_W \xi[\phi_\xi]\, d\xi\, d\varsigma$$

$$M_x = \frac{1}{4\pi}\iint_W \varsigma[\phi_\xi]\, d\xi\, d\varsigma, \qquad \Gamma = \int_W [\phi_\xi]\, d\xi.$$

As in two-dimensional $(M_\infty < 1)$ theory the lift contribution dominates the expansion in the far field. In this case for $r \gg 1$ the leading order term is an elementary horshoe vortex at the origin. Higher order lift contributions consist of a dipole at the origin and, for a wing not symmetric in z, quadrupoles distributed along the axis.

4.3.4 Unsteady Two-Dimensional Flow

The far field for an oscillating two-dimensional transonic $(K > 0)$ airfoil can be found under the assumption that, as in the steady $(K > 0)$ case, circulation dominates [4.3.5]. Then in the far field the flow looks like that from a concentrated vortex at the origin.

The unsteady transonic equation (3.11.13)

$$\left(K - (\gamma + 1)\phi_x\right)\phi_{xx} + \phi_{\hat{y}\hat{y}} - 2\phi_{x\hat{t}} = 0$$

approaches the linearized equation

$$K\phi_{xx} + \phi_{\hat{y}\hat{y}} - 2\phi_{x\hat{t}} = 0 \qquad (4.3.26)$$

as $r = \sqrt{x^2 + K\tilde{y}^2} \to \infty$ since $\phi_x \to 0$. Thus our goal is to solve the system

$$Ku_x + v_{\hat{y}} = 2u_{\hat{t}} \qquad (4.3.27a)$$

$$u_{\hat{y}} - v_x = \tilde{\Gamma}\delta(x)\delta(\tilde{y})e^{i\Omega t} \qquad (4.3.27b)$$

where $u = \phi_x$, $v = \phi_{\tilde{y}}$ and we have assumed that the forced frequency is dominant, that is that the induced harmonics die out faster. Once we have solved the system, then

$$\Gamma(\tilde{t}) = \int u\,dx + v\,d\tilde{y} = \iint_A (u_{\tilde{y}} - v_x)\,dx\,d\tilde{y} = \tilde{\Gamma}e^{i\Omega t} .$$

We have assumed complex notation, later the real part will be taken in accordance with boundary condition (3.11.14).

It is convenient to first eliminate time from equation (4.3.27a). Thus let

$$t^* = \tilde{t} + \frac{2}{K}x ,$$

$$x^* = x ,$$

so that the system (4.3.27) becomes

$$K u_{x^*} + v_{\tilde{y}} = 0 \tag{4.3.28a}$$

$$u_{\tilde{y}} - v_{x^*} - \frac{2}{K}v_{t^*} = \tilde{\Gamma}\delta(x^*)\delta(\tilde{y})e^{i\Omega t^*} . \tag{4.3.28b}$$

Now a stream function ψ can be defined,

$$u = \psi_{\tilde{y}} , \quad v = -K\psi_{x^*} .$$

Under the additional change of variables

$$y^* = \sqrt{K}\,\tilde{y}$$

equation (4.3.28) becomes

$$\psi_{y^*y^*} + \psi_{x^*x^*} + \frac{2}{K}\psi_{x^*t^*} = \frac{\tilde{\Gamma}}{K}\delta(x^*)\delta(y^*)\sqrt{K}e^{i\Omega t^*} . \tag{4.3.29}$$

Finally separating out the time variation so that

$$\psi = \Psi(x^*, y^*)e^{i\Omega t^*} ,$$

we have

$$\nabla^{*2}\Psi + \frac{2i\Omega}{K}\Psi_{x^*} = \frac{\tilde{\Gamma}}{\sqrt{K}}\delta(x^*)\delta(y^*) , \tag{4.3.30}$$

where

$$\nabla^* = \frac{\partial}{\partial x^*} , \frac{\partial}{\partial y^*} .$$

Then letting

$$\Psi = e^{\gamma x}\Lambda, \quad \text{where} \quad \gamma = \frac{-i\Omega}{K} \, ,$$

we have

$$\nabla^{*2}\Lambda + \frac{\Omega^2}{K^2}\Lambda = \frac{\tilde{\Gamma}}{\sqrt{K}}\delta(x^*)\delta(y^*) \, . \tag{4.3.31}$$

For x^*, $y^* \neq 0$ this is just the Helmholtz equation. The solution to equation (4.3.31) is symmetric about the origin hence $\Lambda(r^*)$ and

$$\Lambda_{r^*r^*} + \frac{1}{r^*}\Lambda_{r^*} + \frac{\Omega^2}{K^2}\Lambda = \frac{\tilde{\Gamma}}{\sqrt{K}}\frac{\delta(r^*)}{2\pi r^*} \, , \tag{4.3.32}$$

where $r^* = \sqrt{x^{*2} + y^{*2}}$.

For $\Omega = 0$, steady flow, the solution to equation (4.3.32) is $\Lambda = c\log r^*$, as expected from steady analysis. For $\Omega \neq 0$ the solution is

$$\Lambda = BH_0^{(2)}\left(\frac{\Omega r^*}{K}\right) . \tag{4.3.33}$$

Here $H_0^{(2)} = J_0 - iY_0$, the companion solution $H_0^{(1)}$ has been excluded by the radiation condition that waves must not run in from infinity. For large r^*, $H_0^{(2)}(\Omega r^*/K) \sim (\sqrt{\frac{2K}{\pi\Omega r^*}})e^{-i((\Omega r^*/K)-(\pi/4))}$, so that

$$\psi \sim B\sqrt{\frac{2K}{\pi\Omega r^*}}e^{-\frac{i\Omega}{K}x+i\Omega t^*-i\left(\frac{\Omega r^*}{K}-\frac{\pi}{4}\right)} \, .$$

or in terms of the original coordinates

$$\psi \sim B\sqrt{\frac{2K}{\pi\Omega r}}e^{+i\frac{\pi}{4}+i\Omega\tilde{t}-i\frac{\Omega}{K}(r-x)} \, .$$

Note that the phase is zero if

$$K\tilde{t} = r - x$$

or

$$K\tilde{t}^2 + 2\tilde{t}x = \tilde{y}^2 \, ,$$

which is identical to the equation of the characteristic surface given by (3.11.22) since $K^* \rightarrow K$ as $r \rightarrow \infty$. Thus asymptotically the waves travel outward on the characteristic cone.

To determine B note that from (4.3.31) Λ is the response to a source at $r^* = 0$, so

$$\Lambda \sim \frac{\tilde{\Gamma}}{\sqrt{K}} \frac{\log r^*}{2\pi} \; .$$

From (4.3.33)

$$\Lambda \sim B\left(-\frac{2i}{\pi} \log\left(\frac{\Omega r^*}{K}\right)\right) \; ,$$

so that

$$B = \frac{\tilde{\Gamma} i}{4\sqrt{K}} \; .$$

Thus

$$\Lambda = \frac{i\tilde{\Gamma}}{4\sqrt{K}} H_0^{(2)}\left(\frac{\Omega r^*}{K}\right) \; ,$$

and

$$\psi = \frac{i\tilde{\Gamma}}{4\sqrt{K}} e^{i\Omega\left(\tilde{t}+\frac{x}{K}\right)} H_0^{(2)}\left(\frac{\Omega r^*}{K}\right) \; .$$

Since these are only valid approximations for our original problem for large r, our only interest is in their asymptotic behavior,

$$\psi \underset{\infty}{\sim} \frac{i\tilde{\Gamma}}{4} \sqrt{\frac{2}{\pi\Omega r}} e^{i\Omega\left(\tilde{t}-\frac{r}{K}+\frac{x}{K}\right)} e^{i\frac{\pi}{4}} \; ,$$

To find the potential ϕ note that $\phi_x = \psi_y$, so $\phi = \int_{-\infty}^{x} \psi_y \, dx$. Thus

$$\phi_x = \frac{-i\tilde{\Gamma}}{4} \sqrt{\frac{2}{\pi r}} \frac{\tilde{y}}{r^{\frac{3}{2}}} e^{i\Omega\left(\tilde{t}-\frac{r}{k}+\frac{x}{K}\right)} e^{i\frac{\pi}{4}+\cdots} \; .$$

so that

$$\phi \sim \frac{-i\tilde{\Gamma}}{4} \sqrt{\frac{2}{\pi\Omega}} \tilde{y} e^{i\Omega\tilde{t}} e^{i\frac{\pi}{4}} \int_{-\infty}^{x} \tilde{r}^{-\frac{3}{2}} e^{+i\frac{\Omega}{K}(-r^*+x)} \, dx$$

or

$$\phi \underset{\infty}{\sim} \frac{\tilde{\Gamma}}{4} K \sqrt{\frac{2}{\pi\Omega}} \frac{\tilde{y}(|x|)^{\frac{1}{2}}}{r(|x|+r)} \cos\left(\Omega\tilde{t} - \frac{\Omega}{K}(r+|x|) - \frac{\pi}{4}\right) \; . \qquad (4.3.34)$$

Note that the potential in the far field lags behind that at the origin.

References

[4.3.1] R. Finn, and Gilberg, D., Asymptotic Behavior and Uniqueness of Plane Subsonic Flows, *Comm. Pure and Appl. Math.*, **10**, 1957, pp. 26-63.

[4.3.2] G. S. S. Ludford, The Behavior at Infinity of the Potential Function of a Two Dimensional Subsonic Compresible Flow, *J. Math. and Phys.*, **25**, (1951), pp. 117-130.

[4.3.3] Klunker, E. B., Contribution to Methods for Calculating the Flow about Thin Lifting Wings at Transonic Speeds - Analytic Expressions for the Far Field, NASA TMD-6530, Nov. 1971.

[4.3.4] Murman, E., private communication.

[4.3.5] Krupp, J. A. and J. D. Cole, Unsteady Transonic Flow, Studies in Transonic Flow IV, UCLA School of Engineering and Applied Science Report, UCLA-ENG-76104, October 1976.

5 Transonic Airfoil Theory

In this section transonic airfoil theory is studied. Certain special features of small-disturbance theory such as the flow near the nose and shock wave near a wall are considered first. Then a numerical method is discussed (Section 5.4) for subsonic flow in the physical plane and sonic flow in the hodograph (Section 5.5). Finally a discussion is given of the Transonic Stabilization Law which describes flow near $M_\infty = 1$ in terms of flow at $M_\infty = 1$.

5.1 Problem Formulation

The expansion procedure of Section 3.1 leads to a mathematical problem for the disturbance potential $\phi(x, \tilde{y})$ which is summarized here.

$$\left(K - (\gamma + 1)\phi_x\right)\phi_{xx} + \phi_{\tilde{y}\tilde{y}} = 0, \quad \text{K-G equation}, \tag{5.1.1a}$$

or

$$\left(K\phi_x - \frac{\gamma + 1}{2}\phi_x^2\right)_x + (\phi_{\tilde{y}})_{\tilde{y}} = 0, \quad \text{conservation form}. \tag{5.1.1b}$$

(i) The boundary condition of tangent flow is applied at $\tilde{y} = 0$ in the transonic plane $(-\frac{1}{2} \leq x \leq \frac{1}{2})$,

$$\phi_{\tilde{y}}(x, 0\pm) = F'_{u,\ell}(x) = \pm t'(x) + c'(x) - A, \tag{5.1.2}$$

where $F_{u,\ell}(x)$ is the shape function of the airfoil upper and lower surfaces respectively; alternatively the shape for a fixed airfoil is expressed by $t(x)$ the thickness distribution, $c(x)$ the camber line, $A = \frac{\alpha}{\delta} = \text{constant}$ where α is the geometrical angle of attack. (See Figure 5.1.1)

(ii) The Kutta- Joukowski condition must be satisfied at the trailing edge. Since the pressure must balance at the trailing edge and

$$c_p = -2\delta^{\frac{2}{3}}\phi_x,$$

the K-J condition becomes

$$[\phi_x]_{\text{TE}} = \phi_x\left(\frac{1}{2}+, 0+\right) - \phi_x\left(\frac{1}{2}+, 0-\right) = 0. \tag{5.1.3}$$

(iii) The far-field boundary condition is in the first instance that the perturbation vanish at (upstream) infinity $\phi_x, \phi_{\tilde{y}} \to 0$. The more detailed analysis of

Section 4.3.1 shows that the far field has an expansion of the form (4.3.8)

$$\phi = -\frac{\Gamma}{2\pi}\theta + \left(\frac{\Gamma}{2\pi}\right)^2 \frac{\gamma+1}{4}\frac{\log r}{r}\cos\theta - \left(\frac{\Gamma}{2\pi}\right)^2 \frac{\gamma+1}{16}\frac{\cos 3\theta}{Kr}$$

$$+ \frac{D\cos\theta}{2\pi r} + \frac{E\sin\theta}{2\pi r} + \cdots \qquad (5.1.4)$$

where $r = \sqrt{x^2 + K\tilde{y}^2}$, $\theta = \tan^{-1}\dfrac{\sqrt{K}\tilde{y}}{x}$. Here Γ is the circulation around the airfoil

$$\Gamma = \oint \phi_x\,dx + \phi_{\tilde{y}}\,d\tilde{y} = [\phi]_{\mathrm{TE}}\,. \qquad (5.1.5)$$

We can thus consider that there is a jump in ϕ in the wake, that is across $\tilde{y} = 0$, $x > \frac{1}{2}$,

$$[\phi]_{\mathbf{wake}} = \phi(x,0+) - \phi(x,0-) = \Gamma\,. \qquad (5.1.6)$$

Further, expressions have been worked out for the doublet strengths D and E (cf. 4.3.15 and 4.3.1b) and these involve integrals over the solution and the airfoil. These doublet strengths are more useful for nonlifting cases.

(iv) The shock jump conditions should be satisfied on any shock waves that arise in the solution. As explained in Section 3.6.2 the shock relations are contained in the conservative form (5.1.1b) of the basic TSD equation. Integration across the jumps shows that

$$\left\{ \begin{array}{c} \left[K\phi_x - \dfrac{\gamma+1}{2}\phi_x^2\right]d\tilde{y}_s + [\phi_{\tilde{y}}]\,dx_s = 0 \\[2mm] [\phi_{\tilde{y}}]\,d\tilde{y}_s + [\phi_x]\,dx_s = 0 \end{array} \right\}, \qquad (5.1.7)$$

where

$$[\,\,] = (\ )_b - (\ )_a\,,$$

equivalent to (3.6.21). Also, of course, only compression jumps are allowed,

$$[\phi_x] < 0\,.$$

The basic equation plus the conditions (i)-(iv) define, presumably, a unique[*] solution to the problem of flow past a given airfoil. The Kutta condition is essential for fixing the circulation for flows which are subsonic at the trailing edge just as for subsonic flows. Where the flow is supersonic at the trailing edge the condition is satisfied automatically by the wave system which develops at the trailing edge.

---

[*] Recent calculations have indicated the possibility of nonuniqueness for a narrow range of subsonic Mach numbers.

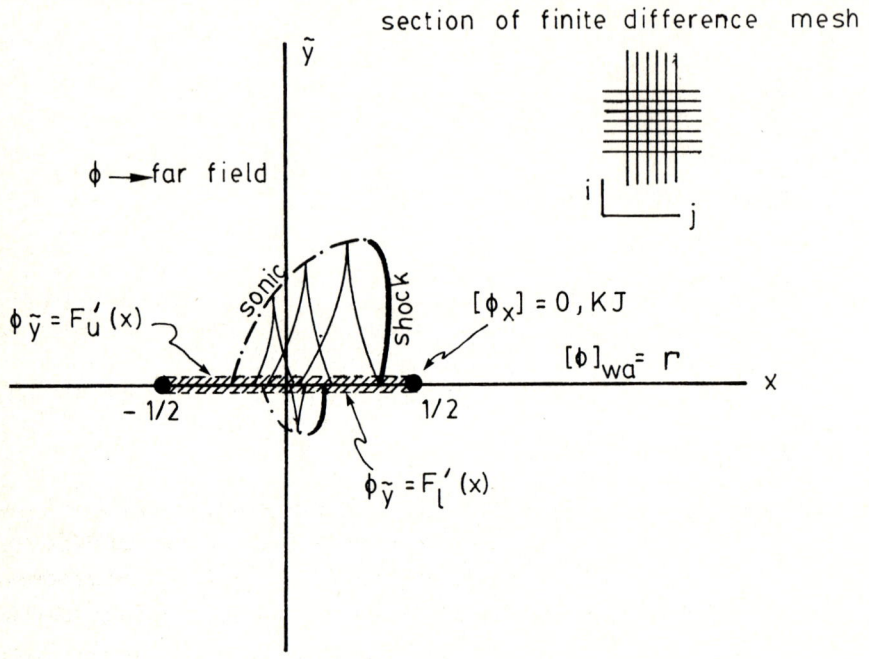

Figure 5.1.1
Typical Subsonic Airfoil Problem, Transonic Plane.

The problem formulation for a supersonic freestream is basically the same except the far-field conditions are different, but are not central to the calculations. All that is required is a uniform flow upstream of the detached or attached bow shock.

5.2 Nose Singularity

In any numerical calculation it is desirable to know the nature of the singularities that appear in the solution. In small disturbance theory the stagnation point in the nose region of an airfoil appears as a singularity. The nature of this singularity can be discovered by a local study of the flow and has the form of a similarity solution for a wide class of cases. In linearized subsonic theory the lifting part has a nose singularity which is a leading edge thrust; the thickness part has a singularity depending on the nature of the leading edge. In the transonic case the non-linearity is essential for characterization of the singularity. The lift and thickness problems can not be split.

Only the case of a finite radius of curvature at the nose is considered here, although the same treatment applies to a range of power law nose shapes. For a parabolic nose

$$F_{u,\ell}(x) = \pm a\sqrt{x}\{1 + \cdots\}; \quad = \sqrt{2R_c}, \ R_c = \text{scaled radius of curvature} \quad (5.2.1)$$

The boundary condition of tangent flow near the nose is thus (cf. 5.1.2)

$$\phi_{\tilde{y}}(x, 0\pm) = \pm \frac{a_0}{2\sqrt{x}}\{1 + \cdots\} \quad \text{for } x > 0. \tag{5.2.2}$$

The basic equation (cf. 5.1.10) can be written

$$(\gamma + 1)\phi_x\phi_{xx} - \phi_{\tilde{y}\tilde{y}} = K\phi_{xx}. \tag{5.2.3}$$

Since a singularity is expected at the nose the dominant terms should be the non-linear terms and $\phi_{\tilde{y}\tilde{y}}$. The effect of the RHS of (5.2.3) should be small.

Thus we can look for a similarity solution of the type discussed in Section 4, as in (4.1.10), (4.1.11),

$$(\gamma + 1)\phi(x, \tilde{y}) = \tilde{y}^{3k-2}f(\xi) + \tilde{y}^\sigma f_1(\xi) + \cdots, \quad \xi = \frac{x}{\tilde{y}^k}. \tag{5.2.4}$$

(See Figure (5.2.1)).

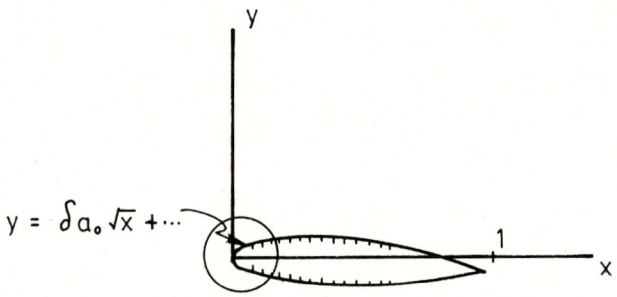

a. Physical Plane

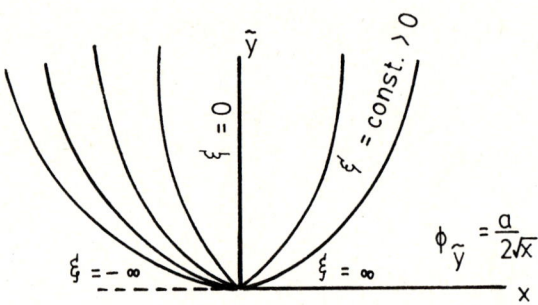

b. Transonic Plane

Figure 5.2.1

Only the first term of (5.2.4) is considered here. Note that

$$(\gamma + 1)\phi_x = \tilde{y}^{2k-2} f'(\xi), \quad (\gamma + 1)\phi_{xx} = \tilde{y}^{k-2} f''(\xi),$$

$$(\gamma + 1)\phi_{\tilde{y}} = \tilde{y}^{3k-3}\left\{(3k-2)f - k\xi f'\right\},$$

$$(\gamma + 1)\phi_{\tilde{y}\tilde{y}} = \tilde{y}^{3k-4}\left\{k^2\xi^2 f'' + 5k(1-k)\xi f' - 3(1-k)(3k-2)f\right\}.$$

(5.2.5)

Thus, considering $\tilde{y} \to 0$ with ξ fixed in (5.2.3)

$$\text{LHS} \sim \tilde{y}^{3k-4}, \quad \text{RHS} \sim \tilde{y}^{k-2},$$

so that the LHS terms domonate for $k < 1$; the solution near the nose is in the first instance independent of the similarity parameter K. $f(\xi)$ is found as a solution of equation (4.1.14).

Consider first the upper half-plane. There ξ goes to infinity as $\tilde{y} \to 0$, $x > 0$. From (4.1.14) we find the expansion

$$(\gamma + 1)\phi = \tilde{y}^{3k-2} \left\{ b_0 \xi^{3-\frac{2}{k}} + b_1 \xi^{3-\frac{3}{k}} + \cdots \right\}, \tag{5.2.6}$$

$$(\gamma + 1)\phi_{\tilde{y}} = \tilde{y}^{3k-3} b_1 \xi^{3-\frac{3}{k}} + \cdots$$
$$= b_1 x^{3-\frac{3}{k}} + \cdots, \quad \tilde{y} \to 0, \, x > 0. \tag{5.2.7}$$

Comparing with the boundary condition (5.2.2) we see that

$$3 - \frac{3}{k} = -\frac{1}{2},$$

or

$$k = \frac{6}{7} < 1. \tag{5.2.8}$$

Further the constant b_1 is identified,

$$b_1 = (\gamma + 1)\frac{a_0}{2}.$$

b_0 is arbitrary, or not determined locally. The general form of the solution near the nose is thus

$$(\gamma + 1)\phi(x, \tilde{y}) = \tilde{y}^{\frac{4}{7}} f(\xi), \quad \xi = \frac{x}{\tilde{y}^{\frac{6}{7}}}. \tag{5.2.9}$$

Now the solution must be continued to the axis ahead ($\xi' = -\infty$) and then to the lower side. In general the whole solution is expressed more simply in terms of the hodograph variables and solutions, which easily connect upper and lower surfaces. In general similarity curves in the physical plane map to similarity curves in the hodograph, the curves $\dfrac{w}{\vartheta^{\frac{2}{3}}} = $ constant. (cf. Section 4.1) This can be seen from (5.2.5) with $w = (\gamma + 1)\phi_x$, $\vartheta = (\gamma + 1)\phi_{\tilde{y}}$,

$$\frac{w}{\vartheta^{\frac{2}{3}}} = \frac{f'}{\left((3k - 2)f - k\xi f' \right)^{\frac{2}{3}}} = fn(\xi). \tag{5.2.10}$$

It also follows that along a similarity curve $\tilde{y} \sim w^{\frac{1}{2k-2}}$. For the nose singularity $\tilde{y} \sim w^{-\frac{7}{2}}$. The hodograph similarity solutions are expressed by equation (4.1.57). The variables are

$$\rho = \sqrt{\vartheta^2 - \frac{4}{9}w^3}, \quad \text{the canonical elliptic distance,} \qquad (5.2.11)$$

and

$$\sin \alpha = \frac{\vartheta}{\rho}. \qquad (5.2.12)$$

It is clear that $\alpha = $ constant represents a similarity curve. For the nose singularity $\tilde{y} \sim \rho^{-7/3}$, so that in (4.1.57) $\beta = \dfrac{13}{6}$. Thus

$$\tilde{y}_I = \rho^{\frac{-7}{3}} F\left(\frac{7}{6}, -1; \frac{1}{2}; \sin^2 \alpha\right) = \rho^{\frac{-7}{3}}\left(1 - \frac{7}{3}\sin^2 \alpha\right), \qquad (5.2.13)$$

$$\tilde{y}_{II} = \rho^{\frac{-7}{3}} \sin \alpha F\left(\frac{5}{3}, -\frac{1}{2}; \frac{3}{2}; \sin^2 \alpha\right). \qquad (5.2.14)$$

The singular solution $\tilde{y}_N$ near the nose, which is equivalent to the far-field behavior ($w \to -\infty$) in the hodograph plane, can thus be represented as

$$\tilde{y}_N = c_1 \rho^{\frac{-7}{3}}\left\{c_2\left(1 - \frac{7}{3}\sin^2 \alpha\right) + \sin \alpha F\left(\frac{5}{3}, -\frac{1}{2}; \frac{3}{2}; \sin^2 \alpha\right)\right\}, \qquad (5.2.15)$$

$$c_1, c_2 = \text{constant}.$$

On the boundary curves representing the upper and lower surfaces of the airfoil $\tilde{y} \to 0\pm$, $x > 0$, and on the axis ahead of the airfoil $\tilde{y} \to 0$, $x < 0$. For a given constant c_2, values of α for which $\tilde{y}_N = 0$ can be found. When there are three roots $\alpha_3 > \alpha_2 > \alpha_1$ the structure is correct to represent the nose flow, α_3 and α_1 specify the upper and lower boundary curves and α_2 the image of the dividing streamline. The course of the hypergeometric functions is shown in Figure 5.2.2 and a sketch of the solution in the canonical hodograph ($\tau = \dfrac{2w^{\frac{3}{2}}}{3}, \vartheta$) and transonic planes appears in Figure 5.2.3. The unsymmetry corresponds to flow around the nose as in a lifting case.

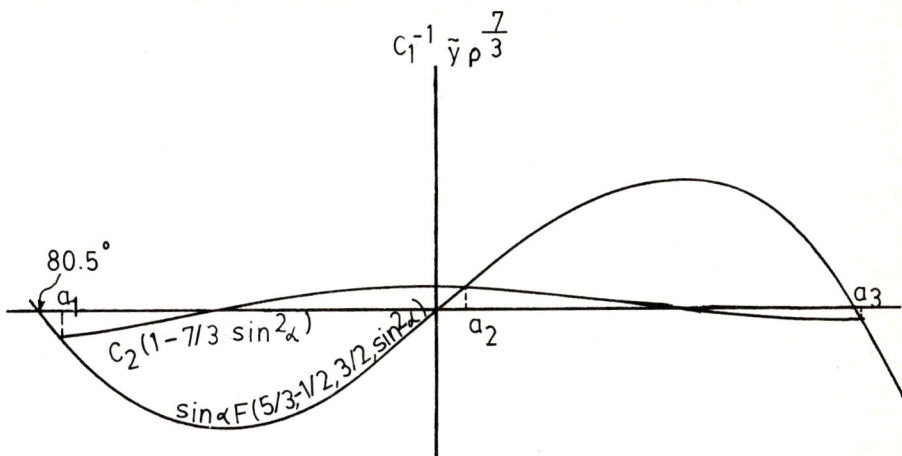

Figure 5.2.2

Sketch of the two hypergeometric functions of the $\tilde{y}_N$ solution

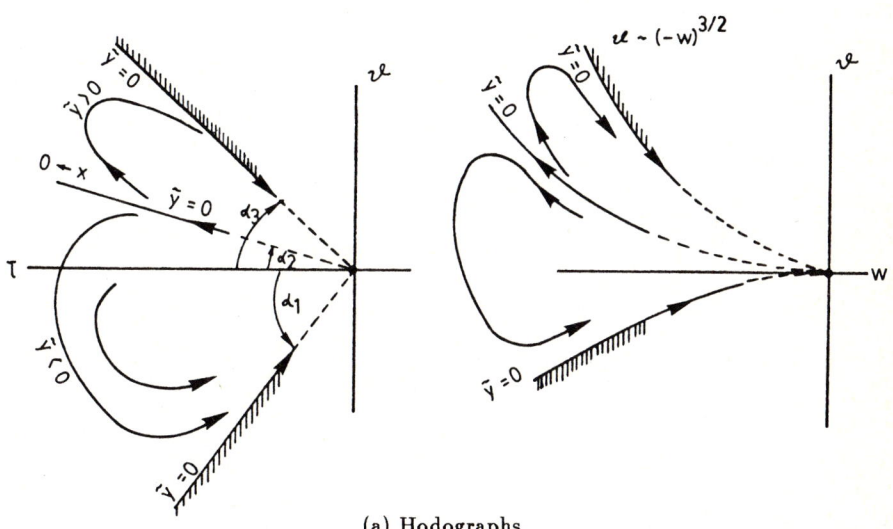

(a) Hodographs

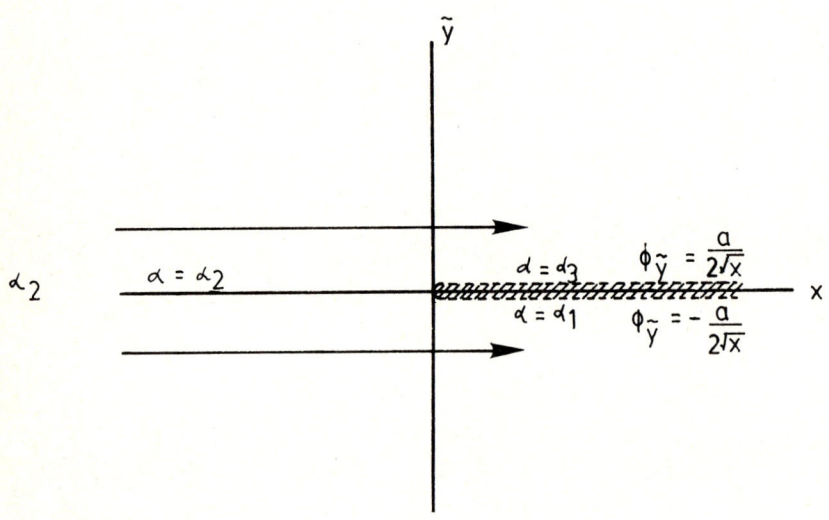

(b) Transonic Plane

Figure 5.2.3

Nose Region Flows

The boundary curves and dividing streamline $\tilde{y}_N = 0$ correpond to $\alpha = $ constant or

$$\vartheta = \tau \tan \alpha, \quad \tau = \frac{2}{3}(-w)^{\frac{3}{2}},$$

$$\text{or} \quad \vartheta = \left(\frac{2}{3}(-w)^{\frac{3}{2}} \tan \alpha\right). \tag{5.2.16}$$

The boundary curves and dividing streamline are straight lines in the ϑ, τ plane and semi-cubical parabolas in the usual hodograph (ϑ, w). For non-lifting airfoils with a potential $\phi(x, \tilde{y})$ symmetric in $\tilde{y}$ the constant c_2 in the nose solution $\tilde{y}_N$ (5.2.15) is zero. Then $\alpha_2 = 0$ and the boundary streamlines near $w = -\infty$ are given by

$$\alpha_3 = 80.5°, \quad \alpha_1 = -80.5°.$$

The constant c_1, of course, fixes the scale of the solution near the nose. For lifting airfoils $\alpha_2 > 0$ and $c_2 < 0$. We also must have $\alpha_1 > -90°$, $\alpha_3 < 90°$ so that $-.2629 < c_2 < 0$, $80.5° < \alpha_3 < 90°$, $-68.3° < \alpha_1 < -80.5°$.

The transonic small disturbance force on the nose region is finite. For instance the nose drag (per unit span) up to $x = \epsilon$ is

$$D_N = \rho_\infty \frac{U^2}{2} C \int_0^\epsilon \left\{ c_{p_u} \delta F'_u(x) - c_{p_\ell} \delta F_\ell(x) \right\} dx, \qquad (5.2.17)$$

where C = characteristic length = airfoil chord, and where c_p = pressure coefficient = $-2\delta^{\frac{2}{3}} \phi_x$.

It follows from (5.2.9) that

$$(\gamma + 1)\phi_x(x, \tilde{y}) = \tilde{y}^{-\frac{2}{7}} f'(\xi). \qquad (5.2.18)$$

Thus, (cf. 5.2.6)

$$(\gamma + 1)\phi_x(x, 0+) \rightarrow b_0 \left(\frac{2}{3} \right) x^{-\frac{1}{3}}. \qquad (5.2.19)$$

This shows the nature of the pressure singularity at the nose. $F'_{u,\ell}(x) \sim x^{-\frac{1}{2}}$ and

$$D_N \sim \int_0^\epsilon \frac{dx}{x^{\frac{5}{6}}} \sim \epsilon^{\frac{1}{6}}. \qquad (5.2.20)$$

The lift force on the nose region is thus also finite and vanishes as $\epsilon \rightarrow 0$.

These considerations can be generalized to the class of power law nose shapes

$$F_{u,\ell}(x) = \pm a x^\mu \left\{ 1 + \cdots \right\}, \qquad \mu < 1. \qquad (5.2.21)$$

Then analogous to (5.2.8)

$$3 - \frac{3}{k} = \mu - 1, \qquad k = \frac{3}{4 - \mu}. \qquad (5.2.22)$$

The general similarity form is

$$(\gamma + 1)\phi = \tilde{y}^{\frac{1+2\mu}{4-\mu}} f(\xi), \qquad \xi = \frac{x}{\tilde{y}^{\frac{3}{4-\mu}}}.$$

and correspondingly the pressure singularity is given by

$$(\gamma + 1)\phi_x(x, 0+) \rightarrow b_0 \left(1 + \frac{2\mu}{3} \right) x^{-\frac{2(1-\mu)}{3}} + \cdots. \qquad (5.2.23)$$

Analogous to (5.2.20) there is

$$D_N \sim \int_0^\epsilon x^{-\frac{2}{3}(1-\mu)} x^{-(1-\mu)} \, dx$$

$$D_N \sim \int_0^\epsilon \frac{dx}{x^{\frac{5}{3}(1-\mu)}} . \tag{5.2.24}$$

The nose drag is thus finite for $2/5 < \mu < 1$ which gives a limit of bluntness for transonic small disturbance theory.

It is often possible to disregard the nature of this singularity when finite difference calculations are done in $(x, \tilde{y})$ (Section 5.4) but its structure is essential for the hodograph construction of airfoils as in Section 5.5.

5.3 Shock Wave at a Curved Surface

This section works out a few details of the flow near the foot of a strong shock wave (supersonic to subsonic) where it strikes a curved surface. The calculation is planar. Because the equation is elliptic in the subsonic region behind the shock a singularity is expected in the solution at the corner where the normal shock meets the surface and where there is a change in the type of boundary condition. On the surface the condition of tangent flow is enforced and at the shock two jump conditions are enforced linking the flow downstream to that ahead. The nature of the singularity was first worked out by Gadd [5.3.1] in the same inviscid framework as used here. Some further details of the shock shape were given by Oswatitsch and Zierep [5.3.2]. Emmons pointed out that the singularity was of the nature of that at a jump in surface curvature in incompressible flow. Details presented below show that this is so and this results in a logarithmically infinite acceleration along the surface immediately downstream of the shock.

In the following paragraph a slightly more general treatment is given than [5.3.2]. The idea is that there is a normal shock at the wall with conditions ahead $w = w_a$, and behind $w_b = -w_a$, in accord with the shock polar (cf. Section 3.6). Suitable series representations are sought ahead of the shock wave and behind the shock wave in terms of distance from the foot of the shock wave. It is shown how these series can locally satisfy the condition of tangent flow and the shock jump conditions.

The flow near the shock is described conveniently in terms of $\varphi(x, \tilde{y})$ where (cf. Section 5.1)

$$\varphi_x \varphi_{xx} - \varphi_{\tilde{y}\tilde{y}} = 0, \tag{5.3.1}$$

and

$$\varphi_x = w = (\gamma + 1)\phi_x - K,$$
$$\varphi_{\tilde{y}} = \vartheta = (\gamma + 1)\phi_{\tilde{y}}.$$

Here w measures the difference from sonic and ϑ the flow deflection. Consider the mixed flow as in Figure (5.3.1) where the shock terminates the subsonic region at a point x_0.

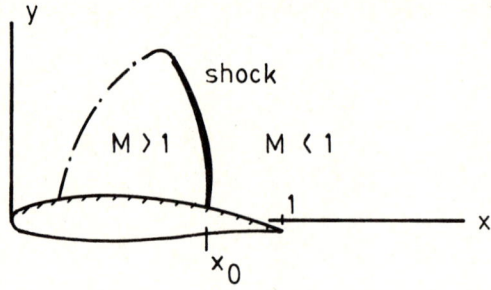

Figure 5.3.1

Shock near curved wall, physical plane

In the transonic plane $(x, \tilde{y})$ (Figure 5.3.2) the shock runs to $x = x_0$ at $\tilde{y} = 0$.

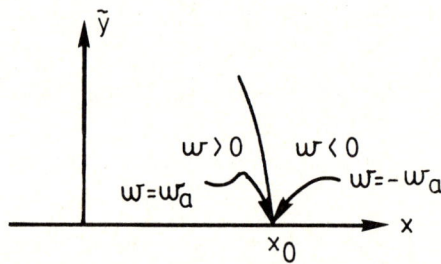

Figure 5.3.2

Shock near curved wall, transonic plane

The boundary condition of tangent flow at $\tilde{y} = 0$ can be expanded about $x = x_0$,

$$\phi_{\tilde{y}}(x,0) = F'(x) = F'(x_0) + (x - x_0)F''(x_0) + \frac{(x - x_0)^2}{2!}F'''(x_0) + \cdots , \quad (5.3.2)$$

or

$$\vartheta(x,0) = \vartheta_0 + (x - x_0)(\gamma + 1)F''(x_0) + \frac{(x - x_0)^2}{2!}(\gamma + 1)F'''(x_0) + \cdots , \quad (5.3.3)$$

where $\vartheta_0 = (\gamma + 1)F'(x_0)$.

Because of the invariance of the original system of equations under translation in x and in ϑ we can write

$$\vartheta - \vartheta_0 \to \vartheta,$$

$$x - x_0 \to x,$$

(5.3.4)

and consider the shock at the origin over a wall which is horizontal at $x = 0$, (Figure 5.3.3). The dominant term of the boundary condition is

$$\vartheta(x,0) = \varphi_{\tilde{y}}(x,0) = -b^2 x, \quad -b^2 = (\gamma + 1) F''(x_0),$$

(5.3.5)

where $b^2 > 0$ for a convex surface.

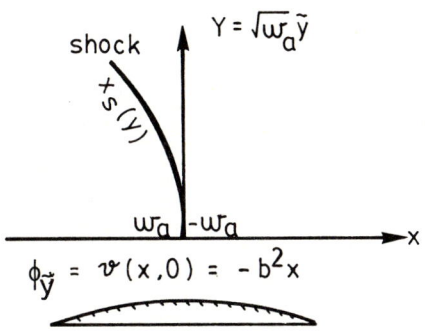

Figure 5.3.3

Shock at a wall

Next equation (5.3.1) is transformed to an appropriate form ahead and behind the shock so that differences from the states $\varphi_x = \pm w_a$ can be easily expressed. Let

$$\varphi(x,\tilde{y}) = w_a\{x + \tilde{\Phi}(x,Y)\} \quad \text{ahead},$$

$$= w_a\{-x + \Phi(x,Y)\} \quad \text{behind},$$

(5.3.6)

where $Y = \sqrt{w_a}\,\tilde{y}$.

Then

$$\varphi_x = w_a\{1 + \tilde{\Phi}_x\} \text{ ahead}, \qquad \varphi_{\tilde{y}} = w_a^{\frac{3}{2}}\tilde{\Phi}_Y \text{ ahead},$$

$$= w_a\{-1 + \Phi_x\} \text{ behind}, \qquad \varphi_{\tilde{y}} = w_a^{\frac{3}{2}}\Phi_Y \text{ behind}.$$

The resulting equations are

$$\tilde{\Phi}_{xx} - \tilde{\Phi}_{YY} = -\tilde{\Phi}_x\tilde{\Phi}_{xx} \quad \text{ahead},$$

(5.3.7)

$$\Phi_{xx} + \Phi_{YY} = \Phi_x\Phi_{xx} \quad \text{behind}.$$

(5.3.8)

The solutions ahead and behind respectively are represented as

$$\tilde{\Phi} = \tilde{\Phi}_w + \tilde{\Phi}_p, \tag{5.3.9}$$

$$\Phi = \Phi_h + \Phi_p, \tag{5.3.10}$$

where $\tilde{\Phi}_w$ is a solution to the wave operator on the left hand side of (5.3.7) and $\tilde{\Phi}_p$ is a particular solution produced by the effect of the non-linear right hand side (caused by $\tilde{\Phi}_w$). This formulation is useful as $x, Y \to 0$. Similarly Φ_h is a harmonic function satisfying the Laplace operator on the left hand side of (5.3.8) and Φ_p is a particular solution produced by the effect of the non-linear right hand side in (5.3.8). Both solutions must satisfy the condition of tangent flow,

$$\tilde{\Phi}_Y(x, 0) = \Phi_Y(x, 0) = -\kappa^2 x, \kappa^2 = \frac{b^2}{w_a^{\frac{3}{2}}}, \tag{5.3.11}$$

and the shock jump conditions which follow. Denote the shock curve by

$$x = x_s(Y) \quad \text{shock curve}, \tag{5.3.12}$$

where $x_s(0) = 0$, $\dfrac{dx_s}{dY}(0) = 0$ (normal shock at origin). Since the potential φ is continuous across the shock we can write

$$x_s(Y) + \tilde{\Phi}(x_s, Y) = -x_s(Y) + \Phi(x_s, Y),$$

or

$$x_s(Y) = \frac{1}{2} \Big\{ \Phi(x_s, Y) - \tilde{\Phi}(x_s, Y) \Big\}. \tag{5.3.13}$$

The second condition follows from equation (3.6.24) for the shock angle

$$\left(\frac{dx}{d\tilde{y}} \right)_s = \mp\sqrt{\langle w \rangle}, \quad \begin{array}{l} - \text{ for } [\vartheta] < 0, \\ + \text{ for } [\vartheta] > 0. \end{array}$$

In terms of (x, Y) thus

$$\frac{dx_s}{dY} = \mp\sqrt{\frac{\Phi_x(x_s, Y) + \tilde{\Phi}_x(x_s, Y)}{2}}. \tag{5.3.14}$$

Equations (5.3.13) and (5.3.14) provide the necessary shock jump conditions along the unknown shock location $x_s, (Y)$. Assume now smooth behavior of the solution ahead of the shock wave. For $\tilde{\Phi}_w$ we have

$$\tilde{\Phi}_w = f(x - Y) + g(x + Y), \quad \text{where } f, g \text{ are arbitrary.}$$

f, g must start with quadratic terms since the velocity perturbations $\tilde{\Phi}_x, \tilde{\Phi}_Y$ must vanish as $(x, Y) \to 0$. The boundary condition of tangent flow (5.3.11) must be satisfied for $x < 0$. So,

$$\tilde{\Phi}_w = \alpha(x - Y)^2 + \beta(x + Y)^2 . \tag{5.3.15}$$

Taking account of the boundary condition this becomes

$$\tilde{\Phi}_w = \mu Y^2 - \kappa^2 xY + \mu x^2 , \quad \mu = \alpha + \beta . \tag{5.3.16}$$

Then $\tilde{\Phi}_{w_x} = -\kappa^2 Y + 2\mu x$, $\tilde{\Phi}_{w_{xx}} = 2\mu = $ acceleration of flow (scaled). We expect that in the neighborhood of the shock $x_s \ll Y$, so that the dominant term of the right hand side is

$$\tilde{\Phi}_{w_x} \tilde{\Phi}_{w_{xx}} = -2\mu\kappa^2 Y + \cdots .$$

This produces a particular solution

$$\tilde{\Phi}_p = -\frac{1}{3}\mu\kappa^2 Y^3 + O(xY^2) .$$

Thus, for the flow ahead of the shock,

$$\tilde{\Phi}(x, Y) = \mu Y^2 - \kappa^2 xY - \frac{1}{3}\mu\kappa^2 Y^3 + \mu x^2 + O(xY^2) , \tag{5.3.17}$$

$$\tilde{\Phi}_x(x, Y) = -\kappa^2 Y + 2\mu x + O(Y^2) , \tag{5.3.18}$$

$$\tilde{\Phi}_Y(x, Y) = 2\mu Y - \kappa^2 x - \mu\kappa^2 Y^2 + O(xY) . \tag{5.3.19}$$

Note that the boundary condition (5.3.11) is satisfied. Downstream of the shock we can write, using the complex potential,

$$\Phi_h(x, Y) + i\Psi_h(x, Y) = Az^2 \log z + \left(B + i\frac{\kappa^2}{2}\right)z^2 + Cz^3 \log^2 z$$
$$+ (D + iD^*)z^3 \log z + (E + iE^*)z^3 + \cdots , \tag{5.3.20}$$

for $x > x_s(Y)$, where

$$z = x + iY = re^{i\theta} ,$$

and the constants A, B, C, D, D^*, E, E^* are real. The dominant term of the boundary condition of tangent flow is satisfied by the term $i\dfrac{\kappa^2}{2}$; the higher order terms D^*, E^* can fix up additions due to Φ_p and $O(x^2)$ terms in the boundary condition (not considered here). Note that

$$
\begin{aligned}
\Phi_{h_z} - i\Phi_{hY} = {}& 2Az \log z + (A + 2B + i\kappa^2)z + 3Cz^2 \log^2 z \\
& + (2C + 3D + i3D^*)z^2 \log z \\
& + (D + iD^* + 2E + i2E^*)z^2 + \cdots .
\end{aligned}
\tag{5.3.21}
$$

Thus when $Y = 0$, $x > 0$,

$$
\Phi_{hY}(x,0) = -\kappa^2 x + O(x^2 \log x).
\tag{5.3.22}
$$

The nature of the largest singularity in (5.3.20) is that of a jump in curvature; two derivatives have $\log z$ behavior and $I(\log z)$ jumps as $\theta = \arg z$ goes from 0 to π. In order to complete the argument we need the dominant behavior of Φ_p. Note that

$$
\begin{aligned}
\Phi_{h_z} &= \Re\ell\Big\{ 2Ar(\cos\theta + i\sin\theta)(\log r + i\theta) \\
&\qquad + (A + 2B + i\kappa^2)r((\cos\theta + i\sin\theta) + \cdots) \Big\} \\
&= 2Ar \log r \cos\theta + O(r), \\
\Phi_{h_{zz}} &= \Re\ell\Big\{ 2A \log z + \cdots \Big\} = 2A \log r + O(1).
\end{aligned}
$$

Thus

$$
\Phi_{h_z}\Phi_{h_{zz}} = 4A^2 r^2 \log^2 r \cos\theta + O(r \log r).
$$

Thus considering equation (5.3.8) in the form

$$
\begin{aligned}
\Phi_{p_{rr}} + \frac{1}{r}\Phi_{p_r} + \frac{1}{r^2}\Phi_{p_{\theta\theta}} &= \Phi_{h_z}\Phi_{h_{zz}} + \cdots \\
&= 4A^2 r^2 \log^2 r \cos\theta + O(r^2 \log r),
\end{aligned}
\tag{5.3.23}
$$

we find the particular solution starting

$$
\Phi_p = \frac{A^2}{2} r^3 \log^2 r \cos\theta + O(r^3 \log r),
$$

$$
\Phi_p = \frac{A^2}{2} x(x^2 + Y^2) \log^2 \sqrt{x^2 + Y^2} + \cdots .
\tag{5.3.24}
$$

In summary

$$
\Phi = \Re \Bigg\{ A(x+iY)^2 \log(x+iY) + \left(B + i\frac{\kappa^2}{2} \right)(x+iY)^2
$$
$$
+ C(x+iY)^2 \log^2(x+iY) + (D+iD^*))(x+iY)^3 \log(x+iY)
$$
$$
+ (E+iE^*)(x+iY)^3 + \cdots \Bigg\}
$$
$$
+ \frac{A^2}{2} x(x^2 + Y^2) \log^2 \sqrt{x^2 + Y^2} + \cdots . \tag{5.3.25}
$$

To evaluate this function near the shock note that $x_s(Y) \ll Y$, in fact roughly $O(Y^2)$. (see below) Thus

$$
\log(x+iY) = \log iY + \log\left(1 - i\frac{x}{Y}\right) = \log Y + i\left(\frac{\pi}{2} - \frac{x}{Y}\right) + O\left(\frac{x^2}{Y^2}\right),
$$
$$
\log^2(x+iY) = \log^2 Y + 2i\left(\frac{\pi}{2} - \frac{x}{Y}\right)\log Y - \left(\frac{\pi}{2} - \frac{x}{Y}\right)^2 + \cdots ,
$$

$$
\Phi_x = \Re\ell \Bigg\{ 2A(x+iY)\log(x+iY) + (A+2B+i\kappa^2)(x+iY)
$$
$$
+ 3C(x+iY)^2 \log^2(x+iY) + (2C+3D+i3D^*)(x+iY)^2 \log(x+iY)
$$
$$
+ (D+iD^* + 2E + 2E^*i)(x+iY)^2 + \cdots \Bigg\}
$$
$$
+ \frac{A^2}{2}\{3x^2 + Y^2\} \log^2 \sqrt{x^2 + Y^2} + \cdots . \tag{5.3.26}
$$

Thus the first shock condition (5.3.13) becomes, from (5.3.18) and (5.3.26)

$$
x_s(Y) = \frac{1}{2}\Big\{ -AY^2 \log Y - BY^2 + \cdots - \mu Y^2 + O(x_s Y) \Big\},
$$

or

$$
x_s(Y) = -\frac{1}{2}AY^2 \log Y - \frac{1}{2}(B+\mu)Y^2 + O(Y^3 \log Y). \tag{5.3.27}
$$

Note that if $A < 0$, $x_s(Y) < 0$ as $Y \to 0$. Thus,

$$
\frac{dx_s}{dY} = -AY \log Y - \left(\frac{1}{2}A + B + \mu\right)Y + O(Y^2 \log Y),
$$

and

$$
\left(\frac{dx_s}{dY}\right)^2 = A^2 Y^2 \log^2 Y - (A^2 + 2AB + 2A\mu)(Y^2 \log Y) + O(Y^2). \tag{5.3.28}
$$

Another expression for this quantity can be found from the second shock relation (5.3.14). From (5.3.26)

$$
\begin{aligned}
\Phi_x = \Re\ell\Big\{ & 2A(x+iY)\Big(\log Y + i\Big(\frac{\pi}{2}+\frac{x}{Y}\Big)\Big) + (A+2B+i\kappa^2)(x+iY) \\
& + 3C(x^2 - Y^2 + i2xY)\Big(\log^2 Y + 2i\Big(\frac{\pi}{2}-\frac{x}{Y}\Big)\log Y\Big) \\
& \qquad\qquad - (2C+3D+i3D^*)(-Y^2\log Y + \cdots)\Big\} \\
& + \frac{A^2}{2}Y^2\log^2 Y + \cdots ,
\end{aligned}
\tag{5.3.29}
$$

or

$$
\Phi_x = -(\pi A + \kappa^2 Y) + 2Ax\log Y - 3CY^2\log^2 Y + \frac{A^2}{2}Y^2\log^2 Y + O(Y\log Y).
\tag{5.3.30}
$$

Thus,

$$
\begin{aligned}
\Phi_x(x_s,Y) &+ \tilde\Phi_x(x_s,Y) \\
&= -(\pi A + \kappa^2)Y + 2Ax_s\log Y + \Big(\frac{A^2}{2}-3C\Big)Y^2\log^2 Y + O(Y^2\log Y) \\
&\quad + (-\kappa^2 Y + 2\mu x_s + \cdots) \\
&= -(\pi A + 2\kappa^2)Y + 2Ax_s\log Y + \Big(\frac{A^2}{2}-3C\Big)Y^2\log^2 Y + O(Y^2\log Y).
\end{aligned}
\tag{5.3.31}
$$

(5.3.14) reads

$$
\Phi_x(x_s,Y) + \tilde\Phi(x_s,Y) = 2\Big(\frac{dx_s}{dY}\Big)^2.
$$

Comparing (5.3.31) with (5.3.28) we see that there is no solution unless $\pi A + 2K^2 = 0$,

$$
A = -\frac{2}{\pi}\kappa^2.
\tag{5.3.32}
$$

Then

$$
\begin{aligned}
2A^2Y^2\log^2 Y &- 2(A^2 + 2AB + 2A\mu)Y^2\log Y + O(Y^2) \\
&= 2A\Big(-\frac{1}{2}AY^2\log Y - \frac{1}{2}(B+\mu)Y^2 + \cdots\Big)\log Y \\
&\quad + \Big(\frac{A^2}{2}-3C\Big)Y^2\log^2 Y + O(Y^2\log Y).
\end{aligned}
$$

This gives a value for C to balance the terms $O(Y^2 \log^2 Y)$

$$C = -\frac{5}{6}A^2 . \tag{5.3.33}$$

The terms $O(Y^2 \log Y)$ can be balanced by choosing the constant D etc., B is not determined in this process and presumably depends on global considerations. For a convex wall the constant A is negative and proportional to the curvature κ^2. The shock bends upstream for a convex surface; for a concave wall the shock bends downstream. cf. Figure (5.3.4).

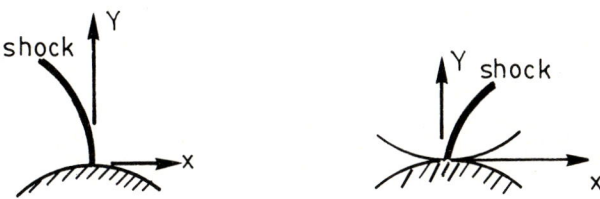

Figure 5.3.4

Shock shape near a wall

Note that the curvature of the shock wave,

$$\frac{d^2 x_s}{dY^2} \sim \log Y \to \infty$$

at the foot of the shock. Downstream of the shock as $x \to 0+$, $Y = 0$, the acceleration $\varphi_{xx} \sim -\kappa^2 \log x \to +\infty$ (convex) and the pressure coefficient $c_p \sim -\varphi_x \sim \kappa^2 x \log x$ showing an expansion downstream of the shock. These features have been observed in careful calculations and early experiments of Ackeret.

References

[5.3.1] Gadd, George E., The Possibility of Normal Shock Waves on a body with convex surfaces in Inviscid Transonic Flow, *Zeit. Ang. Math. and Phys.* **11**, 1960, pp 51-55.

[5.3.2] Oswatitsch, K. and Zierep, J., Das Problem des Senkrechten Stosses an einer gekrümmten Wand. *Zeit. Ang. Math. and Mech.*, 1960, Supplement, pp 143-144.

5.4 Numerical Methods, Physical Plane, Steady Flow

Because of the impossibility of finding sufficiently general analytic solutions to
the non-linear and mixed type problem of (5.1.1) the best solutions that can be
obtained for flow past airfoils and wings are based on finite difference methods.
Early calculations by Yoshihara and Magnus treated a (pseudo) time dependent
version of (5.1.1) of the full potential equation and achieved a steady state as time
approached infinity. The basic idea, due to Lax and Wendroff, of using artificial
viscosity to smooth out shock jumps was incorporated in their work. In an
effort to reduce computation time and increase the flexibility of the calculations,
relaxation solutions for the steady flow were considered [5.4.1]. Shocks again were
expected to be calculated automatically by the finite difference scheme.

Relaxation schemes are well adapted to the elliptic or subsonic part of the
flow field, since they depend essentially on a mean-value like property of the elliptic operator. When elliptic type differencing is used for supersonic or hyperbolic
points, the local domain of dependence is violated and instability of the numerical
scheme results. For example, if a centered x-difference operator is used in the
hyperbolic region there is local upstream influence of the downstream point. If
this happens at even one hyperbolic point the relaxation scheme diverges. Thus
a type-sensitive difference is required, a typical hyperbolic scheme for supersonic
points and a typical elliptic scheme for subsonic points. Since the nature of the
flow (supersonic or subsonic) at various points is not known until the problem is
solved an iterative procedure has to be used. Thus a test has to be devised to
decide if a computational point is elliptic or hyperbolic and the appropriate difference scheme used at each step of the iteration. Further analysis of this plan shows
that there are two additional types of point on the finite mesh, sonic points, typically on a sonic line, and shock points, typically in the (computational) interior
of a shock wave (cf. [5.4.2]).

In the following paragraphs we present first an analysis of the finite difference
scheme and then some discussion of the implementation for the problem of Section
(5.1). This is followed by a few examples of calculations of the flow field, drag
and lift.

A fully conservative scheme should be used so that the difference equations
contain the shock relations. The conservative form (5.1.1b) will be analysed in
terms of

$$\left(\frac{w^2}{2}\right)_x - \vartheta_{\tilde{y}} = 0, \tag{5.4.1}$$

where $\quad w = \varphi_x = (\gamma + 1)\phi_x - K , \quad \vartheta = \varphi_{\tilde{y}} = (\gamma + 1)\phi_{\tilde{y}} .$

Consider a uniform finite difference mesh $(\Delta x, \Delta \tilde{y})$ constructed in the $(x, \tilde{y})$ plane with points labelled by (i, j). The results are easily generalized to a variable mesh. The equation (5.4.1) can be expressed in conservative flux form for a box centered on a mesh point at (i, j) as in Figure 5.4.1.

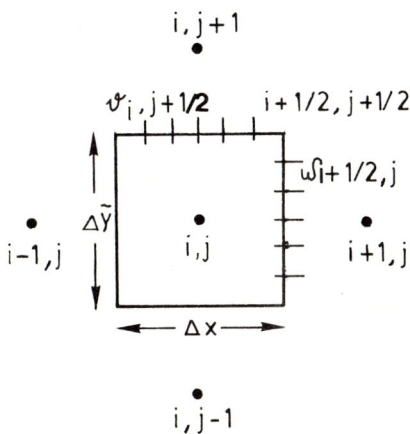

Figure 5.4.1

Control Box

Thus

$$\left\{\left(\frac{w^2}{2}\right)_{i+\frac{1}{2},j} - \left(\frac{w^2}{2}\right)_{i-\frac{1}{2},j}\right\}\Delta\tilde{y} - \left(\vartheta_{i,j+\frac{1}{2}} - \vartheta_{i,j-\frac{1}{2}}\right)\Delta x = 0 , \qquad (5.4.2)$$

or, in factored form

$$\frac{1}{2}\left\{w_{i+\frac{1}{2},j} + w_{i-\frac{1}{2},j}\right\}\left\{w_{i+\frac{1}{2},j} - w_{i-\frac{1}{2},j}\right\}\Delta\tilde{y} - \left\{\vartheta_{i,j+\frac{1}{2}} - \vartheta_{i,j-\frac{1}{2}}\right\}\Delta x = 0 . \tag{5.4.3}$$

$\vartheta_{i,j+\frac{1}{2}}$ is always calculated from a centered expression and if we use a centered expression to calculate $w_{i\pm\frac{1}{2},j}$ then we obtain a difference form of the differential equation that is suitable for elliptic points, a precise definition of which will be given in a moment. That is

$$\vartheta_{i,j+\frac{1}{2}} \equiv (\varphi_{\tilde{y}})_{i,j+\frac{1}{2}} = \frac{\varphi_{i,j+1} - \varphi_{i,j}}{\Delta\tilde{y}} , \vartheta_{i,j-\frac{1}{2}} = \frac{\varphi_{i,j} - \varphi_{i,j-1}}{\Delta\tilde{y}} , \qquad (5.4.4)$$

and

$$w^{(c)}_{i+\frac{1}{2},j} \equiv (\varphi_x)_{i+\frac{1}{2},j} = \frac{\varphi_{i+1,j} - \varphi_{i,j}}{\Delta x}, w^{(c)}_{i-\frac{1}{2},j} = \frac{\varphi_{i,j} - \varphi_{i-1,j}}{\Delta x}. \qquad (5.4.5)$$

Thus (5.4.3) becomes

$$\left(\frac{\varphi_{i+1,j} - \varphi_{i-1,j}}{2\Delta x}\right)\left(\frac{\varphi_{i+1,j} - 2\varphi_{i,j} + \varphi_{i-1,j}}{(\Delta x)^2}\right) - \left(\frac{\varphi_{i,j+1} - 2\varphi_{i,j} + \varphi_{i,j-1}}{(\Delta \tilde{y})^2}\right) = 0$$

$$(5.4.6)$$

(Subsonic) Elliptic Difference Form

This differential operator is a second-order accurate centered expression, in conservative form, as might be written for an elliptic operator. The equation is of elliptic type and stable if a centered approximation to the perturbation from sonic $\varphi^{(c)}_x$ as expressed by the first bracket in (5.4.6) is negative.

$$\varphi^{(c)}_x \equiv \frac{\varphi_{i+1,j} - \varphi_{i-1,j}}{2\Delta x}, \quad \varphi^{(c)}_x < 0 \quad \text{for elliptic stability}. \qquad (5.4.7)$$

The computational star involved in (5.4.6) is shown in Figure (5.4.2) where the central (i,j) point is to be calculated from all its nearest neighbors.

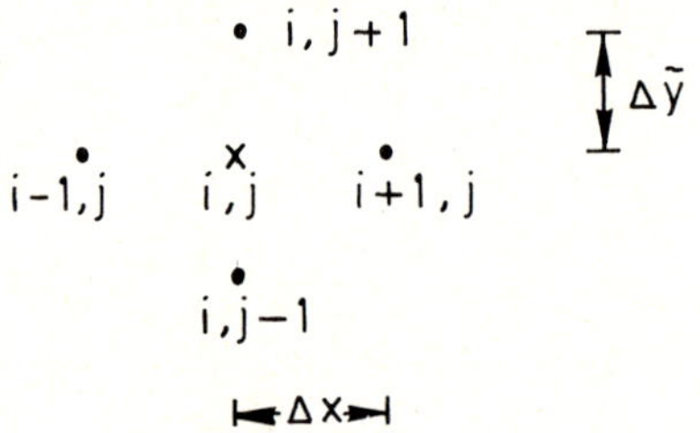

Figure 5.4.2

Elliptic Computational Star

This star is not suitable for a hyperbolic point since there is local upstream influence of point $i+1, j$ on i, j, but the points could be relabelled with $i+1 \to i$. The downstream point can be calculated explicitly from the four upstream points. But, the well known stability criterion (CFL) for explicit hyperbolic schemes demands that the computational domain of dependence include the limiting continuous domain of dependence,

$$\frac{\Delta \tilde{y}}{\Delta x} > \left(\frac{d\tilde{y}}{dx}\right)_{\textbf{characteristic}} = \frac{1}{\sqrt{\varphi_x}}. \qquad (5.4.8)$$

Near the sonic line where $\varphi_x = 0$ this scheme appears awkward since Δx will have to be very small for a given $\Delta \tilde{y}$. Implicit hyperbolic schemes, however, are unconditionally stable and fit in nicely with a line relaxation procedure. Their form can be achieved by shifting the x-difference operator upstream by one unit so that three unknown points $(i, j+1)$, (i, j), $(i, j-1)$ are connected to two upstream ones. This form is achieved if a backward (or upstream) expression for the x-fluxes is used.

$$w_{i+\frac{1}{2},j}^{(b)} = \frac{\varphi_{i,j} - \varphi_{i-1,j}}{\Delta x}, \quad w_{i-\frac{1}{2},j}^{(b)} = \frac{\varphi_{i-1,j} - \varphi_{i-2,j}}{\Delta x}. \tag{5.4.9}$$

Then (5.4.3) becomes

$$\left(\frac{\varphi_{i,j} - \varphi_{i-2,j}}{2\Delta x} \right) \left(\frac{\varphi_{i,j} - 2\varphi_{i-1,j} + \varphi_{i-2,j}}{(\Delta x)^2} \right) - \left(\frac{\varphi_{i,j+1} - 2\varphi_{i,j} + \varphi_{i,j-1}}{(\Delta \tilde{y})^2} \right) = 0 \tag{5.4.10}$$

(Supersonic) Hyperbolic Difference form

The computational star involved in (5.4.10) is shown in Figure (5.4.3). All the points on the i-line are unknown.

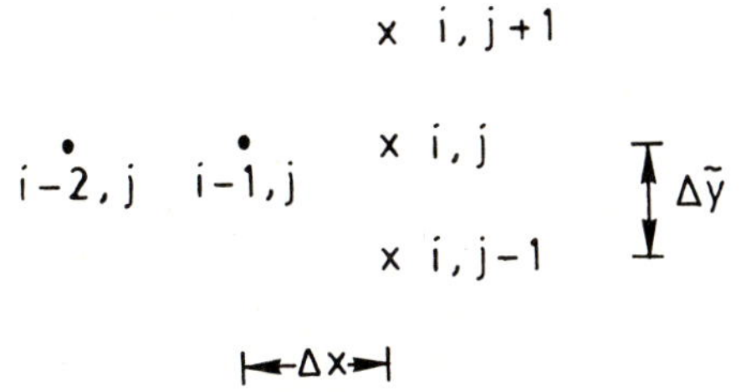

Figure 5.4.3
Hyperbolic Computational Star

Unconditional linear stability holds for this scheme as long as $\phi_x^{(b)}$ as expressed by the first factor in (5.4.10) is positive.

$$\varphi_x^{(b)} \equiv \frac{\varphi_{i,j} - \varphi_{i-2,j}}{2\Delta x}, \quad \varphi_x^{(b)} > 0 \text{ for hyperbolic stability.} \tag{5.4.11}$$

As long as $\varphi_x^{(c)}$, $\varphi_x^{(b)}$ are in agreement it is clear which form (5.4.6) or (5.4.10) to use but there can be (ij) points of disagreement when the flow is accelerating through sonic (sonic points) or decelerating through sonic (shock points). For these points two other finite difference operators are introduced as indicated in the following table.

Table 5.4.1

$\varphi_x^{(c)}$	$\varphi_x^{(b)}$	operator
< 0	< 0	elliptic (5.4.6)
> 0	> 0	hyperbolic (5.4.10)
> 0	< 0	sonic (5.4.13)
< 0	> 0	shock (5.4.14)

At an (i,j) point for which the inequalities of the sonic point operator hold neither (5.4.6) or (5.4.10) is formally stable. It is reasonable to use a backward flux approximation on the downstream face and a centered flux approximation on the upstream face of the control box

$$w^{(b)}_{i+\frac{1}{2},j} = \frac{\varphi_{i,j} - \varphi_{i-1,j}}{\Delta x} , \quad w^{(c)}_{i-\frac{1}{2},j} = \frac{\varphi_{i,j} - \varphi_{i-1,j}}{\Delta x} . \tag{5.4.12}$$

Thus these terms cancel in (5.4.3) and result in

$$\frac{\varphi_{i,j+1} - \varphi_{i,j} + \varphi_{i,j+1}}{(\Delta \tilde{y})^2} = 0 \tag{5.4.13}$$

sonic point difference form.

The computational star for the (i,j) sonic point is simply that shown in Figure 5.4.4.

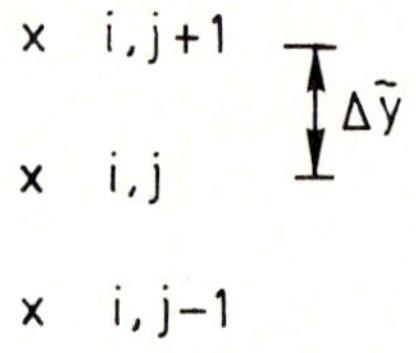

x i,j+1

$\uparrow \downarrow \Delta\tilde{y}$

x i,j

x i,j-1

Figure 5.4.4

Computational Star for Sonic Point Operator

As the mesh size is refined it is clear that this difference operator converges to $\varphi_{\tilde{y}\tilde{y}} = 0$ which should be a good approximation to $\varphi_x \varphi_{xx} - \varphi_{\tilde{y}\tilde{y}} = 0$ near the sonic line where φ_x is close to zero. φ is, of course, a continuous function and its discrete approximation near a sonic point is sketched in Figure 5.4.5.

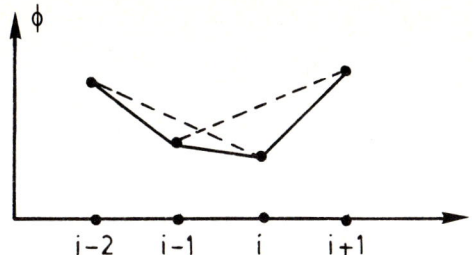

Figure 5.4.5

Qualitative behavior of φ near a sonic point (i, j)

On the other hand the shock point operator is designed for deceleration from supersonic to subsonic. Thus the flux at $i - \dfrac{1}{2}$ can be approximated by a backward formula and the flux at $i + \dfrac{1}{2}$ by a centered formula

$$w^{(b)}_{i-\frac{1}{2},j} = \frac{\varphi_{i-1,j} - \varphi_{i-2,j}}{\Delta x}, \quad w^{(c)}_{i+\frac{1}{2},j} = \frac{\varphi_{i+1,j} - \varphi_{i,j}}{\Delta x}.$$

From (5.4.3) this results in the difference operator

$$\left(\frac{\varphi_{i+1,j} - \varphi_{i,j} + \varphi_{i-1,j} - \varphi_{i-2,j}}{2\Delta x} \right) \left(\frac{\varphi_{i+1,j} - \varphi_{i,j} - \varphi_{i-1,j} + \varphi_{i-2,j}}{(\Delta x)^2} \right)$$

$$\times \left(-\frac{\varphi_{i,j+1} - 2\varphi_{i,j} + \varphi_{i,j-1}}{(\Delta \tilde{y})^2} \right) = 0 \qquad (5.4.14)$$

Shock Point Difference Operator

This can also be interpreted as the addition of both elliptic and hyperbolic x-difference operators in (5.4.6) and (5.4.10). That is if the fluxes are expressed as

$$\left(\frac{w^2}{2} \right)^{(c)}_{i+\frac{1}{2},j} - \left(\frac{w^2}{2} \right)^{(c)}_{i-\frac{1}{2},j} + \left(\frac{w^2}{2} \right)^{(c)}_{i-\frac{1}{2},j} - \left(\frac{w^2}{2} \right)^{(b)}_{i-\frac{1}{2},j} =$$

$$\left(\frac{w^2}{2} \right)^{(c)}_{i+\frac{1}{2},j} - \left(\frac{w^2}{2} \right)^{(c)}_{i-\frac{1}{2},j} + \left(\frac{w^2}{2} \right)^{(b)}_{i+\frac{1}{2},j} - \left(\frac{w^2}{2} \right)^{(b)}_{i-\frac{1}{2},j}$$

since

$$\left(\frac{w^2}{2} \right)^{(b)}_{i+\frac{1}{2},j} \equiv \left(\frac{w^2}{2} \right)^{(c)}_{i-\frac{1}{2},j}.$$

The first part is the elliptic flux and the second part is the hyperbolic flux. Thus the computational star for the shock point is shown in Figure 5.4.6

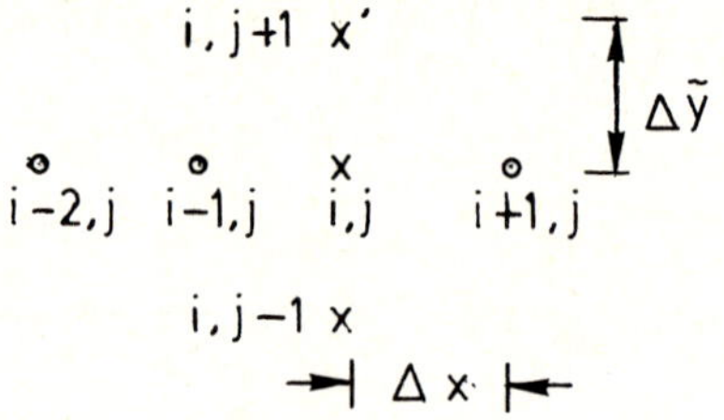

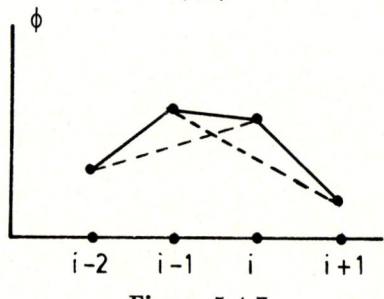

Figure 5.4.6

Computational Star for Shock Point (i, j)

The course of φ near the shock point (i, j) is shown below in Figure 5.4.7.

Figure 5.4.7

Qualitative behavior of φ near a shock point

If the continuous approximations are used in (5.4.14), for example,

$$\varphi_{i,j} = \varphi(x, \tilde{y}) \, ,$$

$$\varphi_{i+1,j} = \varphi(x + \Delta x, \tilde{y}) = \varphi(x, \tilde{y}) + \Delta x \varphi_x(x, \tilde{y}) + \frac{\Delta x^2}{2} \varphi_{xx}(x, \tilde{y}) + \cdots \, ,$$

then the limiting form of (5.4.14) at the shock point is

$$2\varphi_x \varphi_{xx} - \varphi_{\tilde{y}\tilde{y}} = 0 \qquad (5.4.15)$$

But this is an adequate representation to the equation

$$\varphi_x \varphi_{xx} - \varphi_{\tilde{y}\tilde{y}} = 0 \quad \text{(TSD Equation)}$$

when φ_x is sufficiently small $\varphi_x = O(\Delta x)$. According to the tests the shock point operator would also be used for smooth decelerations. More important is the way that the shock point operator calculates shocks. Some analysis of this is given now.

The normal wave solution $w_b = -w_a$ is obtained from the shock point operator at (i, j) in the form

$$\left(\frac{w^2}{2}\right)^{(c)}_{i+\frac{1}{2},j} = \left(\frac{w^2}{2}\right)^{(b)}_{i-\frac{1}{2},j} \tag{5.4.16}$$

together with the hyperbolic operator for $(i - 1, j)$ and the elliptic operator for $(i + 1, j)$.

$$\left(\frac{w^2}{2}\right)^{(b)}_{i-\frac{1}{2},j} = \left(\frac{w^2}{2}\right)^{(b)}_{i-\frac{3}{2},j} \qquad \text{hyperbolic,}$$

$$\left(\frac{w^2}{2}\right)^{(c)}_{i+\frac{3}{2},j} = \left(\frac{w^2}{2}\right)^{(c)}_{i+\frac{1}{2},j} \qquad \text{elliptic.}$$

A solution is (using centered formulas)

$$w_b = w_{i+\frac{3}{2}} = w_{i+\frac{1}{2}} = -w_{i-\frac{1}{2}} = -w_{i-\frac{3}{2}} = -w_a . \tag{5.4.17}$$

The shock is spread over three mesh points here. If now an oblique wave is considered, the shock jump as a solution can also be verified by using the discrete conservation properties of the formulation. A uniform oblique wave is considered as in Figure 5.4.8. Consider the discrete representation on a special mesh such that

$$N\frac{\Delta \tilde{y}}{\Delta x} = \tan \theta_s , \qquad \theta_s = \text{shock angle}, \qquad N = \text{integer} .$$

Thus φ has the periodicity $\varphi(x + \Delta x, \tilde{y} + N\Delta \tilde{y}) = \varphi(x, \tilde{y})$. It is sufficient to consider the flow in a band from $j = 1, N$ as in the figure because of the assumed periodicity, which can be expressed as a boundary condition

$$\varphi_{i,j+\frac{1}{2}} = \varphi_{i+1,j+\frac{1}{2}+N} . \tag{5.4.18}$$

The upstream flow is supersonic $\varphi_x^{(c)}$, $\varphi_x^{(b)} > 0$ and the hyperbolic operator (5.4.10) is used to march downstream toward the shock. This is done until some points are reached where $\varphi_x^{(b)} > 0$ but $\varphi_x^{(c)} < 0$, assuming now subsonic flow downstream. These points are the shock points, shown for a typical case in Figure 5.4.8. At these points the shock point operator (5.4.14) is used. Downstream of these points the flow is subsonic and the elliptic operator (5.4.6) is used.

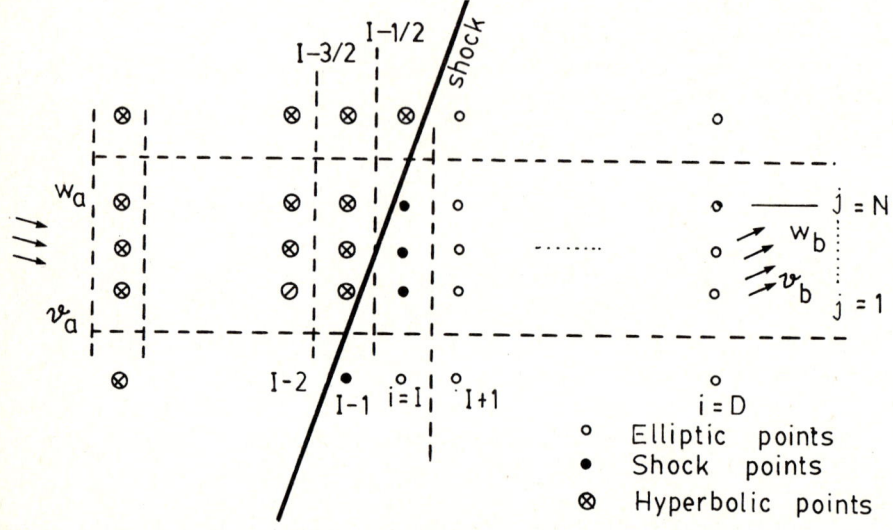

Figure 5.4.8

Oblique Shock Wave

Consider all the hyperbolic points in the band $j = 1, N$ and sum all the fluxes as

$$\sum_{i=1}^{I-1} \sum_{j=1}^{N} \left\{ \frac{1}{2} w_{i+\frac{1}{2},j}^{2(b)} - \frac{1}{2} w_{i-\frac{1}{2},j}^{2(b)} \right\} \Delta \tilde{y} - \left(\vartheta_{i,j+\frac{1}{2}} - \vartheta_{i,j-\frac{1}{2}} \right) \Delta x = 0. \qquad (5.4.19)$$

or

$$\sum_{i=1}^{I-1} \sum_{j=1}^{N} \left\{ \frac{1}{2} w_{i-\frac{1}{2},j}^{2(c)} - \frac{1}{2} w_{i-\frac{3}{2},j}^{2(c)} \right\} \Delta \tilde{y} - \left\{ \vartheta_{i,j+\frac{1}{2}} - \vartheta_{i,j-\frac{1}{2}} \right\} \Delta x = 0. \qquad (5.4.20)$$

Here $i = 1$ on the discrete points representing upstream infinity. Note that

$$\sum_{i=1}^{I-1} \left(w_{i-\frac{1}{2},j}^{2} - w_{i-\frac{3}{2},j}^{2} \right) = \sum_{i=1}^{I-2} w_{i-\frac{1}{2},j}^{2} + w_{I-\frac{3}{2},j} - \sum_{i=2}^{I-1} w_{i-\frac{3}{2},j}^{2} - w_{-\frac{1}{2},j}^{2}$$

$$= w_{I-\frac{3}{2},j} - w_{-\frac{1}{2},j}^{2}$$

and

$$\sum_{j=1}^{N} \left(\vartheta_{i,j+\frac{1}{2}} - \vartheta_{i,j-\frac{1}{2}} \right) = \sum_{j=1}^{N-1} \vartheta_{i,j+\frac{1}{2}} - \sum_{j=2}^{N} \vartheta_{i,j-\frac{1}{2}} + \vartheta_{i,N+\frac{1}{2}} - \vartheta_{i,\frac{1}{2}}$$

$$= \vartheta_{i,N+\frac{1}{2}} - \vartheta_{i,\frac{1}{2}}.$$

Thus the discrete divergence theorem for the hyperbolic points (5.4.20) is

$$\frac{1}{2}\sum_{j=1}^{N}\left\{w_{I-\frac{3}{2},j}^2\Delta\tilde{y}\right\} - N\frac{w_a^2}{2}\Delta\tilde{y} - \sum_{i=1}^{I-1}\left\{\vartheta_{i,N+\frac{1}{2}} - \vartheta_{i,\frac{1}{2}}\right\}\Delta x = 0.$$

But, using the periodicity (5.4.18)

$$\sum_{i=1}^{N}\left\{\vartheta_{i,N+\frac{1}{2}} - \vartheta_{i,\frac{1}{2}}\right\} = \sum_{i=2}^{I-1}\vartheta_{i,N+\frac{1}{2}} - \sum_{i=1}^{I-2}\vartheta_{i,\frac{1}{2}} + \vartheta_{i,N+\frac{1}{2}} - \vartheta_{I-1,\frac{1}{2}}$$

$$= \vartheta_a - \vartheta_{I-1,\frac{1}{2}}.$$

That is, hyperbolic points sum to

$$\frac{1}{2}\sum_{j=1}^{N}\left\{w_{I-\frac{3}{2},j}^2\Delta\tilde{y}\right\} - \frac{1}{2}Nw_a^2\Delta\tilde{y} - \left(\vartheta_a - \vartheta_{I-1,\frac{1}{2}}\right)\Delta x = 0. \qquad (5.4.21)$$

Next consider all the elliptic points between $(j = 1, N)$, that is $I + 1$ to $i = D$ the discrete point representing downstream infinity. We have

$$\sum_{j=1}^{N}\sum_{i=I+1}^{D}\frac{1}{2}\left\{w_{i+\frac{1}{2},j}^2 - w_{i-\frac{1}{2},j}^2\right\}\Delta\tilde{y} - \left\{\vartheta_{i,j+\frac{1}{2}} - \vartheta_{i,j-\frac{1}{2}}\right\}\Delta x = 0.$$

In a similar way as above, we find

$$\frac{1}{2}Nw_b^2\Delta\tilde{y} - \frac{1}{2}\sum_{j=1}^{N}\left\{w_{I+\frac{1}{2},j}^2\Delta\tilde{y}\right\} - \left(\vartheta_{I+1,N+\frac{1}{2}} - \vartheta_b\right)\Delta x = 0. \qquad (5.4.22)$$

The shock point operators at the level $i = I$ can also be summed to give

$$\sum_{j=1}^{N}\frac{1}{2}\left\{w_{I+\frac{1}{2},j}^2 - w_{I-\frac{3}{2},j}^2\right\}\Delta\tilde{y} - \left\{\vartheta_{I,j+\frac{1}{2}} - \vartheta_{I,j-\frac{1}{2}}\right\}\Delta x = 0,$$

or

$$\sum_{j=1}^{N}\frac{1}{2}\left\{w_{I+\frac{1}{2},j}^2 - w_{I-\frac{3}{2},j}\right\}\Delta\tilde{y} - \left(\vartheta_{I,N+\frac{1}{2}} - \vartheta_{I,\frac{1}{2}}\right)\Delta x = 0. \qquad (5.4.23)$$

Adding all these flux equations (5.4.21,22,23) the interior terms all cancel and the relation between the upstream and downstream states is obtained

$$\frac{1}{2}N(w_b^2 - w_a^2)\Delta\tilde{y} + (\vartheta_b - \vartheta_a)\Delta x = 0. \qquad (5.4.24)$$

(5.4.24) is just a discrete version of the shock polar (cf. 5.1.7) since here

$$\tan \theta_s = \left(\frac{d\tilde{y}}{dx}\right)_s = \frac{N\Delta\tilde{y}}{\Delta x}.$$

A similar analysis using only the hyperbolic operators can be carried out if all the points are supersonic. Again the correct shock polar is obtained.

Calculations presented later show that supersonic-subsonic shocks are spread over only three or four mesh points. However supersonic-supersonic shocks tend to spread out as they diffuse from their origin and may reach 6-10 mesh points in width. If sharp supersonic-supersonic shocks are required some modifications of the method must be introduced.

A brief discussion is now given of the numerical formulation of the boundary value problem of Section 5.1 and a solution algorithm. An easy way to connect the unknowns to the boundary conditions is to put the first row of mesh points one half mesh spacing above the airfoil ($\tilde{y} = 0$) (cf. Figure 5.4.9). For a typical box such as $(i, 1)$ over the airfoil the flux equation (5.4.2) reads

$$\left\{\left(\frac{w^2}{2}\right)_{i+\frac{1}{2},1} - \left(\frac{w^2}{2}\right)_{i-\frac{1}{2},1}\right\}\Delta\tilde{y} - \left\{\vartheta_{i,\frac{3}{2}} - \vartheta_{i,\frac{1}{2}}\right\}\Delta x = 0. \tag{5.4.25}$$

But $\vartheta_{i,\frac{1}{2}}$ is the value of ϑ on the upper side of the airfoil and is given by the tangent flow boundary condition (5.4.26)

$$\vartheta_{i,\frac{1}{2}} = \vartheta(x, 0+) = \vartheta_u(x) = (\gamma + 1)F_u'(x). \tag{5.4.26}$$

A similar treatment holds on the lower side of the airfoil. With this treatment the values of ϕ are found first at mesh points one half a mesh spacing above the airfoil. Simple extrapolation provides the values of ϕ, ϕ_x on the surface of the airfoil. For vertical rows ahead of the airfoil there is a straightforward connection between points in the upper half-plane and those in the lower half-plane. For vertical rows behind the airfoil the jump in ϕ, equal to the circulation Γ, must be taken into account, but of course ϑ is continuous. The far-field boundary condition (5.1.4) is applied at the edge of the computational mesh.

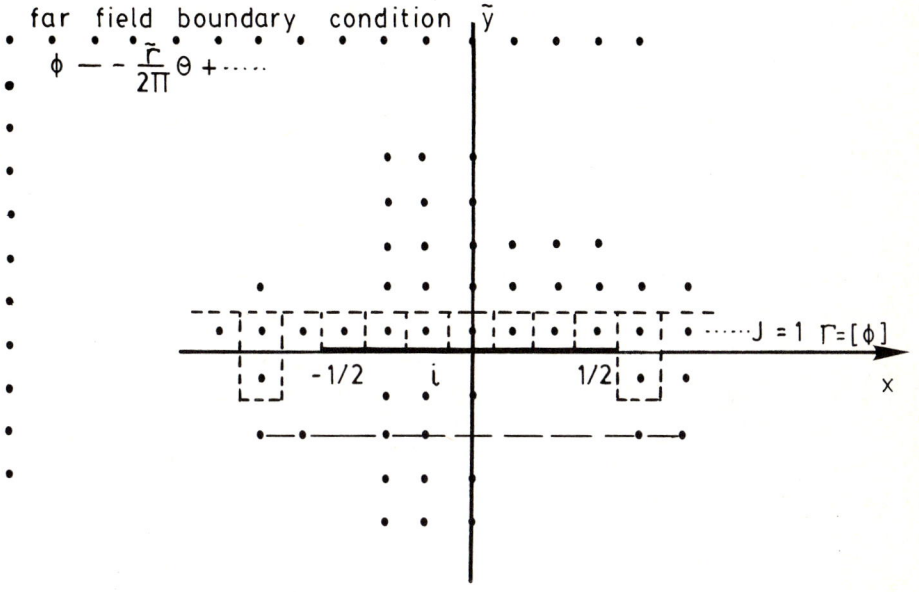

far field boundary condition $\tilde{y}$

$$\phi - - \frac{\tilde{\Gamma}}{2\Pi} \theta + \cdots$$

$J = 1 \quad \Gamma = [\phi]$

$-1/2 \qquad i \qquad 1/2$

x

Figure 5.4.9
Finite Difference Mesh

An iterative line relaxation algorithm, solving (by direct elimination) for all the points along a vertical line has proved effective. The elliptic and hyperbolic operators (5.4.6) and (5.4.10) connect the solution points at the adjacent vertical lines e.g. $i - 1$, i, $i + 1$ or $i - 2$, $i - 1$, i. The sonic point operator concerns only one vertical line and the shock point operator concerns four adjacent lines $(i - 2, i - 1, i, i + 1)$. The iteration proceeds from an initial guess. ϕ values from a previous solution for different K, for example, are initially distributed on the mesh. There is a corresponding initial circulation for the far field. The tests (5.4.7) and (5.4.11) for $\varphi_x^{(c)}$, $\varphi_x^{(b)}$ are applied at each point of the mesh and the appropriate difference scheme chosen. The values of φ on each vertical line are solved for by direct elimination (using quasi-linearization). $x = $ constant lines are solved successively marching downstream. For example in (5.4.10) the "coefficient" is left at the previous iteration step. For the nth step on a line

$$\left(\frac{\varphi_{i,j}^{(n-1)} - \varphi_{i-2,j}^{(n-1)}}{2\Delta x}\right)\left(\frac{\hat{\varphi}_{i,j}^{(n)} - 2\hat{\varphi}_{i-1,j}^{(n)} + \hat{\varphi}_{i-2,j}^{(n)}}{(\Delta x)^2}\right)$$

$$-\left(\frac{\hat{\varphi}_{i,j+1}^{(n)} - 2\hat{\varphi}_{i,j}^{(n)} + \hat{\varphi}_{i,j-1}^{(n)}}{(\Delta \tilde{y})^2}\right) = 0,$$

but wherever possible latest values are used. This means that the tests are reconsidered at each line. The matrix to be solved is tridiagonally dominant. The new values are found from $\hat{\varphi}$ by over-relaxation at elliptic points and by under or no relaxation at hyperbolic points. The marching starts at the upstream boundary and proceeds downstream until the trailing edge is reached. Then the circulation Γ in the far field and wake is readjusted as the calculations proceed in the downstream direction. Sweeps through the mesh starting from the upstream boundary are repeated until convergence. The method is stable and produces well-defined supersonic zones and shock waves. The Kutta condition $[\phi_x] = 0$ at the trailing edge is satisfied automatically with this algorithm. Basically the same method is used for supersonic free stream flows with a far-field boundary condition of zero disturbance upstream.

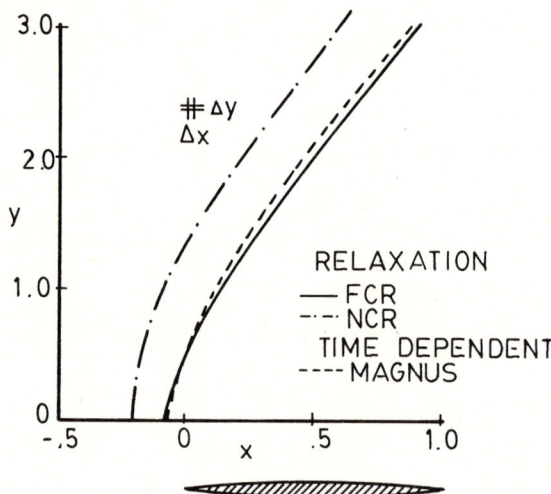

Figure 5.4.10

Comparison of computational methods for 6% parabolic arc with detached bow wave; $M_\infty = 1.15$.

A few computed examples illustrating some details of the method are now given. First from [5.4.2] supersonic flow past a 6% parabolic airfoil, $M_\infty = 1.15$ is considered to show how the shock wave appears with and without the shock point operator. A picture of the detached bow wave (physical coordinates) as calculated by three different methods appears in Figure 5.4.10. FCR denotes the fully conservative relaxation scheme using the shock point operator as outlined. NCR is the result obtained using the elliptic operator at shock points. Magnus (pseudo) time dependent calculations are from [5.4.3].

These calculations were all done with the same equations, boundary conditions and mesh sizes. $(\Delta x = \Delta y = .025)$, $(-.2875 \leq x \leq 1.125, 0 \leq y \leq 3.0)$. Essentially the same result is obtained by Magnus' and FCR method. NCR gives incorrect weak shock jumps. The shock here is defined by (discrete) φ_{xx} maximum. Plots of the discrete solution in the neighborhood of the shock wave against a standardized scaled shock polar appear in Figure 5.4.11.

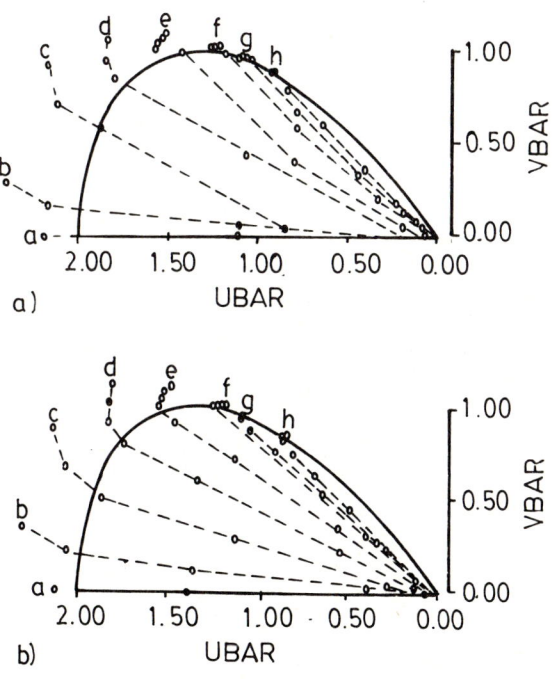

Figure 5.4.11a,b

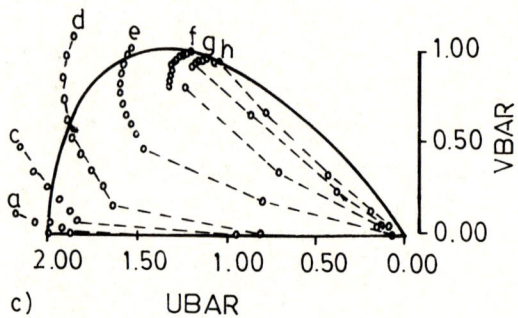

c)   UBAR

Figure 5.4.11c

Hodograph plots for detached bow wave

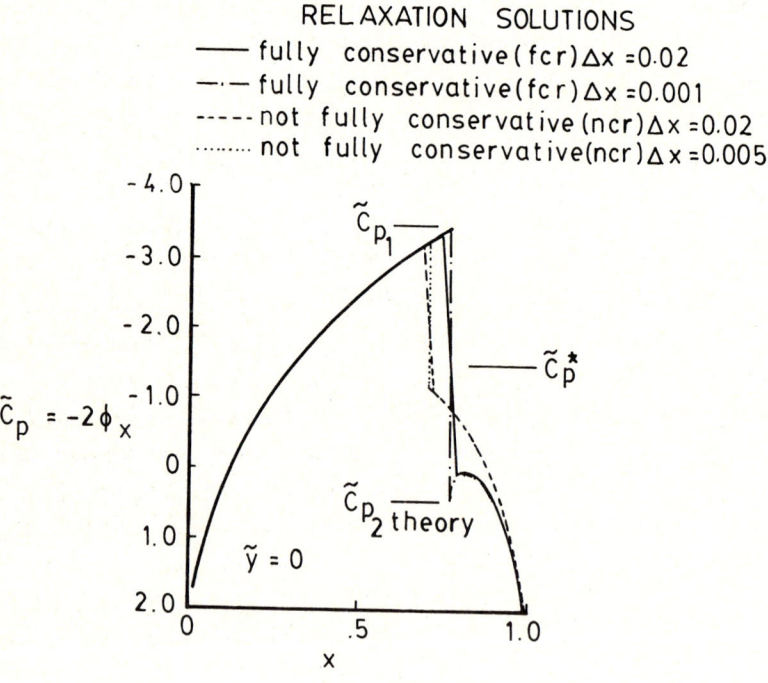

Figure 5.4.12

Pressure distribution for parabolic arc airfoil, $K = 1.8$.

From the plots it is clear that the FCR and Magnus' method make the required shock jumps in a few mesh points but NCR fails. The FCR solution near the y-axis is actually improved with mesh refinement. An example of flow past an airfoil with a supersonic zone and embedded shock was calculated for the parabolic arc at zero incidence with $K = 1.8$. The basic equation was used in the form (5.1.1b). The results appear in Figure 5.4.12. There is a stagnation point (wedge singularity) at the nose and the flow expands aft reaching sonic at approximately 30% of the chord. The flow continues to expand supersonically and terminates in a strong shock. The FCR solution has the shock further aft than the NCR solution and satisfies the correct normal shock jump condition. Further, the (expansion) singularity at the foot of the shock discussed in Section 5.3 is resolved in the FCR calculation. After the shock the flow recompresses smoothly to the tail. details of the flow appear in Figure 5.4.13. The shape of the supersonic zone, the characteristics or Mach waves and the terminating shock are shown. The shock appears to be formed by an envelope of compression waves reflected from the sonic line. The structure agrees with that suggested by Guderley [5.4.4]. The local region near the shock tip in Figure 5.4.14 shows that there is smooth compression for $\tilde{y} > .14$. The shock strengthens continuously as the airfoil surface is approached. Figure 5.4.14 shows the discrete shock structure for this flow plotted on the scaled shock polar at different j levels. The shock traces all values from the normal shock to the weak supersonic-supersonic shock. Near the surface the jump is effectively carried out in four mesh points. But, the weak portions are somewhat smeared out.

Airfoils which have a shock-free supersonic zone can be designed. These can be found by using hodograph methods or by modifying the airfoils which have shocks by successive computations in the physical plane. The physical principle is favourable interference of the Mach waves from the body upon reflection from the sonic line; no envelopes or shocks are formed. Shock free flow for a given shape and attitude can exist only for a definite design Mach number (or similarity parameter K). At off-design condition, for both higher and lower Mach numbers, shock waves appear.

If K is reduced (e.g. M_∞ increased toward unity, δ fixed) the shock wave moves to the tail and then as K is reduced still further an oblique shock (supersonic-supersonic) forms at the tail followed by a wake shock downstream. The configuration appears in Figure 5.4.15.

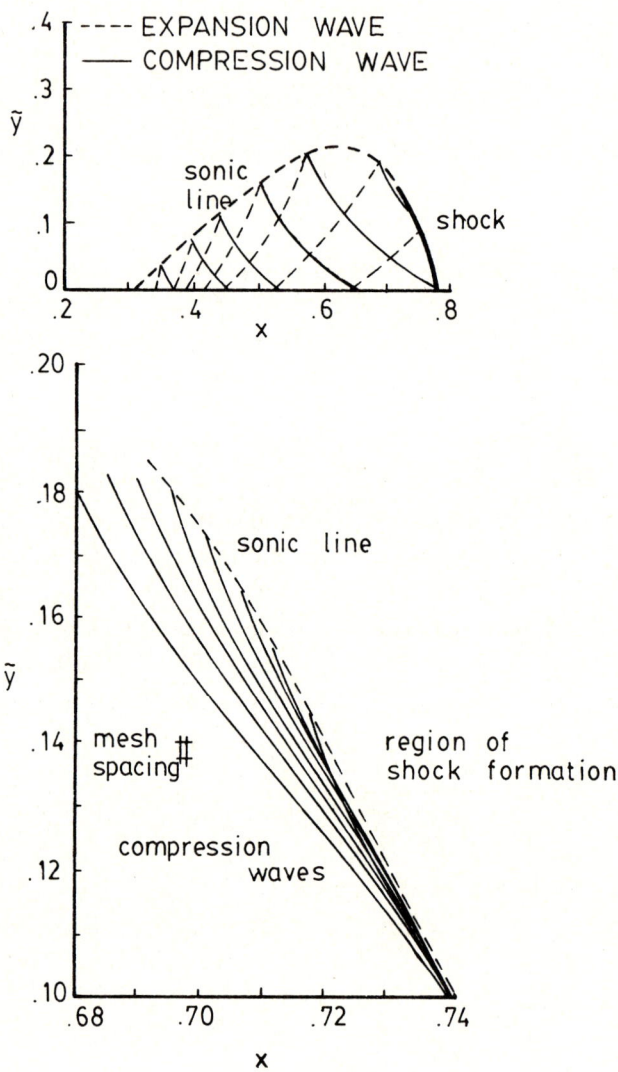

Figure 5.4.13

Structure of embedded shock region, Parabolic arc airfoil, $K = 1.8$.
(a) over-all (b) detail near shock tip

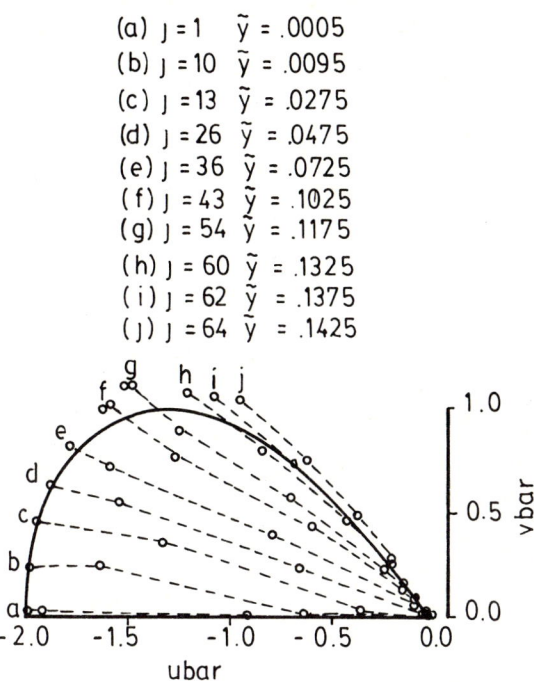

(a) J = 1   ỹ = .0005
(b) J = 10   ỹ = .0095
(c) J = 13   ỹ = .0275
(d) J = 26   ỹ = .0475
(e) J = 36   ỹ = .0725
(f) J = 43   ỹ = .1025
(g) J = 54   ỹ = .1175
(h) J = 60   ỹ = .1325
(i) J = 62   ỹ = .1375
(J) J = 64   ỹ = .1425

Figure 5.4.14

Hodograph plot for embedded shock. Parabolic arc airfoil, $K = 1.8$.

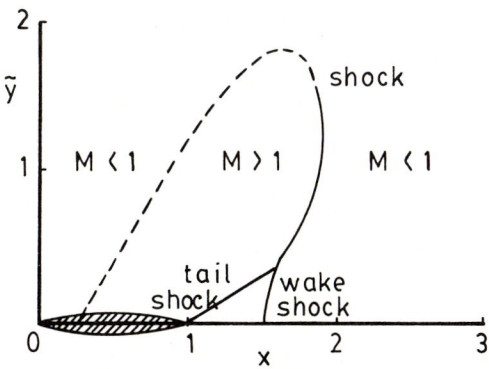

Figure 5.4.15

Flowfield features - parabolic arc airfoil

The wake shock and tail shock intersect and the main shock is oblique, normal and oblique again. There is some smooth supersonic recompression ahead of the wake shock. As K decreases still further the wake shock should move further aft and the tail shock extend in length and curve. Some lifting cases using the same method have been calculated by R. D. Small [5.4.5]. Some earlier results using NCR are reported by Krupp and Murman [5.4.6]. A non-uniform grid was used with 90 x points and 60 y points. A comparison with Sells' results [5.4.7], essentially exact, appears in Figure 5.4.16. There a series of calculations for the NACA-00XX series at different Mach numbers and angle of attack appears in Figure 5.4.17a,b,c,d,e. These surface pressure distributions show increasing lift coefficient with rearward motion of the shock as M_∞ increases, K decreases.

A comparison is given of surface pressure distributions (Figure 5.4.18) with early calculations of Magnus and Yoshihara [5.4.8] and experiments of Stivers [5.4.9] for a 64A410 airfoil. The agreement is good except for the shock location. The calculations were for free flight, the experiments in a wind tunnel with closed walls. Considerable wall interference and viscous effects are likely since the tests Reynolds number was relatively low, about 2×10^6.

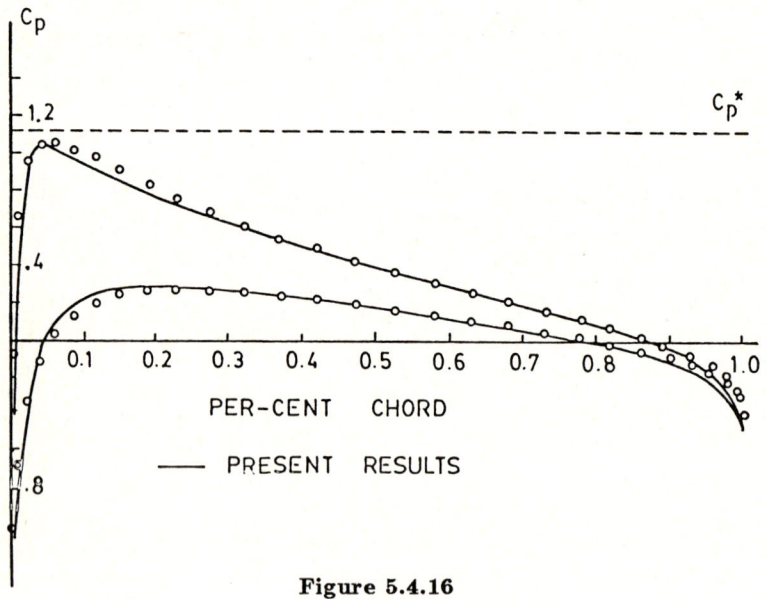

Figure 5.4.16

Pressure distribution, NACA-0012 airfoil, $M_\infty = 0.63$, $\alpha = 2°$.

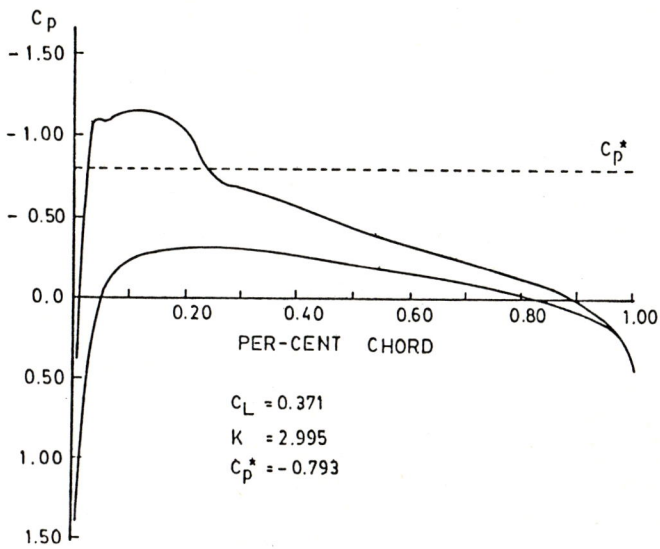

Figure 5.4.17a

Pressure distribution, NACA-0012 airfoil, $M_\infty = .70$, $\alpha = 2°$

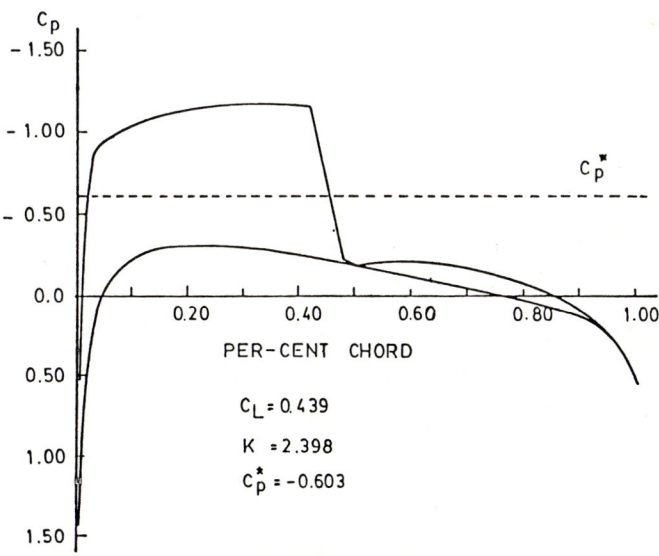

Figure 5.4.17b

Pressure distribution, NACA-0012 airfoil, $M_\infty = .75$, $\alpha = 2°$

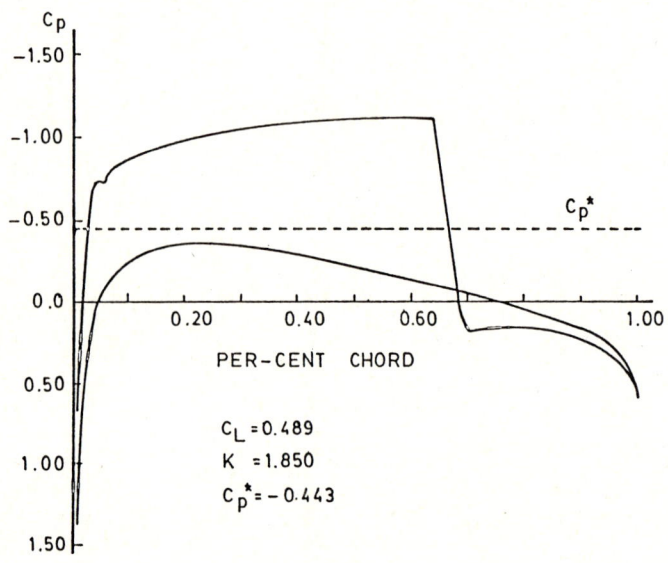

Figure 5.4.17c

Pressure distribution, NACA-0012 airfoil, $M_\infty = .80$, $\alpha = 2°$

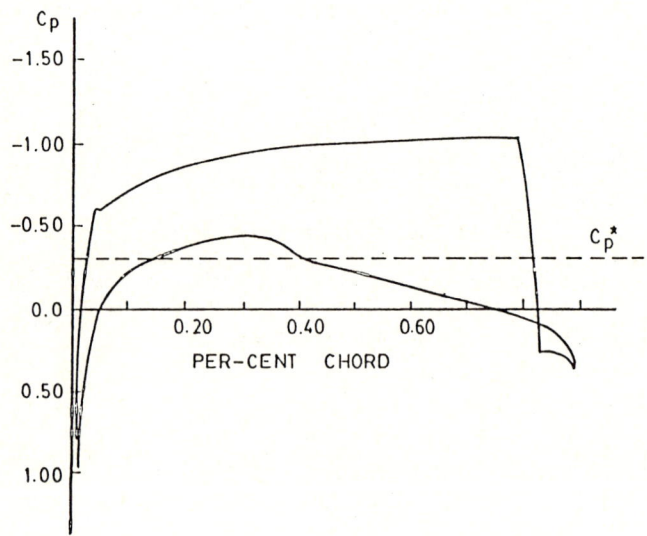

Figure 5.4.17d

Pressure distribution, NACA-0012 airfoil, $M_\infty = .85$, $\alpha = 2°$

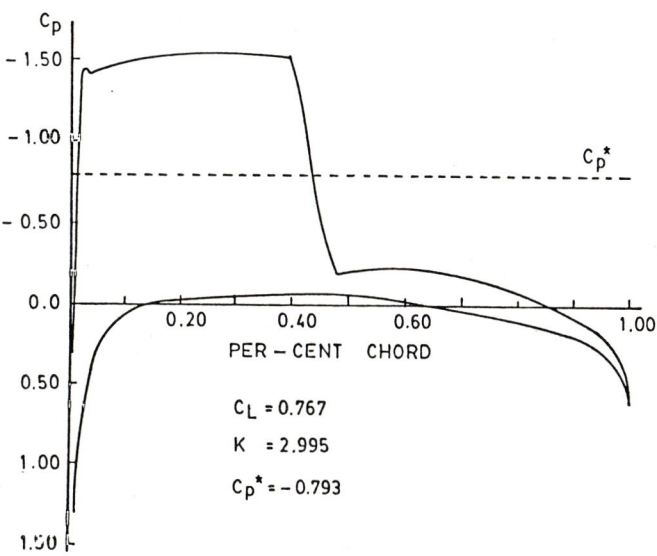

Figure 5.4.17e

Pressure distribution, NACA-0012 airfoil, $M_\infty = .70$, $\alpha = 4°$

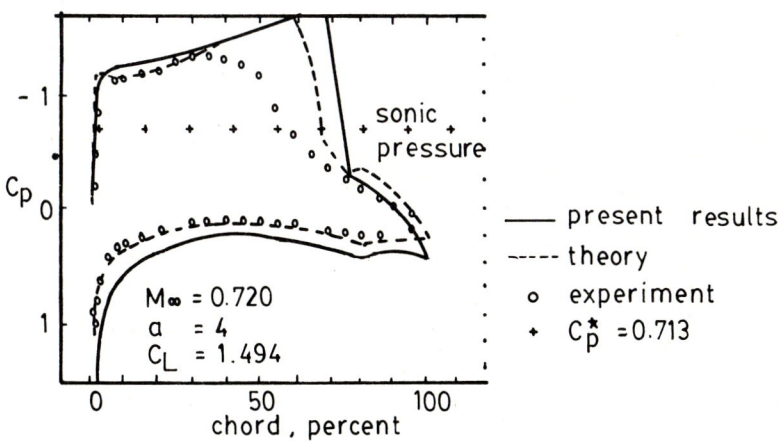

Figure 5.4.18

Comparison with calculations of Magnus and Yoshihara

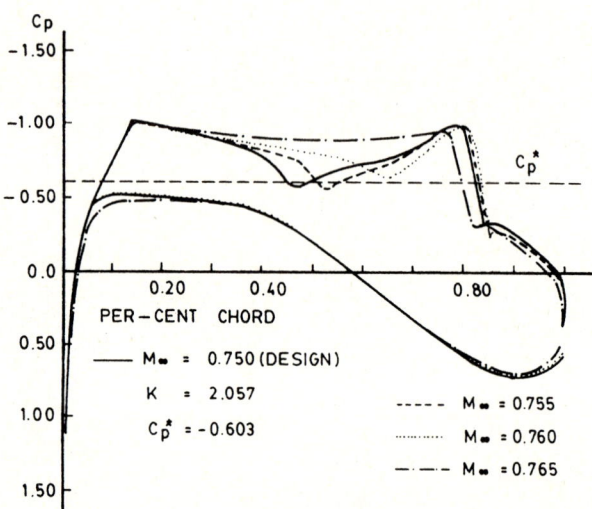

Figure 5.4.19a

Pressure distribution, shock free airfoil, $\delta = 0.151$, $\alpha = 0°$

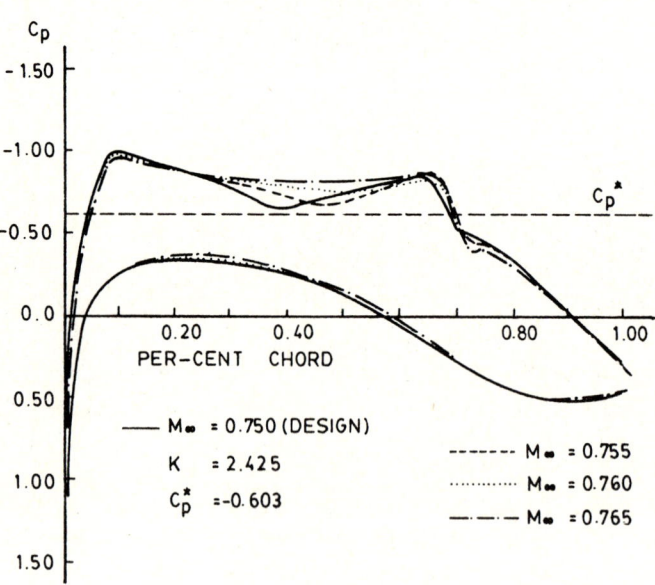

Figure 5.4.19b

Pressure distribution, shock free airfoil, $\delta = 0.151$, $\alpha = 0°$

Also a series of test cases was done on some shock free airfoil designs proposed by Korn, Garabedian et al. The first airfoil had a thickness ratio $\delta = .151$ [5.4.10] and the second had a thickness ratio of $\delta = .118$ [5.4.11]. The computed pressure distributions for the shock free airfoils are shown in Figure 5.4.19a,b. These results agree with the hodograph generated results quite well but not at the proposed design Mach number. For both these airfoils the flow is designed as shock free for $M_\infty = .750$, $\alpha = 0°$. In both cases better agreement is obtained at nominal Mach number $M_\infty = .765$.

Finally in this section we give some numerical results associated with a study of the drag calculated from small-disturbance theory. These results are taken from the report of Murman and Cole [5.4.12]. Drag, being relatively small, is a difficult quantity to determine in this theory. Different ways of calculating the drag by using the theorems of Section 3.10 are discussed here.

Transonic Wave Drag

For comparison with hodograph solutions and to approximate $C_p^\bullet$ the following empirical changes in the scaling definitions are used in the remainder of this section.

$$K = \frac{1 - M_\infty^2}{M_\infty \delta^{\frac{2}{3}}}$$

$$C_p = \frac{\delta^{\frac{2}{3}}}{M_\infty^{\frac{4}{5}}} \bar{C}_p$$

$$(\tilde{y}, \tilde{z}) = \delta^{\frac{1}{3}} M_\infty^{\frac{1}{2}}(y, z)$$

$$C_L = \frac{\delta^{\frac{2}{3}}}{M_\infty^{\frac{4}{5}}} \bar{C}_L, \quad C_D = \frac{\delta^{\frac{5}{3}}}{M_\infty^{\frac{4}{5}}} \bar{C}_D$$

(5.4.27)

For $M_\infty > 1$ Spreiter's scaling laws are used [5.4.13].

The drag integral derived in Section 3.10 is rewritten here including the contribution on the control surface which is shown in Figure 5.4.20.

$$
\begin{aligned}
C_D(\delta, M_\infty, b) &= \delta^{\frac{5}{3}} \bar{C}_D(K, B) \\
&= \frac{\delta^{\frac{5}{3}}}{B} \left\{ 2 \iint_C \left(\left(\frac{v^2 + w^2}{2} - \frac{\gamma + 1}{3} u^3 - K \frac{u^2}{2} \right) \tilde{n}_x - u \mathbf{v}_T \cdot \tilde{n}_T \right) dA \right.\\
&\qquad\qquad \left. - \frac{\gamma + 1}{6} \iint_{S \supset C} [u]_s^3 \, d\tilde{y} d\tilde{z} \right\}, \quad (5.4.28)
\end{aligned}
$$

where $v_T = (v, w)$. This is done because it is useful to calculate the drag with various control surfaces.

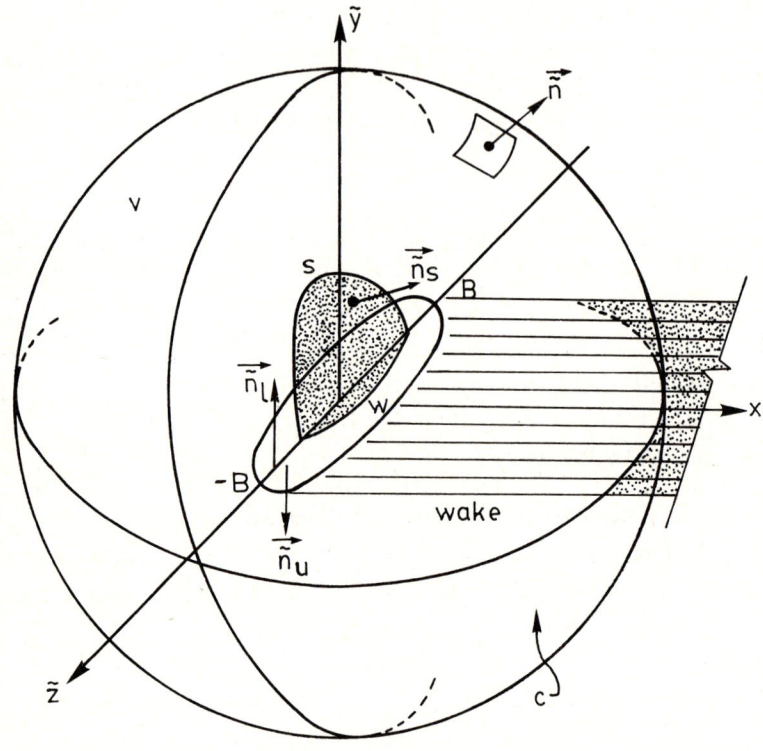

Figure 5.4.20

Contour C, enclosing volume V, shock waves S, wing W and vortex wake.

The conclusion of Section 3.10 is that the drag may be computed analytically by integrating along any convenient contour and enclosed shock waves. The question naturally arises as to whether this result may be realized in a numerical solution of finite-difference equations. The drag is usually a small quantity and hence is sensitive to numerical errors. A basic element in the derivation of (5.4.28) is the use of the divergence theorem. Analogous divergence theorems may be established for finite-difference equations when they are written in conservative form. It is postulated that the effect of truncation errors on the evaluation of (5.4.28) along various contours will be minimized when conservative forms of the finite difference equations are used. To illustrate this, a divergence theorm is established for the difference equations that represent the TSD equations (5.1.1) and the corresponding three-dimensional equation.

Consider first the situation when the flow is completely subsonic so that the governing equation is elliptic throughout the domain. Then, using standard finite-difference notation, the second-order accurate elliptic difference equation for (5.1.1b) may be written as

$$\frac{\bar{u}_{i+\frac{1}{2}} - \bar{u}_{i-\frac{1}{2}}}{\Delta x} + \frac{v_{j+\frac{1}{2}} - v_{j-\frac{1}{2}}}{\Delta \tilde{y}} + \frac{w_{k+\frac{1}{2}} - w_{k-\frac{1}{2}}}{\Delta z} = \left(\nabla \cdot \psi\right)_{ijk} = R_\epsilon \approx 0 \,,$$
(5.4.29)

where

$$\psi = (\bar{u}, v, w)$$
(5.4.30)

$$\bar{u} = K\phi_x - \frac{\gamma + 1}{2}\phi_x^2 \,; \quad v \equiv \phi_{\tilde{y}}\,; \quad z \equiv \phi_{\tilde{z}}\,,$$
(5.4.31)

and

$$\left. \begin{aligned}
\bar{u}_{i+\frac{1}{2}} &\equiv K\left(\frac{\phi_{i+1,j,k} - \phi_{i,j,k}}{\Delta x}\right) \\
&\quad - \frac{\gamma + 1}{2}\left(\frac{\phi_{i+1,j,k} - \phi_{i,j,k}}{\Delta x}\right)^2 . \\
v_{j+\frac{1}{2}} &\equiv \frac{\phi_{i,j+1,k} - \phi_{i,j,k}}{\Delta \tilde{y}} . \\
w_{k+\frac{1}{2}} &\equiv \frac{\phi_{i,j,k+1} - \phi_{i,j,k}}{\Delta \tilde{z}} .
\end{aligned} \right\}$$
(5.4.32)

with analogous expressions for $u_{i-\frac{1}{2}}$, $v_{j-\frac{1}{2}}$, and $w_{k-\frac{1}{2}}$. The term R_ϵ is the residual or the accuracy to which the difference equations are solved on the computer. It is assumed that R_ϵ is less than any truncation errors and is set to zero in the remainder of the analysis. Equation (5.4.29) is multiplied by $\Delta x \Delta \tilde{y} \Delta \tilde{z}$ to put it in flux form:

$$\left(\bar{u}_{i+\frac{1}{2}} - \bar{u}_{i-\frac{1}{2}}\right)\Delta \tilde{y}\Delta \tilde{z} + \left(v_{j+\frac{1}{2}} - v_{j-\frac{1}{2}}\right)\Delta x \Delta \tilde{z}$$
$$+ \left(w_{k+\frac{1}{2}} - w_{k-\frac{1}{2}}\right)\Delta x \Delta \tilde{y} = \left(\nabla \cdot \psi\right)_{ijk}\Delta x \Delta \tilde{y}\Delta \tilde{z} = 0. \quad (5.4.33)$$

Consider now a contour that encloses a group of mesh points, as shown in Figure 5.4.21 for a two-dimensional case. Summing the difference equations for the enclosed mesh points yields

$$\sum_j \Delta \tilde{y} \sum_k \Delta \tilde{z} \sum_i \left(\bar{u}_{i+\frac{1}{2}} - \bar{u}_{i-\frac{1}{2}}\right) + \sum_i \Delta x \sum_k \Delta \tilde{z} \sum_j \left(v_{j+\frac{1}{2}} - v_{j-\frac{1}{2}}\right)$$

$$+ \sum_i \Delta x \sum_j \Delta \tilde{y} \sum_k (w_{k+\frac{1}{2}} - w_{k-\frac{1}{2}})$$

$$= \sum_i \sum_j \sum_k (\nabla \cdot \psi)_{ijk} \Delta x \Delta \tilde{y} \Delta \tilde{z} . \qquad (5.4.34)$$

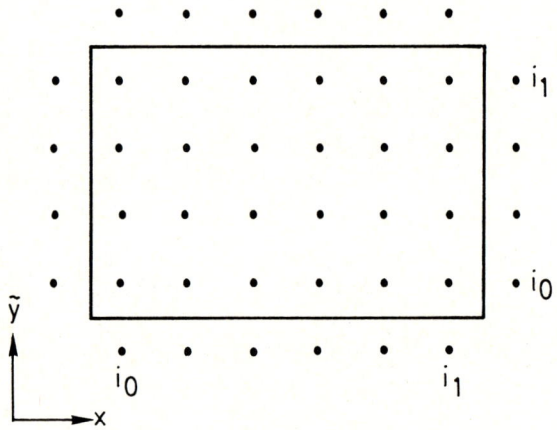

Figure 5.4.21

Mesh Point Notation for Summation Formulae.

But, because of the conservative form of the difference equations,

$$\left. \begin{aligned}
\sum_i (\bar{u}_{i+\frac{1}{2}} - \bar{u}_{i-\frac{1}{2}}) &= \bar{u}_{i_1+\frac{1}{2}} - \bar{u}_{i_0-\frac{1}{2}} , \\
\sum_j (v_{j+\frac{1}{2}} - v_{j-\frac{1}{2}}) &= v_{j_1+\frac{1}{2}} - v_{j_0-\frac{1}{2}} , \\
\sum_k (w_{k+\frac{1}{2}} - w_{k-\frac{1}{2}}) &= w_{k_1+\frac{1}{2}} - w_{k_0-\frac{1}{2}} ,
\end{aligned} \right\} \qquad (5.4.35)$$

and all interior fluxes together with their truncation errors identically cancel on the left hand side of equation (5.4.34). The result is

$$\sum_j \sum_k (\bar{u}_{i_1+\frac{1}{2}} + \bar{u}_{i_0-\frac{1}{2}}) \Delta \tilde{y} \Delta \tilde{z} + \sum_i \sum_k (v_{j_1+\frac{1}{2}} - v_{j_0-\frac{1}{2}}) \Delta x \Delta \tilde{z}$$

$$+ \sum_i \sum_j (w_{k_1+\frac{1}{2}} - w_{k_0-\frac{1}{2}}) \Delta x \Delta \tilde{y} = \sum_i \sum_j \sum_k (\nabla \cdot \psi)_{ijk} \Delta x \Delta \tilde{y} \Delta \tilde{z} = 0 . (5.4.36)$$

The error between the true contour integral and the finite summation formula on the left hand side of (5.4.36) is caused by the truncation error terms on the contour boundary, which are $O\left[(\Delta x)^2, (\Delta \tilde{y})^2, (\Delta \tilde{z})^2\right]$. All interior fluxes identically cancel so that there is no accumulation of truncation error terms (representing sources or sinks) as the contour expands. This is a basic feature of conservative difference formulas. Equation (5.4.36) represents a divergence theorem for the difference equation. The finite summation formula is invariant with choice of contour.

If C completely encloses regions of mesh points for which the flow is supersonic, the above result is unaltered. The FCR difference equations are constructed so that all interior flux terms cancel. If, however, the equation is hyperbolic at one or more of the mesh points on an $x = $ constant plane of C, the above formula must be modified to yield an invariant expression. For example, let the flow be supersonic at all mesh points i_1, j, k; then the first summation in (5.4.36) must be replaced by

$$\sum_j \sum_k \left(\bar{u}_{i_1 - \frac{1}{2}} - \bar{u}_{i_0 - \frac{1}{2}}\right)\Delta \tilde{y}\Delta \tilde{z} \qquad (5.4.37)$$

to account for the backward difference formula for hyperbolic points. The finite summation formula remains invariant, but now represents the line integral with an accuracy $O\left[\Delta x, (\Delta \tilde{y})^2, (\Delta \tilde{z})^2\right]$, at least for the supersonic portions of the contour.

It would be desirable to establish a finite-difference divergence theorem analogous to (5.4.36) for (3.10.6), but this has not yet been accomplished. Computed examples below show, however, that when the difference equations are in conservative form, the drag may be calculated for any convenient contour and the result is essentially invariant for a given solution if $[u]^3$ can be accurately defined.

For the calculated two-dimensional results reported here the integrals along the contour and enclosed shock waves are evaluated numerically by using the trapezoidal rule. The velocities u and v are calculated by centered difference formulas, except for hyperbolic mesh points on an $x = $ constant line. At these points, backward difference formulas for u are used, as suggested by the divergence theorem for the difference equations. A calculation using Simpson's rule for the integrations yielded the same results.

The drag may also be calculated by using the surface-pressure integration equation (3.10.23). Here, again, the trapezoidal formula is used, with u given by centered difference formulas and v defined by the boundary condition $F'(x)$. In the calculations, the leading and trailing edges lie midway between mesh points in the x-direction. Consequently, the interval of integration extends from one mesh point ahead of the leading edge $i_0 - 1$ to one mesh point behind the trailing edge $i_1 + 1$ (Figure 5.4.22).

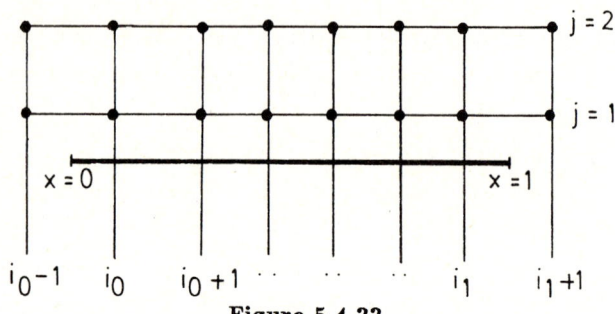

Figure 5.4.22

Mesh Point Notation Near Body

The resulting formula for the surface-pressure integral is then (for a nonlifting symmetric two-dimensional body);

$$\bar{C}_D = 2 \sum_{i_0}^{i_1+1} \left\{ -2 \left(\frac{u_i F_i' + u_{i-1} F_{i-1}'}{2} \right) (x_i - x_{i-1}) \right\} \tag{5.4.38}$$

where

$$u_i \equiv \left(\frac{u_{i+\frac{1}{2}} + u_{i-\frac{1}{2}}}{2} \right)_{y=0} \equiv \frac{1}{2} \left(\frac{\phi_{i+1} - \phi_i}{x_{i+1} - x_i} + \frac{\phi_i - \phi_{i-1}}{x_i - x_{i-1}} \right)_{y=0}. \tag{5.4.39}$$

Since the first j mesh point is a half mesh interval above the body, the values of ϕ on $y = 0$ are obtained by the extrapolation formula

$$(\phi_i)_{y=0} = \frac{3}{2} \phi_{i,1} - \frac{1}{2} \phi_{i,2}. \tag{5.4.40}$$

After some simple algebra, (5.4.38) may be rewritten as

$$\bar{C}_D = 2 \sum_{i_0}^{i_1} \left(-2 u_i F_i' \left(\frac{x_{i+1} - x_{i-1}}{2} \right) \right), \tag{5.4.41}$$

since $F_{i_0-1}' = F_{i_1+1}' = 0$. Simpson's rule was also tried for the surface-pressure integral, but the results were inferior. In addition, both numerical integration formulas were applied over the interval i_0 to i_1 with leading- and trailing-edge contributions included by analytical formulas for the stagnation point singularities. The results were not as accurate as the procedure stated above.

The jump of u across the shock may be readily defined for most cases when the downstream velocity is subsonic. For shocks with supersonic downstream velocities, both the shock location and jump are difficult to evaluate because of the dissipation effects of the first-order, hyperbolic difference equation. It is thus advantageous to select contours that exclude such shocks. Sample contours incorporated in the computer program are shown in Figure 5.4.23.

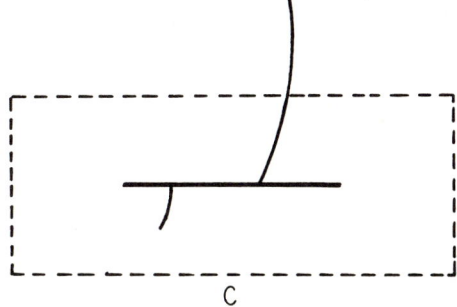

C

Figure 5.4.23a

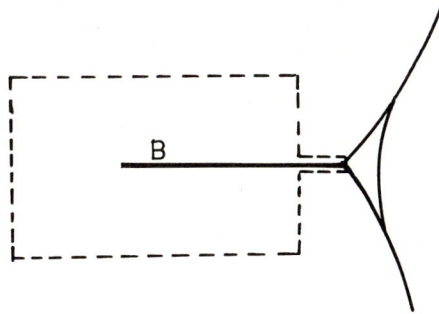

B

Figure 5.4.23b

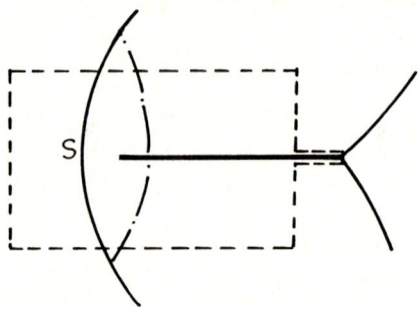

Figure 5.4.23c

Typical contours C for evaluation of drag integral enclosing body
mean surface B and shock waves S.

Comparison of integrals along shock and body

The parabolic arc shape is a simple geometry that has been used extensively
for theoretical studies. A pressure distribution for a supercritical subsonic flow
is shown in Figure 5.4.24. The corresponding function that must be integrated
along the body surface to obtain the pressure drag is shown in Figure 5.4.25.
Thrust contributions T must be subtracted from the drag contributions D to
obtain a small remainder noted by the average value $\frac{1}{2}\bar{C}_D$ in the figure. Such
a procedure is sensitive to numerical integration errors, particularly in the nose
and tail regions and near the shock-wave discontinuity. As mentioned previously,
the trapezoidal rule seems to treat these regions most accurately. The function
that must be integrated along the length of the shock is shown in Figure 5.4.26
for three different mesh spacings. In this case, the drag is the area under a single
curve and its evaluation is insensitive to the numerical integration procedure.
For a 6% thick section ($\tau = 0.06$), $K = 1.8$ corresponds to $M_\infty = 0.87$ and $\bar{C}_D$
$= 0.0028$. Figure 5.4.27 shows another example for the same airfoil at a higher
Mach number.

Effect of conservative difference formulas

Several computations have been completed to investigate the importance of
having the difference equations in conservative form. Two separate situations
should be considered; namely, shock-free flows and flows with shock waves. For

shock-free flows, finite-difference equations in nonconservative form that are consistent with the differential equations are also consistent with the integral forms of the equations. However, internal fluxes between mesh cells do not cancel identically, but do cancel to within a truncation error.

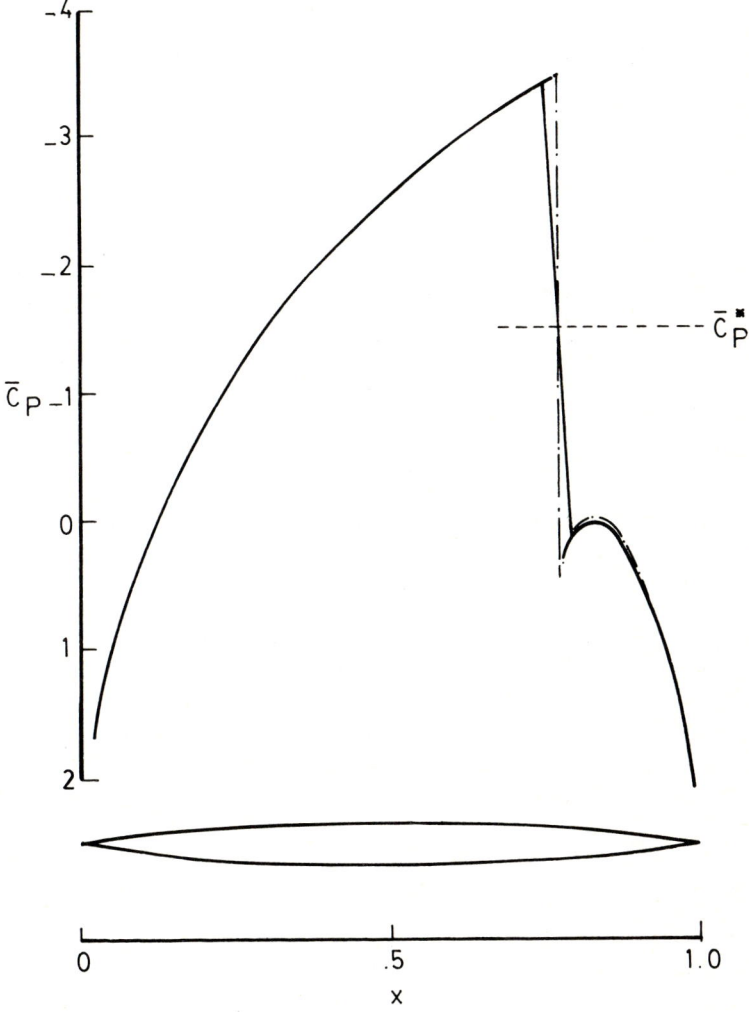

Figure 5.4.24

Pressure distribution for parabolic arc, $K = 1.8$.

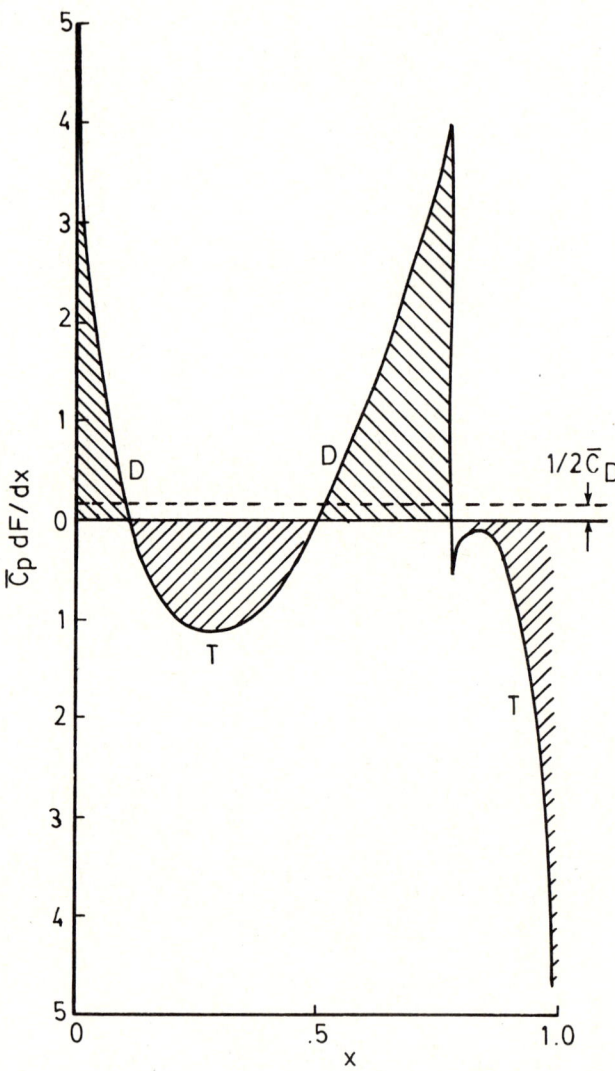

Figure 5.4.25

Distributions of drag D and thrust T along parabolic arc. $C_D = .319$.

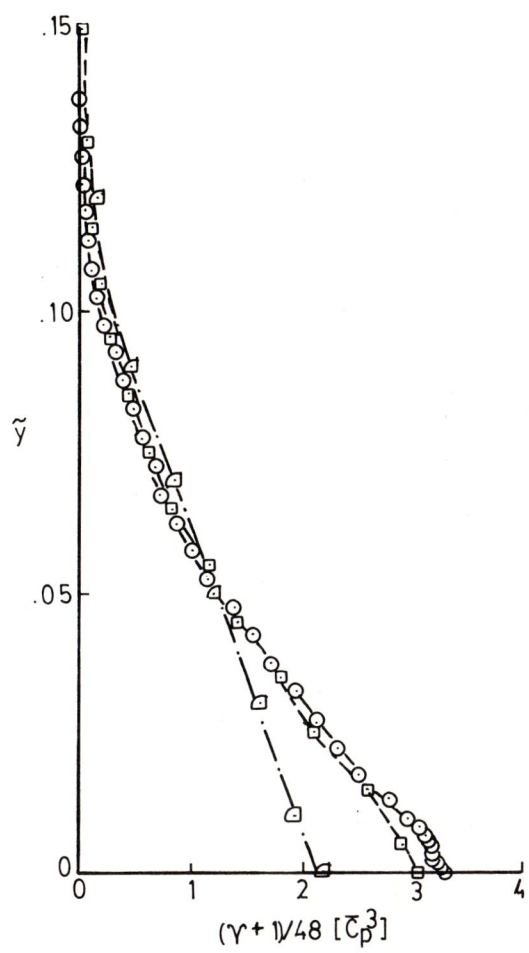

Figure 5.4.26

Distribution of drag along shock wave of parabolic arc, $K = 1.8$. Coarse mesh, $C_D = .263$, medium mesh, $C_D = .293$, fine mesh, $C_D = .297$.

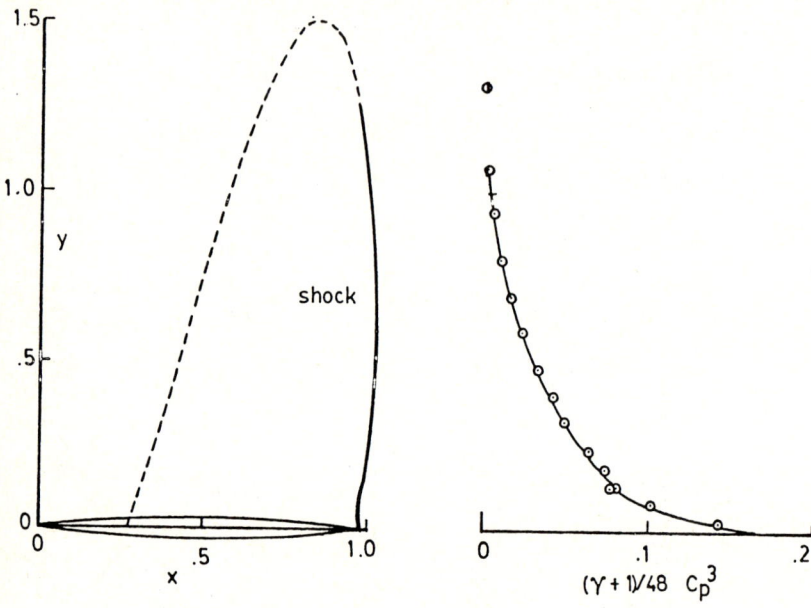

Figure 5.4.27

Embedded supersonic region, shock wave, and distribution of drag along shock wave for parabolic arc, $M = .909$. From surface pressure integration $C_D = 0.0315$. From integration along shock, $C_D = 0.0320$.

Thus the divergence expression (5.4.36) will not equal zero, but will equal a truncation error term that becomes smaller as the mesh is refined or as the accuracy of the difference equations is increased. Then, the drag might not be invarient with choice of contour. This is verified by calculations for a (nearly) shock-free airfoil in Figure 5.4.28. The three calculations represent difference equations in fully conservative form (FCR) [5.4.1], not fully conservative form (NCR) [5.4.2] (where the equations are in nonconservative form at the downstream sonic line), and the method of Garabedian and Korn [5.4.14], and Jameson [5.4.15] (GKJ), where the first derivatives are approximated by central differences in the hyperbolic regions (nonconservative). All methods are formally first-order accurate in the supersonic flow. The term C_{D_B} was calculated by integrating the pressure along the surface, $C_{D_{\sigma_1}}$ was obtained by a contour integral two mesh points above the body and well ahead and behind the body, and $C_{D_{\sigma_2}}$ corresponds to a contour well away from the body and embedded supersonic zone.

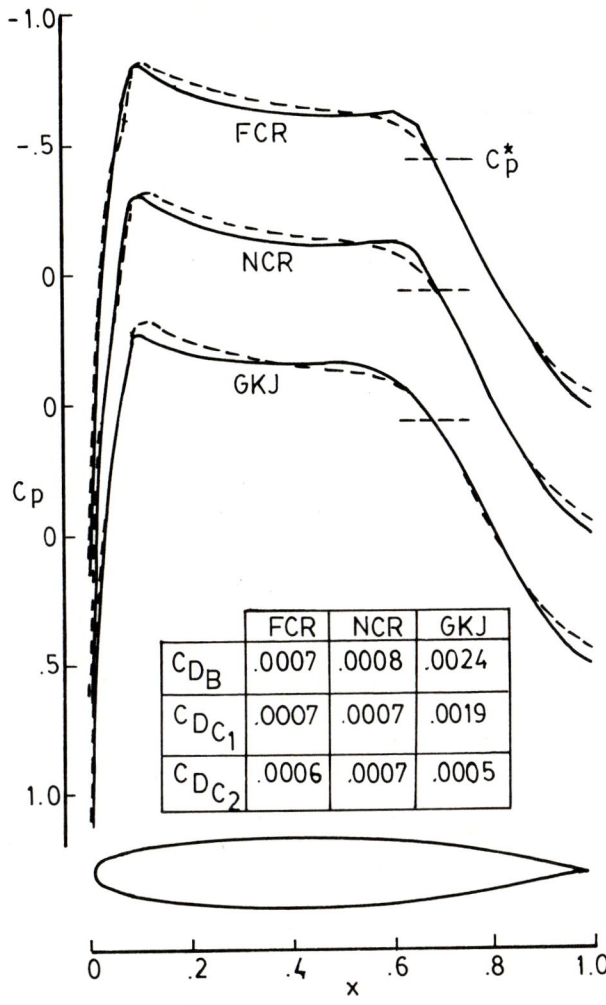

Figure 5.4.28

Pressure distributions and frag coefficients for Korn airfoil at design conditions $M = .80$, $\alpha = 0°$, - - - exact solution, $C_D = 0$. —— numerical solution using various finite difference equations.

For flows with shock waves, the conservative form must be used to calculate the correct shock jumps. The error introduced at shock waves on the right-hand side of (5.4.36) by nonconservative equations is not a truncation error in that it does not disappear as the mesh is refined. Calculated results at an off-design condition for the same airfoil illustrate this effect in the drag calculation (Figure 5.4.29). The airfoil which shows considerable drag is off-design by only .01 Mach number.

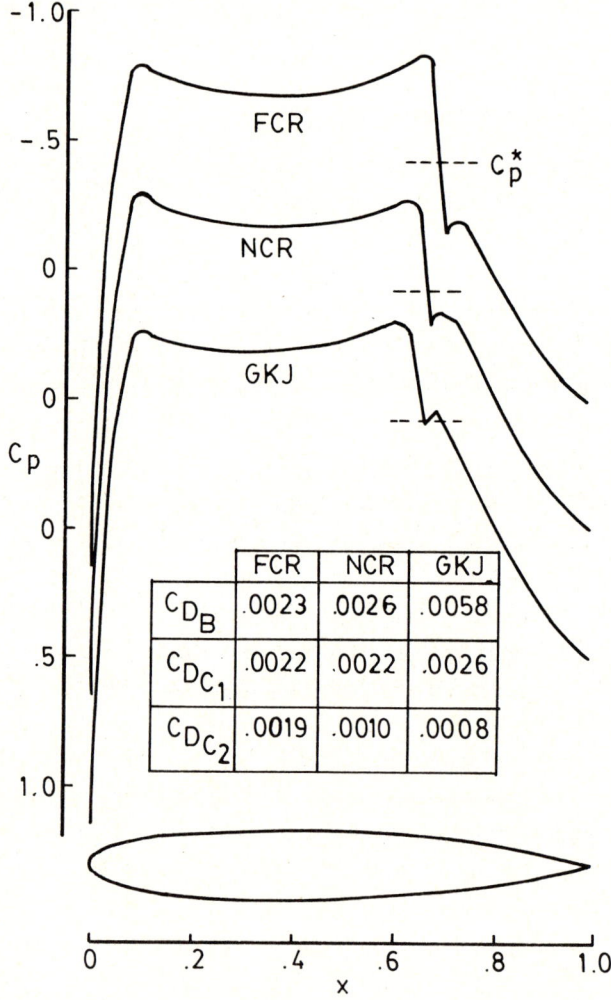

Figure 5.4.29

Pressure distribution and drag coefficients for Korn airfoil at off-design condition $M = .81$, $\alpha = 0°$ using various finite difference equations.

Accuracy

Several factors in addition to those already mentioned affect the accuracy of the calculated drag. For the results reported here, the drag was calculated both by integrating the surface pressures C_{D_B} and by integrating along a contour C_{D_O} in the flow field. The two results generally agree to within $\Delta C_D = 0.0005$. For subcritical flows C_{D_O} is zero to four significant figures while C_{D_B} may be as large as -0.0005. Thus, calculating the drag by two different procedures provides a useful internal check on the accuracy of a given solution.

A source of inaccuracy is the evaluation of $[u]^3$ across the shock. A mesh spacing of $\Delta x = 0.01$ to 0.02 at the foot of the shock seems sufficient for a reliable result. Occasionally, C_{D_B} and C_{D_O} differ by an amount larger than 0.0005, probably as a result of poor mesh spacing relative to the shock wave somewhere in the flow field. This has not proved to be a serious problem. For a few cases, a large variation between C_{D_B} and C_{D_O} resulted, which could always be traced to some unusual shock system (e.g., two shocks very close together).

The choice of mesh spacing near the nose is important, particularly for blunt-nosed bodies with strong shocks. Essentially three different meshes were used in the present study (Table 5.4.2). For the parabolic arc case of Figs. 5.4.24-26, no essential change was noted in C_D as the mesh was refined in x or y. For the Korn airfoil of Figure 5.4.28 (FCR), the drag changed by about 0.0003 between meshes 1 and 2 with only small changes in the pressure distribution. A subcritical flow past an NACA 0012 suggested by Lock for a standard check case is shown in Figure 5.4.30. Again, the change in C_D with mesh was slight. However, for the same airfoil at a supercritical condition with a relatively strong shock, the change in mesh spacing has a more pronounced effect on the value of C_D (Figure 5.4.31). For most calculations here, mesh 1 was used. Refinement from mesh 1 to mesh 3 increased the drag by at most ten to fifteen percent. Generally, the dominant factor in this change was the x mesh spacing near the nose. The y mesh spacing near the body had little effect.

Table 5.4.2

mesh 1	mesh 2	mesh 3
-1.60	-1.60	-1.70
.	.	.
.	.	.
.	.	.
-0.015	-0.006	-0.003
-0.005	-0.002	-0.001
0.005	-0.002	0.001
0.015	0.006	0.003
0.025	0.010	0.005
0.035	0.015	0.007
0.045	0.020	0.010
0.055	0.025	0.014
0.065	0.030	0.018
0.075	0.035	0.022
0.085	0.0425	0.026
0.095	0.050	0.030
0.105	0.0575	0.035
.	0.065	0.040
.	0.075	0.045
.	0.085	0.050
.	0.095	0.055
.	0.105	0.060
.	.	0.065
.	.	0.070
.	.	0.0775
.	.	0.085
$\Delta x = 0.02\text{-}0.04$ over remainder of		0.095
body with $\Delta x = 0.01$ near shock		0.105
		.
		.
0.01	0.002	0.002
0.03	0.006	0.006
0.05	0.010	0.010
0.07	0.014	0.014
.	.	.
.	.	.
.	.	.

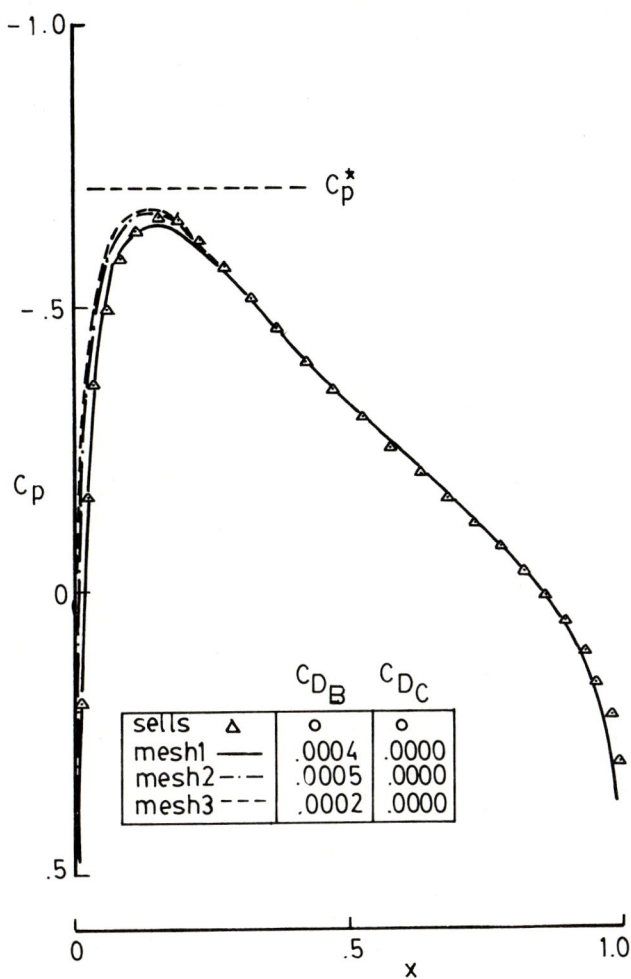

Figure 5.4.30

Effect of mesh spacing on drag coefficient and pressure distribution, NACA 0012 airfoil, $\alpha = 0°$. Subcritical $M = .72$.

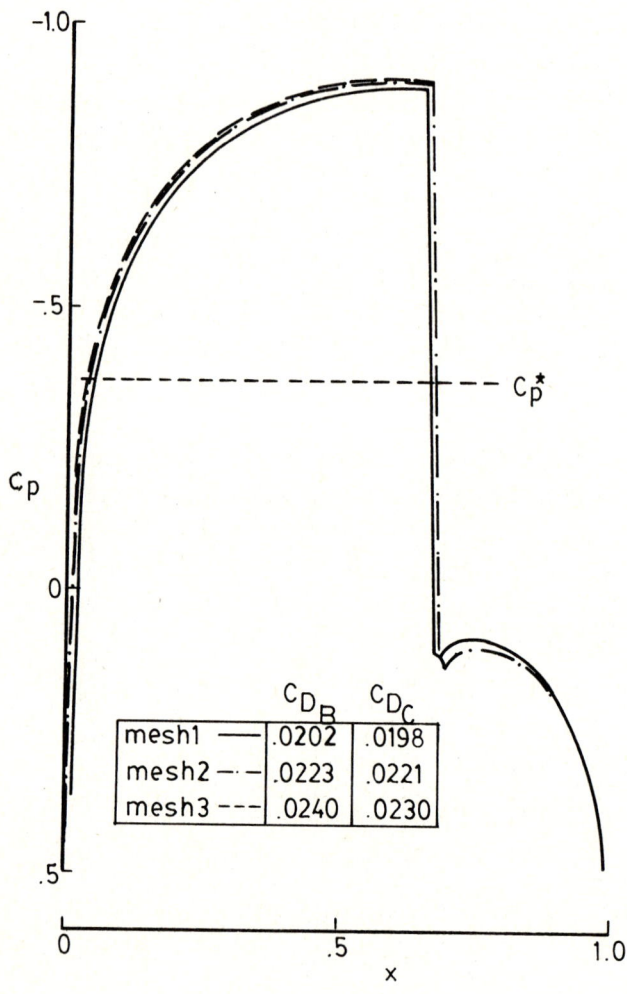

The table shown within the figure:

		c_{D_B}	c_{D_C}
mesh1	——	.0202	.0198
mesh2	—·—	.0223	.0221
mesh3	–––	.0240	.0230

Figure 5.4.31

Effect of mesh spacing on drag coefficient and pressure distribution, NACA 0012 airfoil, $\alpha = 0°$. Supercritical $M = .825$.

Tunnel wall interference drag

An interesting application of the basic drag integral (5.4.28) is the prediction of an interference drag due to momentum flux through a wind-tunnel wall. Consider a model inside a wind tunnel of infinite extent in the upstream and downstream directions. If the contour enclosing the body is extended as far as possible away from the model, the situation shown in Figure 5.4.32 results. The contributions to the integral along the tunnel walls at the end boundaries may not be zero as would be the case in free air. These contributions represent net momentum fluxes out of the wind tunnel, and the drag force felt on the model is basically different from the wave drag. For a straight, solid-wall tunnel $(v = 0)$, a free-jet $(u = 0)$, and an ideal slotted tunnel $(u \sim \partial v/\partial x)$ wall boundary condition, the drag is zero. However, for a perforated wall-boundary condition, $u \sim v$, the integrand uv is positive definite and the drag will be nonzero. For more complicated boundary conditions representative of true tunnel walls, there may be, in general, a nonzero drag. A similar result is found from subsonic linear theory [5.4.16] where the drag is identified as a bouyancy force that is due to a pressure gradient over the length of the model. The current theory represents an extension of the bouyancy drag to transonic speeds. In addition to the bouyancy drag force, the wave drag in free air and in a tunnel may also differ.

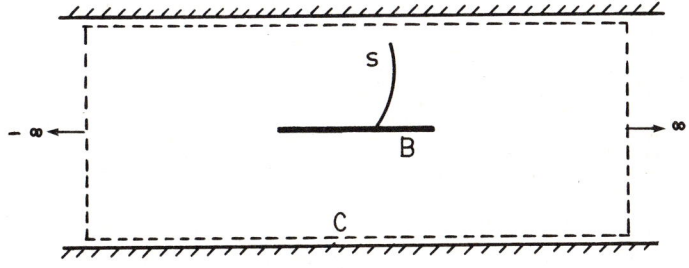

Figure 5.4.32

Contour C enclosing body B and shock S inside a wind tunnel.

To illustrate these tunnel interference drag effects, a test arrangement used at ONERA (Ponteziere and Bernard-Guelle [5.4.17]) was selected. Three NACA 0012 airfoils of different chord lengths were tested in a perforated wind tunnel with various open area ratios σ (σ = area of perforations/tunnel wall area) for Mach numbers from 0.4 to 0.95. It was found that, at the zero lift condition, the shock-wave position for all three models was the same for $\sigma = 12.5\%$ at a given

Mach number, provided the Reynolds number was greater than 3×10^6. It was concluded that the $\sigma = 12.5\%$ condition was interference-free (i.e., zero blockage or Mach number correction) and that the other conditions ($\sigma = 0$, 3.1, 5.5%) interference effects were occurring. A correction procedure based on matching shock locations was developed for the latter conditions, and porosity parameters were deduced for the three nonzero values of σ.

In the present study, calculations were performed using the method of Murman [5.4.18] and the reported porosity parameters for $\sigma = 5.5$ and 12.5%. The intermediate size airfoil ($C = 80$ mm) was selected, yielding a tunnel semi-height/chord ratio of 1.25 (4.8% blockage). The results are shown in Figure 5.4.33 and 5.4.34. In Figure 5.4.33a, the calculated shock positions for the two tunnel and free air calculations are plotted together with several sets of data. The calculations show qualitatively that the shock location for the $\sigma = 12.5\%$ condition is the same as for the free air location, while downstream displacement of the shock occurs for the $\sigma = 5.5\%$ condition. There is, however, substantial disagreement in the measured and theoretical shock locations. Two other independent sets of data (Osborne 1971 and Stivers, private communication, May 1974) at approximately the same Reynolds number are shown. The disagreement apparently is not due solely to two-dimensional, shock-wave, boundary-layer effects.

The calculated pressure drag for $\sigma = 12.5\%$ is compared with the free air values in Figure 5.4.33b. For the tunnel condition, the total drag and wave drag are shown separately, the difference being the "transonic bouyancy" or tunnel wall drag. These values are also listed in Table 5.4.3, together with the bouyancy drag correction predicted with linear subsonic theory, [5.4.16].

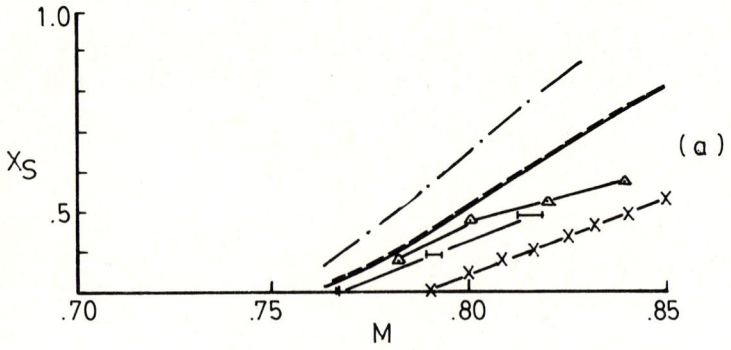

Figure 5.4.33a

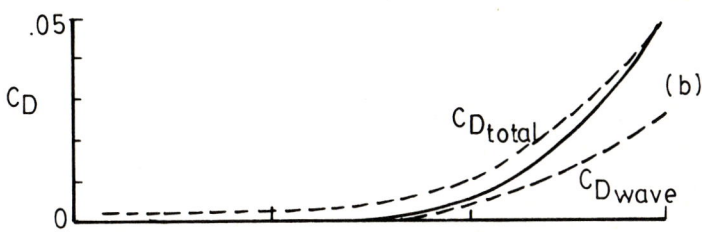

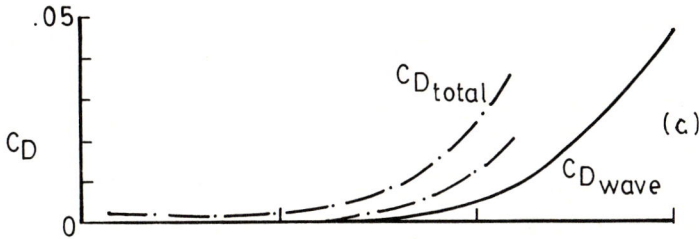

Figure 5.4.33b,c

Calculations for NACA 0012 airfoil, $\alpha = 0°$, inside a perforated wall tunnel $\text{- - -}\ \sigma = 12.5\%$, and $\text{— - —}\ \sigma = 5.5\%$, and in free air ——. (a) Shock wave position compared with data, $\times$ ONERA, $\text{Re} = 3.6 \times 10^6$,   RAE, $\text{Re} = 3.6 \times 10^6$, $\triangle$ NASA-Ames, $\text{Re} = 4 \times 10^6$. (b) Drag coefficients for $\sigma = 12.5\%$. (c) Drag coefficients for $\sigma = 5.5\%$.

Table 5.4.3

Values of C_D for NACA 0012 in perforated tunnel ($\sigma = 12.5\%$) and free air

	C_D				
	Transonic Theory				Subsonic Theory
M	Free Air Wave C_D	Total C_D	Wind Tunnel Wall C_D	Wave C_D	
0.706	0	0.0022	0.0022	0	0.0022
0.763	0.0001	0.0032	0.0032	<0.0001	0.0031
0.780	0.0010	0.0049	0.0043	0.0006	0.0036
0.804	0.0069	0.0124	0.0072	0.0052	0.0044
0.810	0.0099	0.0155	0.0084	0.0071	0.0046
0.827	0.0215	0.0273	0.0131	0.0142	0.0055
0.840	——	0.0390	0.0185	0.0205	0.0063
0.850	0.0482	0.0482	0.0211	0.0249	0.0071

Several points are evident. First, although the shock waves are in the same location in the tunnel and in free air, the wave drags are different and the discrepancy increases markedly with Mach number. Figure 5.4.34 is a plot of the pressure distribution, embedded supersonic zone, and distribution of drag along the shock for $M = 0.81$. Second, at subcritical conditions, the tunnel interference drag equals the bouyancy drag from linear theory. Third, it is not sufficient to simply subtract the tunnel interference drag from the total drag to correct the tunnel data. The nonlinearity of the flow processes precludes using such a superposition principle. The drag results calculated for $\sigma = 5.5\%$ are shown in Figure 5.4.33c, and the already observed features are more noticeable. In all of the results shown in Figure 5.4.33b and c, the shock waves did not reach the tunnel walls. Finally, note that, for some of the cases shown in Figures 5.4.33 and 5.4.34 and table 5.4.3, the mesh spacing near the shock wave exceeded $\Delta x = 0.01$, the required value, stated above, for good accuracy. Consequently, the calculated wave drags may be ten to fifteen percent below the actual value, which will not significantly affect the tunnel interference drag, shock position, or any of the general conclusions.

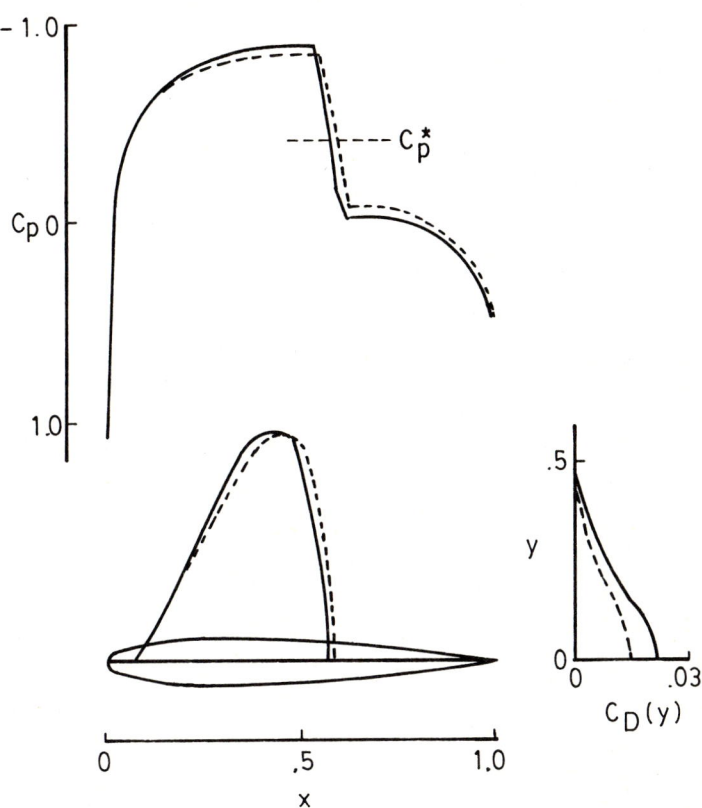

Figure 5.4.34

Pressure distribution, Embedded supersonic zone, and shock losses
for NACA 0012 airfoil, $\alpha = 0^0$, $M = .81$ in free air —— and
perforated wall tunnel - - - $\sigma = 12.5\%$.

Inviscid wake profiles

A set of calculations was completed for a 64A010 airfoil for both free air and ideal slotted tunnel wall boundary conditions. Detailed experimental data for the airfoil were obtained by Stivers (private communication, May 1974) in the Ames 2- by 2-foot Wind Tunnel. For the experiments, a 15.25-cm (6-in.) chord model instrumented for surface-pressure measurements spanned the test section. Detailed impact and static-pressure measurements were obtained in the wake 1.8 chord lengths behind the model. Section drag coefficients are obtained by integrating the wake data in the standard manner. The chord Reynolds number for the experiments is 3 to 4×10^6, and the boundary layer is untripped. The upper and lower walls of the 2- by 2-Foot Wind Tunnel have longitudinal slots filled with triangular inserts. For the airfoil tests, the side walls are solid.

The variation of drag coefficient with Mach number for the experimental results and three theoretical calculations is shown in Figure 5.4.35.

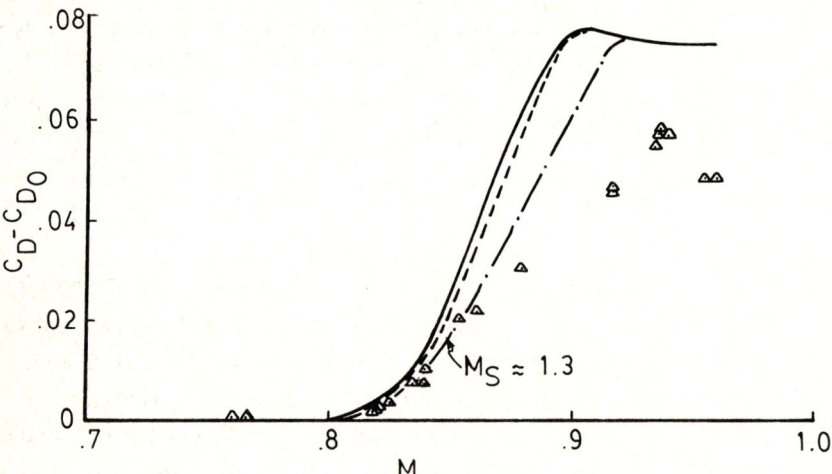

Figure 5.4.35

Comparison of theory with experiment for transonic wave drag, 64A010 airfoil, $\alpha = 0°$, Δ Stivers data, $C_D = 0.0080$. Transonic theory, $C_D = 0$, —— free air, - - - thick slotted wall, — – — thin slotted wall.

For subcritical conditions, a nearly constant profile drag C_{D_0} of 0.0080 was measured. This value has been subtracted from all the experimental data which are uncorrected for tunnel-wall effects. The drag rise is predicted adequately by the inviscid wave drag calculations up to the point where the shock Mach number reaches approximately 1.3. Beyond this point, the theory and experiment deviate due to a combination of increasing viscous-inviscid interaction, shock-wave entropy production, and tunnel interference effects. The true boundary condition for the tunnel walls is not known, and the two ideal slotted tunnel wall models (Gothert [5.4.19]) can only give an indication of the actual wall effects.

Surface-pressure distributions for selected Mach numbers are shown in Figure 5.4.36 and are compared with free air theory and the tunnel wall condition that gives best agreement with the preceeding drag data. A clear pattern with increasing Mach number is apparent. Ahead of the shock wave, the experiment and theory are in excellent agreement. As the shock wave strengthens, the effects of wall interference and viscous interaction alter the shock-wave location. At the highest Mach numbers shown, the pressures downstream of the shock wave are in poor agreement. There is no indication of shock-wave/boundary-layer separation for any of the data.

The total pressure loss through the shock, or local inviscid drag coefficient $C_D(y)$, may be computed from the entropy jump in the wake. As shown in Equation (3.10.15) the drag is connected to the entropy rise across shocks. The distribution of entropy produced at the shock remains unchanged with downstream distance so that the entropy wake at $x \to \infty$ is, to the TSD approximation

$$S(\infty, \tilde{y}, \tilde{z}) = \sum_{\text{shocks}}' \left[S(\tilde{y}, \tilde{z}) \right]_S . \tag{5.4.42}$$

These results are compared with the wake surveys in Figure 5.4.37. The shock losses are quite evident in the wake and are well predicted by the theory. It is interesting that, even at the higher Mach numbers where the surface pressures show poor agreement with theory, the detailed distributions of losses through the shock above the airfoil surface are in good agreement with the theory. This would indicate that only the foot of the shock wave is being altered in strength and location by the viscous effects. The upper portion of the shock, which is formed by coalescing compression waves generated from the forward portion of the airfoil, seems to be basically unaltered by the shock-wave/boundary-layer interaction. This feature is expected to disappear when massive separation is induced by the shock wave. Also, for lifting cases, the altered pressure distribution downstream of the shock will significantly change the circulation. Thus the good agreement noted above would not be expected unless comparisons are made at the same C_L.

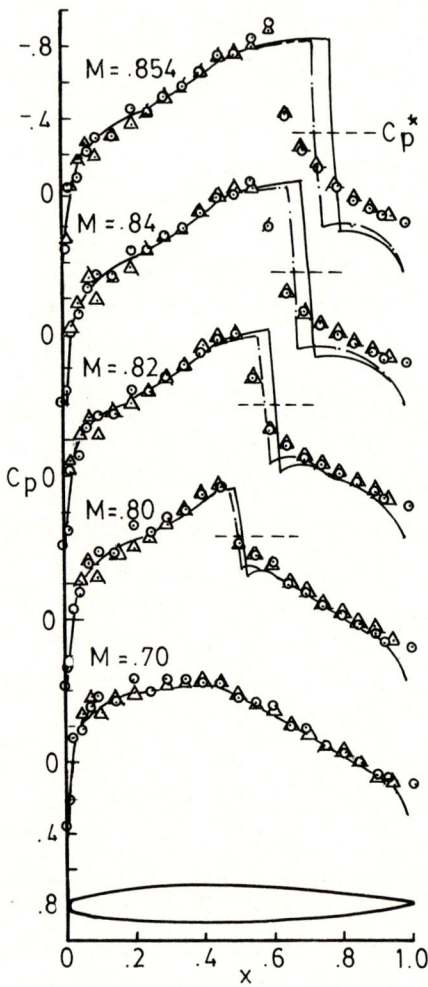

Figure 5.4.36

Surface pressure distributions for 64A010 airfoil, $\alpha = 0°$.
△   Stivers data, ——— Free air theory, —— · —— ideal slotted tunnel theory (thin wall).

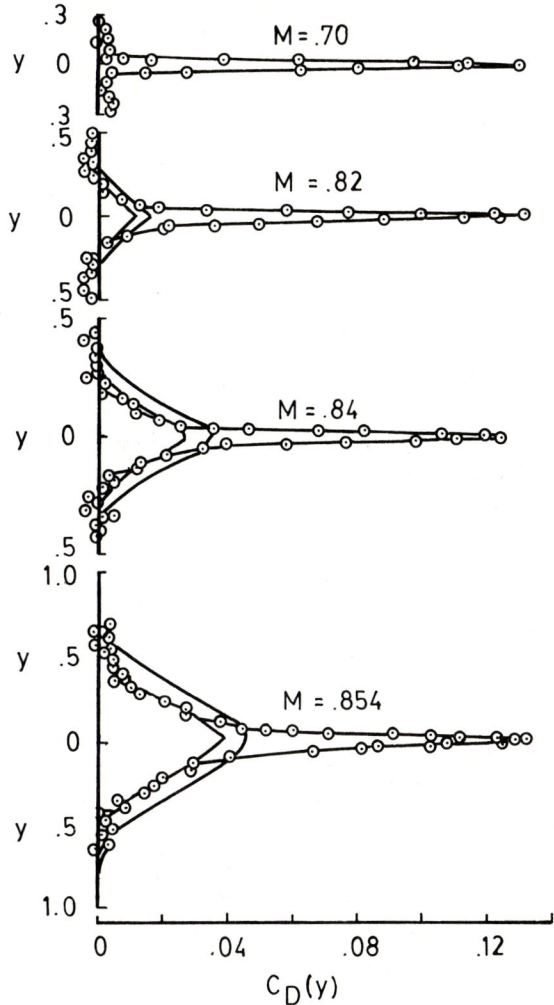

Figure 5.4.37

Wake surveys for 64A010 airfoil, $\alpha = 0°$. $\triangle$ Stivers data,
_____ free air theory, _ . _ ideal slotted tunnel theory (thin wall).

To conclude this example, it is interesting that the viscous wake profile does not change substantially with increasing Mach number. The principal changes are due to the inviscid wake created by the shock losses; This is reflected in the good agreement with the drag rise curves in Figure5.3.35. Clearly, the shock losses can be predicted accurately from the small-disturbance theory if the shocks are not too strong. The pressure drag on the body has unambiguous interpretations in terms of wave drag, induced drag, and wind-tunnel-wall interference drag. The nonlinearity of the transonic equations and the accompanying shock jump solutions permit the correct first-order entropy production to be calculated by the first-order theory. In addition, moderately blunt-nosed shapes may be treated without encountering a singular drag force at the leading edge. The effects of wind-tunnel walls can modify the measured body drag in a manner basically different from that of classical tunnel-wall interference theory. Furthermore, with a proper choice of the finite difference equations and other computational considerations, the pressure drag may be computed to acceptable accuracy for aeronautical applications. The important element is the use of conservative difference formulas, not only to give the correct shock jumps, but also to provide a divergence theorem for the difference equations. It is felt that inaccurcies in the drag calculation arising from mesh spacing and the evaluation of $[u]^3$ are probably within the overall accuracy of the small-disturbance theory.

This section has concentrated mainly on the developments of the authors and their co-workers. The whole subject of transonic calculation has boomed in the last ten years and space does not permit a complete review here.

The methods outlined in this section have recently been enlarged in scope to include refinements such as more efficient computation algorithms, embedded refined grids, primitive viscous interactions, and more complex equations. Theories of type-sensitive schemes and shock point operators are used. Second-order (mostly inconsistent) transonic, full potential, and Euler equations have been attacked. Nevertheless small-disturbance theory remains a useful tool due to its properties of simplicity and internal consistency.

References

[5.4.1] Murman, E. M. and Cole, J. D., Calculation of Plane Steady Transonic Flows, *AIAAJ.* **9**, 1971, pp 114-121.

[5.4.2] Murman, E. M. Analysis of Embedded Shock Waves Calculated by Relaxation Methods, *AIAAJ*, **12**, 1974, pp 626-632.

[5.4.3] Magnus, R. M. The Direct Comparison of the Relaxation Method and the Pseudo-Unsteady Finite-Difference Method for Calculating Steady Planar Transonic Flow, TN-73-SP03, General Dynamics, Convair Aerospace Division, San Diego, CA, 1973.

[5.4.4] Guderley, K. G., *The Theory of Transonic Flow*, Pergamon Press, London, 1962.

[5.4.5] Small, R. D. Numerical Solutions for Transonic Flow. Part 1. Plane Steady Flow over Lifting Airfoils, TAE Report No. 273, Technical Dept. of Aeronautical Engineering, Haifa, Isreal, 1976.

[5.4.6] Krupp, J. A. and Murman, E. M. Computation of Transonic Flows past Lifting Airfoils and Slender Bodies, *AIAAJ* **10**, 1972, pp880-886.

[5.4.7] Sells, C. C. L., Plane Subcritical Flow past a Lifting Airfoil, RAE TR 67146, Royal Aircraft Establishment, England, 1967.

[5.4.8] Magnus, R. and Yoshihara, H., Inviscid Flow Over Airfoils, AIAA Paper 70-47, N.Y., Jan. 1970.

[5.4.9] Stivers, L., Effects of Subsonic Mach Number on the Forces and Pressure Distribution of Four NACA 64A series Airfoil Sections, NACA TN 3162, 1954.

[5.4.10] Korn, D. Computations of Shock-Free Transonic Flows for Airfoil Design, *NYU Rept.* 1480-125, NYU, NY, NY, 1969.

[5.4.11] Bauer, F., Garabedian, P., and Korn, D. *Supercritical Wing Sections*, Springer-Verlag, 1972.

[5.4.12] Murman, E. M. and Cole, J. D. Inviscid Drag at Transonic Speeds, *Studies in Transonic Flow* III. UCLA-Eng.-7603, School of Engineering and Applied Science, Univ. of California, Los Angeles, 1975.

[5.4.13] Spreiter, J. R. On the Application of Transonic Rules to Wings of Finite Span NACA TR 1153, 1953.

[5.4.14] Garabedian, P. R. and Korn, D. G. Analysis of Transonic Airfoils, *Comm. Pure and Appl. Math* **24**, 1971, pp841-8751.

[5.4.15] Jameson, A. Transonic Flow Calculations for Airfoils and Bodies of Revolution, Grumman Aerodynamics Rept. 390-71-1, Dec., 1971.

[5.4.16] Pindzola, M. and Lo, C. F. Boundary Interference at Subsonic Speeds in Wind Tunnels with Ventilated Walls, AEDC-TR-69-47, 1969.

[5.4.17] Ponteziere, J. and Bernarde-Guelle, R., A Critique of Transonic Airfoil

Testing Techniques, Pt. II, Experimental Study of Wall Corrections in R1Ch, *L'Aeronautique et l'Astronautique*, **22**, 1971, pp 9-20.

[5.4.18] Murman, E. M. Computation of Wall Effects in Ventilated Transonic Wind Tunnels, AIAA paper 72-1007, 1972.

[5.4.19] Gothert, B. Transonic Wind Tunnel Testing, AGARD, 49, 1961.

5.5 Airfoils at Sonic Velocity

Flight at the speed of sound provides a decisive dividing case between subsonic and supersonic speeds. There are not many solutions available for this regime. We can note the wedge as an exact solution, and some special airfoils designed by Guderley [5.5.1]. Therefore some preliminary calculations were carried out by Tse based on the hodograph representations in order to find the sonic flow about realistic airfoil shapes (Tse [5.5.2]). Some details of the method were improved by Ziegler [5.5.3] and Rimbey [5.5.4]. The basic idea is to choose a boundary curve in the hodograph and to construct the flow and resulting airfoil numerically in the hodograph The method is thus an inverse or design method. Only for airfoils with special shapes, such as wedges, where the flow has a prescribed direction, is the boundary curve in the hodograph known in advance for mixed subsonic supersonic flows. In this section an outline is given of the hodograph method and a few results are presented.

The free-stream Mach number $M_\infty = 1$, and the airfoil shape is as usual

$$y = \delta F_{u,\ell}(x), \quad 0 < x < 1. \tag{5.5.1}$$

The qualitative features of flow at $M_\infty = 1$ are shown in Figure 5.5.1. The flow approaching the airfoil initially slows down and then speeds up to again reach sonic over the airfoil. It then expands supersonically until the tail where a tail shock is formed in such a way that the flow leaves the trailing edge smoothly. The sonic line that starts at the airfoil runs to infinity. This pattern is the limit of the growth of the finite supersonic zone over the airfoil as $M_\infty \to 1-$. The calculations of the previous section show also that the shock reaches the tail of the airfoil, in general, before $M_\infty = 1$. The flow behind the tail shock turns out to be supersonic and the flow then compresses downstream smoothly to $M_\infty = 1$.

The expansion Mach waves (characteristic lines) that originate from the airfoil surface near the sonic line run to the sonic line and are reflected to the surface as compression waves (cf. Figure 5.5.1).

Those expansion waves that leave near the tail run into the tail shock. There is one distinguished Mach wave, the limit Mach wave, that divides these two families and runs to infinity. It is asymptotic to both the sonic line and the tail shock at infinity. The structure of this far field was discussed in Section 4.1. Domains of dependence can be defined in terms of mutual influence.

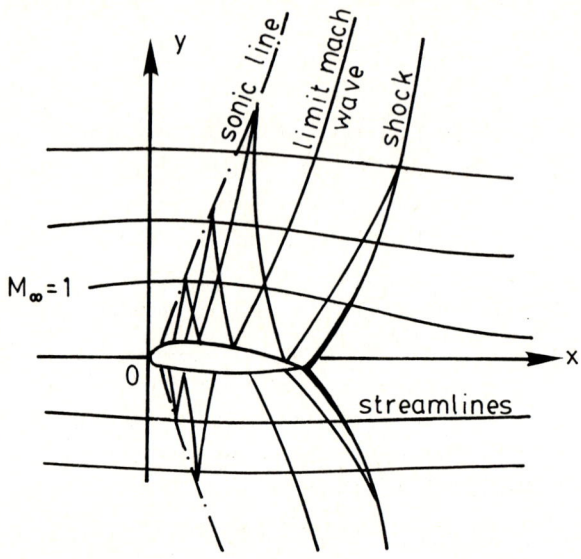

Figure 5.5.1

Typical sonic flow pattern over an airfoil

A change at the boundary ahead of the limit Mach wave can send a signal to the sonic line and influence the entire upstream region. Any change downstream of the limit Mach line can not affect this upstream region. Thus the flow is effectively divided into two regions ahead of the tail shock. The region upstream of the limit Mach wave has the elliptic property of mutual influence and has to be calculated all at once. The regions downstream of the limit Mach wave can be calculated step by step marching in the downstream direction once the flow in the upstream region is known. However, since the problem in the physical plane is non-linear, the location of this limit Mach wave in the physical plane is not known until the solution to the problem is found. As discussed in Section 3.5 the images of characteristics in the physical plane are characteristics in the hodograph. Thus the limit Mach wave is known in advance in the hodograph plane and this is one of the factors suggesting that a treatment in the hodograph plane is useful. The problem is now summarized in the physical and hodograph planes. The TSD equation is to be solved in $(x, \tilde{y})$ with the similarity parameter $K = 0$,

$$(\gamma + 1)\phi_x\phi_{xx} - \phi_{\tilde{y}\tilde{y}} = 0 \,, \tag{5.5.2}$$

with

$$\phi_{\tilde{y}}(x, 0\pm) = F'_{u,\ell}(x), \quad 0 < x < 1, \tag{5.5.3}$$

as the boundary condition of tangent flow. The disturbances must die out at infinity. Further the form of the far field can be given as studied in Section 4.1.

$$\phi(x, \tilde{y}) = \frac{1}{A^3} \tilde{y}^{\frac{2}{5}} f(A\xi), \tag{5.5.4}$$

where

$$\xi = \text{similarity coordinate} = \frac{x}{\tilde{y}^{\frac{4}{5}}},$$

A = scaling parameter, connected to properties of the airfoil.

This form is for a solution in which ϕ is symmetric in $\tilde{y}$, $\phi(x, \tilde{y}) = \phi(x, -\tilde{y})$ and, as shown in Section 4.1, this is the correct form even if the airfoil is carrying lift. It is also shown in Section 4.1 that a closed form (parametric) expression exists for (5.5.4) with a corresponding hodograph representation.

The problem is studied in the hodograph since the nature of the singularities and the computational domain are known there if one is willing to find the airfoil shape as part of the solution. The hodograph equations are linear. Further, the solution for the front part can be continued to the rear part easily if provision is made for two sheets in the hodograph to cover upper and lower sides of the supersonic region. By continuing the flow on these sheets the airfoil can be made to close.

Some details are now given. The hodograph equations and some general properties are discussed in Section(3.5). The basic hodograph equations for $(x, \tilde{y})$ in terms of $w = (\gamma + 1)\phi_x$, $\vartheta = (\gamma + 1)\phi_{\tilde{y}}$, are

$$\left\{ \begin{array}{c} w\tilde{y}_\vartheta - x_w = 0 \\ x_\vartheta - \tilde{y}_w = 0 \end{array} \right\}, \tag{5.5.5}$$

or, for the lines $\tilde{y}$ = constant (approximate streamlines) we must construct a solution of the Tricomi equation

$$w\tilde{y}_{\vartheta\vartheta} - \tilde{y}_{ww} = 0. \tag{5.5.6}$$

A sketch of the domains of the flow in the hodograph and transonic planes appears in Figure 5.5.2.

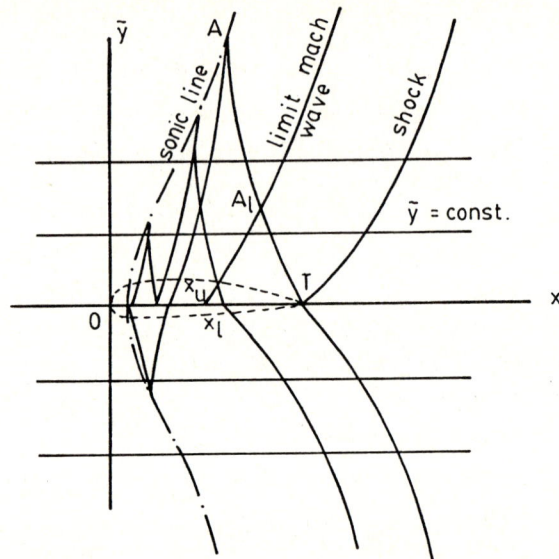

Figure 5.5.2a

Transonic Physical Plane

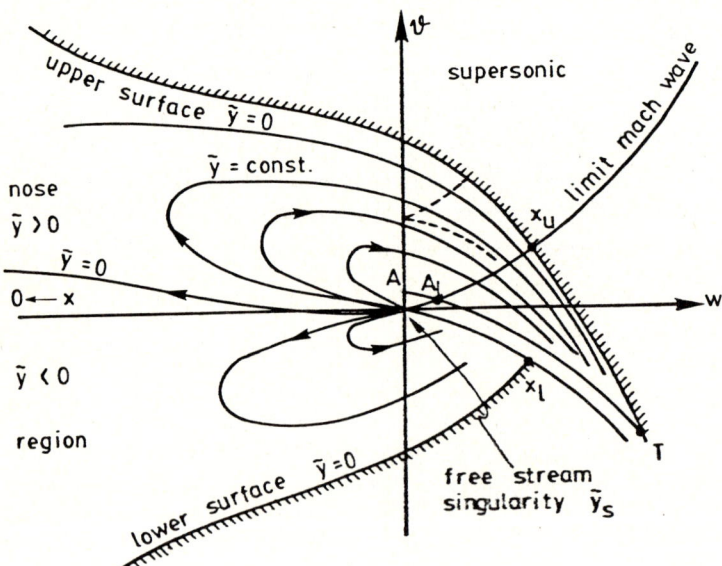

Figure 5.5.2b

Hodograph Plane

All lines $\tilde{y}$ = constant run into a singular point in the hodograph which corresponds to the free-stream where disturbances die out, $w = \vartheta = 0$. The streamline $\tilde{y} = 0$ runs from the singular point toward the nose of the airfoil. As shown in Section 5.2 the stagnation region at the nose ($x \to 0$) of an airfoil corresponds, in small disturbance theory, to $w \to -\infty$. The nature of the flow at the nose in the hodograph is also worked out in Section 5.2. At the nose the $\tilde{y} = 0$ streamline divides and passes over the upper and lower airfoil surfaces. The arrows on $\tilde{y}$ = constant in Figure 5.5.2b show the direction of x-increasing. The image of the upper surface crosses the sonic line ($w = 0$) and proceeds to higher supersonic velocities as it runs toward the tail point T. The sonic line ($w = 0$) in the hodograph runs from the free-stream singularity to the body surface. The limit characteristics on the upper and lower sides respectively are

$$\vartheta = \pm \frac{2}{3} w^{\frac{3}{2}} . \tag{5.5.7}$$

These connect the free-stream singularity to the upper and lower airfoil surfaces. The limit Mach wave leaves the upper and lower surfaces at $x_{u,\ell}$. Other characteristics run from the body to the sonic line and then reflect and run to the limit characteristic. The image of the body streamline $\tilde{y} = 0$ has to satisfy some constraints spelled out more explicitly below. But, the image of the upper surface continues past the limit Mach wave into the region of negative ϑ. Similarly, the image of the lower surface continues to positive ϑ so that two sheets are needed in the hodograph to represent the flow downstream of the limit Mach waves. For $\tilde{y} > 0$, ϑ is less than its value on the limit Mach wave $\vartheta = \frac{2}{3} w^{\frac{3}{2}}$ along $\tilde{y}$ = constant, and for $\tilde{y} < 0$, ϑ is greater than its value on $\vartheta = -\frac{2}{3} w^{\frac{3}{2}}$ along $\tilde{y}$ = constant. Since the two sheets are calculated separately it is not difficult to carry out this separation.

Once the boundary curve

$$\vartheta = \vartheta_b(w) \tag{5.5.8}$$

is chosen the domain for the solution of the Tricomi equation is specified. The subsonic region between the boundary surfaces and that part of the supersonic region up to the limit characteristic must be solved for first. The boundary condition is that

$$\tilde{y} = 0 \quad \text{on} \quad \vartheta = \vartheta_b \tag{5.5.9}$$

with no boundary condition prescribed on the limit characteristic. A sketch appears in Figure 5.5.3 and 5.5.4.

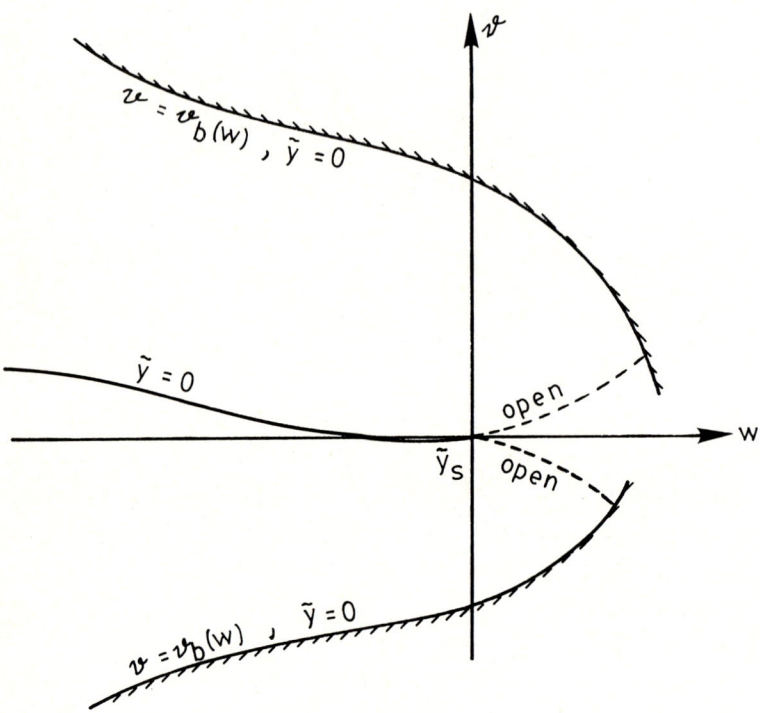

Figure 5.5.3

Domain of calculation of the front part

The solution is not simply $\tilde{y} = 0$ because there is a singularity at the origin whose strength is Q, say. After the solution in $\tilde{y}$ is known from the front part it is known along the limit Mach wave and provides an initial condition on a characteristic. The solution in the hyperbolic region downstream of the limit characteristic can be calculated by the method of characteristics. It is convenient also to carry out this calculation in the hodograph. The region of interest (for $\tilde{y} > 0$) is also bounded by the last compression wave $A_L T$ that runs from the limit Mach wave to the tail. No condition is prescribed along $A_L T$ either, and indeed it is not really found until the tail point T is located. However, the boundary conditions $\tilde{y} = 0$ on the surface and $\tilde{y}$ given on the limit characteristic serve to define the solution in the region up to $\vartheta = -\frac{2}{3} w^{\frac{3}{2}}$ for $\tilde{y} > 0$. This becomes clearer once the details of the boundary shape are explained below.

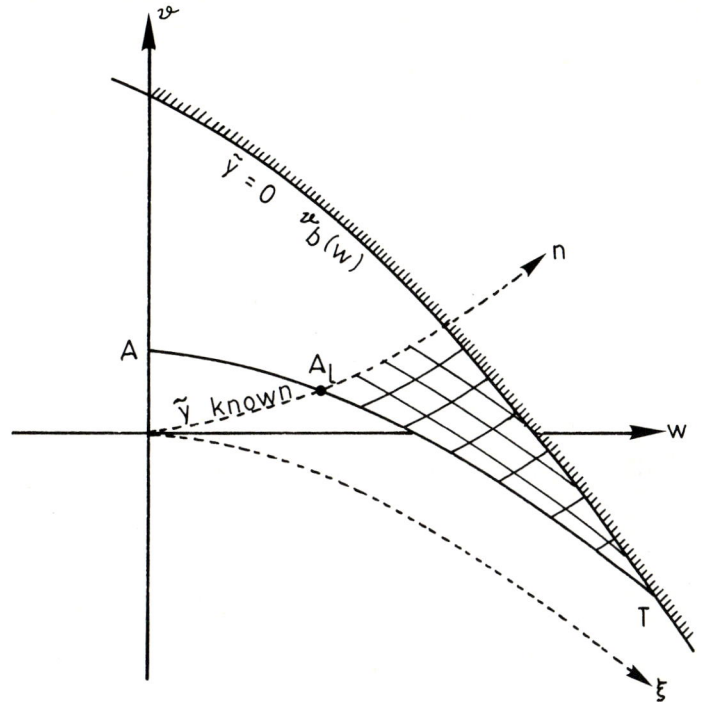

Figure 5.5.4

Domain of calculation of the rear part

Once $\tilde{y}(w, \vartheta)$ is known then $x(w, \vartheta)$ can be found by integration of one of the equations of (5.5.5). The choice of the "source strength" Q of the free-stream singularity determines the scale of the solution in the physical plane, but because of linearity does not affect it qualititatively.

Now some necessary properties of the boundary curve $\vartheta_b(w)$ are ennumerated. As discussed in Section 5.2 the solution $\tilde{y}_N$ (Equation 5.2.15) describes the flow in the neighborhood of a rounded nose with a finite radius of curvature.

$$\tilde{y}_N(w, \vartheta) = c_1 \rho^{\frac{-7}{3}} \left\{ c_2 \left(1 - \frac{7}{3} \sin^2 \alpha \right) + \sin \alpha F \left(\frac{5}{3}, -\frac{1}{2}; \frac{3}{2}; \sin^2 \alpha \right) \right\}, \quad (5.5.10)$$

where

$$\rho^2 = \vartheta^2 - \frac{4}{9} w^3, \quad \sin \alpha = \frac{\vartheta}{\rho}.$$

This means that the boundary curve $\vartheta_b(w)$ on which $\tilde{y} = 0$ appears on a line $\alpha = $ constant, or $\vartheta_b(w) \sim (-w)^{\frac{3}{2}}$. The same remark applies to the dividing streamline $\tilde{y} = 0$. Also as worked out in Section 5.2, $-.2621 < c_2 < 0$ so that three real roots $\alpha_{1,2,3}$ exist. (cf. Figure 5.2.3). For symmetric flow near the nose $c_2 = 0$. Along the airfoil we would like the flow to accelerate to the tail. Since

$$dx = x_w \, dw + x_\vartheta \, d\vartheta$$
$$d\tilde{y} = \tilde{y}_w \, dw + \tilde{y}_\vartheta \, d\vartheta = 0 \quad \text{on} \quad \vartheta = \vartheta_b(w), \tag{5.5.11}$$

we can write one expression for the change in x along the boundary (upper surface, say).

$$dx = \left(x_w + x_\vartheta \vartheta_b'(w)\right) dw = \left(w \tilde{y}_\vartheta + \tilde{y}_w \vartheta_b(w)\right) dw.$$

According to the sketch of Figure 5.5.5 $\tilde{y}_w < 0$, $\tilde{y}_\vartheta > 0$ on the boundary.

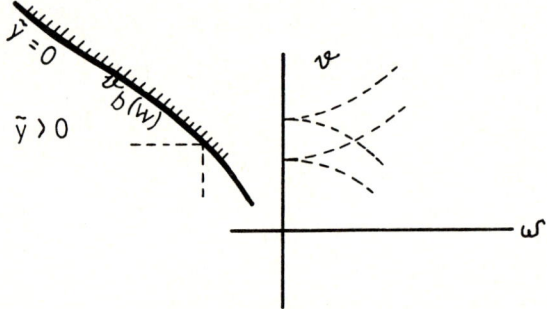

Figure 5.5.5

Flow Near Boundary

Therefore $dx \geq 0$ for $dw > 0$ when $\vartheta_b'(w) < 0$ and we will consider upper surface shapes that have this property. That is, the flow inclination along the upper surface of the profile is a monotonic decreasing function of arc-length along the surface, and correspondingly along the lower surface. This represents a fairly wide class of flows. Furthermore the flow must remain vortex-like as it passes into the supersonic region. As long as the flow remains vortex-like the image of the streamline $\tilde{y} = 0$ in the hodograph is not tangent to a characteristic and a limit line does not occur in the physical plane. (cf. Sections 3.6 and 3.7). This means that the characteristics running from the boundary to the sonic line are expansion Mach waves and those reflected from the sonic line are compression

Mach waves. Further, the maximum flow speed is attained on the boundary. In terms of formulas, for vortex like flow we have

$$|\vartheta_b'(w)| \equiv \left.\frac{d\vartheta}{dw}\right|_{\tilde{y}=0} < \left.\frac{d\vartheta}{dw}\right|_{\text{char}}, \qquad \left(\frac{d\vartheta}{dw}\right)_{\text{char}} = \pm\sqrt{w}.$$

Thus, from (5.5.11), (see Figure 5.5.6)

$$-\frac{\tilde{y}_w}{\tilde{y}_\vartheta} < -\sqrt{w}.$$

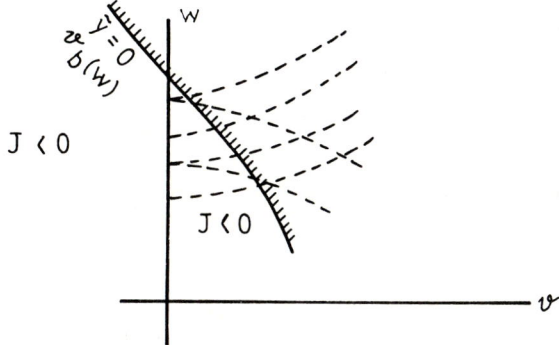

Figure 5.5.6

Vortex-like flow

For the supersonic region the jacobian J is thus negative $(J < 0)$,

$$J = \frac{\partial(w, \vartheta)}{\partial(\tilde{y}, x)} = \tilde{y}_w x_\vartheta - x_w \tilde{y}_\vartheta = w\tilde{y}_w^2 - \tilde{y}_\vartheta^2,$$

of course $J < 0$ in the subsonic region. The vortex-like property guarantees $J < 0$ everywhere in the flow; the total mapping is order reversing.

Two examples of upper boundary curves are used in [5.5.2]. These are chosen arbitrarily.

Example 1

$$\begin{aligned} \vartheta_b(w) &= \frac{2}{3}c_w(w_c - w)^{\frac{3}{2}} + \vartheta_c & w \le w_1 \le w_c < 0, \\ &= a_s w + b_s & w_1 \le w \le w_2, \\ &= a_p w^2 + b_p w + d_p & w > w_2. \end{aligned} \qquad (5.5.12)$$

that is, the boundary curve consists of a semi cubical parabola, joined to a parabola by a straight segment. The requirement of continuity of ϑ_b, $\vartheta'_b(w)$ at the joins is enforced. There are four relations among the original ten constants so that a six-parameter family of upper surfaces, and of lower surfaces is used. However, upper and lower surfaces are not completely independent. As pointed out in Section 5.1, certain relations must be satisfied among the parameters to generate a nose flow.

Example 2

The boundary curve is given parametrically by

$$w_b(t) = \frac{b_1 t}{(1+t)^2}, \quad \vartheta_b(t) = \frac{a_2 + b_2 t + c_2 t^2}{(1+t)^3}, \quad b_1 > 0, \ -1 < t < 1.$$

$$(5.5.13)$$

For the upper surface $a_2 > 0$, $b_2, c_2 < 0$; lower surface $a_2 < 0$, $b_2, c_2 > 0$. The parameter ranges are chosen to meet the restrictions outlined above.

Once the boundary curves are specified the solution and computational procedures can be outlined. Since the hodograph equation and the Tricomi equation are linear superposition can be used. The free-stream singularity (Eqn 4.1.71, 4.1.72) can be used to represent the flow far from the airfoil. Here these singular solutions are multiplied by a source strength Q. This singularity is at the origin $(w = \vartheta = 0)$,

$$\left(\begin{array}{l} \tilde{y}_s = Q\rho^{-3}\left\{ (\rho + \vartheta)^{\frac{1}{3}}(3\vartheta - \rho) + (\rho - \vartheta)^{\frac{1}{3}}(3\vartheta + \rho) \right\} \\ x_s = -\frac{2}{3}Qw^2\rho^{-3}\left\{ (\rho + \vartheta)^{-\frac{2}{3}}(3\vartheta + 2\rho) + (\rho - \vartheta)^{-\frac{2}{3}}(3\vartheta - 2\rho) \right\} \end{array} \right). \quad (5.5.14)$$

This singular solution can be added to other solutions, computational or analytic. However, as shown in Section 4.1, as $\rho \to \infty$, α fixed, then

$$\tilde{y}_s \sim \rho^{-\frac{5}{3}}. \quad (5.5.15)$$

Compared with the nose solution (5.5.10) this decay is too slow so that two negative images, out of the computational domain are used to get a sufficiently rapid decay as $\rho \to \infty$. The analytic part of the solution $\tilde{y}_a$ is defined as

$$\tilde{y}_a = Q\left\{ \tilde{y}_s(w, \vartheta) - \frac{1}{2}\tilde{y}_s(w, \vartheta - \vartheta_0) - \frac{1}{2}\tilde{y}_s(w, \vartheta + \vartheta_0) \right\}, \quad (5.5.16)$$

then as $\rho \to \infty$,

$$\tilde{y}_a \sim \rho^{-\frac{8}{3}}. \tag{5.5.17}$$

ϑ_0 lies above the intersection of the upper boundary curve and the sonic line, $-\vartheta_0$ below the lower. The total solution is represented as the sum of the analytic $(\tilde{y}_a)$ and computational $(\tilde{y}_c)$ parts,

$$\tilde{y} = \tilde{y}_a + \tilde{y}_c. \tag{5.5.18}$$

The boundary condition for the computational part is that

$$\tilde{y}_c = -\tilde{y}_a \quad \text{on the boundaries} \quad \vartheta_b(w).$$

(as in Figure 5.3.3). Also the computational domain is truncated at $w = w_0$ (negative) at a sufficiently large distance from the origin that the nose region solutions should apply. On that boundary, simply,

$$\tilde{y}_c = -\tilde{y}_a + \tilde{y}_N.$$

(cf. Equation 5.5.10).

In the elliptic region a rectangular grid is convenient to use for the finite difference solution of the Tricomi equation (5.5.6) for $\tilde{y}_c(w, \vartheta)$. The finite difference form is found from a simple centered scheme in the elliptic region. In the hyperbolic region the solution, for the front part of the flow, needs to be continued up to the limit characteristics. The limit characteristics are open boundaries with no boundary condition prescribed. It is natural to use a (ξ, η) characteristic coordinate system in the hyperbolic region. The solution can be obtained by marching along the other set of characteristics toward the limit characteristic. Here

$$\xi = -\vartheta + \frac{2}{3} w^{\frac{3}{2}},$$

$$\eta = \vartheta + \frac{2}{3} w^{\frac{3}{2}},$$

and the Tricomi equation becomes,

$$\tilde{y}_{\xi\eta} + \frac{1}{6(\xi + \eta)}(\tilde{y}_\xi + \tilde{y}_\eta) = 0. \tag{5.5.19}$$

This equation is of the general form of the Euler-Poisson-Darboux equations, studied in the classical theory of Partial Differential Equations. In the supersonic region backward difference schemes should be used, along the streamlines $\tilde{y} =$ constant, to express the proper local domain of dependence. The situation is analogous to that in the previous section 5.4 except that now the hyperbolic and elliptic domains are known in advance, due to linearity.

The computational stars used to set up the difference schemes are sketched in Figure 5.5.7. (1) is the centered star for the elliptic region, (4) is the backward six point scheme for the hyperbolic region. (2) and (3) are used for points near the sonic line. Details are omitted here. (cf. [5.5.2]) More recent work [5.5.4] shows that a simpler four point scheme works somewhat better for the hyperbolic region, at least for jet flows. As in Section 5.4 line relaxation is used in alternating directions here to solve for the unknown $\tilde{y}_c$ values at the mesh points. For the horizontal lines (cf. Figure 5.5.7)

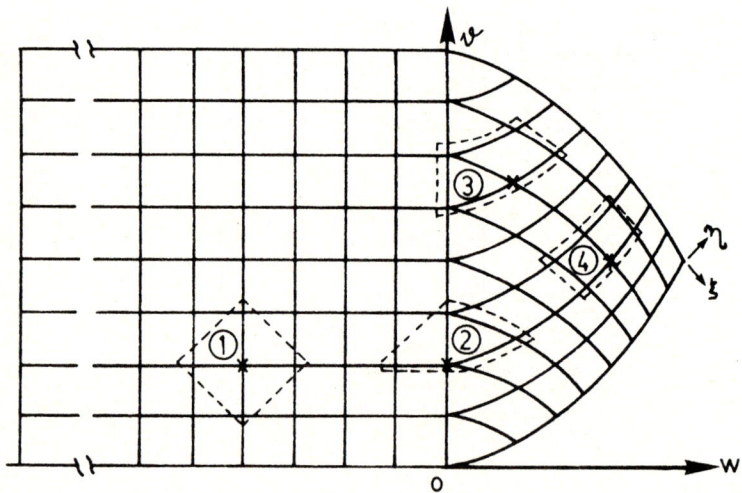

Figure 5.5.7

Various schemes used in the computation

in the elliptic region the line is continued along a $\xi =$ constant characteristic for the upper half plane and along an $\eta =$ constant characteristic for the lower half plane. All the points on this line are solved for at once in terms of latest values. Sweeps are carried out in the direction indicated in Figure 5.5.8. In order to connect the solutions in upper and lower half planes relaxation is also carried out on vertical lines in alternate sweeps. These latter sweeps include only subsonic points.

Once $\tilde{y}_c$ up the the limit characteristics is found $\tilde{y}$ is known and the solution can be continued to the rear part of the airfoil.

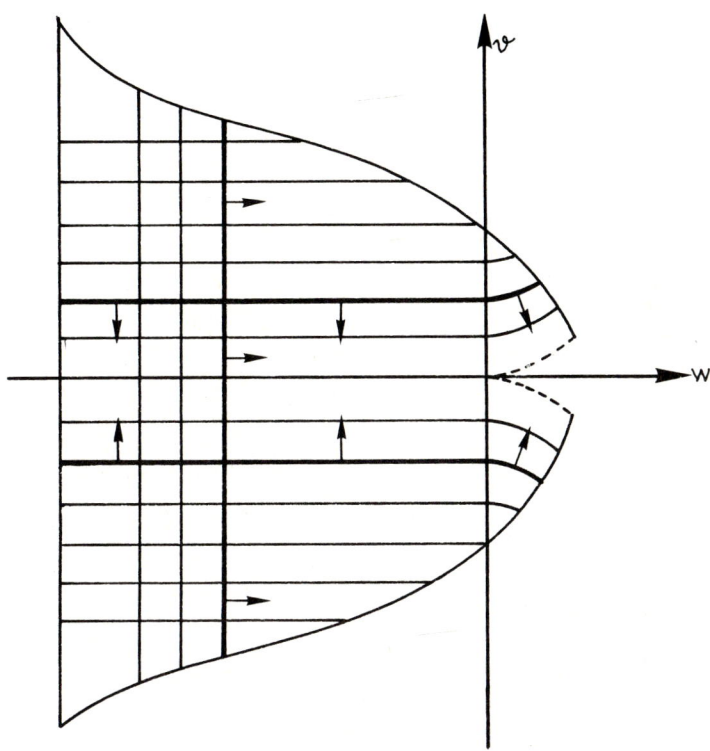

Figure 5.5.8
Direction of line relaxation

The upper and lower rear part are done similarly and independently. For the upper part the solution $\tilde{y}$ is known on the limit characteristic $\xi = 0$ and $\tilde{y} = 0$ on the boundary $\vartheta_b(w)$ (cf. Figure 5.5.9) Using the four point scheme, and the boundary data, the solution can be marched explicitly to fill in the characteristic net. This is carried out until a large negative ϑ is reached. When a complete solution for $\tilde{y}(w, \vartheta)$ is known the corresponding $x(w, \vartheta)$ is found by integration.

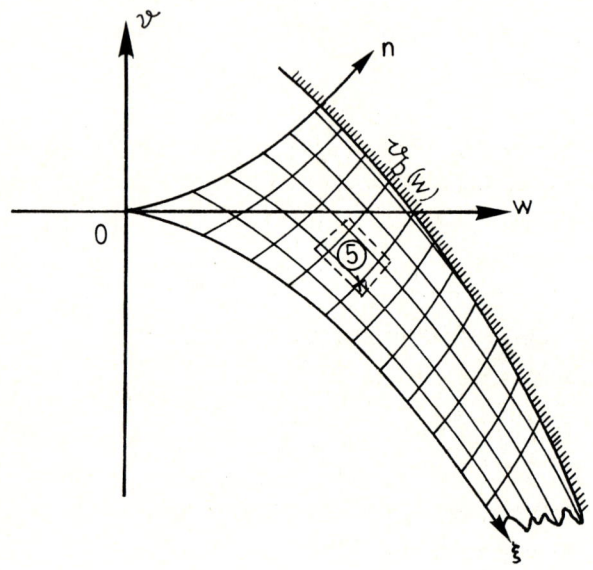

Figure 5.5.9

Domain of calculation for $\tilde{y}$ in the rear part

In the elliptic region Equations (5.5.5) are used and can be integrated, for example, in small steps near the boundary. In the hyperbolic region we have correspondingly

$$\left\{ \begin{array}{l} -x_\xi = \sqrt{w}\,\tilde{y}_\xi \\ x_\eta = \sqrt{w}\,\tilde{y}_\eta \end{array} \right\}.$$

(5.5.20)

The type of grid actually used is given in Figure 5.5.9a.

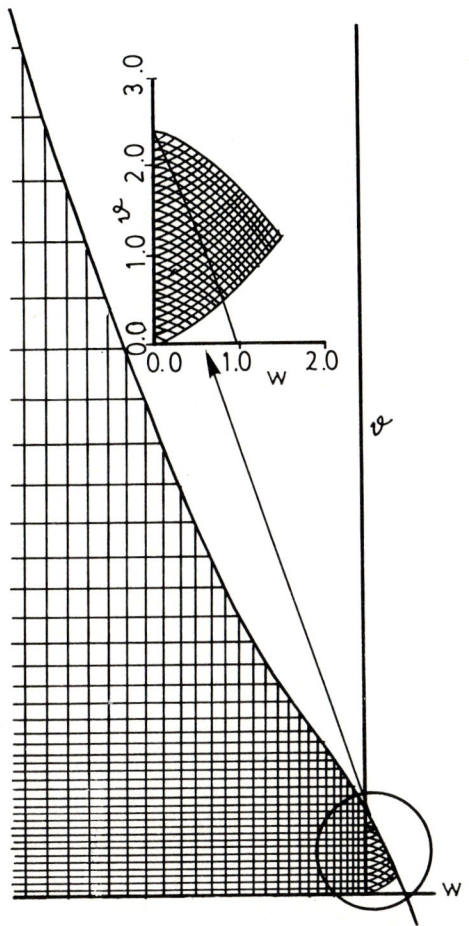

Figure 5.5.9a

Simple grid (actual grid is twice as fine)

In order to recover the shape of the airfoil we use

$$F_{u,\ell}(x) = \int_0^x \phi_{\hat{y}}(x',0\pm)\,dx' = \frac{1}{\gamma+1}\int_0^x \vartheta_{u,\ell}(x')\,dx'. \qquad (5.5.21)$$

The trailing edge is located at that point $x = x_t$ where

$$F_u(x_t) = F_\ell(x_t). \qquad (5.5.22)$$

In this way the airfoil chord $c = x_t$ is not controlled but comes out depending on the free-stream singularity strength Q and the airfoil shape. Pressure coefficient, lift, and drag coefficients are found from

$$C_p^* = \frac{C_p}{\delta^{\frac{2}{3}}} = -\frac{2}{\gamma+1} w \quad \text{on} \quad \vartheta_b,$$ (5.5.23)

$$C_L^* = \frac{C_L}{\delta^{\frac{2}{3}}} = -\frac{2}{\gamma+1} \left\{ \frac{1}{c} \int_0^c w_u dx_u - \frac{1}{c} \int_0^c w_\ell dx_\ell \right\},$$ (5.5.24)

$$C_D^* = \frac{C_D}{\delta^{\frac{5}{3}}} = -\frac{2}{\gamma+1} \left\{ \frac{1}{c} \int_0^c w_u F_u' dx_u - \frac{1}{c} \int_0^c w_\ell F_\ell' dx_\ell \right\}.$$ (5.5.25)

A geometrical angle of attack can also be defined as

$$\alpha = -\delta \frac{F(x_t)}{x_t}.$$ (5.5.26)

The calculated solutions can be rescaled as follows. For any solution

$$\tilde{y} = Y(w, \vartheta), \quad x = X(w, \vartheta),$$

a rescaled solution

$$\tilde{y} = Y^+(w^+, \vartheta^+), \quad x = X^+(w^+, \vartheta^+),$$

can be defined where

$$Y^+ = bY(w^+, \vartheta^+), \quad X^+ = aX(w^+, \vartheta^+),$$

and

$$w^+ = \frac{a^2}{b^2} w, \quad \vartheta^+ = \frac{a^3}{b^3} \vartheta.$$

If $F(x)$ is the shape generated by the original solution the scaled solution generates a similar shape according to

$$F^+(X^+) = \frac{1}{\gamma+1} \int_0^{X^+} \vartheta^+ dX^+ = \frac{a^4}{b^3} \frac{1}{\gamma+1} \int_0^X \vartheta \, dX = \frac{a^4}{b^3} F(X).$$

Choose $a = 1$ and the chord to be unity; then

$$F^+(x) = \frac{1}{b^3} F(x).$$ (5.5.27)

If the calculated shape $F_u(x) - F_\ell(x)$ has a thickness t then the solution

$$X^+ = X(w^+, \vartheta^+),$$
$$Y^+ = t^{\frac{1}{3}} Y(w^+, \vartheta^+),$$ (5.5.28)

has a shape $F^+(x)$ with a unit thickness.

Now a few numerical results are given. Guderley and Yoshihara calculated the sonic flow toward a wedge in [5.5.5]. Since $\vartheta_b = $ constant the boundary curve is known in the hodograph. The sonic point occurs at the corner. (cf. Figure 5.5.10).

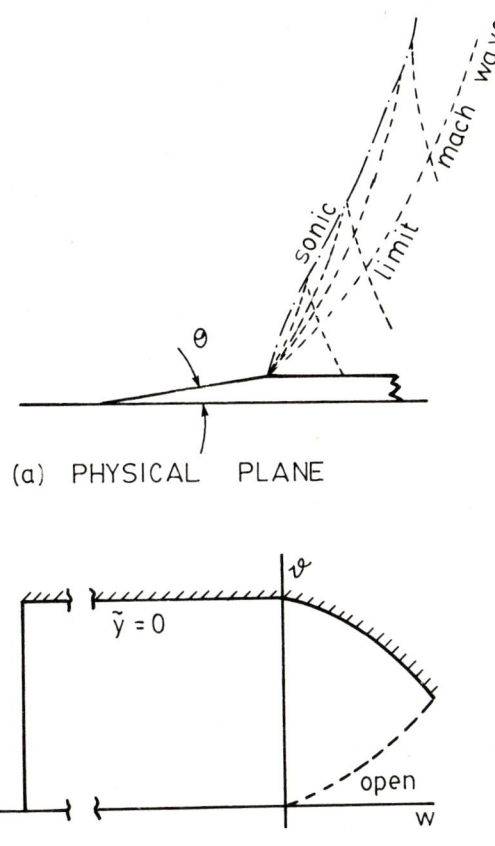

(a) PHYSICAL PLANE

(b) HODOGRAPH PLANE

Figure 5.5.10

Sonic flow over a wedge

The calculation uses an eigenfunction expansion and a numerical method to satisfy the boundary conditions in the supersonic part. The sonic flow was calculated using the method just outlined. A comparison of the resulting pressure distribution is shown in Figure 5.5.11.

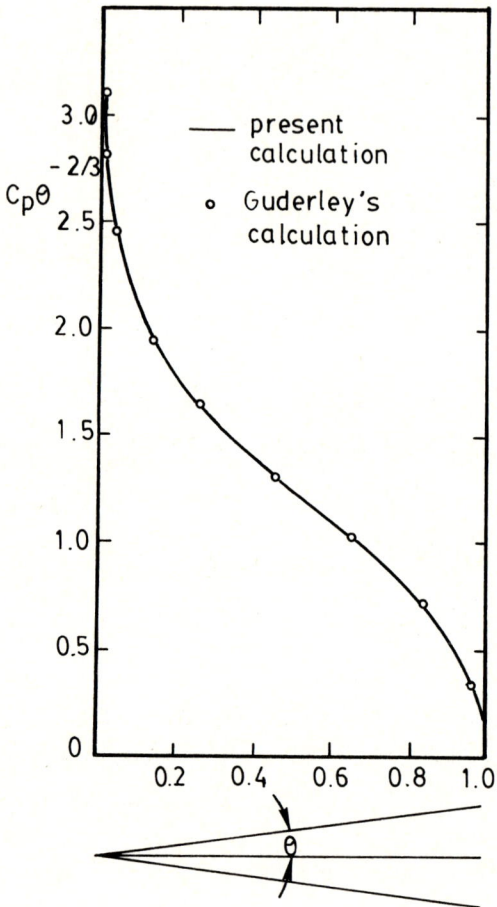

Figure 5.5.11

Pressure distribution over the wedge

A few airfoils were designed with $Q = 1$ and the parametric form of the boundary curve (5.5.13). A typical airfoil shape and pressure distribution appears in Figure 5.5.12.

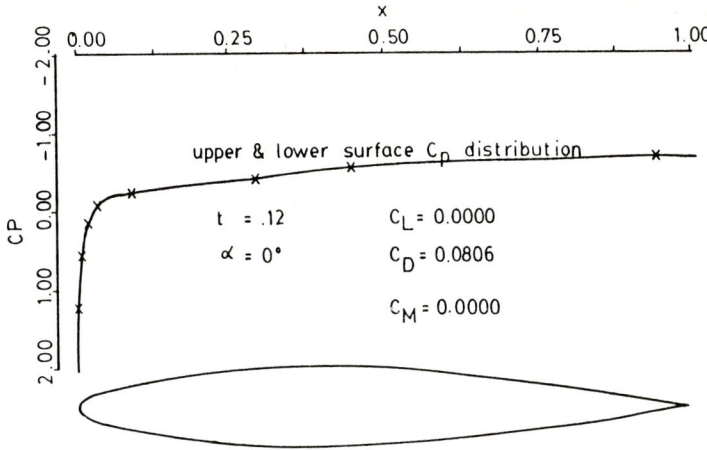

Figure 5.5.12

C_p distribution and profile of airfoil B-1

Another case using the boundary curve of (5.5.12) is given in Figure 5.5.13.

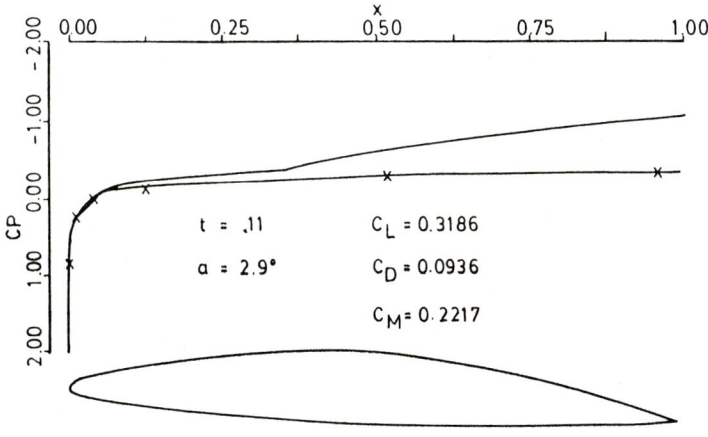

Figure 5.5.13

C_p distribution and profile of airfoil C-9

The difference in pressure on the upper and lower surfaces accounts for the lift.
The airfoil shapes are reasonable. A comparison of drag coefficients with some

calculations of Guderley for the wedge and cusped noses with rear part of minimum drag [5.5.5] shows that these smooth airfoils have somewhat less wave drag. An order of magnitude comparison of C_D and C_L/α is made with experiments of Vincenti et al [5.5.6] on .07 thickness double wedge and appears in Figs. 5.5.14 and 5.5.15.

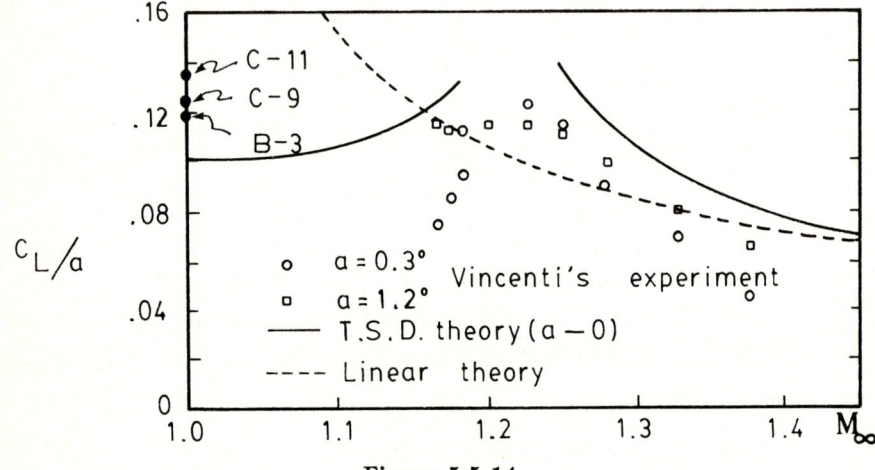

Figure 5.5.14

Order-of-magnitude comparison of C_L

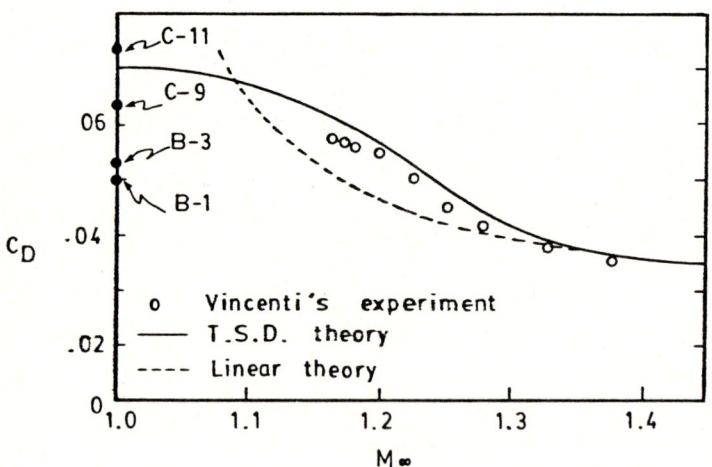

Figure 5.5.15

Order-of-magnitude comparison of C_D

These smooth airfoils have more lift and less drag in general. Better calculations can be produced with some refinement of the difference scheme. Also, the same scheme can be applied to Chaplygins' exact equation. (Some preliminary results have already been obtained). The computation time for the implementation of this scheme is minimal, a few seconds on an IBM 91/360. Thus it should be possible to search for optimum airfoils maximising C_L for a given C_D with some constraints. In view of the Law of Stabilization discussed in Section 5.6 these sonic flows have added significance for flows near sonic.

References

[5.5.1] Guderley, K. G. *Theory of Transonic Flow*, Addison Wesley, 1962.

[5.5.2] Tse, Egbert, *Airfoils at Sonic velocity*, PhD. Thesis, School of Engineering and Applied Science, University of California, Los Angeles 1980.

[5.5.3] Ziegler, Frederick J., *Finite Span Wings at Sonic Speed*, PhD. Thesis, Department of Matematics, University of California, Los Angeles 1981.

[5.5.4] Rimbey, S., Private Communication.

[5.5.5] Guderley, K. G. and Yoshihara, H., Flow on a Wedge Profile at Mach Number One, *J. Aero. Sci.*, **17** no. 11, Nov. 1950 pp. 723-735.

[5.5.6] Vincenti, W. G., Dugan, D. W. and Phelps, E. R., An Experimental Study of the Lift and Pressure Distribution on a Double Wedge Profile at Mach Number near Shock Attachment, *NACA TN* 3325, 1954.

5.6 The Stabilization Law

In this section transonic small disturbance flows are analyzed for their dependence on the free-stream Mach number in a neighborhood of one. ($M_\infty \to 1$, $K \to 0$.) Of special interest is the "law of stabilization" or "freezing" of the local Mach number near the body in front of the shock as the free stream Mach number approaches one. The main results of this section are the determination of the order of the correction to sonic flow due to variations in the Mach number and the formulation of the boundary value problem which describes the correction.

The full boundary value problem (see Section 3.1) consists of the small disturbance equation

$$\left(K - (\gamma + 1)\phi_x\right)\phi_{xx} + \phi_{\tilde{y}\tilde{y}} = 0 \,, \tag{5.6.1}$$

in the plane slit along $\tilde{y} = 0$, $x > 0$, and the boundary conditions of

tangent flow,

$$\phi_{\tilde{y}}|_{\tilde{y}=0} = F'_{u,\ell}(x)\,, \quad 0 < x < 1\,, \tag{5.6.2}$$

no disturbance at upstream infinity,

$$\phi_x \to 0\,, \tag{5.6.3}$$

no pressure jump across the wake,

$$\left[\phi_x\right]_{\tilde{y}=0} \quad \text{for} \quad x > 1\,, \tag{5.6.4}$$

the shock jump conditions

$$\left[K\phi_x - \frac{(\gamma + 1)}{2}\phi_x^2\right]_s \left[\phi_x\right]_s + \left[\phi_{\tilde{y}}\right]_s^2 = 0\,, \tag{5.6.5}$$

$$\left[\phi\right]_s = 0\,, \tag{5.6.6}$$

and for $M_\infty < 1$, the Kutta condition

$$\left[\phi_x\right]_{x=1} = 0\,. \tag{5.6.7}$$

Here we have assumed that the airfoil is located, in small disturbance coordinates, along $\tilde{y} = 0$, $0 \le x \le 1$.

This boundary value problem has already been examined for K strictly greater than zero ($M_\infty > 1$), for K precisely zero ($M_\infty = 1$), and for K strictly less than zero ($M_\infty > 1$). However, the results so obtained do not hold uniformly as M_∞ actually passes through one. This can be seen quite clearly from the form of the far fields as well as from the form of the equation. For $K > 0$ the equation is elliptic in the entire field. For $K = 0$ the equation changes from elliptic to hyperbolic in the far field. Fot $K < 0$ the equation is hyperbolic in the entire far field. Thus as $K \to 0$ the boundary value problem is of singular perturbation type.

The nonuniformity can be seen explicitly from the $M_\infty < 1$ far field expansion (4.3.14),

$$\phi \underset{r \to \infty}{\sim} -\frac{\Gamma \theta}{2\pi} + \frac{(\gamma + 1)}{4\pi} \left(\frac{\Gamma}{2\pi} \right)^2 \frac{\log r}{r} \cos \theta - \frac{\gamma + 1}{16K} \left(\frac{\Gamma}{2\pi} \right)^2 \frac{\cos 3\theta}{r}$$

$$+ \frac{D \cos \theta}{2\pi r} + \frac{E \sin \theta}{2\pi r} + \cdots ,$$

where $r = (x^2 + K\tilde{y}^2)^{1/2}$, $\theta = \tan^{-1} \frac{\sqrt{K}\tilde{y}}{x}$ as in (4.3.1). This expansion does not remain valid as $K \to 0$ (unless at a similar rate $r \to \infty$). Furthermore this expansion is dominated by a circulation (nonsymmetric) term whereas the sonic flow far field expansion is (4.1.102),

$$(\gamma + 1)\phi \underset{\tilde{y} \to \infty}{\sim} \tilde{y}^{\frac{2}{5}} \frac{f(a\xi)}{a^3} + c_0 + c_1 \frac{\tilde{y}^{-\frac{1}{5}} f_1(a\xi)}{a^3} + \cdots ,$$

where $\xi = \tilde{y}^{4/5}$ c_0, c_1, a are determined in section 4.1.1. As $M_\infty \to 1^-$ the effect of lift dies out faster in the far field than the effect of thickness.

In order to clarify the difficulty as K passes through zero, and to clarify the limiting processes to be described, a picture of the flow patterns is given below. (Figure (5.6.1)).

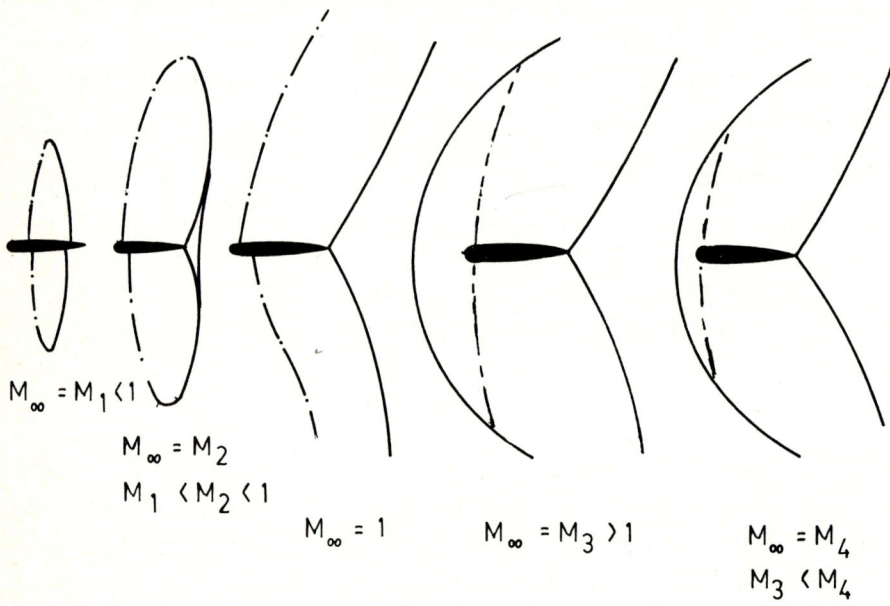

$M_\infty = M_1 < 1$

$M_\infty = M_2$
$M_1 < M_2 < 1$

$M_\infty = 1$ $M_\infty = M_3 > 1$ $M_\infty = M_4$
$M_3 < M_4$

Figure 5.6.1

Flow Patterns as M_∞ Passes Through One

Note that as M_∞ approaches one from below the shock moves toward the trailing edge of the airfoil. At the same time the supersonic zone grows until, at $M_\infty = 1$, it extends to infinity. The shock actually hits the tail for some $M_\infty < 1$, thereafter a second fishtail shock forms. As M_∞ approaches one, this second shock moves off to infinity. As the free stream Mach number increases even further a bow shock appears upstream of the body and moves, with increasing Mach number, towards the body.

$K \to 0+ :$

As K approaches zero from above the supersonic zone grows until at $K = 0$ it extends to infinity. Thus analysis of the flow lends itself nicely to two limiting procedures, the inner limit in which x, $\tilde{y}$ are fixed as $K \to 0$, and the outer limit in which x, $\tilde{y}$ are stretched as $K \to 0$. In the inner limit the flow looks sonic to leading order since the supersonic zone grows to infinity, hence passing any fixed $\tilde{y}$ station. In the outer limit the flow looks subsonic with a small embedded supersonic zone since the x, $\tilde{y}$ coordinates stretch to infinity at least as fast as the termination of the supersonic zone. The expansions in these two regions are now found and matched. Through this matching the correction to the sonic flow in the inner region is found.

Inner Expansion: $(x, \tilde{y}$ fixed, $K \to 0)$.

In the inner region the expansion has the form

$$\phi = \phi_0(x, \tilde{y}) + K\phi_1(x, \tilde{y}) + \mu_2(K)\phi_2(x, \tilde{y}) + \cdots , \qquad (5.6.8)$$

where

$$\mu_2(K) \ll K .$$

The $O(K)$ correction arises from the equation (5.6.1), and matching guarantees that no other terms need to be inserted. As will be seen, this term contributes only a uniform translation to the velocity and hence does not contribute to the lift. The determination of μ_2 and the formulation of the boundary value problem for ϕ_2 are the desired results.

Substitution of the expansion (5.6.8) into the boundary value problem (5.6.1) - (5.6.7) generates parts of the boundary value problem for the ϕ_i. They are, in front of the shock,

$$-(\gamma + 1)\phi_{0x}\phi_{0xx} + \phi_{0\tilde{y}\tilde{y}} = 0 , \qquad (5.6.9)$$

$$\phi_{0\tilde{y}}|_{\tilde{y}=0} = F'_{u,\ell}(x) , \qquad (5.6.10)$$

$$(\gamma + 1)\phi_0 \underset{\tilde{y}\to\infty}{\sim} \frac{\tilde{y}^{\frac{2}{5}}}{a^3} f(a\xi) + c_0 + c_1 \frac{\tilde{y}^{-\frac{1}{5}}}{a^3} f_1(a\xi) + \cdots , \qquad (5.6.11)$$

where $\xi = \dfrac{x}{\tilde{y}^{\frac{4}{5}}}$ and the far field behavior was obtained in section 4.1.1.,

$$-(\gamma + 1)(\phi_{0x}\phi_{1x})_x + \phi_{1\tilde{y}\tilde{y}} = -\phi_{0xx} ,$$

$$\phi_{1\tilde{y}}|_{\tilde{y}=0} = 0 ,$$

and

$$-(\gamma + 1)(\phi_{0x}\phi_{2x})_x + \phi_{2\tilde{y}\tilde{y}} = 0 ,$$

$$\phi_{2\tilde{y}}|_{\tilde{y}=0} = 0 ,$$

where the far fields of ϕ_1, ϕ_2 are determined by matching with the outer expansion. Note that the ϕ_0 problem is the sonic flow problem. A particular solution of the ϕ_1 problem is

$$\phi_1^P = \frac{x}{\gamma + 1}.$$

Matching will determine that ϕ_1^P is the full ϕ_1 solution, and this in fact determines the form of the ϕ_2 equation. Matching also determines μ_2 and the far field of ϕ_2. Note that the equation for ϕ_2 actually has a forcing term if $\mu_2 = K^2$, but since, as we shall see, $\phi_{1xx} = 0$, the equation is homogeneous for all $\mu_2 \gg K^2$ or $= K^2$.

Outer Expansion: ($\bar{x} = \alpha(K)x$, $\tilde{y} = \beta(K)\hat{y}$ fixed as $K \to 0$, $\alpha, \beta \ll 1$).

In the outer region the $x, \tilde{y}$ variables are stretched so that

$$\bar{x} = \alpha(K)x, \quad \bar{y} = \beta(K)\hat{y}, \tag{5.6.12}$$

are held fixed as $K \to 0$, where α, β are to be determined. The outer expansion has the form

$$\phi = \nu_0(K)\varphi_0(\bar{x}, \bar{y}) + \nu_1(K)\varphi_1(\bar{x}, \bar{y}) + \nu_2(K)\varphi_2(\bar{x}, \bar{y}) + \cdots, \tag{5.6.13}$$

where $\nu_1 \ll \nu_0$, ν_0 and ν_1 are to be determined.

In order to determine α, β we require that similarity curves go into similarity curves so that matching can be carried out with the inner expansion. Thus,

$$\xi = \frac{\bar{x}}{\bar{y}^{\frac{4}{5}}} = \frac{\alpha(K)x}{\beta(K)^{\frac{4}{5}}\hat{y}^{\frac{4}{5}}} = \frac{x}{\tilde{y}^{\frac{4}{5}}},$$

or

$$\alpha(K) = \beta(K)^{\frac{4}{5}}. \tag{5.6.14}$$

Also we require that when equation (5.6.1) is rewritten in outer variables,

$$\left(K - (\gamma + 1)\nu_0(K)\alpha(K)\varphi_{0\bar{x}}\right)\alpha^2(K)\varphi_{0\bar{x}\bar{x}} + \beta^2 \varphi_{0\bar{y}\bar{y}} = 0, \tag{5.6.15}$$

the full equation be retained as $K \to 0$ so that the subsonic nature of the flow is retained while at the same time matching can be accomplished with the sonic (inner) flow. Thus

$$\alpha^2 K = \beta^2 = \nu_0 \alpha^3. \tag{5.6.16}$$

From (5.6.14) and (5.6.16),

$$\beta = K^{\frac{5}{2}}, \quad \alpha = K^2, \quad \nu_0 = \frac{1}{K}.$$

Substitution of these values into the expansion (5.6.13) and then substitution into (5.6.1), (5.6.3) gives parts of the boundary value problems to be satisfied by the φ_i;

$$\left(1 - (\gamma + 1)\varphi_{0\bar{x}}\right)\varphi_{0\bar{x}\bar{x}} + \varphi_{0\bar{y}\bar{y}} = 0,$$

$$\varphi_0 \underset{\bar{y}\to 0}{\sim} \frac{\bar{y}^{\frac{2}{5}} f(a\xi)}{a^3(\gamma + 1)} + \frac{\bar{x}}{\gamma + 1} + \frac{C_1 \bar{y}^{\frac{6}{5}}}{a^3} g_1(a\xi) + \cdots,$$

$$\varphi_{0\bar{x}} \underset{\bar{y}\to\infty}{\sim} 0,$$

and

$$\left(\left(1 - (\gamma + 1)\varphi_{0\bar{x}}\right)\varphi_{1\bar{x}}\right)_{\bar{x}} + \varphi_{1\bar{y}\bar{y}} = 0,$$

$$\varphi_1 \underset{\bar{y}\to 0}{\sim} D_1 \frac{\bar{y}^p}{a^3} h_p(a\xi) + \cdots,$$

$$\varphi_{1\bar{x}} \underset{\bar{y}\to\infty}{\sim} 0,$$

and

$$\left(\left(1 - (\gamma + 1)\varphi_{0\bar{x}}\right)\varphi_{2\bar{x}}\right)_{\bar{x}} + \varphi_{2\bar{y}\bar{y}} = \begin{cases} 0 & \text{if } \nu_2 \gg K\nu_1^2, \\ (\gamma + 1)\varphi_{1\bar{x}}\varphi_{1\bar{x}\bar{x}} & \text{if } \nu_2 = K\nu_1^2 \end{cases}$$

$$\varphi_2 \underset{\bar{y}\to 0}{\sim} D_2 \frac{\bar{y}^\lambda}{a^3} h_\lambda(a\xi),$$

$$\varphi_{2\bar{x}} \underset{\bar{y}\to\infty}{\sim} 0.$$

Note that the first matching of the inner and outer expansion has already been carried out in order to determine the first term in the near field of φ_0. The next term, $\dfrac{\bar{x}}{\gamma + 1}$, in the near field of φ_0 is a forced term since, as $\bar{x}, \bar{y} \to 0$, $\varphi_{0\bar{x}} \gg 1$, so that the equation behaves like the sonic equation with a small forcing term e.g.

$$-(\gamma + 1)\varphi_{0\bar{x}}\varphi_{0\bar{x}\bar{x}} + \varphi_{0\bar{y}\bar{y}} = -\varphi_{0\bar{x}\bar{x}}.$$

The remaining terms in the near field of φ_0 as well as the first term in the near field of φ_1 are homogeneous similarity solutions of the variational sonic equation. The form of these solutions was found in Section (4.1.1) to be $y^\alpha g(\xi)$ where α has the values $-\dfrac{2}{5}n, \ \dfrac{2}{5} + \dfrac{6}{5}n, \ -\dfrac{2}{5}n - \dfrac{1}{5}, \ \dfrac{6}{5}n + 1$, where n is an integer. The value of the exponent p (or n) in the first term in the near field of φ_1 is determined through matching. The term y^1 is missing from the near field of ϕ_0 since we expect the solution to be even in $\bar{y}$. There can be no circulation to this order since that would imply $\Gamma = O(1/K)$.

Matching

Summarizing so far we have for the

Inner Expansion:

$$\phi \sim \phi_0(x,\tilde{y}) + K\phi_1(x,\tilde{y}) + \mu_2(K)\phi_2(x,\tilde{y}) + \cdots,$$

where

$$(\gamma+1)\phi_0 \underset{\tilde{y}\to\infty}{\sim} \frac{\tilde{y}^{\frac{2}{5}}}{a^3}f(a\xi) + c_0 + c_1\frac{\tilde{y}^{-\frac{1}{5}}}{a^3}f_1(a\xi) + \cdots,$$

$$\phi_1 = \frac{x}{\gamma+1}, \qquad\qquad (5.6.17)$$

$$\phi_2 \underset{\tilde{y}\to\infty}{\sim} d_2\frac{\tilde{y}^s}{a^3}\ell(a\xi) + \cdots,$$

where we have assumed a solution of the sonic variational equation for the far field of ϕ_2 and the power of s will be determined by matching, and for the

Outer Expansion:

$$\phi \sim \frac{1}{K}\varphi_0(\bar{x},\bar{y}) + \nu_1(K)\varphi_1(\bar{x},\bar{y}) + \nu_2(K)\varphi_2(\bar{x},\bar{y}) + \cdots, \qquad (5.6.19)$$

where

$$\bar{x} = K^2 x, \quad \bar{y} = K^{\frac{5}{2}}\tilde{y}, \qquad\qquad (5.6.20)$$

and

$$\varphi_0 \underset{\bar{y}\to 0}{\sim} \frac{\bar{y}^{\frac{2}{5}}f(a\xi)}{a^3(\gamma+1)} + \frac{\bar{x}}{\gamma+1} + \frac{C_1\bar{y}^{\frac{6}{5}}}{a^3}g_1(a\xi) + \cdots, \qquad (5.6.21)$$

$$\varphi_1 \underset{\bar{y}\to 0}{\sim} D_1\frac{\bar{y}^p}{a^3}h_p(a\xi) + \cdots, \qquad\qquad (5.6.22)$$

$$\varphi_2 \underset{\bar{y}\to 0}{\sim} D_2\frac{\bar{y}^\lambda}{a^3}h_\lambda(a\xi) + \cdots. \qquad\qquad (5.6.23)$$

Matching can be accomplished by writing both expansions in an intermediate variable or more simply by writing the far field of the outer expansion in inner variables and equating terms with the near field of the inner expansion. The result of matching to this order is that

$$D_1 = \frac{c_0}{(\gamma + 1)}, \quad p = 0, \quad \nu_1(K) = 1, \quad D_2 = \frac{c_1}{(\gamma + 1)}, \quad \lambda = -\frac{1}{5}, \quad \nu_2(K) = K^{\frac{1}{2}},$$

and

$$\mu_2 = K^3, \quad d_2 = C_1, \quad s = \frac{8}{5}.$$

The boundary value problem to be solved for the $O(K^3)$ correction to the sonic flow is,

$$-(\gamma + 1)(\phi_{0x}\phi_{2x})_x + \phi_{2\tilde{y}\tilde{y}} = 0, \tag{5.6.24}$$

$$\phi_{2\tilde{y}}|_{\tilde{y}=0} = 0, \tag{5.6.25}$$

$$\phi_2 \underset{\tilde{y}\to\infty}{\sim} C_1 \frac{\tilde{y}^{\frac{8}{5}}}{a^3} g_1(a\xi), \tag{5.6.26}$$

where the constant C_1 is determined by solving the outer boundary value problem which is,

$$\left(1 - (\gamma + 1)\varphi_{0\bar{x}}\right)\varphi_{0\bar{x}\bar{x}} + \varphi_{0\bar{y}\bar{y}} = 0, \tag{5.6.27}$$

$$\varphi_{0\bar{x}} \underset{\bar{y}\to\infty}{\sim} = 0, \tag{5.6.28}$$

$$\varphi_0 \underset{\substack{\bar{y}\to 0 \\ \xi \text{ fixed}}}{\sim} \frac{\bar{y}^{\frac{2}{5}} f(a\xi)}{a^3(\gamma + 1)} + \frac{\bar{x}}{\gamma + 1} + C_1 \frac{\bar{y}^{\frac{8}{5}}}{a^3}\ell(a\xi). \tag{5.6.29}$$

Note that φ_0 sees a shock in its near field since the function f' has a jump at ξ_s. The constant a is determined from the ϕ_0 problem (5.6.1)-(5.6.11) as in Section 4.1.1. The constant C_1 can probably be determined in a similar manner to a using conservation laws.

Once C_1 is known, numerical evaluation of the solution to the boundary value problem (5.6.24)-(5.6.26) can be carried out. In fact this boundary value problem governing ϕ_2 is the same as that governing the lifting line correction to sonic 2-D flow which is discussed in Section 6.3. Numerical results are, therefore, available for specially designed airfoils.

Since the lift can be found as $\oint_{\text{body}} \phi_x$, in the case that the shock is on the tail, the lift correction is $O(K^3)$. If the shock is in front of the tail the $O(K^3)$ correction applies only in front of the shock and hence is not the full lift. Further analysis must be carried out to determine the order of the correction behind the shock. This then is the stabilization law. Close to sonic flow the flow along the body deviates weakly, $O(K^3)$, from sonic. (Figure 5.6.2).

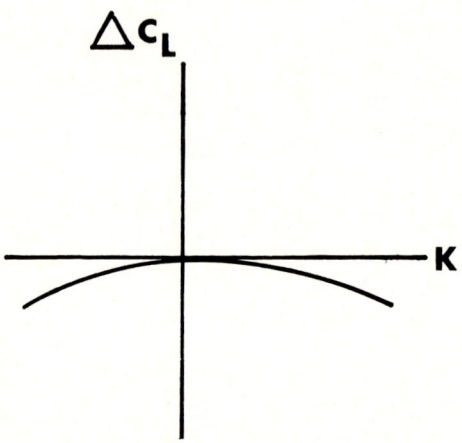

Figure 5.6.2

Stabilization: Lift coefficient variation from sonic

These results agree in order of magnitude with those obtained earlier by other means [5.6.1], [5.6.2], [5.6.3], [5.6.4] and [5.6.5] for $K < 0$. The approach taken here is simpler and more systematic. Moreover, an explicit boundary value problem is formulated for the correction potential.

$K \to 0^-$:

The behavior of ϕ as K approaches zero from below is easily visualized in the hodograph plane. The flow in the physical and in the hodograph plane is sketched in Figure (5.6.3a,b). The second shock in the $K > 0$ flow moves to $x = +\infty$ as $K \to 0$, and reappears then in front of the airfoil for $K < 0$. In front of this shock there is uniform flow. Behind the shock the flow is subsonic and then speeds up again to supersonic. We first compute the flow in the region between the shock and the characteristic CB. Then we can complete the flow by continuing it stepwise from CB through the purely hyperbolic region.

In the hodograph plane the shock polar surrounds the origin. The image of the body, DC is unknown a priori. The boundary value problem is given by (3.6)

$$w\tilde{y}_{\vartheta\vartheta} - \tilde{y}_{ww} = 0, \tag{5.6.30}$$

with the boundary conditions

$$\tilde{y} = 0 \quad \text{on} \quad w_b(\vartheta), \tag{5.6.31}$$

$$\tilde{y}_w \pm \frac{7w - K}{5w - 3K} \frac{\sqrt{w - K}}{\sqrt{2}} \tilde{y}_\vartheta = 0, \tag{5.6.32}$$

on the shock polar which is given by

$$\vartheta = \pm \frac{(w + K)\sqrt{w - K}}{\sqrt{2}}. \tag{5.6.33}$$

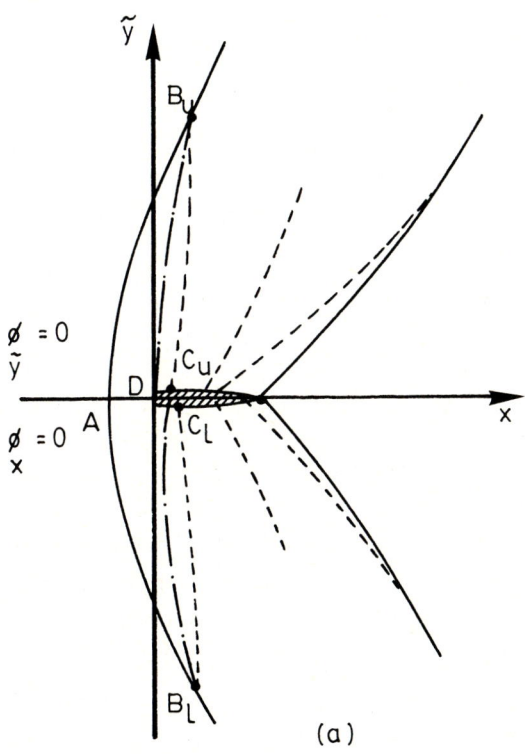

(a)

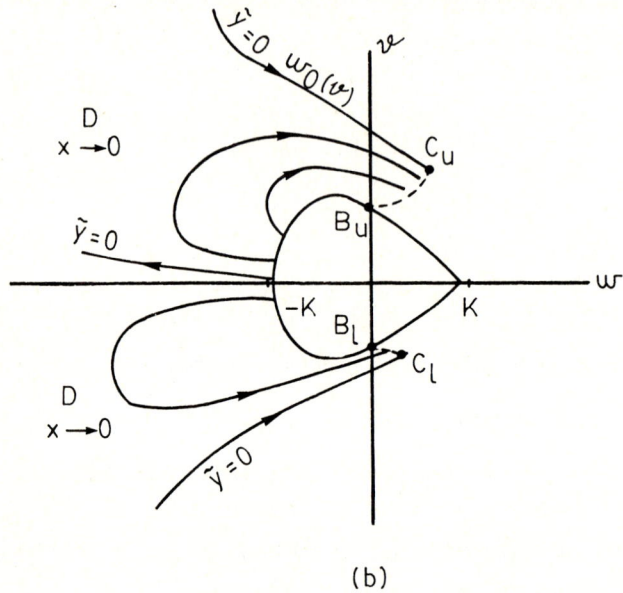

(b)

Figure 5.6.3

Flow patterns as $M_\infty > 1$ in (a) the transonic small disturbance plane; (b) the hodograph plane

Also,

$$x = x_b(\vartheta) \quad \text{is given on} \quad w_b(\vartheta). \tag{5.6.34}$$

Once $\tilde{y}$ is determined x can be found through the relations

$$w\tilde{y}_\vartheta = x_w,$$
$$x_\vartheta = \tilde{y}_w. \tag{5.6.35}$$

Note that in the hodograph plane the location of the curve which is the image of the body depends on K, that is $w_b(\vartheta; K)$, or in other words ϕ_x along the body changes with K. This is the quantity we want to find. However $x_b(\vartheta)$ is independent of K, that is $\phi_{\tilde{y}}$ is given on $\tilde{y} = 0$.

As $K \to 0$, and the shock polar shrinks to a point singularity the flow near the boundary changes little, that near the shock polar changes more drastically. Thus the flow yields nicely to two limiting procedures, the inner limit (valid near the body or the boundary $w_b(\vartheta))$ in which w, ϑ are fixed as $K \to 0$, and the outer limit (valid near the shock polar) in which w, ϑ are stretched as $K \to 0$. Thus in the inner limit the shock polar shrinks to a point singularity, in the outer limit the shock polar remains $O(1)$ but the body recedes to infinity.

Inner Expansion: w, ϑ fixed, $K \to 0$.

In the inner limit x, $\tilde{y}$ have the expansions

$$\tilde{y} \sim \tilde{y}_0(\vartheta, w) + \nu_1(K)\tilde{y}_1(\vartheta, w) + \cdots, \tag{5.6.36}$$

$$x \sim x_0(\vartheta, w) + \nu_1(K)\tilde{x}_1(\vartheta, w) + \cdots, \tag{5.6.37}$$

where $\nu_1 \ll 1$ is to be determined, and the boundary has the expansion

$$w_b(\vartheta; K) \sim w_0(\vartheta) + \nu_1(K)w_1(\vartheta) + \cdots. \tag{5.6.38}$$

One might assume a different guage function for w_b, however substitution into the boundary condition would confirm that it must be $\nu_1(K)$.

Substituting (5.6.36,37) into the governing equation (5.6.30), and equating coefficients of corresponding gauge functions gives the equations to be satisfied by $\tilde{y}_0$, y_1, x_0, x_1,

$$w\tilde{y}_{0\vartheta\vartheta} - \tilde{y}_{0ww} = 0,$$
$$w\tilde{y}_{1\vartheta\vartheta} - \tilde{y}_{1ww} = 0, \tag{5.6.39}$$

and

$$x_{0\vartheta} = \tilde{y}_{0w}, \quad x_{1\vartheta} = \tilde{y}_{1w}$$
$$x_{0w} = w\tilde{y}_{0\vartheta}, \quad x_{1w} = w\tilde{y}_{1\vartheta} \tag{5.6.40}$$

Along the boundary $w_b(\vartheta)$ one has, by substitution of (5.6.38) into (5.6.31) and (5.6.34) and Taylor series expanding about $w_0(\vartheta)$,

$$\tilde{y}(\vartheta, w_b(\vartheta)) = \tilde{y}_0(\vartheta, w_0(\vartheta))$$
$$+ \nu_1(K)\left\{ \tilde{y}_1(\vartheta, w_0(\vartheta)) + w_1(\vartheta)\frac{\partial \tilde{y}_0}{\partial w}(\vartheta, w_0(\vartheta)) \right\} + \cdots = 0,$$

$$x(\vartheta, w_b(\vartheta)) = x_0(\vartheta, w_0(\vartheta))$$
$$+ \nu_1(K)\left\{ x_1(\vartheta, w_0(\vartheta)) + w_1(\vartheta)\frac{\partial x_0}{\partial w}(\vartheta, w_0(\vartheta)) \right\} + \cdots = x_b(\vartheta).$$

Thus equating coefficients of corresponding gauge functions,

$$\tilde{y}_0(\vartheta, w_0(\vartheta)) = 0, \tag{5.6.41}$$

$$\tilde{y}_1(\vartheta, w_0(\vartheta)) + w_1(\vartheta)\frac{\partial \tilde{y}_0}{\partial w}(\vartheta, w_0(\vartheta)) = 0, \tag{5.6.42}$$

and

$$x_0\left(\vartheta, w_0(\vartheta)\right) = x_b(\vartheta)\,, \tag{5.6.43}$$

$$x_1\left(\vartheta, w_0(\vartheta)\right) + w_1(\vartheta)\frac{\partial x_0}{\partial w}\left(\vartheta, w_0(\vartheta)\right) = 0\,. \tag{5.6.44}$$

We can eliminate the unknown $w_1(\vartheta)$ from (5.6.42) by substituting (5.6.44), and then we can eliminate x_1 from the result by using the relations (5.6.40) and by differentiating. Thus,

$$\left(\tilde{y}_1 - \frac{x_1}{x_{0w}}\tilde{y}_{0w}\right)\Bigg|_{\left(\theta, w_0(\theta)\right)} = 0\,,$$

or

$$x_1\left(\theta, w_0(\theta)\right) = \left(\frac{x_{0w}\tilde{y}_1}{\tilde{y}_{0w}}\right)\Bigg|_{\left(\theta, w_0(\theta)\right)} = \left(\frac{w\tilde{y}_{0\vartheta}\tilde{y}_1}{\tilde{y}_{0w}}\right)_{\left(\theta, w_0(\theta)\right)}.$$

Thus

$$\left(x_{1\vartheta} + x_{1w}w_0'\right)\Big|_{\left(\theta, w_0(\theta)\right)} = \frac{d}{d\vartheta}\left\{\left(\frac{w_0\tilde{y}_{0\vartheta}\tilde{y}_1}{\tilde{y}_{0w}}\right)_{w = w_0(\vartheta)}\right\}\,,$$

or

$$\left(\tilde{y}_{1w} + w_0 w_0' \tilde{y}_{1\vartheta}\right)\Big|_{\left(\theta, w_0(\theta)\right)} = -\frac{d}{d\vartheta}\left(w_0 w_0' \tilde{y}_1\right)\,.$$

since from (5.6.38) $\tilde{y}_{0\vartheta}\left(\theta, w_0(\theta)\right) + w_0'\tilde{y}_{0w}\left(\theta, w_0(\theta)\right) = 0$. Collecting terms we have that the boundary condition for $\tilde{y}_1$ can be written as,

$$\tilde{y}_{1w}\left(1 + w(w')^2\right) + \tilde{y}_{1\vartheta}2ww' + \tilde{y}_1\left((w')^2 + ww''\right) = 0 \quad \text{on} \quad w = w_0(\vartheta)\,. \tag{5.6.45}$$

In this limit the shock polar has shrunk to a point singularity. The singular solutions of Tricomi's equation were examined in Section (4.1.1) and thus we find the behavior of $\tilde{y}_0$.

Summarizing, we have the following boundary value problems for $\tilde{y}_0$, $\tilde{y}_1$,

$$-w\tilde{y}_{0\vartheta\vartheta} + \tilde{y}_{0ww} = 0\,, \tag{5.6.46}$$

$$\tilde{y}_0\left(\theta, w_0(\theta)\right) = 0\,, \tag{5.6.47}$$

$$\tilde{y}_0 \underset{\substack{\vartheta \to 0 \\ \frac{\vartheta}{\rho}\text{ fixed}}}{\sim} K\rho^{-\frac{5}{3}}\left\{\left(1 + \frac{\vartheta}{\rho}\right)^{\frac{1}{3}}\left(3\frac{\vartheta}{\rho} + 1\right) + \left(1 - \frac{\vartheta}{\rho}\right)^{\frac{1}{3}}\left(3\frac{\vartheta}{\rho} - 1\right)\right\}$$

$$+ b_1\rho^{-\frac{2}{3}}f_1\left(\frac{\vartheta}{\rho}\right) + b_2\,, \tag{5.6.48}$$

where f_1 is a known hypergeometric function, (4.1.104) and $\rho = \vartheta^2 - \frac{4}{9}w^3$,

$$-w\tilde{y}_{1\vartheta\vartheta} + \tilde{y}_{1ww} = 0\,,$$

$$\tilde{y}_{1w}\left(1 + w(w')^2\right) + \tilde{y}_{1\vartheta}2ww' + y_1\left((w')^2 + ww''\right) = 0 \quad \text{on} \quad w = w_0(\vartheta)\,,$$

$$\tilde{y}_1 \underset{\substack{\vartheta\to 0\\ \frac{\vartheta}{\rho}\text{ fixed}}}{\sim} e_0\rho^{\lambda_1}g_0\left(\frac{\vartheta}{\rho}\right) + e_1\rho^{\lambda_2}g_1\left(\frac{\vartheta}{\rho}\right) + \cdots\,.$$

Note that the $\tilde{y}_0$ problem is that for exactly sonic flow, the constants a, b_1, b_2 are determined as in Section 4.1. The unknowns e_0, λ_1 are determined by matching.

Outer Expansion: $W = \frac{w}{-K}$, $\Theta = \frac{\vartheta}{(-K)^{\frac{3}{2}}}$ $K \to 0$.

In the outer limit w, ϑ are scaled so that K drops out of the problem (5.6.30) and (5.6.32, 5.6.33). In the new variables equation (5.6.30) becomes

$$W\tilde{y}_{\Theta\Theta} - \tilde{y}_{WW} = 0\,, \tag{5.6.49}$$

and the shock conditions (5.6.32) and (5.6.33) become

$$\tilde{y}_W \pm \frac{7W + 1}{5W + 3}\frac{\sqrt{W + 1}}{\sqrt{2}}\tilde{y}_\Theta = 0\,, \tag{5.6.50}$$

on the shock polar

$$\Theta = \pm\frac{(W - 1)\sqrt{W + 1}}{\sqrt{2}}\,. \tag{5.6.51}$$

The outer expansion is then

$$\tilde{y}(\vartheta, w) = \mu_0(K)Y_0(W, \Theta) + \mu_1(K)Y_1(W, \Theta) + \cdots\,. \tag{5.6.52}$$

The boundary condition on the body, $w_b(\vartheta)$, is lost in the limit and is replaced by the condition of matching with the inner expansion.

Substitution of (5.6.52) into (5.6.49) and (5.6.50) then gives the various boundary value problems for the Y_i. All satisfy Tricomi's equation and the full shock conditions.

The first matching determines that

$$Y_0 \underset{\substack{\Theta\to\infty\\ \frac{P}{\Theta}\text{ fixed}}}{\sim} KP^{-\frac{5}{3}}\left\{\left(1 + \frac{\Theta}{P}\right)^{\frac{1}{3}}\left(3\frac{\Theta}{P} + 1\right) + \left(1 - \frac{\Theta}{P}\right)^{\frac{1}{3}}\left(3\frac{\Theta}{P} - 1\right)\right\} + \cdots\,,$$

where

$$P^2 = \Theta^2 - \frac{4}{9}W^2 ,$$

(Note that $\dfrac{W^{\frac{3}{2}}}{\vartheta} = \dfrac{w^{\frac{3}{2}}}{\theta}$ so that similarity curves go into similarity curves as $K \to$

0), and

$$\mu_0(K) = (-K)^{-\frac{5}{2}} . \tag{5.6.53}$$

Thus

$$\tilde{y} \sim (-K)^{-\frac{5}{2}} Y_0 ,$$

which is the same $\tilde{y}$ stretching as was found in the $M_\infty < 1$ stabilization section.

Summarizing the boundary value problems for Y_0, Y_1 we have

$$\tilde{y} \sim (-K)^{-\frac{5}{2}} Y_0(\Theta,W) + \mu_1(K)Y_1(\Theta,W) + \cdots ,$$

where

$$WY_{0\Theta\Theta} - Y_{0WW} = 0, \tag{5.6.54}$$

$$Y_{0W} \pm \frac{7W-1}{5W-3}\sqrt{\frac{W-1}{2}}Y_{0\Theta} = 0 \quad \text{on} \quad \Theta = \pm\frac{(W-1)\sqrt{W+1}}{\sqrt{2}} , \tag{5.6.55}$$

and

$$Y_0 \underset{\substack{\Theta\to\infty \\ \frac{P}{\Theta}\text{ fixed}}}{\sim} KP^{-\frac{5}{3}}f\left(\frac{\Theta}{P}\right) + B_1 P^{-\frac{11}{3}} F_1\left(\frac{\Theta}{P}\right) + \cdots , \tag{5.6.56}$$

where F_1 is the hypergeometric function, $\dfrac{\Theta}{P}F\left(\dfrac{7}{3}, -\dfrac{7}{6}, \dfrac{3}{2}, \dfrac{\Theta^2}{P^2}\right)$. (See 4.1.57).

Note that the second term in the far field of Y_0 was chosen such that the exponent to which P is raised is -11/3. Examination of the solutions to (5.5.54) indicates there is also a solution with exponent -8/3. (See (4.1.103), (4.1.57)). However, the boundary value problem given by (5.6.54), (5.6.55), and the first terms in (5.6.56) clearly defines an odd function Y, hence the $P^{-\frac{8}{3}}$ term is missing.

And, for Y_1

$$WY_{1\Theta\Theta} - Y_{1WW} = 0,$$

$$Y_{1W} \pm \frac{7W-1}{5W-3}\frac{\sqrt{W-1}}{\sqrt{2}}Y_{1\Theta} = 0 \quad \text{on} \quad \Theta = \pm\frac{(W-1)\sqrt{W+1}}{\sqrt{2}} ,$$

$$Y_1 \sim E_1 P^{\sigma_1} G_0\left(\frac{\Theta}{P}\right) + \cdots .$$

Matching:

Matching is accomplished through intermediate limits however it is simpler to write the near field of the outer expansion in inner variables and compare it with the far field of the inner expansion. Thus,

Outer Expansion:

$$\tilde{y} \sim (-K)^{-\frac{5}{2}} Y_0 + \mu_1 Y_1 \underset{\substack{\Theta \to \infty \\ \frac{P}{\Theta} \text{ fixed}}}{\sim} K P^{-\frac{5}{3}} f\left(\frac{\Theta}{P}\right) + B_1 P^{-\frac{11}{3}} F_1\left(\frac{\Theta}{P}\right),$$

and

Inner Expansion:

$$\tilde{y} \sim \tilde{y}_0 + \nu_1(K)\tilde{y}_1$$

$$\underset{\substack{\Theta \to \infty \\ \frac{P}{\Theta} \text{ fixed}}}{\sim} K \rho^{-\frac{5}{3}} f\left(\frac{\vartheta}{p}\right) + b_1 \rho^{-\frac{2}{3}} f_1\left(\frac{\vartheta}{\rho}\right) + \cdots + \nu_1 e_0 \rho^{\lambda_1} g_0\left(\frac{\vartheta}{\rho}\right) + \cdots .$$

Here μ_1, E_0, σ_1, ν_1, e_0, λ_1 are the unknowns. Thus,

$$\delta_1 = -\frac{2}{3}, \quad \mu_1 = (-K)^{-1}, \quad E_0 = b_1, \quad \nu_1 = (-K)^3, \quad \lambda_1 = -\frac{11}{3}, \quad e_0 = B_1.$$

Summarizing we have that

$$\tilde{y} = \tilde{y}_0 - K^3 \tilde{y}_1 + \cdots, \tag{5.6.57}$$

$$\tilde{y} \sim (-K)^{-\frac{5}{2}} Y_0 - K^{-1} Y_1 + \cdots, \tag{5.6.58}$$

where $\tilde{y}_0$ satisfies the sonic boundary value problem (5.6.46)-(5.6.48), see Section 4.1, $\tilde{y}_1$ satisfies the boundary value problem (see Figure 5.6.4)

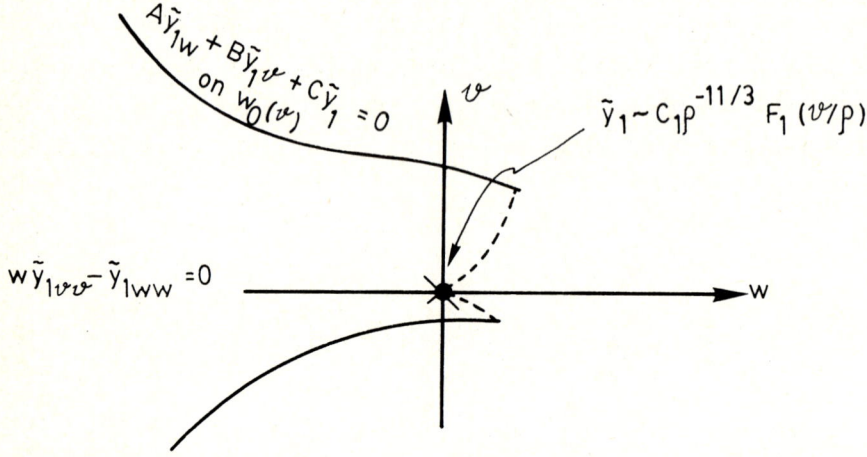

Figure 5.6.4

Boundary value problem for $\tilde{y}_1$

$$-w\tilde{y}_{1\vartheta\vartheta} + \tilde{y}_{1ww} = 0\,,$$

$$\tilde{y}_{1w}\left(1 + w(w')^2\right) + \tilde{y}_{1\vartheta}(2ww') + \tilde{y}_1\left((w')^2 + ww''\right) = 0 \quad \text{on} \quad w = w_0(\vartheta)\,,$$

$$\tilde{y}_1 \underset{\substack{\theta \to \infty \\ \frac{P}{\theta} \text{ fixed}}}{\sim} B_1\rho^{-\frac{11}{3}} F_1\left(\frac{\theta}{\rho}\right) + \cdots\,.$$

where

$$F_1 = \frac{\theta}{\rho} F\left(\frac{7}{3}, -\frac{7}{6}, \frac{3}{2}, \frac{\theta^2}{\rho^2}\right)\,,$$

$w_0(\vartheta)$ is known, and B_1 must be found by solving (5.6.63), (5.6.64), (5.6.65).

Once $\tilde{y}_1(\vartheta, w)$ is known $w_1(\vartheta)$ can be computed from (5.6.51)

$$(\tilde{y}_1 + w_1\tilde{y}_{0w})_{w=w_0(\vartheta)} = 0\,.$$

Then

$$w_b(\vartheta) = w_0(\vartheta) - K^{+3}w_1(\vartheta) + \cdots\,. \tag{5.6.59}$$

This then is the stabilization law. Near sonic flow conditions the flow on the body deviates weakly $O(K^3)$ from sonic flow. (see Figure (5.6.2).)

To determine the explicit solutions the value of B_1 must be found. This can be done using conservation laws as in Section(4.1.1). The conservation law for this case comes from setting $\beta = 7/2$ in the family of conservation laws (4.1.108) so that

$$Y = 2^{-4}\left\{(P - \Theta)^{\frac{1}{3}}(7P^3 + 45P^2\Theta - 27P\Theta^2 - 81\Theta^3)\right.$$
$$\left. - (P + \Theta)^{\frac{1}{3}}(7P^3 - 45P^2\Theta - 27P\Theta^2 + 81\Theta^3)\right\}, \qquad (5.6.60)$$

and

$$X = -2^{-\frac{10}{3}}3^{\frac{1}{3}}11^{-1}\left\{(P + \Theta)^{\frac{2}{3}}(-64P^3 + 171P^2\Theta + 270P\Theta^2 - 405\Theta^3)\right.$$
$$\left. + (P - \Theta)^{\frac{2}{3}}(-64P^3 - 171P^2\Theta + 270P\Theta^2 + 405\Theta^3)\right\}, (5.6.61)$$

where $P^2 = \Theta^2 - \frac{4}{9}W^3$. In this case we integrate in the hodograph plane along the path shown in Figure 5.6.5, that is the path of integration extends from the point at which the polar is intersected a $P = 0$ characteristic below around to the corresponding point above then out to the limiting characteristics and closes at infinity.

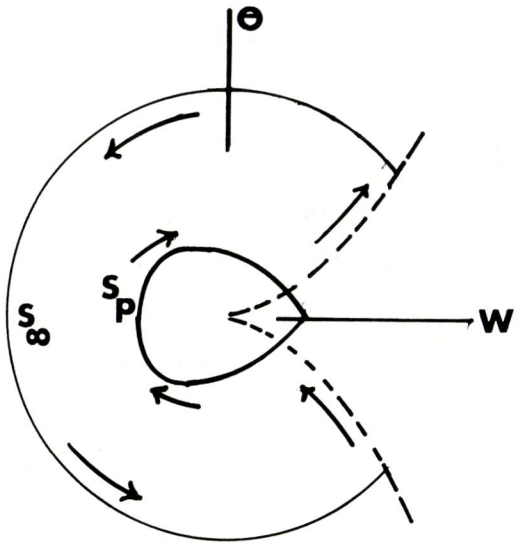

Figure 5.6.5

Contour of Integration for B_1

The final result obtained by Ziegler [5.6.6] is that

$$B_1 = (.0260) \int_{S_p} (Y X_{0W} + XY_{0W})dW + (YX_{0\Theta} + XY_{0\Theta})d\Theta, \qquad (5.6.62)$$

where $X(W,\Theta)$, $Y(W,\Theta)$ are given in (5.6.60) and (5.6.61), and $Y_0(W,\Theta)$ is the solution to the boundary value problem given by (5.6.54)-(5.6.56), see Figure 5.6.6, and X_0 is related to Y_0 through the Tricomi relations. Thus B_1 links the near field of the outer expansion to the flow on the shock polar. If the velocity components are known on the shock polar, then B_1 can be computed.

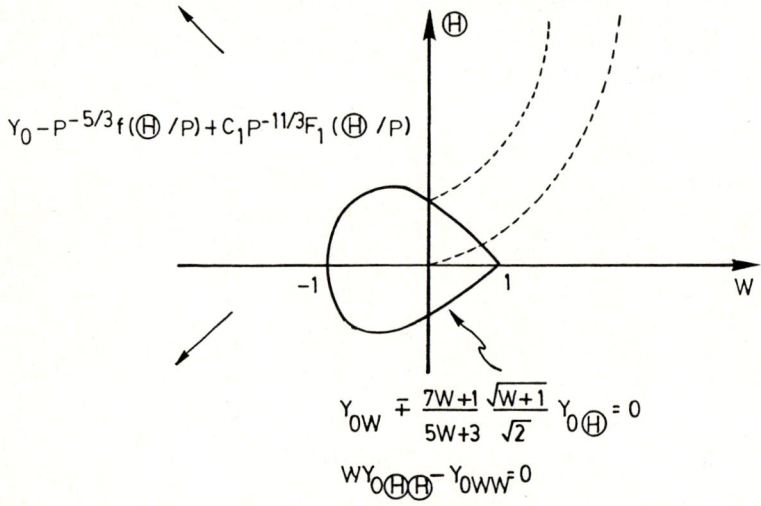

Figure 5.6.6

Boundary value problem for Y_0

Note that the hodograph plane analysis has an advantage over the physical plane analysis in that the equation governing the zeroth order (y_0, sonic) problem and the first order correction near the airfoil (y_1) are the same, Tricomi's equation. In the physical plane formulation the zeroth order (ϕ_0) and correction (ϕ_1) potentials near the airfoil satisfy different equations.

References

[5.6.1] Diesperov, V. N., Lifschitz, Yu. B., Ryzhov, O. S., The Stabilization Law for Transonic Flows, *Archives of Mechanics (Archwum Mechaniki Stasowanej)* **26**. No. 3, Warsaw 1974, pp. 511-521.

[5.6.2] Diesperov, V. N., Lipschitz, Yu. B., Rhyzov, O. S., Stabilization and Drag in Transonic Range of Velocities, *Symposium Transsonicum*, ed. Oswatitch and Rues, Springer, N.Y. 1976.

[5.6.3] Guderley, Gottfried, Two-dimensional flow patterns with a free-stream Mach number close to one. AF Tech. Rpt. #6343, May 1951, Wright-Patterson Air Force Base, Dayton, Ohio.

[5.6.4] Lipschitz, Yu. B. and Ryzhov, O. S., Transonic Flow around a Carrying Profile, English Translation, *Fluid Dynamics*, V. 13, #1, 1978, pp. 78-84.

[5.6.5] Guderley, Gottfried, Two Dimensional Bodies at Slightly Supersonic Mach Numbers, *WADC Tech. Rpt.* 53-454, Nov. 1953, Wright-Patterson Air Force Base, Dayton, Ohio.

[5.6.6] Cook, L. Pamela and Fred Ziegler, The Stabilization Law for Transonic Flow, to appear in *SIAM J. Appl. Math.*

6 Three Dimensional Wings

In this section lifting-line theory is applied to describe the flow about lifting, high aspect ratio wings, at transonic ($M_\infty < 1$ and $M_\infty = 1$) speeds. Lifting-line theory provides the mechanism for integrating the two-dimensional flow results up to three dimensions.

Classical lifting-line theory for incompressible flow, due to Prandtl, gives the aspect ratio corrections to the two dimensional cross section flow at each spanwise station as an induced downwash. This leads to an integral equation which had to be solved for the circulation. More recently [6.1] it was recognized that the problem of computing the incompressible flow about an unswept wing could be considered as a singular perturbation problem with the inverse of the aspect ratio as the small parameter. Asymptotic expansions were constructed in various regions and then matched. The advantage of this approach was that aspect ratio corrections were obtained explicitly, without recourse to integral equations.

To apply the methods of matched asymptotic expansions to three dimensional *transonic* flows we examine the dependence of the solution $\phi(x, y, z; K, A, B)$ on the reduced aspect ratio parameter $B = b\delta^{\frac{1}{3}}$, as $B \to \infty$. [6.2]. Two distinguished limits of the boundary value problem exist, one valid near the airfoil, the other valid far from the airfoil. The limiting procedures differ in the $M_\infty < 1$ ($K \neq 0$) case from the $M_\infty = 1$ ($K = 0$) case. In fact, the $M_\infty = 1$ flow can not be obtained by taking the $M_\infty \to 1$ limit of the $M_\infty < 1$ expansion. In that limit the expansions are nonuniform. This is to be expected since for $M_\infty < 1$ the supersonic (hyperbolic) zone is a finite bubble. Hence, the flow on the rear of the wing or behind the wing influences the flow in front. At $M_\infty = 1$ the limiting Mach surface extends to infinity so that the flow in the rear of the wing, or behind the wing, does not influence the flow in front. The front part of the flow at $M_\infty = 1$ can be calculated independently of the rear part.

The boundary value problems describing the flows in these inner (near the airfoil) and outer (far from the airfoil) regions are formulated completely through the matching conditions. The result obtained is an explicit representation of the boundary value problem governing the first aspect ratio correction to the two dimensional flow. The equation governing the correction is linear, but involves as coefficients the solution of the nonlinear two dimensional transonic problem, hence its solution must be evaluated numerically.

In what follows we first discuss lifting line theory for $M_\infty < 1$. In this case the equations governing the outer flow are linear and basically those of incompressible flow. The outer solution can thus be found in closed form. The simplest case is that of an unswept wing which is considered first. The swept wing solution is similar to the unswept except that new terms arise in the asymptotic expansions, that is the downwash is of larger order. This case was first considered by Thurber [6.3], and later independently and more completely by Cheng and Meng [6.4] and by Cook [6.5]. After the swept wing ($M_\infty < 1$) is considered, unswept wings in sonic flow ($M_\infty = 1$) are dealt with. These flows are governed by compressibility effects (nonlinear) even far from the airfoil. Explicit closed form solutions cannot be found even in the outer flow region. [6.6]

The lifting-line results generate a boundary value problem for the aspect ratio correction at each spanwise station. This gives qualitative results for the effect of finite span, but is not efficient numerically since the boundary value problem must be solved at each spanwise (z) station. In the special case that the wing has similar sections, the boundary value problem (with the exception of the skewed wing) can be scaled to be independent of spanwise coordinate, hence only one computation need be done. This scaling is considered within each relevant section.

Finally, some discussion of the uniqueness of the solutions of the newly formulated boundary value problems is given. In particular a proof of the uniqueness of the first aspect ratio correction for $M_\infty < 1$, in the case that the cross-sectional flow is shock free, is given.

References

[6.1] Van Dyke, M. D., *Perturbation Methods in Fluid Mechanics*, Parabolic Press, Stanford, Ca, 1975.

[6.2] Cook, L. Pamela and Julian D. Cole, Lifting Line Theory for Transonic Flow, *SIAM J Appl. Math.* **35** (2) 1978, pp 209-228.

[6.3] Thurber, James K., An asymptotic Method for Determining the Lift Distribution of a Swept-Back Wing of Finite Span, *Comm. Pure and Appl. Math.*, **18**, 1965, pp733-756.

[6.4] Cheng, H. K. and S. Y. Meng, The Oblique Wing as a Lifting-Line Problem in Transonic Flow, *J. F. M.* **97**, #3, (1980), pp 531-566; also Univ. of Southern Ca. Dept of Aerospace Eng. Rept. #136.

[6.5] Cook, L. Pamela, Lifting-Line Theory for a Swept Wing at Transonic

Speeds, *Q. Appl. Math.*, **37**, #2, 1979 pp 177-202.

[6.6] Cole, J. D., L. Pamela Cook, F. Ziegler, Finite Span-Wings at Sonic Speed, *Mech. Rsch. Comm.*, **1**, #4, (1980), pp 253-260.

6.1 Unswept Wings, $M_\infty < 1$.

The boundary value problem describing the transonic $(M_\infty < 1)$ small disturbance flow about a lifting finite aspect ratio wing, (with spanwise coordinate z),
was formulated in Section 3.1. The governing equation is

$$\left(K - (\gamma + 1)\phi_x\right)\phi_{xx} + \phi_{\tilde{y}\tilde{y}} + \phi_{\tilde{z}\tilde{z}} = 0. \tag{6.1.1}$$

In this approximation the wing surface and the vortex sheet both lie in the $\tilde{y} = 0$
plane for $-B < \tilde{z} < B$. The wing surface is defined by $x_{\text{LE}}(\tilde{z}/B) < x < x_{\text{TE}}(\tilde{z}/B)$,
where $x_{\text{LE}}(1) = x_{\text{LE}}(-1) = x_{\text{TE}}(1) = x_{\text{TE}}(-1) = 0$, i.e. the wing is unswept.
The vortex sheet is defined by $x > x_{\text{TE}}(\tilde{z}/B)$, $-B < \tilde{z} < B$, (see Figure 6.1.1).

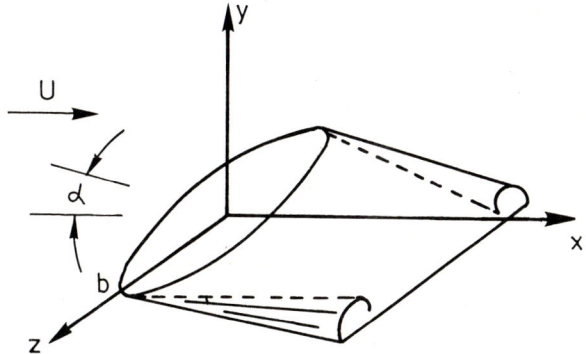

Figure 6.1.1a

Three dimensional lifting wing in physical coordinates.

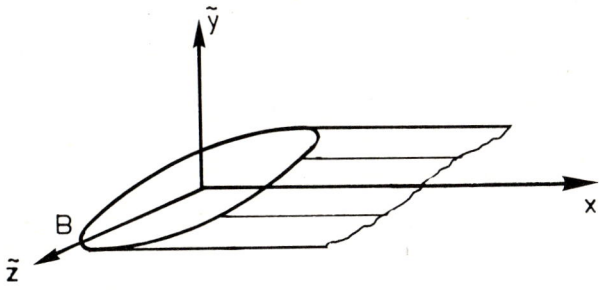

Figure 6.1.1b

Three dimensional lifting wing in transonic coordinates.

The boundary conditions on these surfaces are;
tangent flow on the body,

$$\phi_{\tilde{y}}(x, 0\pm, \tilde{z}) = \frac{\partial F_{u,\ell}}{\partial x}\left(x, \frac{\tilde{z}}{B}\right) \quad \text{on} \quad x_{\text{LE}}\left(\frac{\tilde{z}}{B}\right) < x < x_{\text{TE}}\left(\frac{\tilde{z}}{B}\right), \quad |\tilde{z}| < B,$$
$$(6.1.2)$$

the Kutta-Joukowsky condition that the flow leave the trailing edge smoothly,

$$[\phi_x]_{\text{TE}} = 0, \tag{6.1.3}$$

and no pressure jump across the wake,

$$[\phi_x]_{\text{vs}} = 0. \tag{6.1.4}$$

There is no disturbance at upstream infinity,

$$\phi_x, \phi_{\tilde{y}}, \phi_{\tilde{z}} \to 0 \quad \text{as} \quad x \to -\infty, \tag{6.1.5}$$

and pressure disturbances die out downstream,

$$\phi_x \to 0 \quad \text{as} \quad x \to +\infty. \tag{6.1.6}$$

The specification of the problem is completed by the addition of the shock jump conditions. For the shock surface given by

$$S(x, \tilde{y}, \tilde{z}) = x - g(\tilde{y}, \tilde{z}) = 0, \tag{6.1.7}$$

the conditions are;

no jump of tangential velocity across the shock,

$$[\phi] = 0, \tag{6.1.8a}$$

conservation of mass,

$$\left[K\phi_x - \frac{\gamma + 1}{2}\phi_x^2\right]_s - [\phi_{\tilde{y}}]_s g_{\tilde{y}} - [\phi_{\tilde{z}}]_s g_{\tilde{z}} = 0. \tag{6.1.8b}$$

From (6.18a,b) the shock polar can be found,

$$\left[K\phi_x - \frac{\gamma + 1}{2}\phi_x^2\right]_s [\phi_x]_s + [\phi_{\tilde{y}}]_s^2 + [\phi_{\tilde{z}}]_s^2 = 0. \tag{6.1.9}$$

There are two basic limiting procedures which correspond to distinguished limits of the boundary value problem (6.1.1)-(6.1.9). In the inner limit x, $\tilde{y}$, $z^* = \tilde{z}/B$ are held fixed as $B \to \infty$. That is, the x and $\tilde{y}$ coordinates are measured relative to the chord, the $\tilde{z}$ coordinate relative to the span. For fixed z^*, $\tilde{z} \to \infty$ as $B \to \infty$, so that the wing approaches infinite span, and the boundary value problem is essentially two dimensional. In the outer limit $x^* = x/B$, $y^* = \tilde{y}/B$, $z^* = \tilde{z}/B$ are held fixed as $B \to \infty$. That is, all coordinates are measured with respect to the span. In this limit for x fixed, $x^* \to 0$ so that the image of the wing shrinks to a line. The nonlinearity of the flow is lost to first order in this limit, corresponding to the fact that it is valid far from the airfoil, but the three dimensionality of the flow is preserved.

Inner Expansion: $(x, \tilde{y}, z^* = \dfrac{\tilde{z}}{B}$ fixed as $B \to \infty)$.

This expansion has the form

$$\phi(x, \tilde{y}, \tilde{z}; B) = \phi_0(x, \tilde{y}; z^*) + \frac{1}{B}\phi_1(x, \tilde{y}; z^*) + \cdots . \tag{6.1.10}$$

The corresponding approximate forms of the basic transonic equation (6.1.1), obtained by substituting (6.1.10) into (6.1.1) and equating coefficients of corresponding powers of B, are,

$$\left(K - (\gamma + 1)\phi_{0x}\right)\phi_{0xx} + \phi_{0\tilde{y}\tilde{y}} = 0, \tag{6.1.11}$$

$$\left(K - (\gamma + 1)\phi_{0x}\right)\phi_{1xx} - (\gamma + 1)\phi_{1x}\phi_{0xx} + \phi_{1\tilde{y}\tilde{y}} = 0. \tag{6.1.12}$$

In these equations the spanwise coordinate z^* appears only as a parameter. The tangent flow boundary conditions on $|z^*| < 1$, $x_{\mathrm{LE}}(z^*) < x < x_{\mathrm{TE}}(z^*)$, from (6.1.2) are,

$$\phi_{0\tilde{y}}(x, 0\pm, z^*) = \frac{\partial F_{u,\ell}(x, z^*)}{\partial x}, \tag{6.1.13}$$

$$\phi_{1\tilde{y}}(x, 0\pm, z^*) = 0. \tag{6.1.14}$$

The Kutta-Joukowsky condition must be satisfied for each term of the inner expansion,

$$\left[\phi_{0x}\right]_{\mathrm{TE}} = 0, \tag{6.1.15}$$

$$\left[\phi_{1x}\right]_{\mathrm{TE}} = 0. \tag{6.1.16}$$

The conditions at infinity must be thought of in terms of asymptotic matching to the near field of the outer expansion. The shock jump conditions are obtained by substituting (6.1.10) into (6.1.8) and (6.1.9) and expanding about the zeroth order shock locus. That is, the shock locus (6.1.7) is expanded as

$$x - g(\tilde{y}, \tilde{z}/B; B) = x - \left(g_0(\tilde{y}; z^*) + \frac{1}{B}g_1(\tilde{y}; z^*) + \cdots\right) = 0.$$

Then for any quantity $f(x, \tilde{y}, \tilde{z}; B)$,

$$[f]_s = \left(f\left(\left(g_0 + \frac{1}{B}g_1 + \cdots\right)^+, \tilde{y}, \tilde{z}; B\right) - f\left(\left(g_0 + \frac{1}{B}g_1 + \cdots\right)^-, \tilde{y}, \tilde{z}; B\right)\right),$$

where $+(-)$ means as x approaches the value from above (below) respectively. Expanding f in its expansion in terms of f_i for large B, and then the f_i in their right and left Taylor series about g_0,

$$[f]_s = [f_0]_{s_0} + \frac{1}{B}\left\{[f_1]_{s_0} + g_1[f_{0x}]_{s_0}\right\} + \cdots,$$

where

$$s_0 : \quad x - g_0(\tilde{y}) = 0.$$

Applying this to (6.1.8) and (6.1.9) gives,

$$\left[K\phi_{0x} - \frac{\gamma+1}{2}\phi_{0x}\right]_{s_0}[\phi_{0x}]_{s_0} + [\phi_{0\tilde{y}}]_{s_0}^2 = 0, \tag{6.1.17}$$

$$[\phi_0]_{s_0} = 0, \tag{6.1.18}$$

and, to first order,

$$\left[K\phi_{0x} - \frac{\gamma+1}{2}\phi_{0x}^2\right]_{s_0}[\phi_{1x}] + [\phi_{0x}]_{s_0}[K\phi_{1x} - (\gamma+1)\phi_{0x}\phi_{1x}]_{s_0} + 2[\phi_{0\tilde{y}}]_{s_0}[\phi_{1\tilde{y}}]_{s_0}$$

$$= g_1\left\{[\phi_{0xx}]_{s_0}\left[K\phi_{0x} - \frac{\gamma+1}{2}\phi_{0x}^2\right]_{s_0} + [\phi_{0x}]_{s_0}\cdot\right.$$

$$\left.[K\phi_{0xx} - (\gamma+1)\phi_{0x}\phi_{0xx}]_{s_0} + 2[\phi_{0\tilde{y}}]_{s_0}[\phi_{0x\tilde{y}}]_{s_0}\right\}, \tag{6.1.19}$$

$$[\phi_1]_{s_0} = -g_1[\phi_{0x}]_{s_0}. \tag{6.1.20}$$

The zeroth order conditions (6.1.17), (6.1.18) are the usual two-dimensional shock conditions (3.1.23), (3.1.24). The first order conditions involve three unknowns, $[\phi_{1x}]_{s_0}$, $[\phi_{1\tilde{y}}]_{s_0}$, $g_1(\tilde{y})$. More will be said about this later.

In fact, the problem for ϕ_0, (6.1.11), (6.1.13), (6.1.15), (6.1.17), (6.1.18) is the two-dimensional transonic problem for the flow past a lifting airfoil at a given spanwise station of the wing. The ϕ_1 problem appears to have solution zero, however we have not yet specified the conditions that it must satisfy at infinity. It is this far field condition, which will be determined from matching, which defines the problem for ϕ_1, the first aspect ratio correction to the two-dimensional cross-section flow.

Outer Expansion: $(x^* = \dfrac{x}{B}, y^* = \dfrac{\tilde{y}}{B}, z^* = \dfrac{\tilde{z}}{B}$ fixed as $B \to \infty)$.

In this limit all lengths are measured relative to the span. The effect of the wing, which has shrunk to a line, is represented by a line-vortex shedding its vorticity downstream (Figure 6.1.2) plus higher order singularities.

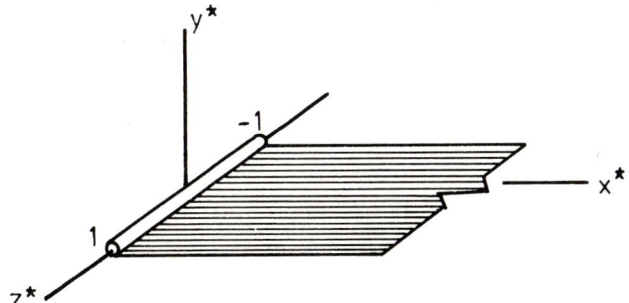

Figure 6.1.2

Bound vortex line and trailing vortex sheet

The outer expansion has the form

$$\phi(x, \tilde{y}, \tilde{z}; B) = \varphi_0(x^*, y^*, z^*) + \frac{\log B}{B}\varphi_1(x^*, y^*, z^*)$$
$$+ \frac{1}{B}\varphi_2(x^*, y^*, z^*) + \cdots .$$

(6.1.21)

The $\log(B)/B$ term must be introduced for matching, as will be seen. Substitution into (6.1.1) gives the equations governing the ϕ_i, which are valid at some distance from the wing,

$$\left. \begin{array}{l} K\varphi_{0,1x^*x^*} + \varphi_{0,1y^*y^*} + \varphi_{0,1z^*z^*} = 0, \\ K\varphi_{2y^*y^*} + \varphi_{2y^*y^*} + \varphi_{2z^*z^*} = (\gamma+1)\varphi_{0x^*}\varphi_{0x^*x^*} . \end{array} \right\}$$

(6.1.22)

and on one further change of varibles,

$$\rho = \frac{\tau}{y^*} \quad \text{for} \quad y^* > 0\,,$$

we have

$$\varphi_0 = \frac{1}{4\pi}\left\{\int_0^\infty h_1(y^*\rho)f(\rho)\,d\rho + \int_0^\infty h_2(y^*\rho)f(\rho)\,d\rho\right\}$$

where

$$h_{1,2}(y^*\rho) = \gamma_0(z^* \mp y^*\rho)H(1 \pm z^* - y^*\rho)\,,$$

$$f(\rho) = \frac{1}{1+\rho^2}\left(1 + \frac{A}{\sqrt{E+\rho^2}}\right),$$

where H is the Heaviside unit function, and $A = \dfrac{x^{*2}}{y^*}$, $E = K + \dfrac{x^{*2}}{y^{*2}}$, are fixed in the limit $y^* \to 0$.

Now the last integral has the form for which in asymptotic expansions as $y^* \to 0$ can be found by Mellin Transforms [6.1.1]. That is, if M denotes the Mellin transform, then since

$$h_j(y^*\rho) \underset{y^*\rho\to 0+}{\sim} \gamma_0(z^*) \mp y^*\rho\gamma_0'(z^*) + O\big((y^*\rho)^2\big)\,,$$

$$f(\rho) \underset{\rho\to\infty}{\sim} \frac{1}{\rho}\,,$$

we have that

$$M\big[h_j; s\big] = \int_0^{\frac{1\pm z^*}{y^*}} \rho^{s-1}\gamma_0(1 \mp y^*\rho)d\rho$$

is analytic for $\Re s > 0$ and its analytic continuation to $\Re s > -2$ is analytic with the exception of poles at the nonpositive integers, and

$$M\big[f : s\big] = \int_0^\infty \frac{\rho^{s-1}}{1+\rho^2}\left(1 + \frac{A}{\sqrt{E+\rho^2}}\right)d\rho$$

is analytic for $0 < \Re s < 2$ and its analytic continuation to $\Re s < 0$ is analytic with the exception of a pole at $s = 2$. Also,

$$M\big[hj : 1 - s\big]M\big[f : s\big] \to \infty \quad \text{as} \quad |\ln s| \to \infty\,.$$

The boundary condition at the airfoil surface is replaced by a matching condition. φ_0 is represented as the (linearized) flow due to a line distribution of lift producing singularities. According to the results familiar from low speed flow [6.3] the correct solution is that of a bound-vortex along the z^* axis which trails a vortex sheet-downstream. Considerations of matching with the inner expansion decide about the type of singularities needed in $\varphi_{1,2}$. The essential transonic nonlinear terms appear first in the outer expansion on the right hand side of the equation for φ_2. Matching is carried out here as $(x^*, y^* \to 0, z^*$ fixed) in the outer expansion, that is approaching the line of singularities, and as $x, y^* \to \infty, z^*$ fixed) in the inner expansion, that is in the far field of the airfoil flows. We find

$$\varphi_0 = \frac{y^*}{4\pi} \int_{-1}^{1} \frac{\gamma_0(\varsigma)\,d\varsigma}{y^{*2}+(z^*-\varsigma)^2} \left\{ 1 + \frac{x^*}{\sqrt{x^{*2}+Ky^{*2}+K(z^*-\varsigma)^2}} \right\} d\varsigma, \quad (6.1.23)$$

$$\varphi_1 = \frac{x^*}{4\pi r^{*2}} \frac{\partial}{\partial z^*} \int_{-1}^{1} \frac{D_1(\varsigma)\sqrt{K}(z^*-\varsigma)}{\sqrt{x^{*2}+Ky^{*2}+K(z^*-\varsigma)^2}}\,d\varsigma, \quad (6.1.24)$$

$$\varphi_2 = \frac{x^*}{4\pi r^{*2}} \frac{\partial}{\partial z^*} \int_{-1}^{1} \frac{D_2(\varsigma)\sqrt{K}(z^*-\varsigma)}{\sqrt{x^{*2}+Ky^{*2}+K(z^*-\varsigma)^2}}\,d\varsigma$$

$$+ \frac{y^*}{4\pi r^{*2}} \frac{\partial}{\partial z^*} \int_{-1}^{1} \frac{\mathcal{E}_2(\varsigma)\sqrt{K}(z^*-\varsigma)}{\sqrt{x^{*2}+Ky^{*2}+K(z^*-\varsigma)^2}}\,d\varsigma$$

$$+ \frac{y^*}{4\pi} \int_{-1}^{1} \frac{\gamma_2(\varsigma)}{y^{*2}+(z^*-\varsigma)^2} \left\{ 1 + \frac{x^*}{\sqrt{x^{*2}+Ky^{*2}+K(z^*-\varsigma)^2}} \right\} d\varsigma$$

$$+ \varphi_2^p, \quad (6.1.25)$$

where φ_2^p is a particular solution of (6.1.22) whose behavior is specified as $(x^*, y^*) \to 0$.

Expansions of the integrals in (6.1.23), (6.1.24), (6.1.25) as $r^* \to 0$ can be found by integration by parts or by Mellin transform techniques. Since these techniques will be needed again for treatment of the swept wing, they are first introduced here.

Working first with φ_0, under the change of variable

$$\tau = z^* - \varsigma,$$

we have

$$\varphi_0 = \frac{1}{4\pi} \int_{z^*-1}^{z^*+1} \frac{y^* \gamma_0(z^* - \tau)}{y^{*2}+\tau^2} \left(1 + \frac{x^*}{\sqrt{x^{*2}+Ky^{*2}+\tau^2}} \right) d\tau,$$

Thus,

$$\varphi_0 = \frac{1}{4\pi} \sum_{j=1}^{2} h_j(y^*\rho) f(\rho)\, d\rho$$

$$= \frac{1}{4\pi} \sum_{j=1}^{2} 2\pi i \int_{\nu-i\infty}^{\nu+i\infty} M[h_j : 1-s] M[[f : s]\, ds \qquad \text{for} \quad 0 < \nu < 1,$$

$$= -\frac{1}{4\pi} \sum_{j=1}^{2} \operatorname*{Residue}_{s=1,2} \left\{ M[h_j; 1-s] M[f : s] \right\} + O(y^{*2} \ln y^*) \quad \text{as} \quad y^* \to 0.$$

(See Bleistein and Handelsman [6.1.1]).

Now,

$$M[f; s] = b + O(s-1) \quad \text{as} \quad s \to 1,$$

$$= -\frac{1}{s-2} + a + O(s-2) \quad \text{as} \quad s \to 2,$$

where

$$b = \int_0^\infty \frac{1}{1+\rho^2} \left(1 + \frac{A}{\sqrt{E+\rho^2}} \right) d\rho,$$

$$a = \int_0^\infty \left\{ \frac{\rho}{1+\rho^2} \left(1 + \frac{A}{\sqrt{E+\rho^2}} \right) - \frac{1}{\rho} H(\rho-1) \right\} d\rho,$$

and

$$M[h_j; 1-s] = -\frac{\gamma_0(z^*)}{s-1} + O(1) \quad \text{as} \quad s \to 1,$$

$$= \pm y^* \frac{\gamma_0(z^*)}{s-2} + e_j + O(s-1) \quad \text{as} \quad s \to 2,$$

where

$$e_j = \int_0^{\frac{1+z^*}{y^*}} \rho^{-2} \left\{ \gamma_0(z^* \mp y^*\rho) - \gamma_0(z^*) \pm y^*\rho\gamma_0'(z^*) \right\} d\rho$$

$$\qquad - \gamma_0(z^*) \left(\frac{y^*}{1\pm z^*} \right) \mp y^*\gamma_0'(z^*) \ln \left(\frac{1\pm z^*}{y^*} \right),$$

$$= y^* \oint_{z^*-1}^{z^*+1} \frac{\gamma_0'(z^*-s)}{s}\, ds, \quad \text{by integration-by-parts.}$$

Similar analysis holds for $y^* < 0$, thus combining terms,

$$\varphi_0 = +\frac{1}{2\pi} b\gamma_0(z^*) + y^* \oint_{z^*-1}^{z^*+1} \frac{\gamma_0'(z^*-s)}{s}\, ds + O(y^{*2} \ln y^*)$$

$$= -\frac{\gamma_0(z^*)}{2\pi}\theta + y^* \oint_{-1}^{1} \frac{\gamma_0'(s)}{z^*-s} + O(y^{*2} \ln y^*).$$

The asymptotic behavior of φ_1 and $\varphi_2 - \varphi_2^p$ can be found in a similar manner. For the details of that calculation see Section 6.2.

The results are,

$$\varphi_0 = \frac{\gamma_0 z^*}{2\pi}\theta - \frac{y^*}{2\pi}\int_{-1}^{1}\frac{\gamma_0'(\xi)}{z^* - \xi}\,d\xi + \cdots, \quad \text{as} \quad r^* \to 0, \qquad (6.1.26)$$

$$\varphi_1 = \frac{x^*}{2\pi r^{*2}}D_1(z^*) + O(r^*\log r^*), \qquad (6.1.27)$$

$$\varphi_2 = \varphi_2^p + \frac{1}{2\pi\sqrt{K}}\left\{D_2\frac{x^*}{r^{*2}} + \varepsilon_2\frac{\sqrt{K}y^*}{r^{*2}}\right\} - \frac{\gamma_2 z^*}{2\pi}\theta + \cdots, \qquad (6.1.28)$$

where $\theta = \tan^{-1}\frac{\sqrt{K}y^*}{x^*} = \tan^{-1}\frac{\sqrt{K}\tilde{y}}{x}$, and $r^* = \sqrt{x^{*2} + Ky^{*2}}$. In order to find the asymptotic behavior of φ_2^p as $r^* \to 0$ note that φ_2^p is a particular solution of (6.1.22) which, using (6.1.26), takes the form

$$K\varphi_{2x^*x^*} + \varphi_{2y^*y^*} + \varphi_{2z^*z^*} = -(\gamma+1)\frac{\gamma_0^2}{2\pi^2}K\frac{x^*y^{*2}}{r^{*6}} + \cdots,$$

as $r^* \to 0$. Hence, by successive substitutions

$$\varphi_2^p = \frac{\gamma+1}{4\pi^2 K}\left(\frac{\gamma_0(z^*)}{2}\right)^2\frac{\log r^*}{r^*}\cos\theta - \frac{\gamma+1}{16\pi^2 K}\left(\frac{\gamma_0(z^*)}{2}\right)^2\frac{\cos 3\theta}{r^*}$$
$$+ O(r^*\log r^*). \qquad (6.1.29)$$

Note the term $\frac{\log r^*}{r^*}$ in φ_2^p; it is this term that necessitates the introduction of the $\frac{\log B}{B}$ term in the outer expansion for matching.

Matching

Matching can be carried out with the help of an intermediate limit in each cross-section plane z^* fixed, $|z^*| < 1$. In a class of limits intermediate to the inner and outer limits

$$r_\beta = \frac{r}{\beta(B)} \quad \text{is fixed as} \quad B \to \infty.$$

where $r = \sqrt{x^2 + K\tilde{y}^2} = Br^*$ and $1 \ll \beta \ll B$. Thus in the intermediate limit

$$r = \beta(B)r_\beta \to \infty, \quad r^* = \frac{\beta(B)}{B}r_\beta \to 0.$$

The calculations are simplified if the far-field of the inner expansion is merely written in terms of outer $(\cdot)^*$ coordinates and direct comparison of inner $(r^* \to \infty)$ and outer expansion $(r^* \to 0)$ is made.

For subsonic flow past a lifting airfoil in two dimensions the far-field expansion was worked out in Section 4.3. That is, the far-field behavior of the boundary value problem specified by the nonlinear transonic equation (6.1.11) and its boundary condition, is

$$
\phi_0 = -\frac{\Gamma_0}{2\pi}\theta + \frac{1}{4}\frac{\gamma+1}{K}\left(\frac{\Gamma_0}{2\pi}\right)^2\frac{\log r}{r}\cos\theta + \frac{1}{r}\left\{\frac{D_0}{2\pi K^{\frac{1}{2}}}\cos\theta + \frac{E_0}{2\pi K^{\frac{1}{2}}}\sin\theta\right.
$$
$$
\left. -\frac{1}{16}\frac{\gamma+1}{K}\left(\frac{\Gamma_0}{2\pi}\right)^2\cos 3\theta\right\} + O\left(\frac{\log^2 r}{r}\right), \quad r\to\infty, \quad (6.1.30)
$$

where $r = \sqrt{x^2 + K\tilde{y}^2}$, $\Gamma_0(z^*) = $ circulation at a spanwise station z^*, $D_0(z^*)$, $E_0(z^*) = $ doublet strengths at spanwise station z^*.

The behavior of $\phi_1(x,\tilde{y})$ as $r\to\infty$ is one of the main results of matching. We expect the general form

$$
\phi_1(x,\tilde{y}:z^*) = A_1(z^*)x + B_1(z^*)\tilde{y} + \frac{\Gamma_1(z^*)}{2\pi}\theta + \cdots, \quad (x,\tilde{y}\to\infty). \quad (6.1.31)
$$

The dominant terms represent a possible uniform flow at infinity and the next term in the flow is due to the induced circulation Γ_1. No source terms can appear in the far-field expansion.

Now if we write $r = Br^*$ in (6.1.30) and (6.1.31), then the expansions of (6.1.10) and (6.1.21), with (6.1.26), (6.1.27), (6.1.28) and (6.1.29), take the form:

Inner:

$$
\phi = -\frac{\Gamma_0}{2\pi}\theta + \frac{1}{4}\frac{\gamma+1}{K}\left(\frac{\Gamma_0}{2\pi}\right)^2\frac{\log B + \log r^*}{Br^*}\cos\theta
$$
$$
+\frac{1}{Br^*}\left\{\frac{D_0}{2\pi K^{\frac{1}{2}}}\cos\theta + \frac{E_0}{2\pi K^{\frac{1}{2}}}\sin\theta - \frac{1}{16}\frac{\gamma+1}{K}\left(\frac{\Gamma_0}{2\pi}\right)^2\cos 3\theta\right\}
$$
$$
+\frac{1}{B}\left\{A_1 Bx^* + B_1 By^* + \frac{\Gamma_1(z^*)}{2\pi}\theta + \cdots\right\} + \cdots. \quad (6.1.32)
$$

Outer:

$$
\phi = -\frac{\gamma_0(z^*)}{2\pi}\theta - \frac{y^*}{4\pi}\int_{-1}^{1}\frac{\gamma_0'(\xi)}{z^*-\xi}d\xi + \frac{\log B}{B}\left\{\frac{x^*}{2\pi r^{*2}}D_1(z^*)\right\}
$$
$$
+\frac{1}{B}\left\{\frac{\gamma+1}{4K}\left(\frac{\gamma_0}{2\pi}\right)^2\frac{\log r^*}{r^*}\cos\theta - \frac{\gamma+1}{16K}\left(\frac{\gamma_0}{2\pi}\right)^2\frac{\cos 3\theta}{r^*}\right.
$$
$$
\left. +\frac{1}{2\pi\sqrt{K}}D_2\frac{\cos\theta}{r^*} + \frac{1}{2\pi\sqrt{K}}\mathcal{E}_2\frac{\sin\theta}{r} - \frac{\gamma_2(z^*)}{2\pi}\theta\right\}, \quad (6.1.33)
$$
$$
x^* = r^*\cos\theta, \quad \sqrt{K}y^* = r^*\sin\theta.
$$

Matching is accomplished for all terms shown if

$$\gamma_0(z^*) = \Gamma_0(z^*), \quad 0 = A_1, \quad -\frac{1}{4\pi} \oint_{-1}^{1} \frac{\gamma_0'(\xi)}{z^* - \xi} \, d\xi = B_1(z^*),$$

$$D_1(z^*) = \frac{\gamma + 1}{8\pi K} \Gamma_0^2(z^*), \quad D_2(z^*) = D_0(z^*),$$

$$\mathcal{E}_2(z^*) = E_0(z^*), \quad \gamma_2(z^*) = \Gamma_1(z^*). \tag{6.1.34}$$

The essential matching for completion of the boundary value problem for the second inner potential ϕ_1, defines $B_1(z^*)$ in terms of the first inner circulation distribution $\Gamma_0(z^*)$. The $B_1(z^*)y$ term in the far-field of ϕ_1 corresponds to the induced downwash of the vortex sheet just as in classical lifting line theory.

As a result of this asymptotic matching complete boundary value problems have been formulated for the first and second order inner potentials ϕ_0, ϕ_1. Those results are summarized here. The problem for ϕ_0 corresponds to two-dimensional flow past an airfoil of the same shape and angle of attack as the actual wing at a given spanwise station (z^*). The correction potential corresponds to a perturbed two-dimensional flow past a flat plate with an induced downwash at infinity due to the trailing vortex sheet. There is also a special set of shock conditions to be considered for ϕ_1.

For ϕ_0 we have (cf. (6.1.11), (6.1.13), (6.1.15), (6.1.17), (6.1.18)

$$\left(K - (\gamma + 1)\phi_{0x}\right)\phi_{0xx} + \phi_{0\tilde{y}\tilde{y}} = 0, \tag{6.1.35}$$

in the $(x, \tilde{y})$ plane with boundary conditions at infinity and on the slit $x_L(z^*) < x < x_T(z^*)$,

$$\phi_{0x} \to 0 \quad \text{at infinity,} \tag{6.1.36}$$

$$\phi_{0\tilde{y}}(x, 0\pm, z^*) = \frac{\partial F_{u,\ell}}{\partial x}(x, z^*), \quad \text{(tangent flow).} \tag{6.1.37}$$

In addition the shock wave relations are integral forms of the conservation form of (6.1.35). In polar form,

$$\left[K\phi_{0x} - \frac{\gamma + 1}{2}\phi_{0x}^2\right][\phi_{0x}] + [\phi_{0\tilde{y}}]^2 = 0, \quad [\phi_0] = 0. \tag{6.1.38}$$

The shock geometry is such that

$$g_0'(\tilde{y}) = -\frac{[\phi_{0\tilde{y}}]}{[\phi_{0x}]}. \tag{6.1.39}$$

These shock jump conditions must apply locally across any shock waves that appear in supersonic zones of the solution. The shock locus $g_0(\tilde{y})$ is in general not known in advance and must be found as part of the solution. Finally, the Kutta-Joukowsky condition must be satisfied at the trailing edge:

$$\left[\phi_{0x}\right] = 0 \,.$$

Also, as a consequence, $[\phi_0]_{\text{wake}} = \text{constant} = \Gamma_0(z^*)$. The boundary value problem is indicated in Figure (6.1.3).

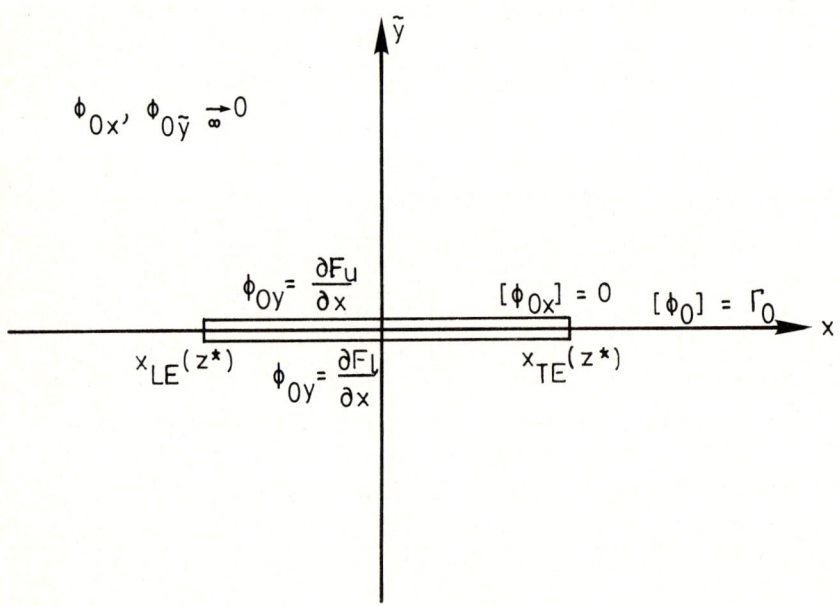

Figure 6.1.3

Boundary value problem for ϕ_0.

For a given airfoil shape it is in general not possible to solve the above boundary value problem analytically. One of the standard computational methods which has the possibility of resolving shock waves has to be used as in Section 5.4.

Now for the correction flow ϕ_1 a new type of boundary value problem must be solved, assuming that ϕ_0, g_0 are now known. The basic equation is:

$$\left(K - (\gamma + 1)\phi_{0x}\right)\phi_{1xx} - (\gamma + 1)\phi_{1x}\phi_{1xx} + \phi_{1\tilde{y}\tilde{y}} = 0 \,, \tag{6.1.40}$$

or, in conservation form

$$\left(K\phi_{1x} - (\gamma + 1)\phi_{0x}\phi_{1x}\right)_x + (\phi_{1\tilde{y}})_{\tilde{y}} = 0. \tag{6.1.41}$$

The boundary condition on the slit $x_{\text{LE}}(z^*) < x < x_{\text{TE}}(z^*)$ is that of a flat plate:

$$\phi_{1\tilde{y}}(x, 0\pm, z^*) = 0. \tag{6.1.42}$$

However, there is a downwash at infinity towards this flat plate as determined by matching in the previous section (cf. (6.1.31), (6.1.34)). We have

$$\phi_1(x, \tilde{y}; z^*) = -\frac{\tilde{y}}{2\pi} \int_{-1}^{1} \frac{\Gamma_0'(\xi)}{z^* - \xi}\, d\xi - \frac{\Gamma_1(z^*)}{2\pi}\theta + \cdots. \tag{6.1.43}$$

$\Gamma_1(z^*)$, the induced circulation, is to be found. The conditions on the trailing edge and wake are identical to those for ϕ_0:

$$\begin{aligned} \left[\phi_{1x}\right]_{\text{TE}} &= 0, \quad \text{Kutta-Joukowsky condition,} \\ \left[\phi_1\right]_{\text{wake}} &= \text{constant} = \Gamma_1(z^*). \end{aligned} \tag{6.1.44}$$

The shock conditions for this linear equation (6.1.40) require special treatment. The jumps are not merely its integrated form. The shock locus $g_0(\tilde{y})$ has been found in the solution for ϕ_0. An expansion of the exact shock jump conditions was made about the locus $g_0(\tilde{y})$ in order to account both for perturbations of the ϕ_0 flow quantities and for the shift in location of the shock wave. The results are

$$\begin{aligned} &\left[K\phi_{0x} - \frac{\gamma + 1}{2}\phi_{0x}^2\right][\phi_{1x}] + [\phi_{0x}]\left[K\phi_{1x} - (\gamma + 1)\phi_{0x}\phi_{1x}\right] + [\phi_{0\tilde{y}}][\phi_{1\tilde{y}}] \\ &= g_1(\tilde{y})\left\{[\phi_{0xx}]\left[K\phi_{0x} - \frac{\gamma + 1}{2}\phi_{0x}^2\right]\right. \\ &\quad \left. + [\phi_{0x}]\left[K\phi_{0xx} - (\gamma + 1)\phi_{0x}\phi_{0xx}\right] + 2[\phi_{0\tilde{y}}][\phi_{0x\tilde{y}}]\right\}. \tag{6.1.45} \end{aligned}$$

$$[\phi_1] = -g_1[\phi_{0x}], \tag{6.1.46}$$

where the shock locus was represented as

$$x = g_0(\tilde{y}; z^*) + \frac{1}{B}g_1(\tilde{y}; z^*) + \cdots,$$

and all the jumps in (6.1.45), (6.1.46) are to be taken on the original shock locus $s_0 : x = g_0(\tilde{y})$. An overall conservation theorem can also be derived to act as a supplementary shock condition. This condition arises by applying Green's theorem to the equation for ϕ_1, which is in divergence form. That is,

$$0 = \iint_D \hat{\nabla} \cdot (-w\phi_{1x}, K\phi_{1\hat{y}}) \, dx \, d\hat{y} = \int (-w\phi_{1x}, K\phi_{1\hat{y}}) \cdot \hat{n} \, dl \,,$$

where $\hat{y} = \sqrt{K}\tilde{y}$, $\hat{\nabla} = (\frac{\partial}{\partial x}, \frac{\partial}{\partial \hat{y}})$, $w = -K + (\gamma+1)\phi_{0x}$, and D is the region bounded by the body B, the zeroth order shock s_0, the wake W and $S_R : x^2 + \hat{y}^2 = R^2$. So,

$$0 = \int_B K\phi_{1\hat{y}} \, dx + \int_W K\phi_{1\hat{y}} \, dx + \int_S (-w\phi_{1x}, K\phi_{1\hat{y}}) \cdot n \, dl$$
$$+ \int_{SR} (-w\phi_{1x}x + K\phi_{1\hat{y}}\hat{y}) \, d\theta \,. \qquad (6.1.47)$$

The integral over the body is zero since $\phi_{1\hat{y}}|_B = 0$. The integral over the wake is zero since $[\phi_{1\hat{y}}]_w = 0$. Since

$$\phi_1 \sim \frac{-\hat{y}}{2\pi\sqrt{K}} \oint \frac{\Gamma_0'(\xi)}{z^* - \xi} \, d\xi + \Gamma_1\theta + O\left(\frac{1}{r}\right),$$
$$- w\phi_{1x}x + K\phi_{1\hat{y}}\hat{y} \sim \frac{\sqrt{K}}{2\pi}\hat{y} \oint \frac{\Gamma_0'(\xi)}{z^* - \xi} + O\left(\frac{1}{r^2}\right). \qquad (6.1.48)$$

So, the integral over S_R is $O(1/r)$. If we let $R \to \infty$, (6.1.47)

$$\int_{S_0} w\phi_{1x} \, d\hat{y} + K\phi_{1\hat{y}} \, dx = 0.$$

Or, since

$$\frac{d\hat{y}}{dx} = -\frac{[\phi_{0x}]_{s_0}}{[\phi_{1\hat{y}}]},$$

$$\int_{s_0} \left\{ [\phi_{1x}] - \frac{[\phi_{0\hat{y}}]}{[\phi_{0x}]}[\phi_{1\hat{y}}] \right\} d\hat{y} = 0. \qquad (6.1.49)$$

This condition guarantees that there are no source terms at infinity in the solution to the boundary value problem just formulated. The boundary value problem is sketched in Figure 6.1.4.

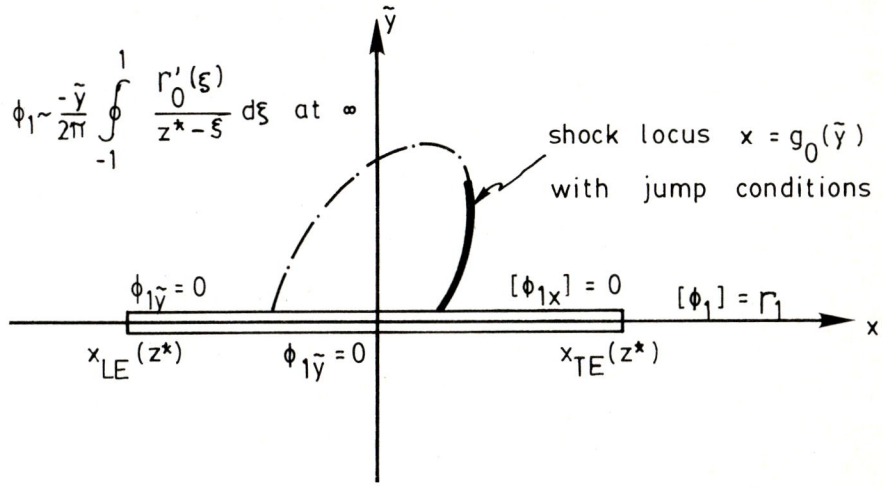

Figure 6.1.4
Boundary value problem for ϕ_1.

The ϕ_1 problem is of mixed type, as was the ϕ_0 problem, but if the ϕ_0 solution is known then the elliptic and hyperbolic regions for ϕ_1 are known in advance. Furthermore, if the flow about the airfoil represented by ϕ_0 is shock free, then the flow represented by ϕ_1 is also shock free. Uniqueness of the ϕ_1 solution, given that the ϕ_0 solution is shock free, will be shown later.

Similar Sections

The planform with similar cross-sections is especially amenable to lifting line analysis since, by suitable scaling, the problems for ϕ_0, ϕ_1 become independent of z^*. Numerical solutions for ϕ_0, ϕ_1 then need be calculated only at one z^* station.

Consider a wing surface given by 6.1 such that the planform has chord $c(z^*)$. Also assume that

$$F_{u,\ell}(x, z^*) = c(z^*) H_{u,\ell}\left(\frac{x}{c(z^*)}\right).$$

Then, if we rescale ϕ_0, x and $\tilde{y}$ by the chord so that

$$\phi_0 = c(z^*)\psi_0(X, Y), \quad X = \frac{x - \frac{1}{2}(x_L + x_T)}{c(z^*)}, \quad Y = \frac{\tilde{y}}{c(z^*)},$$

the problem for ϕ_0 becomes

$$(K - (\gamma + 1)\psi_{0X})\psi_{0XX} + \psi_{0YY} = 0,$$

$$\psi_{0Y}|_{Y=0} = H_X(X), \quad -\frac{1}{2} \le X \le \frac{1}{2},$$

$$\left[\psi_{0X}\right]_{\substack{X=\frac{1}{2} \\ Y=0}} = 0. \qquad \text{Kutta condition}$$

This problem for ψ_0 has no explicit dependence on z^*. The circulation $\Gamma_0(z^*)$ for ϕ_0 is then

$$\Gamma_0(z^*) = c(z^*)\left[\psi_0\right]_{\substack{X=\frac{1}{2} \\ Y=0}}.$$

A similar scaling works for ϕ_1. That is, the downwash

$$\oint_{-1}^{1} \left(\frac{\Gamma_0'(\xi)}{(z^* - \xi)}\right) d\xi = \left[\psi_0\right]_{\substack{X=\frac{1}{2} \\ Y=0}} \int_{-1}^{1} \left(\frac{c'(\xi)}{(z^* - \xi)}\right) d\xi$$

$$= \left[\psi_0\right]_{\substack{X=\frac{1}{2} \\ Y=0}} d(z^*).$$

Now if $\phi_1 = d(z^*)c(z^*)\psi_1(X, Y), X, Y$ as before, the scaled ϕ_1 problem from (6.1.40), (6.1.42), (6.1.43), (6.1.44) can be written

$$\left(K - (\gamma + 1)\psi_{0X}\right)\psi_{1XX} - (\gamma + 1)\psi_{0XX}\psi_{1X} + \psi_{1YY} = 0.$$

$$\psi_{1Y}|_{Y=0} = 0, \quad -\frac{1}{2} \le X \le \frac{1}{2},$$

$$\left[\psi_{1X}\right]_{\substack{X=\frac{1}{2} \\ Y=0}} = 0,$$

$$\psi_1 \sim -\frac{Y}{2}\left[\psi_0\right]_{\substack{X=\frac{1}{2} \\ Y=0}} \quad \text{as} \quad \sqrt{X^2 + Y^2} \to \infty.$$

This ψ_1 problem is z^* independent. The circulation $\Gamma_1(z^*)$ for ϕ_1 is

$$\Gamma_1(z^*) = c(z^*)d(z^*)\left[\psi_1\right]_{\substack{X=\frac{1}{2} \\ Y=0}}.$$

Note that this scaling works whenever the airfoil cross-sections are similar since the ϕ_1 problem is linear. If in addition the planform is elliptic, $(2x_{\text{LE,TE}} - 1)^2 +$

$z^{*2} = 1$, so that $c(z^*) = 2\sqrt{1 - z^{*2}}$, then the downwash for ϕ_1 is independent of z^*. That is

$$\int_{-1}^{1} \left(\frac{\Gamma_0'(\xi)}{(z^* - \xi)} \right) d\xi = \left[\psi_0 \right]_{\substack{X = \frac{1}{2} \\ Y = 0}} \int_{-1}^{1} \left(\frac{c'(z^*)}{(z^* - \xi)} \right) d\xi$$

$$= 2 \left[\psi_0 \right]_{\substack{X = \frac{1}{2} \\ Y = 0}} \int_{-1}^{1} \left(\frac{\xi}{(z^* - \xi)\sqrt{1 - \xi^2}} \right) d\xi .$$

If we let $\sqrt{1 - \xi^2} = \sin \theta$, this is

$$= - \left[\psi_0 \right]_{\substack{X = \frac{1}{2} \\ Y = 0}} \int_{0}^{2\pi} \left(\frac{\cos \theta}{(z^* - \cos \theta)} \right) d\theta$$

$$= 2\pi \left[\psi_0 \right]_{\substack{X = \frac{1}{2} \\ Y = 0}} .$$

so that $d = 2\pi$. Then the circulation $\Gamma_1(z^*)$ for ϕ_1 is

$$\Gamma_1(z^*) = c(z^*) 2\pi \left[\psi_1 \right]_{\substack{X = \frac{1}{2} \\ Y = 0}} .$$

References

[6.1.2] Bleistein, Norman and Richard A. Handelsman, *Asymptotic Expansions of Integrals*, Holt, Rinehart and Winston, N.Y., 1975.

6.2 Swept Wings $M_\infty < 1$.

The boundary value problem describing the transonic $(M_\infty < 1)$ small disturbance flow about a lifting finite aspect ratio wing skewed at a small angle $O(\sqrt{1 - M_\infty^2})$ to the z axis, is similar to that formulated in Section 3.1 for an unswept wing. The wing centerline is given by

$$x = z\delta^{\frac{1}{3}} \tan \beta, \quad -b\cos\beta < z < b\cos\beta,$$

so that the coordinate

$$\sigma = x - z\delta^{\frac{1}{3}} \tan \beta = x - \tilde{z}\tan\beta, \tag{6.2.1}$$

measures the x distance from the centerline of the wing.

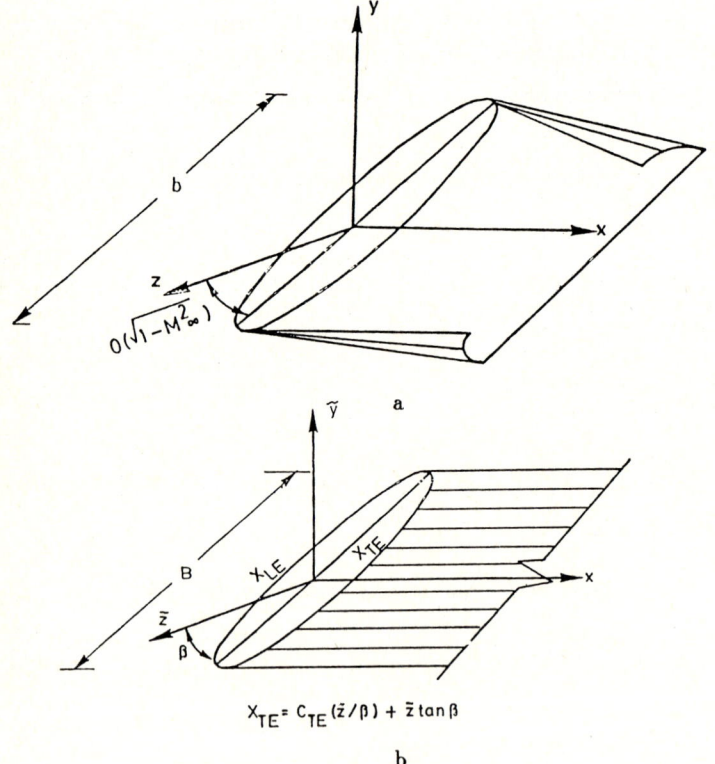

$$X_{TE} = C_{TE}(\tilde{z}/\beta) + \tilde{z}\tan\beta$$

b

Figure 6.2.1

Skewed wing in Physical coordinates (a) and in transonic coordinates (b).

Here $\tilde{z} = z\delta^{\frac{1}{3}}$, $\tilde{y} = y\delta^{\frac{1}{3}}$ and $K = \frac{1-M_\infty^2}{\delta^{\frac{2}{3}}}$ is held fixed as δ, the thickness $\to 0$, as before. That is, the sweep angle is small enough that it does not affect the basic small disturbance derivation obtained previously. (Figure 6.2.1) The wing surface is given by

$$w(x, y, z) = 0 = y - \delta F_{u,\ell}\left(x - z\delta^{\frac{1}{3}}\tan\beta, \frac{z}{b}\right),\qquad(6.2.2)$$

so that the boundary condition of tangent flow on the wing is, in small disturbance coordinates,

$$\phi_{\tilde{y}}(x, 0\pm, \tilde{z}) = \frac{\partial F_{u,\ell}}{\partial x}\left(x - \tilde{z}\tan\beta, \frac{\tilde{z}}{b}\right)\quad\text{on}\quad -B\cos\beta < \tilde{z} < B\cos\beta.\qquad(6.2.3)$$

In order to carry out the lifting-line theory for large aspect ratio wings ($B \gg 1$) the streamwise coordinate, measured from the centerline of the wing, is an important coordinate. Hence, it is convenient to write the problem in terms of the $\sigma, \tilde{y}, \tilde{z}$ coordinates instead of the $x, \tilde{y}, \tilde{z}$ coordinates. In terms of those coordinates the boundary value problem (6.1.1), (6.2.3), (6.1.3-6.1.9) becomes:

$$\left(K^* - (\gamma+1)\phi_\sigma\right)\phi_{\sigma\sigma} + \phi_{\tilde{y}\tilde{y}} - 2\tan\beta\phi_{\sigma\tilde{z}} + \phi_{\tilde{z}\tilde{z}} = 0,\qquad(6.2.4)$$

where

$$K^* = K + \tan^2\beta,\qquad(6.2.5)$$

with the conditions

tangent flow on the body,

$$\phi_{\tilde{y}}\left(\sigma, 0\pm, \frac{\tilde{z}}{B}\right) = \frac{\partial F_u}{\partial \sigma}\left(\sigma, \frac{z}{B}\right)\quad\text{for}\quad -B\cos\beta < \tilde{z} < B\cos\beta,\qquad(6.2.6)$$

the Kutta-Joukowsky condition at the trailing edge,

$$[\phi_\sigma]_{\text{TE}} = 0,\qquad(6.2.7)$$

no pressure jump across the wake,

$$[\phi_\sigma]_{\text{VS}} = 0,\qquad(6.2.8)$$

no disturbance at upstream infinity,

$$\phi_\sigma, \phi_{\tilde{y}}, \phi_{\tilde{z}} \underset{\sigma \to -\infty}{\to} 0, \tag{6.2.9}$$

and

pressure disturbances die out downstream,

$$\phi_\sigma \underset{\sigma \to \infty}{\to} 0. \tag{6.2.10}$$

Note that in $\sigma, \tilde{y}, \tilde{z}$ coordinates the body is located at

$$\sigma_{\text{LE}} = -\sigma_{\text{TE}} < \sigma < \sigma_{\text{TE}}, \quad \tilde{y} = 0, \quad -B\cos\beta < \tilde{z} < B\cos\beta,$$

and the vortex sheet is represented by

$$\sigma > \sigma_{\text{TE}}, \quad \tilde{y} = 0, \quad -B\cos\beta < \tilde{z} < B\cos\beta.$$

The problem is completed by the addition of the shock jump conditions (6.1.8a, 6.1.8b). In the $\sigma, \tilde{y}, \tilde{z}$ coordinates the shock polar is

$$\left[K^* \phi_\sigma - \frac{+1}{2}\phi_\sigma^2 \right][\phi_\sigma]_s + [\phi_{\tilde{y}}]_s^2 + [\phi_{\tilde{z}}]^2 - 2\tan\beta[\phi_{\tilde{z}}]_s[\phi_\sigma]_s = 0, \tag{6.2.11}$$

where the shock locus is given by

$$\sigma = G(\tilde{y}, \tilde{z}; B). \tag{6.2.12}$$

Note that the governing equation (6.2.4) changes types when

$$(\gamma + 1)\phi_x \equiv (\gamma + 1)\phi_\sigma = K^* \equiv K + \tan^2\beta.$$

That is, the effect of sweep is to lower the flow component perpendicular to the wing.

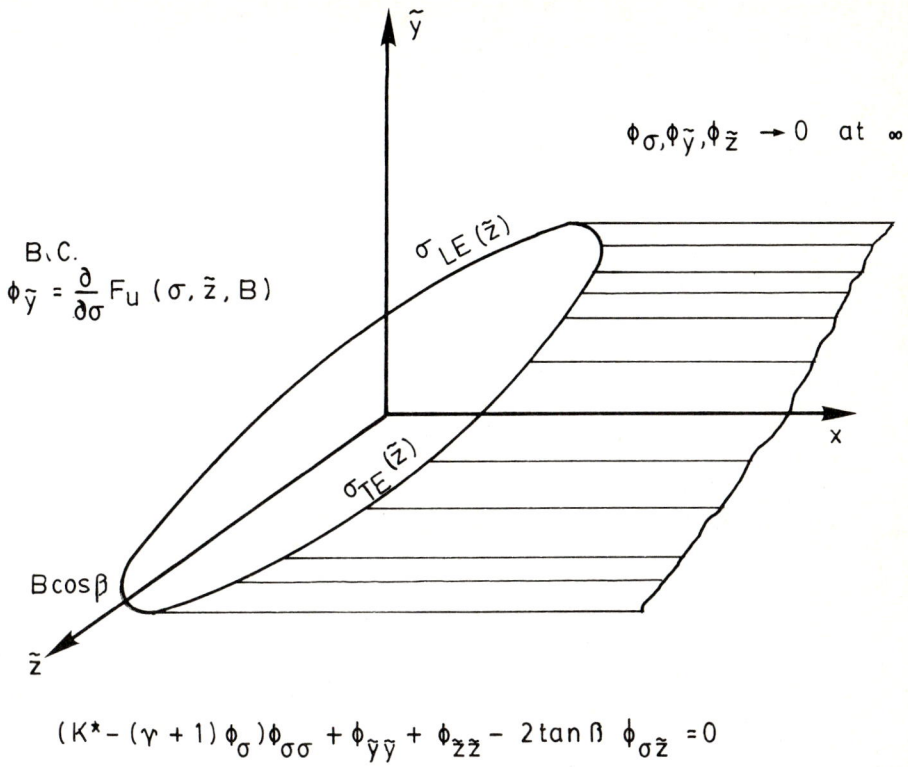

Figure 6.2.2
Boundary value problem in $\sigma, \tilde{y}, \tilde{z}$ coordinates.

In order to study the behavior of $\phi(x, y, z; K, B)$ as the aspect ratio $B \to \infty$ we consider the outer and inner (as in Section 6.1) expansions of ϕ. In the outer limit all lengths are measured relative to the span so that $x^* = \frac{x}{B}(\sigma^* = \sigma/B)$, $y^* = \tilde{y}/B$, $z^* = \frac{\tilde{z}}{B}$ are held fixed as $B \to \infty$. In this limit, as before, (Section 6.1), the wing shrinks to a line (of singularities). In the σ^*, y^*, z^* coordinates the line is along the z^* axis, but in the x^*, y^*, z^* coordinates the line of singularities is skewed at an angle β. In the inner limit all lengths are measured relative to the chord except the spanwise coordinate which, once again, is measured relative to the span. Hence in the inner limit $\sigma = x - z^* \tan \beta$, $\tilde{y}$, $z^* = \frac{\tilde{z}}{B}$ are held fixed as $B \to \infty$. In this limit the flow is essentially two dimensional in the (skewed) cross section of the wing.

Inner Expansion: ($\sigma = x - \tilde{z}\tan\beta$, $\tilde{y}$, $z^* = \frac{z}{B}$ fixed as $B \to \infty$).

The inner expansion has the form

$$\phi(x, y, z; B) = \phi_0(\sigma, y; z^*) + \frac{\log B}{B}\phi_1(\sigma, \tilde{y}z^*) + \frac{1}{B}\phi_2(\sigma, \tilde{y}; z^*) + \cdots, \quad (6.2.13)$$

The correction term ϕ_1, which is $O(\frac{\log B}{B})$, did not appear in the unswept wing case (Section 6.1) and would not normally be placed in the expansion. However, as we shall see, this term is necessary for matching with the outer flow. Substituting (6.2.13) into the boundary value problem (6.2.4)-(6.2.10), and equating cooefficients of corresponding powers of B gives part of the boundary value problems to be solved for the ϕ_i. The missing portion is the behavior of the ϕ_i at infinity, which is dictated by matching with the outer flow.

The equations governing the ϕ_i are

$$\left(K^* - (\gamma + 1)\phi_{0\sigma}\right)\phi_{\sigma\sigma} + \phi_{\tilde{y}\tilde{y}} = 0, \quad (6.2.14)$$

$$\left(K^* - (\gamma + 1)\phi_{0\sigma}\right)\phi_{1\sigma\sigma} - (\gamma + 1)\phi_{0\sigma\sigma}\phi_{1\sigma} + \phi_{1\tilde{y}\tilde{y}} = 0, \quad (6.2.15)$$

$$\left(K^* - (\gamma + 1)\phi_{0\sigma}\right)\phi_{2\sigma\sigma} - (\gamma + 1)\phi_{0\sigma\sigma}\phi_{2\sigma} + \phi_{2\tilde{y}\tilde{y}} = 2\tan\beta\phi_{0\sigma z^*}. \quad (6.2.16)$$

The tangent flow boundary conditions from (6.2.6) are:

$$\phi_{0\tilde{y}}(\sigma, 0\pm; z^*) = \frac{\partial F_{u,\ell}}{\partial\sigma}(\sigma, z^*), \quad (6.2.17)$$

$$\phi_{1\tilde{y}}(\sigma, 0\pm; z^*) = 0, \quad \phi_{2y}(\sigma, 0\pm; z^*) = 0, \quad (6.2.18)$$

on $-\cos\beta < z^* < \cos\beta$, $\sigma_{\text{LE}}(z^*) < \sigma < \sigma_{\text{TE}}(z^*)$. The Kutta-Joukowsky condition must be satisfied for each term of the inner expansion:

$$\left[\phi_{0\sigma}\right]_{\text{TE}} = 0, \quad \left[\phi_{1\sigma}\right]_{\text{TE}} = 0, \quad \left[\phi_{2\sigma}\right]_{\text{TE}} = 0, \quad (6.2.19)$$

where the trailing edge is given by $-\cos\beta < z^* < \cos\beta$, $\sigma = \sigma_{\text{TE}}(z^*)$. The conditions at infinity are obtained by matching to the near field of the outer expansion.

Note that the problem for ϕ_0 has the form of the usual two-dimensional transonic problem for the flow past a lifting airfoil. At this point the influence of sweep angle is not felt. Corrections for the finite aspect ratio appear in both ϕ_1 and ϕ_2 although more strongly in ϕ_2 since some of the ϕ_2 correction remains even in the case that the sweep angle goes to zero.

Outer Expansion ($\sigma^* = \frac{\sigma}{B} = \frac{x - \tilde{z}\tan\beta}{B}$, $y^* = \frac{\tilde{y}}{B}$, $z^* = \frac{\tilde{z}}{B} = \frac{z}{b}$, fixed as $B \to \infty$).

The outer expansion has the same basic form as for the unswept wing,

$$\phi(\sigma, y, z : B) = \varphi_0(\sigma^*, y^*, z^*) + \frac{(\log B)^2}{B}\varphi_1(\sigma^*, y^*, z^*) + \frac{\log B}{B}\varphi_2(\sigma^*, y^*, z^*)$$

$$+ \frac{1}{B}\varphi_3(\sigma^*, y^*, z^*) + \cdots . \tag{6.2.20}$$

The equations satisfied by the ϕ_i are, from substitution into (6.2.4),

$$K^* \varphi_{i\sigma^*\sigma^*} + \varphi_{iy^*y^*} + \varphi_{iz^*z^*} - 2\tan\beta\varphi_{i\sigma^*z^*} = 0, \tag{6.2.21}$$

for $i = 0, 1, 2$, and

$$K^* \varphi_{3\sigma^*\sigma^*} + \varphi_{3y^*y^*} + \phi_{3z^*z^*} - 2\tan\beta\varphi_{3\sigma^*z^*} = (\gamma+1)\varphi_{0\sigma^*}\varphi_{0\sigma^*\sigma^*}. \tag{6.2.22}$$

Note that if rewritten in terms of the x^*, y^*, z^* variables these are precisely the same equations as (6.1.21, 6.1.22). The nonlinearity first appears in the $O(1/B)$ term and the dependence on boundary conditions arises from the matching condition which is as σ^* (not x^*) $\to 0$.

For a fixed σ, $\sigma^* \to 0$ as $B \to \infty$ hence the image of the wing collapses to a line. Its effect on the outer flow can be represented in the x^*, y^*, z^* coordinates as a *skewed* line vortex shedding its vorticity downstream, plus higher order singularities. The boundary conditions on the φ_i are

$$\varphi_{i\sigma^*} \underset{r^* \to \infty}{\longrightarrow} 0,$$

$$[\varphi_{i\sigma}]_{\text{VS}} = 0,$$

where we use the generalized polar coordinates

$$r^* = (\sigma^{*2} + K^* y^{*2})^{\frac{1}{2}}, \quad \theta = \tan^{-1}\sqrt{\frac{K^* y^*}{\sigma^*}} = \tan^{-1}\left(\sqrt{\frac{K^* \tilde{y}}{\sigma}}\right).$$

Thus we have

$$\varphi_0 = \frac{1}{4\pi}\int_{-1}^{1} \frac{y^*\gamma(s)}{y^{*2} + (z^* - s\cos\beta)^2} \cdot$$

$$\left(1 + \frac{x^* - s\sin\beta}{\sqrt{(x^* - s\sin\beta)^2 + K y^{*2} + K(z^* - s\cos\beta)^2}}\right) ds$$

$$= \frac{1}{4\pi}\int_{-1}^{1} \frac{y^*\gamma(s)}{y^{*2} + (z^* - s\cos\beta)^2} \cdot$$

$$\left(1 + \frac{\sigma^* + (z^* - s\cos\beta)\tan\beta}{\sqrt{K^*(z^* - s\cos\beta)^2 + 2\sigma^*(z^* - s\cos\beta)\tan\beta + K y^{*2} + \sigma^{*2}}}\right) ds \tag{6.2.23}$$

which is the superposition of elementary horseshoe vortices distributed along $x^* = z^* \tan \beta, - \cos \beta < \cos \beta$, and trailing off parallel to the σ^* axis,

$$\varphi_1 = \frac{1}{\pi} \int_{-1}^{1} \mathcal{H}_1 \left(\frac{\sigma^* \tan \beta}{\sqrt{K^*}} + z^* \sqrt{K^*} + is \sqrt{\frac{K}{K^*} (K^* y^{*2} + \sigma^{*2})} \right) \frac{ds}{\sqrt{1 - s^2}}, \quad (6.2.24)$$

$$\varphi_2 = \frac{\sigma^*}{4\pi} \frac{K}{K^*} \int_{-1}^{1} \frac{D_2}{\left(K^*(z^* - s\cos\beta)^2 + 2\sigma^*(z^* - s\cos\beta)\tan\beta + Ky^{*2} + \sigma^{*2} \right)^{\frac{3}{2}}} \, ds$$

$$+ \frac{y^*}{4\pi} \int_{-1}^{1} \frac{\gamma_2(s)}{y^{*2} + (z^* - s\cos\beta)^2} \cdot$$

$$\left\{ 1 + \frac{\sigma^* + (z^* - s\cos\beta)\tan\beta}{\sqrt{K^*(z^* - s\cos\beta)^2 + 2\sigma^*(z^* - s\cos\beta)\tan\beta + Ky^* + \sigma^2}} \right\} ds$$

$$+ \frac{1}{4\pi} \int_{-1}^{1} \frac{\mathcal{G}_2(s)}{\sqrt{K^*(z^* - s\cos\beta)^2 + 2\sigma^*(z^* - s\cos\beta)\tan\beta + Ky^{*2} + \sigma^{*2}}} \, ds$$

$$+ \frac{1}{\pi} \int_{-1}^{1} \mathcal{H}_2 \left(z^* \sqrt{K^*} \frac{\sigma^* \tan \beta}{\sqrt{K^*}} + is \sqrt{\frac{K}{K^*} (K^* y^{*2} + \sigma^{*2})} \right) \frac{ds}{\sqrt{1 - s^2}}, \quad (6.2.25)$$

which is the potential in $\sigma^*, y^* z^*$ coordinates of a distribution of divortices along the lifting line, and

$$\varphi_3 = \frac{\sigma^*}{4\pi} \frac{K}{K^*} \int_{-1}^{1} \frac{D_3(s)}{\left(k^*(z^* - s\cos\beta)^2 + 2\sigma^*(z^* - s\cos\beta)\tan\beta + Ky^{*2} + \sigma^{*2} \right)^{\frac{3}{2}}} \, ds$$

$$+ \frac{y^*}{4\pi} \frac{K}{K^*} \int_{-1}^{1} \frac{\mathcal{E}_3}{\left(K^*(z^* - s\cos\beta)^2 + 2\sigma^*(z^* - s\cos\beta)\tan\beta + Ky^{*2} + \sigma^{*2} \right)^{\frac{3}{2}}} \, ds$$

$$+ \frac{y^*}{4\pi} \int_{-1}^{1} \frac{\gamma_3(s)}{y^{*2} + (z^* - s\cos\beta)^2} \cdot$$

$$\left(1 + \frac{\sigma^* + (z^* - s\cos\beta)\tan\beta}{\sqrt{K^*(z^* - s\cos\beta)^2 + 2\sigma^*(z^* - s\cos\beta)\tan\beta + Ky^{*2} + \sigma^{*2}}} \right) ds$$

$$+ \frac{1}{4\pi} \sqrt{K^*} \cos\beta \int_{-1}^{1} \frac{\mathcal{G}_3(s) \, ds}{\sqrt{K^*(z^* - s\cos\beta)^2 + 2\sigma^*(z^* - s\cos\beta)\tan\beta + Ky^{*2} + \sigma^{*2}}}$$

$$+ \frac{1}{\pi} \int_{-1}^{1} \mathcal{H}_3 \left(z^* + \sigma^* \frac{\tan\beta}{\sqrt{K^*}} + is \sqrt{\frac{K}{K^*} (K^* y^{*2} + \sigma^{*2})} \right) \frac{ds}{\sqrt{1 - s^2}} + \varphi_3^p, \quad (6.2.26)$$

where φ_3^p is a particular solution of (6.2.22) whose behavior is specified as $(\sigma^*, y^*) \to 0$, $\frac{\sigma^*}{y^*}$ fixed.

The expansions of the integral in (6.2.23)-(6.2.26) for $\frac{\sigma^*}{y^*}$ fixed as $r^* \to 0$ can be found using Mellin Transforms [6.2.1].

Working first with φ_0, (6.2.23), change the integration variable to τ where

$$\tau = z^* - s \cos \beta$$

to get

$$
\varphi_0 = \frac{1}{4\pi \cos \beta} \int_{z^* - \cos \beta}^{z^* + \cos \beta} \gamma \left(\frac{z^* - \tau}{\cos \beta} \right) \frac{y^*}{y^{*2} + \tau^2}
$$
$$
\times \left(1 + \frac{\sigma^* + \tau \tan \beta}{(K^* \tau^2 + 2\sigma^* \tau \tan \beta + K y^{*2} + \sigma^{*2})^{\frac{1}{2}}} \right) d\tau
$$

$$
= \frac{1}{4\pi \cos \beta} \left\{ \int_0^{\cos \beta + z^*} \gamma \left(\frac{z^* - \tau}{\cos \beta} \right) \frac{y^*}{y^{*2} + \tau^2} \right.
$$
$$
\times \left(1 + \frac{\sigma^* + \tau \tan \beta}{(K^* \tau^2 + 2\sigma^* \tau \tan \beta + K y^{*2} + \sigma^{*2})^{\frac{1}{2}}} \right) d\tau
$$
$$
+ \int_0^{\cos \beta - z^*} \gamma \left(\frac{z^* + \tau}{\cos \beta} \right) \frac{y^*}{y^{*2} + \tau^2}
$$
$$
\left. \times \left(1 + \frac{\sigma^* - \tau \tan \beta}{(K^* \tau^2 - 2\sigma^* \tau \tan \beta + K y^{*2} + \sigma^{*2})^{\frac{1}{2}}} \right) d\tau \right\}
$$

changing variables once more in order to isolate the small parts of φ_0, let $\rho = \tau/y^*$, for $y^* > 0$; then

$$
\varphi_0 = \frac{1}{4\pi \cos \beta} \left\{ \int_0^\infty h_1(y^* \rho) f_1(\rho) \, d\rho + \int_0^\infty h_2(y^* \rho) f_2(\rho) \, d\rho \right\}, \qquad (6.2.27)
$$

where $h_{\frac{1}{2}}(y^* \rho) = \dfrac{\gamma(z^* \mp y^* \rho)}{\cos \beta} H(\cos \beta \pm z^* - y^* \rho)$, H is the Heaviside unit function, and

$$
f_{\frac{1}{2}}(\rho) = \frac{1}{1 + \rho^2} \left(1 + \frac{A \pm C\rho}{(K^* \rho^2 \pm D\rho + E)^{\frac{1}{2}}} \right),
$$

where $A = \dfrac{\sigma^*}{y^*}$, $C = \tan \beta$, $D = \left(\dfrac{2\sigma^*}{y^*} \right) \tan \beta$, $E = K + \left(\dfrac{\sigma^{*2}}{y^{*2}} \right)$ are all fixed in the limit that $\sigma^*, y^* \to 0$.

Now (6.2.27) has the form for which the asymptotic expansion as $y^* \to 0$ can be found using Mellin transforms [6.2.1]. That is, if M denotes the Mellin transform, then since

$$
h_j(y^* \rho) \underset{y^* \rho \to 0+}{\sim} \gamma \left(\frac{z^*}{\cos \beta} \right) \mp y^* \rho \gamma' \left(\frac{z^*}{\cos \beta} \right) + O((y^* \rho)^2),
$$

$$f_j(\rho) \underset{\rho \to \infty}{\sim} \frac{1}{\rho^2}\left(1 \pm \frac{C}{\sqrt{K^*}}\right) + O(1/\rho^3),$$

where $(\ )'$ means $\dfrac{d(\ )}{dz^*}$, we have that

$$M[h_j : s] = \int_0^{(\cos\beta \pm z^*)/y^*} \rho^{s-1}\gamma\left(\frac{z^* \mp y^*\rho}{\cos\beta}\right)d\rho$$

is analytic for Re $s > 0$, and its analytic continuation to Re $s > -2$ is analytic with the exception of poles at the nonpositive integers, and

$$M[h_j : s] = \int_0^\infty \frac{\rho^{s-1}}{1+\rho^2}\left(1 + \frac{A \pm C_p}{(K^*\rho^2 \pm D\rho + E)^{\frac{1}{2}}}\right)d\rho$$

is analytic for $0 < \mathrm{Re}\,s < 2$, and its analytic continuation for $\mathrm{Re}\,s < 3$ is analytic with the exception of a pole at $s = 2$. Also,

$$M[h_j : 1-s]M[f_j : s] \to 0 \quad \text{as} \quad |\mathrm{Im}s| \to \infty.$$

Thus,

$$\varphi_0 = \frac{1}{4\pi\cos\beta}\sum_{j=1}^2 \int_0^\infty h_j(y^*\rho)f_j(\rho)\,d\rho$$

$$= \varphi_0 = \frac{1}{4\pi\cos\beta}\sum_{j=1}^2 2\pi i\int_{\nu-i\infty}^{\nu+i\infty} M[h_j; 1-s]M[f_j; s]\,ds$$

for $0 < \nu < 1$,

$$= -\frac{1}{4\pi\cos\beta}\sum_{j=1}^2 \operatorname*{Residue}_{s=1,2}\left(M[h_j; 1-s]M[f_j; s] + O(y^{*2}\ln y^*)\right) \qquad (6.2.28)$$

as $y^* \to 0$. The proof can be found in Bleistein and Handelsman [6.2.1].

To find the explicit terms in the expansion note that

$$M[f_j; s] =$$
$$\int_0^\infty \left\{\frac{\rho^{s-1}}{1+\rho^2}\left(1 + \frac{A \pm C\rho}{(K^*\rho^2 \pm D\rho + E)^{\frac{1}{2}}}\right) - \rho^{s-3}\left(1 \pm \frac{C}{\sqrt{K^*}}\right)H(\rho-1)\right\}d\rho$$
$$+ \int_0^\infty \rho^{s-3}\left(1 \pm \frac{C}{\sqrt{K^*}}\right)H(\rho-1)\,d\rho,$$

so that

$$M[f_j; s] = b_j + O(s-1) \quad \text{as} \quad s \to 1,$$

$$M[f_j; s] = \left(1 \pm \frac{C}{\sqrt{K^*}}\right)\frac{1}{s-2} + d_j + O(s-2) \quad \text{as} \quad s \to 2, \tag{6.2.29}$$

where

$$b_j = \int_0^\infty \frac{1}{1+\rho^2}\left(1 + \frac{A \pm C_p}{(K^*\rho^2 \pm D\rho + E)^{\frac{1}{2}}}\right) d\rho,$$

$$d_j = \int_0^\infty \frac{\rho}{1+\rho^2}\left(1 + \frac{A \pm C_p}{(K^*\rho^2 \pm D\rho + E)^{\frac{1}{2}}}\right) - \frac{1}{\rho}\left(1 \pm \frac{C}{\sqrt{K^*}}\right)H(\rho - 1)\,d\rho.$$

Similarly,

$$M[h_j; 1-s] =$$

$$\int_0^{\frac{\cos\beta \pm z^*}{v^*}} \rho^{-s}\left\{\gamma\left(\frac{z^* \mp y^*\rho}{\cos\beta}\right) - \gamma\left(\frac{z^*}{\cos\beta}\right) \pm y^*\rho\gamma'\left(\frac{z^*}{\cos\beta}\right)\right\}d\rho$$

$$+ \gamma\left(\frac{z^*}{\cos\beta}\right)\left(\frac{\cos\beta \pm z^*}{y^*}\right)^{1-s}\frac{1}{1-s}$$

$$\pm y^*\gamma'\left(\frac{z^*}{\cos\beta}\right)\left(\frac{\cos\beta \pm z^*}{y^*}\right)^{2-s}\frac{1}{2-s},$$

so that

$$M[h_j; 1-s] = -\gamma\left(\frac{z^*}{\cos\beta}\right)\frac{1}{s-1} + O(1) \quad \text{as} \quad s \to 1,$$

$$M[h_j; 1-s] = \pm y^*\gamma'\left(\frac{z^*}{\cos\beta}\right)\frac{1}{s-2} + e_j + O(s-2) \quad \text{as} \quad s \to 2, \tag{6.2.30}$$

where

$$e_j = \int_0^{\frac{\cos\beta \pm z^*}{v^*}} \rho^{-2}\left\{\gamma\left(\frac{z^* \mp y^*\rho}{\cos\beta}\right) - \gamma\left(\frac{z^*}{\cos\beta}\right) \pm y^*\rho\gamma'\left(\frac{z^*}{\cos\beta}\right)\right\}d\rho$$

$$- \gamma\left(\frac{z^*}{\cos\beta}\right)\left(\frac{y^*}{\cos\beta \pm z^*}\right) \pm y^*\gamma'\left(\frac{z^*}{\cos\beta}\right)\ln\left(\frac{\cos\beta \pm z^*}{y^*}\right). \tag{6.2.31}$$

So, from (6.2.28)-(6.2.30),

$$\varphi_0 = -\frac{1}{4\pi\cos\beta}\sum_{j=1}^{2}\left\{-b_j\gamma\left(\frac{z^*}{\cos\beta}\right) \pm D_jy^*\gamma'\left(\frac{z^*}{\cos\beta}\right) - \left(1 \pm \frac{C}{\sqrt{K^*}}\right)e_j\right\}$$

$$+ O(y^{*2}\ln y^*) \quad \text{as} \quad y^* \to 0.$$

Integrating by parts once in e_j, using the fact that $\gamma(\pm 1) = 0$, and combining terms, we get

$$\varphi_0 = -\frac{1}{4\pi \cos \beta} \left\{ -c_0 \gamma \left(\frac{z^*}{\cos \beta} \right) - \frac{2 \tan \beta}{\sqrt{K^*}} \gamma' \left(\frac{z^*}{\cos \beta} \right) y^* \ln y^* \right.$$

$$+ y^* \oint_{\frac{z^*-\cos \beta}{v^*}}^{\frac{z^*+\cos \beta}{v^*}} \frac{\gamma' \left(\frac{z^*-y^*\rho}{\cos \beta} \right)}{\rho} \, d\rho$$

$$+ \frac{y^* \tan \beta}{\sqrt{K^*}} \left(2\gamma' \left(\frac{z^*}{\cos \beta} \right) + \gamma' \left(\frac{z^*}{\cos \beta} \right) \ln(\cos^2 \beta - z^{*2}) \right.$$

$$\left. + y^* \int_{-\cos \beta}^{\cos \beta} \frac{\gamma' \left(\frac{s}{\cos \beta} \right) - \gamma' \left(\frac{z^*}{\cos \beta} \right)}{|z^* - s|} \, ds \right) + y^* c_1 \gamma' \left(\frac{z^*}{\cos \beta} \right) \right\} + O(y^{*2} \ln y^*),$$

where

$$c_0 = b_1 + b_2, \quad c_1 = d_1 - d_2.$$

These last two constants, c_0, c_1, can be calculated explicitly to give

$$c_0 = -2 \tan^{-1} \left(\frac{\sqrt{K^*} y^*}{\rho^*} \right),$$

$$c_1 = \ln \left| \frac{\tan \beta - \sqrt{K^*}}{\tan \beta + \sqrt{K^*}} \right| - \frac{\tan \beta}{\sqrt{K^*}} \left\{ \ln \left(K^* + \frac{\sigma^{*2}}{y^{*2}} \right) - \ln \frac{4K^{*2}}{K} \right\}.$$

So finally,

$$\varphi_0 = -\frac{\gamma \left(\frac{z^*}{\cos \beta} \right)}{2\pi \cos \beta} \theta + \frac{\tan \beta}{2\pi \sqrt{K^*}} - \frac{\gamma' \left(\frac{z^*}{\cos \beta} \right)}{2\pi \cos \beta} y^* \ln r^*$$

$$- \frac{y^*}{4\pi \cos \beta} \oint_{-\cos \beta}^{\cos \beta} \frac{\gamma' \left(\frac{s}{\cos \beta} \right)}{z^* - s} \, ds$$

$$- \frac{y^* \tan \beta}{2\pi \sqrt{K^*} \cos \beta} \left\{ \gamma' \left(\frac{z^*}{\cos \beta} \right) + \frac{1}{2} \int_{-\cos \beta}^{\cos \beta} \mathrm{sgn}(z^* - s) \gamma'' \left(\frac{s}{\cos \beta} \right) \ln |s - z^*| \, ds \right.$$

$$\left. + \gamma' \left(\frac{z^*}{\cos \beta} \right) \ln \frac{2K^*}{\sqrt{K^*}} \right\}$$

$$- \frac{y^* \gamma' \left(\frac{z^*}{\cos \beta} \right)}{2\pi \cos \beta} \ln \left(\frac{\sqrt{K^*} - \tan \beta}{\sqrt{K^*} + \tan \beta} \right)^{\frac{1}{2}} + O(y^* \ln y^*),$$

or more simply,

$$\varphi_0 = \frac{-\gamma\left(\dfrac{z^*}{\cos\beta}\right)\theta}{2\pi\cos\beta} + \frac{\gamma'\left(\dfrac{z^*}{\cos\beta}\right)\tan\beta}{2\pi\sqrt{K^*}\cos\beta}\, y^*\ln r^*$$

$$-\frac{y^*}{4\pi\cos\beta}\int_{-\cos\beta}^{\cos\beta}\frac{\gamma'\left(\dfrac{s}{\cos\beta}\right)}{z^*-s}\,ds - y^*J_0^c(z^*)+\cdots,\quad (6.2.32)$$

where

$$J_0^c(z^*) = \frac{\tan\beta}{2\pi\sqrt{K^*}\cos\beta}\left\{\gamma'\left(\frac{z^*}{\cos\beta}\right)\left(1+\ln\frac{2K^*}{\sqrt{K}}\right)\right.$$

$$\left. + \frac{1}{2}\int_{-\cos\beta}^{\cos\beta}\mathrm{sgn}(z^*-s)\gamma''\left(\frac{s}{\cos\beta}\right)\ln|s-z^*|\,ds\right\}$$

$$-\frac{\gamma'\left(\dfrac{z^*}{\cos\beta}\right)}{4\pi\cos\beta}\ln\left(\frac{\sqrt{K^*}-\tan\beta}{\sqrt{K^*}+\tan\beta}\right),\qquad (6.2.33)$$

The other integrals in (6.2.24)-(6.2.26) have the form

$$\psi^0 = \frac{1}{\pi}\int_{-1}^{1}\ell\left(\frac{1}{\sqrt{K^*}}\sigma^*\tan\beta + z^*\sqrt{K^*} + is\sqrt{\frac{K}{K^*}(K^*y^{*2}+\sigma^{*2})}\right)\frac{ds}{\sqrt{1-s^2}},$$

$$(6.2.34)$$

or a multiple of

$$\psi^1 = \int_{-1}^{1}\frac{\ell(s)}{\left(\sigma^{*2} + Ky^{*2} + K^*(z^*-s\cos\beta)^2 + 2\sigma^*(z^*-s\cos\beta)\tan\beta\right)^{\frac{1}{2}}}\,ds,$$

$$(6.2.35)$$

or

$$\psi^2 = \int_{-1}^{1}\frac{\ell(s)}{\left(\sigma^{*2} + Ky^{*2} + K^*(z^*-s\cos\beta)^2 + 2\sigma^*(z^*-s\cos\beta)\tan\beta\right)^{\frac{3}{2}}}\,ds.$$

$$(6.2.36)$$

It is easy to see, since ℓ is a smooth function, by Taylor series expansion about $\sigma = y^* = 0$, that

$$\phi^0 \sim \ell(\sqrt{K^*}z^*) + O(r^*).\qquad (6.2.37)$$

The other two integrals (6.2.35), (6.2.36) can be treated similarly to the integral for φ_0. With $z^* - s\cos\beta = \tau$ and then $\rho = \frac{\tau}{y^*}$, D, E as before we have

$$
\psi^1 = \frac{1}{\cos\beta}\left(\int_0^{\frac{z^*+\cos\beta}{v^*}} \frac{\ell\left(\dfrac{z^*-y^*\rho}{\cos\beta}\right)d\rho}{(K^*\rho^2 + D\rho + E)^{\frac{1}{2}}} + \int_0^{\frac{\cos\beta-z^*}{v^*}} \frac{\ell\left(\dfrac{z^*+y^*\rho}{\cos\beta}\right)d\rho}{(K^*\rho^2 - D\rho + E)^{\frac{1}{2}}}\right)
$$

$$
= \frac{1}{\cos\beta}\sum_{j=1}^{2}\int_0^\infty H_j(y^*\rho)\mathcal{F}_j^1(\rho)\,d\rho\,, \tag{6.2.38}
$$

$$
\psi^2 = \frac{y^{*-2}}{\cos\beta}\left(\int_0^{\frac{z^*+\cos\beta}{v^*}} \frac{\ell\left(\dfrac{z^*-y^*\rho}{\cos\beta}\right)d\rho}{(K^*\rho^2 + D\rho + E)^{\frac{3}{2}}} + \int_0^{\frac{\cos\beta-z^*}{v^*}} \frac{\ell\left(\dfrac{z^*-y^*\rho}{\cos\beta}\right)d\rho}{(K^*\rho^2 + D\rho + E)^{\frac{3}{2}}}\right)
$$

$$
= \frac{y^{*-2}}{\cos\beta}\sum_{j=1}^{2}\int_0^\infty H_{\frac{1}{2}}(y^*\rho)F_j(\rho)\,d\rho\,, \tag{6.2.39}
$$

where

$$
H_{\frac{1}{2}}(y^*\rho) = \ell\left(\frac{z^*\mp y^*\rho}{\cos\beta}\right)H(\cos\beta \pm z^* - y^*\rho),
$$

$$
\mathcal{F}_{\frac{1}{2}}(\rho) = (K^*\rho^2 \pm D\rho + E)^{-\frac{1}{2}},
$$

$$
F_{\frac{1}{2}}(\rho) = (K^*\rho^2 \pm D\rho + E)^{-\frac{3}{2}}.
$$

Then

$$
H_j(y^*\rho) \underset{y^*\rho\to 0+}{\sim} \ell\left(\frac{z^*}{\cos\beta}\right) + O(y^*\rho),
$$

$$
\mathcal{F}_j(\rho) \underset{\rho\to\infty}{\sim} (\sqrt{K^*}\rho)^{-1} + O(\rho^{-2}),
$$

$$
F_j(\rho) \underset{\rho\to\infty}{\sim} (\sqrt{K^*}\rho)^{-3} + O(\rho^{-4}),
$$

so that

$$
M[H_j;\, s] = \int_0^{\frac{\cos\beta\pm z^*}{v^*}} \rho^{s-1}\ell\left(\frac{z^*\mp y^*\rho}{\cos\beta}\right)d\rho
$$

is analytic for Re $s > 0$ and its analytic continuation to the negative half plane is analytic except for poles at the negative integers;

$$
M[\mathcal{F}_j;\, s] = \int_0^\infty \frac{\rho^{s-1}}{(K^*\rho^2 \pm D\rho + E)^{\frac{1}{2}}}\,d\rho
$$

is analytic for $0 < \operatorname{Re} s < 1$ and its analytic continuation to the plane has poles at the integers, and

$$M[F_j : s] = \int_0^\infty \frac{\rho^{s-1}}{(K^* \rho^2 \pm D\rho + E)^{\frac{1}{2}}} \, d\rho$$

is analytic for $0 < \operatorname{Re} s < 3$ and its analytic continuation to the plane has poles at the integers. Finally

$$M[H_j : 1 - s] M[\mathcal{F}_j; s] \underset{|\operatorname{Im} s| \to 0}{\longrightarrow} 0 \,,$$

$$M[H_j; 1 - s] M[F_j; s] \underset{|\operatorname{Im} s| \to 0}{\longrightarrow} 0 \,.$$

So,

$$\psi^1 = \frac{1}{\cos \beta} \left(\sum_{j=1}^2 \int_0^\infty H_j(y^* \rho) \mathcal{F}_j(\rho) \, d\rho \right)$$

$$= \frac{1}{\cos \beta} \sum_{j=1}^2 2\pi i \int_{\nu - i\infty}^{\nu + i\infty} M[H_j; 1 - s] M[\mathcal{F}_j; s] \, ds$$

for $0 < \nu < 1$,

$$= \frac{1}{\cos \beta} \sum_{j=1}^2 \operatorname*{Residue}_{s=1} M[H_j; 1 - s] M[\mathcal{F}_j; s] + O(y^* \ln y^*) \,, \quad (6.2.40)$$

as $y^* \to 0$, and similarly

$$\psi^2 = \frac{y^{*-2}}{\cos \beta} \sum_{j=1}^2 \operatorname*{Residue}_{s=1,2} M[H_j; 1 - s] M[F_j; s] + O(y^* \ln y^*) \,. \quad (6.2.41)$$

Since

$$M[H_j; 1 - s] = \int_0^{\frac{\cos \beta \pm z^*}{\nu^*}} \rho^{-s} \ell\left(\frac{z^* \mp y^* \rho}{\cos \beta} \right) d\rho \,,$$

then

$$M[H_j; 1 - s] = \mp \frac{\ell\left(\frac{z^*}{\cos \beta} \right)}{1 - s} + g_j + O(1) \quad \text{as} \quad s \to 1, \quad (6.2.42)$$

$$M[H_j; 1 - s] = \mp \frac{y^* \ell\left(\frac{z^*}{\cos \beta} \right)}{2 - s} + O(1) \quad \text{as} \quad s \to 2, \quad (6.2.43)$$

where

$$g_j = \int_0^{\frac{\cos \beta \pm z^*}{v^*}} \frac{1}{\rho} \left\{ \ell\left(\frac{z^* \mp y^* \rho}{\cos \beta \cdot} \right) - \ell\left(\frac{z^*}{\cos \beta} \right) \right\} d\rho$$
$$- \ell\left(\frac{z^*}{\cos \beta} \right) \ln\left(\frac{y^*}{\cos \beta \pm z^*} \right). \tag{6.2.44}$$

And,

$$M[\mathcal{F}_j; s] = \int_0^\infty \rho^{s-1} \left\{ \frac{1}{(K^* \rho^2 \pm D\rho + E)} - \frac{1}{\sqrt{K^*}\rho} H(\rho - 1) \right\} d\rho + \frac{1}{\sqrt{K^*}(1-s)}$$
$$= \frac{1}{\sqrt{K^*}(1-s)} + \lambda_j + O(1) \quad \text{as} \quad s \to 1, \tag{6.2.45}$$

$$M[F_j; s] = \frac{4\sqrt{K^* E} \mp 2D}{\sqrt{E}(4K^* E - D^2)} + O(1) \quad \text{as} \quad s \to 1, \tag{6.2.46}$$

$$= \frac{4\sqrt{K^* E} \mp 2D}{\sqrt{K^*}(4K^* E - D^2)} + O(1) \quad \text{as} \quad s \to 2, \tag{6.2.47}$$

where

$$\lambda_j = \int_0^\infty \left\{ \frac{1}{\sqrt{K^* \rho^2 \pm D\rho + E}} - \frac{1}{\sqrt{K^*}\rho} H(\rho - 1) \right\} d\rho$$
$$= \frac{1}{\sqrt{K^*}} \left\{ - \ln(2\sqrt{K^* E} \pm D) + \ln 4K^* \right\}. \tag{6.2.48}$$

So that from (6.2.40), (6.2.42), (6.2.44), (6.2.45)

$$\psi^1 \underset{y^* \to 0}{=} \frac{1}{\sqrt{K^*} \cos \beta} \left\{ f_1 + f_2 + \sqrt{K^*}\ell\left(\frac{z^*}{\cos \beta} \right) (\lambda_1 + \lambda_2) \right\} + O(y^* \ln y^*)$$
$$= \frac{1}{\sqrt{K^*} \cos \beta} \left\{ \int_{-\cos \beta}^{\cos \beta} \text{sgn}(z^* - s) \ln(z^* - s)\ell'\left(\frac{s}{\cos \beta} \right) ds - 2\ell\left(\frac{z^*}{\cos \beta} \right) \ln y^* \right.$$
$$\left. + \ell\left(\frac{z^*}{\cos \beta} \right) \left(-2 \ln r^* + 2 \ln y^* + 2 \ln \frac{K^*}{\sqrt{K}} \right) \right\} + O(y^* \ln y^*)$$
$$= \frac{1}{\sqrt{K^*} \cos \beta} \left\{ \int_{-\cos \beta}^{\cos \beta} \text{sgn}(z^* - s) \ln(s - z^*)\ell'\left(\frac{s}{\cos \beta} \right) ds \right.$$
$$\left. - 2\ell\left(\frac{z^*}{\cos \beta} \right) \ln r^* + 2\ell\left(\frac{z^*}{\cos \beta} \right) \ln \frac{K^*}{\sqrt{K}} \right\} + O(y^* \ln y^*)$$
$$= J_1^t(z^*) - \frac{2\ell\left(\dfrac{z^*}{\cos \beta} \right) \ln r^*}{\sqrt{K^*} \cos \beta} + O(y^* \ln y^*), \tag{6.2.49}$$

where

$$J_1^\ell(z^*) = \frac{1}{\sqrt{K^*}\cos\beta}\left\{\int_{-\cos\beta}^{\cos\beta}\operatorname{sgn}(z^*-s)\ln|s-z^*|\ell'\left(\frac{s}{\cos\beta}\right)ds\right.$$
$$\left. + 2\ell\left(\frac{z^*}{\cos\beta}\right)\ln\frac{K^*}{\sqrt{K}}\right\}(6.2.50)$$

and from (6.2.41), (6.2.42), (6.2.43), (6.2.46), (6.2.47)

$$\psi^2 = \frac{2\sqrt{K^*}}{K\cos\beta}\frac{1}{r^{*2}}\ell\left(\frac{z^*}{\cos\beta}\right) + \frac{\sigma\tan\beta}{K\sqrt{K^*}r^{*2}\cos\beta}\ell'\left(\frac{z^*}{\cos\beta}\right) + O(\ln y^*). \quad (6.2.51)$$

Using these results in the expressions (6.2.23)-(6.2.26) we find that the potentials behave, for $r^* \to 0$, $\frac{x^*}{y^*}$ fixed, as;

$$\varphi_0 = -\frac{\gamma_0\left(\frac{z^*}{\cos\beta}\right)}{2\pi\cos\beta}\theta + \frac{\tan\beta}{\sqrt{K^*}}\left(\frac{\gamma_0'}{2\pi\cos\beta}\right)y^*\ln r^*$$
$$- J_0(z^*)y^* + O(y^*\ln r^*), \quad (6.2.52)$$

$$\varphi_1 = \mathcal{H}_1(\sqrt{K^*}z^*) + O(r^*), \quad (6.2.53)$$

$$\varphi_2 = \frac{\sigma^* D_2\left(\frac{z^*}{\cos\beta}\right)}{2\pi\sqrt{K^*}r^{*2}\cos\beta} - \frac{\mathcal{E}_2\left(\frac{z^*}{\cos\beta}\right)}{2\pi\sqrt{K^*}\cos\beta}\ln r^* - \frac{\gamma_2\left(\frac{z^*}{\cos\beta}\right)\theta}{2\pi\cos\beta} + \frac{1}{4\pi}J_1^{\mathcal{G}_2}(z^*)$$
$$+ \mathcal{H}_2(\sqrt{K^*}z^*) + \frac{\tan\beta(\cos 2\theta + 1)D_2'\left(\frac{z^*}{\cos\beta}\right)}{4\pi(K^*)^{\frac{3}{2}}\cos\beta} + O(y^*\ln r^*), \quad (6.2.54)$$

$$\varphi_3 = \frac{\sigma^* D_3\left(\frac{z^*}{\cos\beta}\right)}{2\pi\sqrt{K^*}r^{*2}\cos\beta} + \frac{y^*\mathcal{E}_3\left(\frac{z^*}{\cos\beta}\right)}{2\pi\sqrt{K^*}r^{*2}\cos\beta} - \frac{\mathcal{G}_3\left(\frac{z^*}{\cos\beta}\right)}{2\pi\sqrt{K^*}\cos\beta}\ln r^* - \frac{\gamma_3\left(\frac{z^*}{\cos\beta}\right)\theta}{2\pi\cos\beta}$$

$$+ \frac{J_1^{\mathcal{G}_3}(z^*)}{4\pi} + \mathcal{H}_3(\sqrt{K^*}z^*) + \frac{\tan\beta(\cos 2\theta + 1)D_3'\left(\frac{z^*}{\cos\beta}\right)}{4\pi(K^*)^{\frac{3}{2}}\cos\beta}$$

$$+ \frac{\tan\beta\sin^2\theta\mathcal{E}_3'\left(\frac{z^*}{\cos\beta}\right)}{4\pi(K^*)^{\frac{3}{2}}\cos\beta} + \phi_3^p + O(y^*\ln r^*), \quad (6.2.55)$$

where

$$J_0(z^*) = \frac{1}{4\pi\cos\beta}\oint_{-\cos\beta}^{\cos\beta}\frac{\gamma_0'\left(\frac{s}{\cos\beta}\right)}{z^*-s}ds + J_0^c(z^*), \quad (6.2.56)$$

and $J_0^c(z^*)$, $J_1^\ell(z^*)$ are given in (6.2.33), (6.2.50) respectively, $(\ )'$ means $\dfrac{d}{dz^*}(\ )$, and

$$
\varphi_3^P = \frac{\gamma+1}{4K^*}\left(\frac{\gamma_0}{2\pi\cos\beta}\right)^2\left\{\frac{\log r^*}{r^*}\cos\theta - \frac{\cos 3\theta}{4r^*}\right\} + \tan\beta\frac{\gamma+1}{4K^*}\left(\left(\frac{\gamma_0}{2\pi\cos\beta}\right)^2\right)'
$$
$$
\times\left\{\frac{+(\cos 2\theta+1)}{2}\log r^* + \frac{(\log r^*)^2}{2} + \frac{\cos 2\theta}{8} - \frac{\cos 4\theta}{8}\right\}. \qquad (6.2.57)
$$

Note that although this analysis was carried out for $y^* > 0$, similar results hold for $y^* < 0$ so that (6.2.52)-(6.2.55) hold for all y^*.

Matching

The boundary value problem described by the nonlinear equation and its associated boundary conditions (6.2.17), (6.2.19), has the form of the boundary value problem for two-dimensional transonic flow past a lifting airfoil. In particular the far field is given by

$$
\phi_0 = -\frac{\Gamma_0\theta}{2\pi} + \frac{(\gamma+1)}{4K^*}\left(\frac{\Gamma_0}{2\pi}\right)^2\frac{\log r}{r}\cos\theta + \frac{1}{r}\left\{\frac{D_0}{2\pi\sqrt{K^*}}\cos\theta + \frac{E_0}{2\pi\sqrt{K^*}}\sin\theta\right.
$$
$$
\left. - \frac{1}{16}\frac{(\gamma+1)}{K^*}\left(\frac{\Gamma_0}{2\pi}\right)^2\cos 3\theta\right\} + O\left(\frac{\log r}{r^2}\right) \quad\text{as}\quad r\to\infty, \qquad (6.2.58)
$$

where $r = Br^* = \sigma^2 + K^*\tilde{y}^2$, and $\Gamma_0(z^*)$ is the circulation at the spanwise station z^*; $D_0(z^*)$ and $E_0(z^*)$ are the doublet strengths at the station z^*.

The behavior of ϕ_1, ϕ_2 as $r^* \to \infty$ is one of the main results of matching. In particular we expect that

$$
\phi_1 = A_1(z^*)\sigma + B_1(z^*)\tilde{y} + \frac{\Gamma_0}{2\pi}\frac{A_1}{4}\sin\theta - \frac{\Gamma_1}{2\pi}\theta + \cdots; \qquad (6.2.59)
$$

that is, a uniform flow, then a circulation term. We do not allow a source term in φ_1. The term with $\sin 2\theta$ arises naturally in equation (6.2.15). We also find that

$$
\phi_2 = \frac{\tan\beta}{\sqrt{K^*}}\frac{\Gamma_0'}{2\pi}\tilde{y}\log r + B_2\tilde{y} + A_2\sigma + C_2\log r - \frac{\Gamma_2\theta}{2\pi} + \frac{\gamma+1}{4}A_2\frac{\Gamma_0'}{2\pi}\sin 2\theta
$$
$$
+ \tan\beta\left\{\left(\frac{\gamma+1}{4K^*}\left(\left(\frac{\Gamma_0}{2\pi}\right)^2\right)'\right)\left(\frac{\log r(\cos 2\theta+1)}{2} + \frac{(\log r)^2}{2} + \frac{\cos 2\theta}{8} - \frac{\cos 4\theta}{8}\right)\right.
$$
$$
\left. + \frac{D_0'}{2\pi(K^*)^{\frac{5}{2}}}\frac{\cos 2\theta+1}{2} + \frac{E_0'}{2\pi(K^*)^{\frac{5}{2}}}\frac{\sin 2\theta}{2}\right\} + O\left(\frac{1}{r}\log r\right). \qquad (6.2.60)
$$

as $r \to \infty$.

The term $O(\tilde{y} \log r)$ arises from the forcing part of the equation, $\phi_{0\sigma z^*}$. In the matching this term combines with the $B_1 \tilde{y}$ term in ϕ_1. The A_2, B_2 terms represent a possible uniform flow, the C_2 term is needed to eliminate the source terms in φ_2, Γ_2 is a circulation, and the $O(1)$ terms multiplied by $\tan \beta$ arise from higher order terms in $\phi_{0\sigma z^*}$ as well as from $\phi_{0\sigma} \phi_{2\sigma\sigma}$ and $\phi_{2\sigma} \phi_{0\sigma\sigma}$.

Matching is carried out in each σ cross-section plane (z^* fixed, $|z^*| < \cos \beta$). It can be carried out through intermediate limits as for the unswept wing, or the calculations can be simplified by writing the far-field of the inner expansion in terms of the outer $(\ )^*$ coordinates and a direct comparison can then be made of the inner, $(r^* \to \infty)$ and outer $(r^* \to 0)$ expansions.

Writing $r = Br^*$ in (6.2 13), (6.2.58), (6.2.59), (6.2.60), gives for the inner expansion,

Inner:

$$
\phi = -\frac{\Gamma_0}{2\pi}\theta + \frac{\tan \beta}{\sqrt{K^*}} \frac{\Gamma_0'}{2\pi} y^* \log r^* + B_2 y^* + A_2 \sigma^*
$$

$$
+ \frac{(\log B)^2}{B} \tan \beta \frac{\gamma+1}{8K^{*2}} \left(\left(\frac{\Gamma_0}{2\pi} \right)^2 \right)'
$$

$$
+ \frac{\log B}{B} \left\{ \frac{\gamma+1}{4K^*} \left(\frac{\Gamma_0}{2\pi} \right)^2 \frac{\cos \theta}{r^*} + \tan \beta \frac{\gamma+1}{4K^{*2}} \left(\left(\frac{\Gamma_0}{2\pi} \right)^2 \right)' \log r^* \right.
$$

$$
\left. - \frac{\Gamma_1 \theta}{2\pi} + C_2 + \tan \beta \frac{\gamma+1}{4K^*} \left(\left(\frac{\Gamma_0}{2\pi} \right)^2 \right)' \frac{\cos 2\theta + 1}{2} \right\}
$$

$$
+ \frac{1}{B} \left\{ \frac{\gamma+1}{4K^*} \left(\frac{\Gamma_0}{2\pi} \right)^2 \frac{\log r^*}{r^*} \cos \theta + C_2 \log r^* - \frac{\gamma+1}{4K^*} \left(\frac{\Gamma_0}{2\pi} \right)^2 \frac{\cos 3\theta}{4r^*} \right.
$$

$$
+ \frac{D_0}{2\pi \sqrt{K^*}} \frac{\cos \theta}{r^*} + \frac{E_0}{2\pi \sqrt{K^*}} \frac{\sin \theta}{r^*} - \frac{\Gamma_2 \theta}{2\pi} + \frac{\gamma+1}{4} A_2 \frac{\Gamma_0'}{2\pi} \sin 2\theta
$$

$$
+ \tan \beta \left(\left(\frac{\gamma+1}{4K^*} \right) \left(\frac{\Gamma_0}{2\pi} \right)^2 \right)'
$$

$$
\times \left(\frac{\log r^* (\cos 2\theta + 1)}{2} + \frac{(\log r^*)^2}{2} + \frac{\cos 2\theta}{8} - \frac{\cos 4\theta}{8} \right)
$$

$$
\left. + \frac{D_0}{2\pi (K^*)^{\frac{3}{2}}} \frac{\cos 2\theta + 1}{2} + \frac{E_0'}{2\pi (K^*)^{\frac{3}{2}}} \frac{\sin 2\theta}{2} \right\} \tag{6.2.61}
$$

where we chose $B_1 = \frac{\tan \beta}{\sqrt{K^*}} \frac{\Gamma_0'}{2\pi}$ in order for the expansion to remain valid that is to eliminate the terms $O(\log B)$. Also we chose $A_1 = 0$; the justification for that will be that the expansions match. The outer expansion from (6.2.52)-(6.2.55) is

Outer:

$$
\phi = -\left(\frac{\gamma_0}{2\pi \cos \beta}\right)\theta + \tan \beta \left(\frac{\gamma_0'}{2\pi \cos \beta}\right)\frac{1}{\sqrt{K^*}} y^* \ln r^* - J_0 y^* + \frac{(\log B)^2}{B} \mathcal{H} \sqrt{K^*}\, z^*
$$

$$
+ \frac{\log B}{B}\left\{\frac{\mathcal{D}_2}{2\pi\sqrt{K^*}\cos \beta}\frac{\cos \theta}{r^*} - \left(\frac{\gamma_2}{2\pi \cos \beta}\right)\theta - \frac{\mathcal{G}_2 \ln r^*}{2\pi\sqrt{K^*}\cos \beta}\right.
$$

$$
\left. + \mathcal{H}_2 + \frac{J_1^{\mathcal{G}_2}(z^*)}{4\pi} + \frac{\mathcal{D}_2'}{4\pi(K^*)^{\frac{3}{2}}}\frac{\tan \beta}{\cos \beta}(\cos 2\theta + 1)\right\}
$$

$$
+ \frac{\mathcal{D}_3}{2\pi\sqrt{K^*}\cos \beta}\frac{\cos \theta}{r^*} + \frac{\mathcal{E}_3}{2\pi\sqrt{K^*}\cos \beta}\frac{\sin \theta}{r^*} - \frac{\gamma_3}{2\pi \cos \beta}
$$

$$
+ \tan \beta \left(\frac{\mathcal{D}_3'(\cos 2\theta + 1)}{4\pi(K^*)^{\frac{3}{2}}} + \frac{\mathcal{E}_3' \sin 2\theta}{4\pi(K^*)^{\frac{3}{2}}}\right) - \frac{\mathcal{G}_3}{2\pi\sqrt{K^*}\cos \beta}\log r^* + \mathcal{H}_3 + J_1^{\mathcal{G}_3}
$$

$$\tag{6.2.62}$$

Matching of these two expansions is accomplished to all orders shown if

$$
\frac{\gamma_0\left(\frac{z^*}{\cos \beta}\right)}{\cos \beta} = \Gamma_0(z^*), \qquad \frac{\mathcal{D}_2}{\cos \beta} = \frac{\gamma+1}{4\sqrt{K^*}}\frac{(\Gamma_0(z^*))^2}{2\pi}, \qquad \frac{\mathcal{E}_3}{\cos \beta} = E_0,
$$

$$
B_2 = -J_0(z^*), \qquad \frac{\gamma_2}{\cos \beta} = \Gamma_1(z^*), \qquad \frac{\gamma_3}{\cos \beta} = \Gamma_2(z^*), \qquad A_2 = 0, \qquad \frac{\mathcal{D}_3}{\cos \beta} = D_0,
$$

$$
\mathcal{H}_1 = \tan \beta \frac{\gamma+1}{8K^{*2}}\left(\left(\frac{\Gamma_0}{2\pi}\right)^2\right)', \qquad \mathcal{G}_2 = -2\pi\sqrt{K^*}\cos \beta \tan \beta \frac{\gamma+1}{4K^{*2}}\left(\left(\frac{\Gamma_0}{2\pi}\right)^2\right)',
$$

$$
\mathcal{H}_2 = C_2 - \frac{J_1^{\mathcal{G}_2}}{4\pi}, \qquad \mathcal{G}_3 = -2\pi\sqrt{K^*}\cos \beta C_2, \qquad \mathcal{H}_3 = -\frac{J_1^{\mathcal{G}_3}}{4\pi}
$$

$$\tag{6.2.63}$$

Through this matching process the terms $A_1(z^*)$, $A_2(z^*)$, $B_1(z^*)$ which dictate the far field behavior of ϕ_1, ϕ_2, have been determined in terms of the leading inner circulation $\Gamma_0(z^*)$. The boundary value problems determining the ϕ_i are now complete, with the addition of the shock jump conditions and the determination of C_2. Note that if $\beta = 0$, $B_1 = 0$, $B_2 = \frac{1}{4\pi}\oint \frac{\gamma_0'(s)}{z^*-s}ds$ which is precisely the result obtained in Section (6.1) for the unswept wing.

In order to write down the complete boundary value problems for ϕ_0, ϕ_1, ϕ_2, some details of the shock jump conditions must be considered. The shock jump conditions in the $\sigma, \tilde{y}, \tilde{z}$ coordinates are

$$\left[K^* \phi_\sigma - \frac{\gamma+1}{2}\phi_\sigma^2 \right][\phi_\sigma] + [\phi_{\tilde{y}}]_2^2 - 2\tan\beta[\phi_{\tilde{z}}]_s[\phi_\sigma]_s = 0, \qquad (6.2.64)$$

$$[\phi]_s = 0, \qquad (6.2.65)$$

where the shock locus is given by

$$\sigma = G\left(\tilde{y}, \frac{\tilde{z}}{B}; B\right) = -\tilde{z}\tan\beta + g\left(y, \frac{\tilde{z}}{B}; B\right). \qquad (6.2.66)$$

As $B \to \infty$, we have

$$G\left(\tilde{y}, \frac{\tilde{z}}{B}, B\right) = G_0(\tilde{y}; z^*) + \frac{\log B}{B}G_1(\tilde{y}; z^*) + \frac{1}{B}G_2(\tilde{y}; z^*) + \cdots, \qquad (6.2.67)$$

$$\phi = \phi_0(\sigma, \tilde{y}; z^*) + \frac{\log B}{B}\phi_1(\sigma, \tilde{y}; z^*) + \frac{1}{B}\phi_2(\sigma, \tilde{y}; z^*) + \cdots, \qquad (6.2.68)$$

and, as in section (6.1) for any function f with an expansion of the form (6.2.68)

$$[f]_s = [f_0]_{s_0} + \frac{\log B}{B}\left\{ G_1[f_{0\sigma}]_{s_0} + [f_1]_{s_0}\right\}$$
$$+ \frac{1}{B}\left\{ G_2[f_{0\sigma}]_{s_0} + [f_2]_{s_0}\right\} + \cdots, \qquad (6.2.69)$$

where s_0 is given by

$$\sigma - G_0(\tilde{y}; z^*) = 0.$$

Substitution of (6.2.69) with (6.2.66), (6.2.67), (6.2.68) into (6.2.64) and (6.2.65) gives

$$\left\{ \left[K^*\phi_{0\sigma} - \frac{\gamma+1}{2}\phi_{0\sigma}^2 \right]_{s_0} + \frac{\log B}{B}\left[K^*\phi_{1\sigma} - (\gamma+1)\phi_{0\sigma}\phi_{1\sigma}\right]_{s_0} \right.$$
$$+ \frac{1}{B}\left[K^*\phi_{2\sigma} - (\gamma+1)\phi_{0\sigma}\phi_{2\sigma}\right]_{s_0}$$
$$+ \left(\frac{\log B}{B}g_1 + \frac{1}{B}g_2\right)\left[K^*\phi_{0\sigma\sigma} - (\gamma+1)\phi_{0\sigma}\phi_{0\sigma}\right]_{s_0} \right\}$$
$$\times \left\{ [\phi_{0\sigma}]_{s_0} + \frac{\log B}{B}[\phi_{1\sigma}] + \frac{1}{B}[\phi_{2\sigma}]_{s_0} + \left(\frac{\log B}{B}g_1 + \frac{1}{B}g_2\right)[\phi_{0\sigma\sigma}]_{s_0} \right\}$$
$$+ [\phi_{0\tilde{y}}]_{s_0}^2 + \frac{2\log B}{B}[\phi_{0\tilde{y}}]_{s_0}[\phi_{1\tilde{y}}]_{s_0} + 2\frac{1}{B}[\phi_{2\tilde{y}}]_{s_0}[\phi_{0\tilde{y}}]_{s_0}$$
$$+ \left(\frac{\log B}{B}g_1 + \frac{1}{B}g_2\right)2[\phi_{0y}]_{s_0}[\phi_{0\sigma y}]_{s_0} - \frac{2\tan\beta}{B}[\phi_{0z^*}]_{s_0}[\phi_{0\sigma}]_{s_0}$$
$$+ O\left(\left(\frac{\log B}{B}\right)^2\right) = 0, \qquad (6.2.70)$$

and

$$[\phi_{0\sigma}]_{s_0} + \frac{\log B}{B}[\phi_1]_{s_0} + \frac{1}{B}[\phi_2]_{s_0} + \left(\frac{\log B}{B}g_1 + \frac{1}{B}g_2\right)[\phi_{0\sigma}]_{s_0}$$

$$+ O\left(\left(\frac{\log B}{B}\right)^2\right) = 0.\qquad(6.2.71)$$

Collecting terms of corresponding powers then gives the shock jump conditions for each of the ϕ_i. These are given below in the appropriate boundary value problem.

As in the unswept case, there is one other shock condition to be checked which arises from applying Greens Theorem to the equations for ϕ_1, ϕ_2 which equations are in divergence form. For ϕ_1,

$$0 = \iint_D \hat{\nabla} \cdot (-\omega^* \phi_{1s}, K^* \phi_{1\hat{y}}) \, d\sigma \, d\hat{y}$$

$$= \int_B K^* \phi_{1\hat{y}} \, d\sigma + \int_w K^* \phi_{1\hat{y}} \, d\sigma + \int_{s_0} (-\omega \phi_{1\sigma}, K^* \phi_{1\hat{y}}) \cdot \hat{n} d\ell$$

$$+ \int_{S_R} (-\omega \phi_{1\sigma}\sigma + K^* \phi_{1\hat{y}}\hat{y}) \, d\theta,$$

where $S_R : \sigma^2 + \hat{y}^2 = R^2$. The integral over the body is zero since $\phi_{1\hat{y}}|_B = 0$; the integral over the wake is zero since $[\phi_{1\hat{y}}]_w = 0$. Since

$$\phi_1 \sim B_1 \tilde{y} - \frac{\Gamma_1}{2\pi}\theta + \cdots,$$

$$\int_{S_R} (-\omega^* \phi_{1\sigma}\sigma + K^* \phi_{1\hat{y}}\hat{y}) \, d\theta \underset{R \to \infty}{=} 0.$$

So,

$$\int_{S_0} \left\{ [\omega^* \phi_{0\sigma}] - \frac{[\phi_{0\hat{y}}]}{[\phi_{0\sigma}]}[K^* \phi_{1\hat{y}}] \right\} d\hat{y} = 0.\qquad(6.2.72)$$

For ϕ_2,

$$0 = \iint_D \hat{\nabla} \cdot \left((-\omega^* \phi_{2\delta} - 2(\tan\beta)\phi_{0z^*}),\ K^* \phi_{2\hat{y}}\right) d\sigma \, d\hat{y}$$

$$= \int_B K^* \phi_{2\hat{y}} \, d\sigma + \int_w K^* \phi_{2\hat{y}} \, d\sigma + \int_{S_0} (-\omega^* \phi_{2\sigma} - 2\tan\beta \phi_{0z^*},\ K^* \phi_{2\hat{y}}) \cdot \hat{n} \, d\ell$$

$$+ \int_{S_R} (-\omega^* \phi_{2\sigma} - 2\tan\beta \phi_{0z^*})\sigma + K^* \phi_{2\hat{y}}\hat{y}) \, d\theta.$$

Again the integrals over the body and wake are zero. Now, since the asymptotic form of ϕ_2 is given by (6.2.60) we find

$$\int_{S_R} \left((-\omega^* \phi_{2\sigma} - 2\tan\beta\phi_{0z^*})\sigma + K^* \phi_{2\hat{y}}\hat{y} \right) d\theta$$

$$\underset{R\to\infty}{=} 2\pi K^* C_2 - \frac{D_0'}{\sqrt{K^*}}\tan\beta - \left(\left(\frac{\Gamma_0}{2\pi}\right)^2 \right)' \frac{(\gamma+1)}{16(K^*)^{\frac{5}{2}}}\tan\beta.$$

Thus to avoid source terms we must choose

$$C_2 = \frac{D_0'}{2\pi(K^*)^{\frac{3}{2}}}\tan\beta + \left(\left(\frac{\Gamma_0}{2\pi}\right)^2 \right)' \frac{\gamma+1}{16(K^*)^{\frac{5}{2}}}\tan\beta. \tag{6.2.73}$$

Then we see

$$\int_{S_0} \left\{ [\omega^* \phi_{2\sigma} + 2\tan\beta\phi_{0z^*}] - \frac{[\phi_{0\hat{y}}]}{[\phi_{0\sigma}]}[K^* \phi_{2\hat{y}}] \right\} d\hat{y} = 0. \tag{6.2.74}$$

The complete boundary value problems for ϕ_0, ϕ_1, ϕ_2 can now be written explicitly.

The problem for ϕ_0 is that for a two-dimensional flow past an airfoil at the same shape and angle of attack as the actual (skewed) wing at a given z^* station. It is (6.2.14), (6.2.17), (6.2.19), (6.2.58), (6.2.70), (6.2.71)

$$\left(\left(K^* - (\gamma+1)\right)\phi_{0\sigma}\right)\phi_{0\sigma\sigma} + \phi_{0\hat{y}\hat{y}} = 0,$$

with the boundary conditions

$$\phi_{0\sigma} \to 0 \quad \text{at infinity,}$$

$$\phi_{0\hat{y}}(\sigma, 0\pm, z^*) = \frac{\partial F_{u,\ell}}{\partial\sigma}(\sigma, z^*),$$

$$[\phi_{0\sigma}]_{TE} = 0,$$

and the shock conditions which are in fact integral forms of the conservation form of

$$\left[K^* \phi_{0\sigma} - \frac{\gamma+1}{2}\phi_{0\sigma}^2 \right] + [\phi_{0\hat{y}}]^2 = 0, \quad [\phi_0] = 0,$$

on the first-order shock locus

$$S_0 : \sigma = G_0(\tilde{y}).$$

The shock geometry is such that

$$G_0'(\tilde{y}) = -\frac{[\phi_{0\hat{y}}]}{[\phi_{0\sigma}]}.$$

These shock wave jump conditions apply locally across any shock waves that appear in supersonic zones of the solution. The shock locus $G_0(\tilde{y})$ is not known in advance and must be found as part of the solution. Figure 6.2.3a shows the boundary problem for ϕ_0. As in Section 6.1 this boundary value problem must be solved numerically.

The $O(\frac{\log B}{B})$ correction, ϕ_1 satisfies a linear boundary value problem, assuming that ϕ_0, G_0 are known. ϕ_1 corresponds to the potential of a perturbed two-dimensional flow past a flat plate with induced downwash at infinity. The boundary value problem is (6.2.15), (6.2.18), (6.2.19), (6.2.59), (6.2.63), (6.2.70), (6.2.71), (6.2.72),

$$\left(K^* - (\gamma+1)\phi_{0\sigma}\right)\phi_{0\sigma\sigma} - (\gamma+1)\phi_{1\sigma}\phi_{0\sigma\sigma} + \phi_{1\tilde{y}\tilde{y}} = 0,$$

with the boundary conditions

$$\phi_{1\tilde{y}}(\sigma, 0\pm, z^*) = 0, \quad \tilde{y} = 0, \quad \sigma_{\text{LE}}(z^*) < \sigma < \sigma_{\text{TE}}(z^*),$$

$$\phi_1 \underset{r\to\infty}{=} -\tilde{y}\tan\beta\frac{\Gamma_0'}{2\pi\sqrt{K^*}} - \frac{\Gamma_1'(z^*)}{2\pi}\theta + O\left(\frac{\ln r}{r}\right),$$

$$[\phi_{1\sigma}]_{\text{TE}} = 0,$$

and the shock conditions, which are not intergral forms of the basic equation,

$$\left[K^*\phi_{0\sigma} - \frac{\gamma+1}{2}\phi_{0\sigma}^2\right][\phi_{1\sigma}] + [\phi_{0\sigma}]\left[K^*\phi_{1\sigma} - (\gamma+1)\phi_0\phi_{1\sigma}\right] + 2[\phi_{0\hat{y}}][\phi_{1\hat{y}}]$$

$$= -g_1(\tilde{y})\left\{[\phi_{0\sigma}]\left[\phi_{0\sigma\sigma} - \frac{\gamma+1}{2}\phi_{0\sigma}^2\right] + [\phi_{0\sigma}]\left[K^*\phi_{0\sigma\sigma} + (\gamma+1)\phi_{0\sigma}\phi_{0\sigma\sigma}\right]\right.$$

$$\left. + 2[\phi_{0\sigma\hat{y}}]\right\}$$

$$[\phi_1] = -g_1[\phi_{0\sigma}].$$

and the conservation condition

$$\int_{S_0}\left\{[\omega^*\phi_{1\sigma}] - \frac{[\phi_{0\hat{y}}]}{[\phi_{0\sigma}]}[\phi_{2\tilde{y}}]\right\}d\tilde{y} = 0.$$

Figure (6.2.3b) illustrates the boundary value problem for ϕ_1. The circulation correction, $\Gamma_1(z^*)$ is given by $[\phi_1]_{vs}$.

The $O(1/B)$ correction, ϕ_2, is a new type of term. It corresponds to the flow past a flat plate with an induced $O(\tilde{y} \ln \tilde{y})$-type behavior at infinity. The boundary value problem for ϕ_2 is also linear and requires that ϕ_1, ϕ_0 be known. From (6.2.16), (6.2.18), (6.2.19), (6.2.60), (6.2.63), (6.2.70), (6.2.71), (6.2.73), we get

$$\left(K^* - (\gamma+1)\phi_{0\sigma}\right)\phi_{2\sigma\sigma} - (\gamma+1)\phi_{0\sigma\sigma}\phi_{2\sigma} + \phi_{2\tilde{y}\tilde{y}} = 2\tan\beta\,\phi_{0\sigma z^*}\,,$$

with

$$\phi_{2\tilde{y}}(\sigma, 0\pm, z^*) = 0 \quad \text{on} \quad y = 0, \quad \sigma_{\text{LE}}(z^*) < \sigma < \sigma_{\text{TE}}(z^*)\,,$$

$$\phi_2 \underset{r\to\infty}{\sim} \tan\beta \frac{\Gamma_0'}{2\pi\sqrt{K^*}}\tilde{y}\log r - J_0(z^*)\tilde{y} + C_2\log r + \tan\beta\,\cdot$$

$$\left\{\left(\frac{\gamma+1}{4K}\right)\left(\left(\frac{\Gamma_0}{2\pi}\right)^2\right)'\left(\frac{(\log r)(\cos 2\theta + 1)}{2} + \frac{(\log r)^2}{2} + \frac{\cos 2\theta}{8} - \frac{\cos 4\theta}{8}\right)\right.$$

$$\left. + \frac{D_0'}{4\pi(K^*)^{\frac{3}{2}}}(\cos 2\theta + 1) + \frac{E_0'}{4\pi(K^*)^{\frac{3}{2}}}\sin 2\theta\right\} - \frac{\Gamma_2(z^*)}{2\pi}\theta + O\left(\frac{\log r}{r}\right)\,,$$

where $\Gamma_2(z^*)$, the induced circulation, is to be found, and

$$J_0(z^*) = +\frac{1}{4\pi}\fint_{-\cos\beta}^{\cos\beta}\frac{\Gamma_0'(s)}{z^* - s}ds - \frac{\tan\beta}{2\pi\sqrt{K^*}}\left\{+\Gamma_0'\left(1 + \ln\left(\frac{2K^*}{\sqrt{K}}\right)\right)\right.$$

$$\left. + \frac{1}{2}\int_{-\cos\beta}^{\cos\beta}\text{sgn}(z^* - s)\gamma''\left(\frac{s}{\cos\beta}\right)\ln(s - z^*)\,ds\right\}$$

$$- \frac{\Gamma_0'}{4\pi\cos\beta}\ln\left(\frac{\sqrt{K^*} - \tan\beta}{\sqrt{K^*} + \tan\beta}\right)\,,$$

and

$$[\phi_{2\sigma}]_{\text{TE}} = 0\,,$$

$$[\phi_2]_{\text{VS}} = \Gamma_2(z^*)\,.$$

The shock conditions

$$\left[K^*\phi_{0\sigma} - \frac{(\gamma+1)}{2}\phi_{0\sigma}^2\right][\phi_{2\sigma}^2] + [\phi_{0\sigma}]\left[K^*\phi_{0\sigma\sigma} - (\gamma+1)\phi_{0\sigma}\phi_{2\sigma}\right] + 2[\phi_{0y}][\phi_{2y}]$$

$$= -g_2(\tilde{y})\left\{\left[K^*\phi_{0\sigma\sigma} - (\gamma+1)\phi_{0\sigma}\phi_{0\sigma\sigma}\right][\phi_{0\sigma}] + [\phi_{0\sigma\sigma}]\left[K^*\phi_{0\sigma} - \frac{\gamma+1}{2}\phi_0^2\right]\right.$$

$$\left. + 2[\phi_{0\tilde{y}}][\phi_{0\sigma\tilde{y}}]\right\} + 2\tan\beta[\phi_{0\sigma z^*}][\phi_{0\sigma}]\,,$$

$$[\phi_2] = -g_2[\phi_{0\sigma}],$$

and the conservation condition

$$\int_{S_0}\left\{[\omega^*\phi_{2\sigma} + 2\tan\beta\phi_{0z^*}] - \frac{[\phi_{0\hat{y}}]}{[\phi_{0\sigma}]}[K^*\phi_{2\hat{y}}]\right\}d\tilde{y} = 0.$$

All jumps are evaluated on the zeroth-order locus $\sigma = G_0(\tilde{y})$. Figure (6.2.3c) illustrates the boundary value problem for ϕ_2.

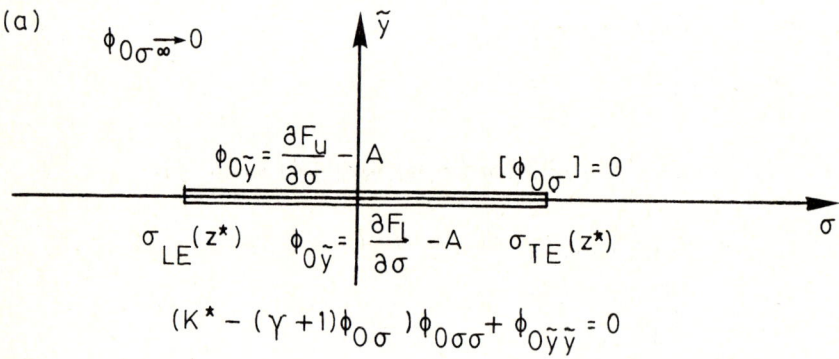

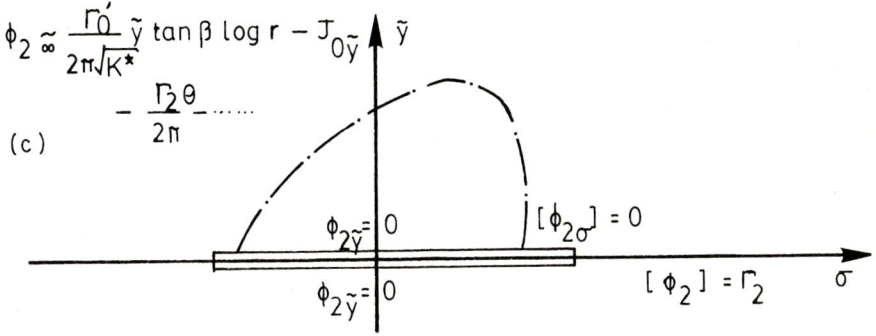

$$(K^* - (\gamma + 1)\phi_{0\sigma})\phi_{2\sigma\sigma} - (\gamma + 1)\phi_{0\sigma\sigma}\phi_{2\sigma} + \phi_{2\tilde{y}\tilde{y}} = 2\tan\beta\,\phi_{0\sigma}z^*$$

Figure 6.2.3

Boundary value problems for: (a) ϕ_0: (b) ϕ_1 and (c) ϕ_2.

Two points should be noted. The first is that in this analysis, as for the unswept wing, if the ϕ_0 problem is shock free, so are the ϕ_1, ϕ_2 problems shock free.

The second is that if the wing has similar sections in the swept coordinates, that is if the shape function is given by

$$F'_{u,\ell}(\sigma, z^*) = H_{u,\ell}\left(\frac{\sigma}{c(z^*)}\right)$$

then the problem can be rescaled to be independent of z^*. The results follow similar to those of Section 6.1. That is if we scale ϕ_0 by the chord,

$$\phi_0 = c(z^*)\psi_0(\Sigma, Y),$$

where

$$\Sigma = \frac{\sigma - \frac{(\sigma_{LE} - \sigma_{TE})}{2}}{c(z^*)}, \quad Y = \frac{\tilde{y}}{c(z^*)},$$

then the boundary value problem for ψ_0 has no explicit dependence on z^*.

We obtain,

$$\left(K^* - (\gamma + 1)\psi_{0\Sigma}\right)\psi_{0\Sigma\Sigma} + \psi_{0YY} = 0,$$

$$Y_{0Y}|_{Y=0} = G'_{e,\ell}(\Sigma), \quad -\frac{1}{2} < \Sigma < \frac{1}{2},$$

$$[\psi_{0\Sigma}]_{\substack{\Sigma=\frac{1}{2} \\ Y=0}} = 0.$$

The circulation $\Gamma_0(z^*)$ for ψ_0 is given by

$$\Gamma_0(z^*) = c(z^*)[\psi_0]_{\substack{\Sigma=\frac{1}{2} \\ Y=0}},$$

and the doublet strength D_0 is given by

$$\frac{D_0}{2\pi\sqrt{K^*}} = -\frac{\gamma+1}{4K^*}\left(\frac{\Gamma_0}{2\pi}\right)^2 \log c + c^2 \tilde{D}_0,$$

where $\tilde{D}_0$ is the corresponding doublet strength for ψ_0 and is independent of z^*.

Similarly, with

$$\phi_1 = c'(z^*)c(z^*)\psi_1,$$

the boundary value problem for ψ_1 is independent of z^*. It is,

$$\left((K^* - (\gamma+1)\psi_{0\Sigma})\psi_{1\Sigma}\right)_\Sigma + \psi_{1YY} = 0,$$

$$\psi_{1Y}|_{Y=0} = 0, \quad -\frac{1}{2} < \Sigma < \frac{1}{2},$$

$$[\psi_{1\Sigma}]_{\substack{\Sigma=\frac{1}{2} \\ Y=0}} = 0,$$

$$\psi_1 \underset{R\to\infty}{\sim} -Y \tan\beta \frac{[\psi_0]}{2\pi\sqrt{K^*}}.$$

The circulation $\Gamma_1(z^*)$ for ψ_1 is

$$\Gamma_1(z^*) = c(z^*)c'(z^*)[\psi_1]_{\substack{\Sigma=\frac{1}{2} \\ Y=0}}$$

Finally for ϕ_2 we write

$$\phi_2 = cc'\psi_2^p + J\psi_2^h + \tilde{L}$$

where the boundary value problems for ψ_2^p, ψ_2^h are,

$$(K^* - (\gamma+1)\psi_{0\Sigma})\psi_{2\Sigma\Sigma}^p - (\gamma+1)\psi_{0\Sigma\Sigma}\psi_{2\Sigma}^p + \psi_{2YY}^p = -2\tan\beta\,\Sigma\psi_{0\Sigma\Sigma}$$

$$(K^* - (\gamma+1)\psi_{0\Sigma})\psi_{2\Sigma\Sigma}^h - (\gamma+1)\psi_{0\Sigma\Sigma}\psi_{2\Sigma\Sigma}^h + \psi_{2YY}^h = 0,$$

$$\psi_{2Y}^{p,h}|_{Y=0} = 0, \qquad \left[\psi_{2\Sigma}^{p,h}\right]_{\substack{\Sigma=\frac{1}{2} \\ Y=0}} = 0,$$

and

$$\psi_2^p \underset{\infty}{\sim} \frac{\tan\beta}{2\pi\sqrt{K^*}} \frac{\Gamma_0'}{c'} Y \log R + \tilde{C}_2 \log R$$

$$+ \tan\beta \left\{ \left(\frac{\gamma+1}{\psi K^{*2}}\right) \left(\left(\frac{\Gamma_0}{2\pi}\right)^2\right)' \frac{1}{cc'} \left(\frac{\log R}{2}(\cos 2\theta + 1) + \frac{(\log R)^2}{2}\right.\right.$$

$$\left. + \frac{\cos 2\theta}{8} - \frac{\cos 4\theta}{8}\right)$$

$$\left. + \frac{\tilde{D}_0'}{2\pi(K^*)^{\frac{5}{2}}} \frac{(\cos 2\theta + 1)}{2} + \frac{\tilde{E}_0'}{2\pi(K^*)^{\frac{5}{2}}} \frac{\sin 2\theta}{2}\right\}$$

$$- \frac{\tilde{\Gamma}_2^p}{2\pi}\theta + \cdots,$$

$$\psi_2^h \underset{\infty}{\sim} Y - \frac{\tilde{\Gamma}_2^h}{2\pi}\theta + \cdots,$$

and shock conditions. Here

$$J = C\left\{-J_0(z^*) + \frac{\tan\beta}{\sqrt{K^*}}\frac{\Gamma_0'}{2\pi}\ln c\right\},$$

$$\tilde{C}_2 = \frac{\left\{C_2 + \frac{\gamma+1}{4\pi}\left(\left(\frac{\Gamma_0}{2\pi}\right)^2\right)' \tan\beta \log c\right\}}{cc'},$$

$$\frac{\tilde{D}_0'}{2\pi\sqrt{K^*}} = \frac{\left\{\frac{D_0'}{2\pi\sqrt{K^*}} + \frac{\gamma+1}{4K^*}\left(\left(\frac{\Gamma_0}{2\pi}\right)^2\right)' \log c\right\}}{cc'},$$

$$\mathcal{E}_0' = \frac{E_0}{cc'}.$$

Note that

$$\tilde{L} = C_2 \log c + \frac{\gamma+1}{8K^*}\left(\left(\frac{\Gamma_0}{2\pi}\right)^2\right)' \tan\beta(\log c)^2,$$

and note also that $\tilde{D}_0$, and hence $\tilde{C}_2$ are independent of z^*. One must also check the shock condition for the similar section pieces. With

$$g_2(\tilde{y}) = cc's^p + Js^h$$

we have

$$\left[\psi_2^p\right]_{S_0} = -s^p \left[\psi_{0\Sigma}\right]_{S_0},$$

$$\left[\psi_2^h\right]_{S_0} = -s^h \left[\psi_{0\Sigma}\right]_{S_0},$$

and

$$\left[K^* \psi_{2\Sigma}^{p,h} - \frac{\gamma+1}{2}\psi_{0\Sigma}^2\right]\left[\psi_{2\Sigma}^{p,h}\right] + \left[\psi_{0\Sigma}\right]\left[K^* \psi_{2\Sigma}^{p,h} - (\gamma+1)\psi_{0\Sigma}\psi_{2\Sigma}^{p,h}\right]$$

$$+ 2\left[\psi_{0\Sigma}\right]\left[\psi_{2\Sigma}^{p,h}\right] = -s^{p,h}\left\{\left[K^* \psi_{0\Sigma\Sigma} - (\gamma+1)\psi_{0\Sigma}\psi_{0\Sigma\Sigma}\right]\left[\psi_{0\Sigma}\right]\right.$$

$$+ \left[\psi_{0\Sigma\Sigma}\right]\left[K^* \psi_{0\Sigma} - \frac{\gamma+1}{2}\psi_{0\Sigma}^2\right] + 2\left[\psi_{2Y}\right]\left[\psi_{0\Sigma Y}\right]\right\}$$

$$+ \begin{cases} 2\tan\beta\left[\Sigma\psi_{0\Sigma}\right]\left[\psi_{0\Sigma}\right] & \text{for } \psi_2^p \\ 0 & \text{for } \psi_2^h \end{cases}$$

This circulation for ϕ_2 is then given by

$$\Gamma_2 = cc'\tilde{\Gamma}_2^p + J\tilde{\Gamma}_2^h.$$

This similarity form was first recognized by Cheng and Meng [6.2.2]. In that paper they did not compute the $\tilde{L}$ term which, in fact, is of minor importance since it does not contribute to the lift.

References

[6.2.1] Bleistein, Norman and Richard A. Handelsman, *Asymptotic Expansions of Integrals*, Holt, Rinehart and Winston, N.Y., 1975.

[6.2.2] Cheng, H. K. and S. Y. Meng, The Oblique Wing as a Lifting-Line Problem in Transonic Flow, U. of Southern Ca. Aerospace Rpt. #136, May 1979.

6.3 Unswept Wings, $M_\infty = 1$.

The order of the first three-dimensional correction to the two-dimensional cross-section flow about a lifting wing at exactly sonic speed, $M_\infty = 1$, differs from that for slightly subsonic speed, $M_\infty < 1$ [6.3.1] [6.3.2]. This is due primarily to the difference in the far field behavior of the two-dimensional $M_\infty = 1$ flow from the $M_\infty < 1$ flow. (Sections 4.1, 4.3). The $M_\infty < 1$ far field expansion is dominated by an $O(1)$ circulation term, whereas the $M_\infty = 1$ far field expansion is dominated by an $O(\tilde{y}^{\frac{2}{5}})$ symmetric flow.

The equation governing the transonic small disturbance flow at $M_\infty = 1$ is (6.1.1) with $K = 0$,

$$(\gamma + 1)\bar{\phi}_x\bar{\phi}_{xx} + \bar{\phi}_{\tilde{y}\tilde{y}} + \bar{\phi}_{\tilde{z}\tilde{z}} = 0\,. \tag{6.3.1a}$$

Here, as usual, the full potential has been expanded as

$$\Phi(x, y, z; b) = a_\infty\left(x + \delta^{\frac{2}{5}}\bar{\phi}(x, \tilde{y}\tilde{z}; B) + \cdots\right)$$

where $\tilde{y}, \tilde{z}, B$ are as in Section 3.1, and the original wing shape was

$$W(x, y, z) = y - \delta F_{u,\ell}\left(x, \frac{z}{b}\right)\,.$$

The equations are simplified if we write

$$\bar{\phi} = (\gamma + 1)\phi\,,$$

so that (6.3.1a) becomes

$$-\phi_x\phi_{xx} + \phi_{\tilde{y}\tilde{y}} + \phi_{\tilde{z}\tilde{z}} = 0\,. \tag{6.3.1b}$$

The boundary conditions for ϕ are:

(i) no disturbance at upstream infinity,

$$\phi_x, \ \phi_{\tilde{y}}, \ \phi_{\tilde{z}} \underset{x \to -\infty}{\longrightarrow} 0\,; \tag{6.3.2}$$

(ii) the flow is tangent to the body,

$$\phi_{\tilde{y}}(x, 0\pm, \tilde{z}) = (\gamma + 1)\frac{\partial}{\partial x}F_{u,\ell}\left(x, \frac{\tilde{z}}{B}\right) \quad \text{for} \quad x_{\text{LE}}\left(\frac{\tilde{z}}{B}\right) < x < x_{\text{TE}}\left(\frac{\tilde{z}}{B}\right)\,. \tag{6.3.3}$$

This set, (6.3.1b), (6.3.2), (6.3.3), constitute a complete boundary value problem for the determination of the front portion of the flow. That is, at exactly sonic speeds the sonic surface extends from the body to infinity. Equation (6.3.1b) remains of mixed type even far from the airfoil. This is unlike the $M_\infty < 1$ flow governed by equation (6.1.1), which is completely subsonic far from the airfoil. Also, at exactly sonic speeds, there is a last characteristic surface, the limiting Mach surface, which extends from the body to infinity and is tangent to the sonic surface at infinity. All characteristic surfaces in front of the limiting Mach surface reflect off the sonic line. All characteristic surfaces which start behind the limiting Mach surface stay behind the limiting Mach surface (Figure 6.3.1). Thus the flow in front of the limiting Mach surface is not influenced by the flow behind the surface and hence the front portion of the flow can be computed independently of the rear portion of the flow. Later, using the front flow, shock jump conditions, and conditions downstream, the rear portion of the flow can be computed. Thus we begin by discussing the three-dimensional correction to the front portion of the flow.

The dependence of $\phi(x, \tilde{y}, \tilde{z}; B)$ on the aspect ratio parameter $B \gg 1$ is obtained by considering the distinguished limits of (6.3.1b) as $B \to \infty$. In the inner limit $x, \tilde{y}, z^* = \tilde{z}/B$ are held fixed as $B \to \infty$, and the problem becomes essentially two-dimensional. In order to determine the stretchings for the outer limit note that all terms of the equation must be retained. Thus $y^* = \tilde{y}/B$, in order that the y^* derivative terms balance the z^* terms. Then, with

$$\phi = \mu(B), \tag{6.3.4}$$

$$x^* = \alpha(B)x, \quad \alpha \ll 1, \tag{6.3.5}$$

equation (6.1.1b) becomes

$$-\mu(B)\alpha^3 \varphi_{x^*}\varphi_{x^*x^*} + \frac{1}{B^2}\varphi_{y^*y^*} + \frac{1}{B^2}\varphi_{z^*z^*} = 0,$$

and the planform shrinks to a line of singularities. Hence, to balance terms,

$$\mu\alpha^3 = \frac{1}{B^2}. \tag{6.3.6}$$

Finally, note that the far field of the (inner) two-dimensional flow is expressed in terms of the similarity parameter

$$\xi = \frac{x}{\tilde{y}^{\frac{4}{5}}} = \frac{x^*}{\alpha(B)y^{*\frac{4}{5}}B^{\frac{4}{5}}} = \frac{x^*}{y^{*\frac{4}{5}}},$$

if

$$\alpha(B) = B^{-\frac{4}{5}} \, . \tag{6.3.7}$$

Then, from (6.3.6)

$$\mu(B) = B^{\frac{2}{5}} \, . \tag{6.3.8}$$

Inner Expansion: $(x, \tilde{y}, z^* = \frac{\tilde{z}}{B}$ fixed, $B \to \infty)$.

One expects the inner expansion to have the form

$$\phi(x, \tilde{y}, \tilde{z}; B) = \phi_0(x, \tilde{y}, z^*) + \nu_1(B)\phi_1(x, \tilde{y}, z^*) + \cdots ,$$

where $\nu_1 \ll 1$, is determined by matching with the outer expansion. In fact for matching purposes it turns out that

$$\phi(x, \tilde{y}z^*, B) = \left(B^{\frac{2}{5}}C_0(z^*)\right) + \phi_0 + \nu_1(B)\phi_1 + \cdots . \tag{6.3.9}$$

This $O(B^{\frac{2}{5}})$ spanwise induced flow is constant with respect to the $x, \tilde{y}$ variations and does not effect the pressure $\hat{c}_p = -2\phi_x$. Furthermore it does not enter as a coupling term in the successive approximate solutions to equation (6.3.1) until $O(B^{-\frac{2}{5}})$.

Substituting (6.3.9) into (6.3.1b), (6.3.2) and (6.3.3), as written in $x, \tilde{y}, z^*$ variables, and equating coefficients of corresponding powers of B shows that ϕ_0, ϕ_1 satisfy the equations

$$-\phi_{0x}\phi_{0xx} + \phi_{0\tilde{y}\tilde{y}} = 0 , \tag{6.3.10}$$

$$-(\phi_{0x}\phi_{1x})_x + \phi_{1\tilde{y}\tilde{y}} = \begin{cases} 0 & \text{if } \nu_1(B) \ll \dfrac{1}{B^2} , \\[2mm] -\phi_{0z^*z^*} & \text{if } \nu_1(B) = \dfrac{1}{B^2} , \end{cases} \tag{6.3.11}$$

and the boundary conditions,

$$\phi_{0\tilde{y}}(x, 0\pm, z^*) = \frac{\partial F_{u,t}}{\partial x}(x, z^*)$$

$$\phi_{1\tilde{y}}(x, 0\pm, z^*) = 0 , \tag{6.3.12}$$

for $x_{LE}(z^*) < x < x_{TE}(z^*)$, $|z^*| < 1$. The boundary conditions at infinity arise from matching to the outer expansion.

Note that the ϕ_0 problem is that for the two-dimensional flow past a lifting airfoil (at a given spanwise station) at sonic speed. The far field expansion for ϕ_0 is

$$\phi_0 \underset{\substack{\tilde{y} \to \infty \\ \xi \text{ fixed}}}{\sim} \frac{\tilde{y}^{\frac{2}{5}}}{a^3} f(a\xi) + c_0 + \tilde{y}^{-\frac{1}{5}} \frac{c_1}{a^3} f_1(a\xi) + \cdots , \tag{6.3.14}$$

where

$$\xi = \frac{x}{\tilde{y}^{\frac{4}{5}}}, \tag{6.3.15}$$

and a, c_0, c_1 are functions af z^*. The properties of f, f_1, as well as the origin of the similarity coordinate ξ, and the parameters a, c, were given in Section 4.1..

The far field expansion for ϕ_1 is the main result of this matching. Assuming that

$$\phi_1 \underset{\tilde{y}\to\infty}{\sim} y^\alpha \frac{g(a\xi)}{a^3},$$

one finds that g satisfies the equation

$$\left(f' - \frac{16}{25}a^2\xi^2\right)g'' + \left(f'' + \frac{4}{5}\left(2\alpha - \frac{9}{5}\right)a\xi\right)g' - \alpha(\alpha - 1)g = 0.$$

The solutions to this equation for which $y^\alpha g$ is smooth from $\xi = 0$ (the negative x axis) through the limit Mach line $\xi_L = \frac{1}{a(z^*)}$ were found in Section 4.1. The α's for which such solutions exist are given in (4.1.101),

$$\alpha = -\frac{2}{5}n, \quad \frac{2 + 6n}{5}, \quad \frac{-2n - 1}{5}, \quad 1 + \frac{6}{5}n.$$

One would expect that $\alpha = 1$, the next (after $2/5$) most singular solution near infinity. In fact, matching will show that this solution is missed and $\alpha = 8/5$. Thus,

$$\phi_1 \underset{\substack{\tilde{y}\to\infty \\ \xi \text{ fixed}}}{\sim} \frac{d_0 \tilde{y}^{\frac{8}{5}} h_0(a\xi)}{a^3} + \frac{d_0 c_1 \tilde{y} h_1(a\xi)}{a^3} + \cdots, \tag{6.3.16}$$

where d_0 is a function of z^* to be determined by matching,

$$h_0(s) = s^{-\frac{4}{5}}(1 - 6s), \tag{6.3.17}$$

c_1 comes from (4.1.120), and h_1 satisfies the forced equation,

$$\left(f' - \frac{16}{25}a^2\xi^2\right)h_1'' + \left(f'' + \frac{4}{25}a\xi\right)h_1' = (f_1'h_0').$$

In terms of the s variable h_1 can be found explicitly,

$$h_1 = \frac{\sqrt{3a_1}\, s^{-\frac{3}{5}}}{3s + 1}. \tag{6.3.18}$$

Outer Expansion: $(x^* = \dfrac{x}{B^{\frac{4}{5}}}, \; y^* = \dfrac{\tilde{y}}{B}, \; z^*$ fixed; $B \to \infty)$.

The outer expansion has the form

$$\phi(x, \tilde{y}, \tilde{z} : B) = B^{\frac{2}{5}} \varphi_0(x^*, y^*, z^*) + \mu_1(B)\varphi_1 + \mu_2(B)\varphi_2 + \cdots, \qquad (6.3.19)$$

from (6.3.4), (6.3.8) where $\mu_2 \ll \mu_1 \ll B^{\frac{2}{5}}$ as $B \to \infty$. The equations satisfied by $\varphi_0, \varphi_1, \varphi_2$ are:

$$-\varphi_{0x^*}\varphi_{0x^*x^*} + \varphi_{0y^*}\varphi_{0y^*y^*} + \varphi_{0z^*}\varphi_{0z^*z^*} = 0, \qquad (6.3.20)$$

$$-(\varphi_{0x^*}\varphi_{1x^*})_{x^*} + \varphi_{1y^*y^*} + \varphi_{1z^*z^*} = 0, \qquad (6.3.21)$$

$$-(\varphi_{0x^*}\varphi_{2x^*})_{x^*} + \varphi_{2y^*y^*} + \varphi_{2z^*z^*} = \begin{cases} 0 & \text{if } \mu_1^2 B^{-\frac{2}{5}} \ll \mu_2, \\ \phi_{1x^*}\phi_{1x^*x^*} & \text{if } \mu_1^2 B^{-\frac{2}{5}} = \mu_2. \end{cases} \qquad (6.3.22)$$

The φ_0 equation is the full nonlinear equation, the φ_1 equation is the linear (variational) equation as is the φ_2 equation which may also hav a forcing term. The boundary conditions at infinity for the φ_i are

$$\varphi_{1x^*}, \; \varphi_{iy^*}, \; \varphi_{iz^*} \underset{x^* \to -\infty}{\longrightarrow} 0. \qquad (6.3.23)$$

The boundary condition near the airfoil surface is replaced by the requirement that the outer and inner expansions match. Note that if (6.3.14) as written in outer variables is to match up with (6.3.19), then necessarily $\mu = 1$ or $\mu \gg 1$ and $\mu_2 = B^{-\frac{1}{5}}$. At any rate the equation governing φ_2 (6.3.22) must be homogeneous.

Now matching dictates that the near field of φ_0 must have the form

$$\varphi_0 \sim y^{*\frac{2}{5}} \frac{f(a\xi)}{a^3} + y^{*\lambda_1} \frac{C_1(z^*)F_1(a\xi)}{a^3} + \cdots. \qquad (6.3.24)$$

Substitution of this expansion into (6.3.20) shows that three dimensional forcing terms (z^* derivarives) do not arise in this expansion until $O(y^{*\frac{12}{5}})$. Thus, by (4.1.101) the possibilities for λ_1 are $1, 8/5, \cdots$. If the solution (6.3.20), (6.3.21) (in fact an axisymmetric far field), and $\varphi_0 \sim y^{*\frac{2}{5}} \dfrac{f(a\xi)}{a^3}$, is unique, then the solution φ_0 is even in y hence $\lambda_1 = \frac{8}{5}$. There is also the possibility of a term $C_0(z^*)$ preceeding the $y^{*\frac{2}{5}}$ term in (6.3.24). Whether or not this purely sidewash term is present must be decided by examining the φ_0 problem more closely. If it is present then for matching a term $B^{\frac{2}{5}}C_0(z^*)$ must be present in the inner expansion (6.3.9). Note that higher order (singular) solutions are not allowable since they are not trivial solutions of (6.3.20) and will affect the other terms so that matching can not be achieved. Hence,

$$\varphi_0 \underset{y^{\bullet} \to 0}{\sim} C_0(z^{\bullet}) + y^{\bullet \frac{2}{5}} \frac{f(a\xi)}{a^3} + y^{\bullet \frac{8}{5}} \frac{C_1(z^{\bullet})h_0(a\xi)}{a^3} + \cdots . \qquad (6.3.25)$$

Similar reasoning, and the fact that we expect φ_1 to be more singular than φ_0 as $y^{\bullet} \to \infty$, leads to

$$\varphi_1 \underset{y^{\bullet} \to 0}{\sim} E_0 + \cdots ,$$

$$\varphi_2 \underset{y^{\bullet} \to 0}{\sim} \frac{D_0(z^{\bullet})y^{\bullet -\frac{1}{5}}f_1(a\xi)}{a^3} + \frac{C_1(z^{\bullet})D_0(z^{\bullet})y^{\bullet}h_1(a\xi)}{a^3} + \cdots . \qquad (6.3.26)$$

Here C_1, E_0, D_0 are determined from matching by the inner expansion. Summarizing we have;

Inner Expansion:

$$\phi = B^{\frac{2}{5}}C_0(z^{\bullet}) + \phi_0(x, \tilde{y}; z^{\bullet}) + B^{-\frac{6}{5}}\phi_1(x, \tilde{y}; z^{\bullet}) + \cdots , \qquad (6.3.27)$$

where

$$\phi_0 \underset{\tilde{y} \to \infty}{=} \tilde{y}^{\frac{2}{5}} \frac{f(a\xi)}{a^3} + c_0(z^{\bullet}) + \frac{c_1(z^{\bullet})\tilde{y}^{-\frac{1}{5}}h_1(a\xi)}{a^3} + O(\tilde{y}^{-\frac{2}{5}}), \qquad (6.3.28)$$

$$\phi_1 \underset{\tilde{y} \to \infty}{=} \tilde{y}^{\frac{8}{5}} \frac{d_0(z^{\bullet})h_0(a\xi)}{a^3} + \frac{c_1(z^{\bullet})d_0(z^{\bullet})\tilde{y}h_1(a\xi)}{a^3} + O(\tilde{y}^{\frac{2}{5}}), \qquad (6.3.29)$$

Outer Expansion:

$$\phi = B^{\frac{2}{5}}\varphi_0 + \mu_1\varphi_1 + \mu_2\varphi_2 + \cdots , \qquad (6.3.30)$$

where

$$\varphi_0 \underset{y^{\bullet} \to 0}{=} C_0(z^{\bullet}) + y^{\bullet \frac{2}{5}} \frac{f(a\xi)}{a^3} + y^{\bullet \frac{8}{5}} \frac{C_1(z^{\bullet})h_0(a\xi)}{a^3} + O(y^{\bullet 2}), \qquad (6.3.31)$$

$$\varphi_1 \underset{y^{\bullet} \to 0}{=} E_0(z^{\bullet}) + O(y^{\bullet \frac{8}{5}}), \qquad (6.3.32)$$

$$\varphi_2 \underset{y^{\bullet} \to 0}{=} D_0(z^{\bullet})\frac{y^{\bullet -\frac{1}{5}}f_1(a\xi)}{a^3} + C_1(z^{\bullet})\frac{D_0(z^{\bullet})y^{\bullet}h_1(a\xi)}{a^3} + O(y^{\bullet \frac{7}{5}}); \qquad (6.3.33)$$

where a is still a function of $z^{\bullet}$.

Matching can be accomplished by taking an intermediate limit or more simply by comparing the near field of the outer expansion as written in inner variables to the inner expansion. Then matching is accomplished if

$$E_0(z^*) = c_0(z^*), \quad D_0(z^*) = c_1(z^*), \quad d_0(z^*) = C_1(z^*).$$

Here $c_1(z^*)$ is known (see (4.1.120)), $C_1(z^*)$ must still be found either numerically or more compactly, by solving the φ_0 problem.

Summarizing then we have that given that $\varphi_0(x, y; z^*)$ is known, φ_1 is given by the solution to the boundary value problem

$$-(\phi_{0x}\phi_{1x}) + \phi_{1\tilde{y}\tilde{y}} = 0, \tag{6.3.34}$$

$$\phi_{1\tilde{y}}|_{\tilde{y}=0} = 0 \quad \text{for} \quad x_{\text{LE}}(z^*) < x < x_{\text{TE}}(z^*), \tag{6.3.35}$$

$$\phi_1 \sim \tilde{y}^{\frac{8}{5}} C_1(z^*) \frac{h_0(a\xi)}{a^3} + \tilde{y} \frac{C_1(z^*)c_1(z^*)h_1(a\xi)}{a^3} + \cdots, \tag{6.3.36}$$

where h_0, h_1 are known functions (6.3.17), (6.3.18), $c_1(z^*)$ is known (4.1.120), and $C_1(z^*)$ is yet to be found. Note that the ϕ_1 equation is linear and the boundary condition on $y^* = 0$ is homogeneous, thus $C_1(z^*)$ is solely a multiplicative factor.

Finally note that for a wing of similar sections the boundary value ϕ_0, (6.3.10), (6.3.12), (6.3.14) and that for ϕ_1 (6.3.34), (6.3.35), (6.3.36), can be scaled to be independent of the chord. That is, the boundary value problem for ϕ_0, ϕ_1 are two dimensional problems at each spanwise station. If the wing planform of chord $c(z^*)$ is given by

$$F_{u,\ell}(x, z^*) = c(z^*)G_{u,\ell}\left(\frac{x}{c(z^*)}\right), \tag{6.3.37}$$

then the ϕ_0, ϕ_1 problems can be scaled to be independent of z^* and hence solutions need only be calculated at one z^* station. This is accomplished by first scaling ϕ_0, x, $\tilde{y}$ by the chord,

$$\psi_0 = \frac{\phi_0}{c(z^*)}, \quad X = \frac{x - \frac{1}{2}(x_{\text{LE}} + x_{\text{TE}})}{c(z^*)}, \quad Y = \frac{\tilde{y}}{c(z^*)}. \tag{6.3.38}$$

Then, since $a \propto c^{-\frac{1}{5}}$, (see 4.1.115), $A = ac^{\frac{1}{5}}$ is independent of the z^* and the ϕ_0 problem becomes

$$-\psi_{0X}\psi_{0XX} + \psi_{0YY} = 0, \tag{6.3.39}$$

$$\psi_{0X}|_{Y=0} = G_X(X), \quad -\frac{1}{2} < x < \frac{1}{2}, \tag{6.3.40}$$

$$\psi_0 \underset{Y \to \infty}{\sim} y^{\frac{2}{5}} \frac{f\left(\frac{AX}{Y^{\frac{4}{5}}}\right)}{A^3}, \tag{6.3.41}$$

independent of z^*. Then, with

$$\phi_1 = c^{\frac{11}{5}}(z^*)C_1(z^*)\phi_1, \tag{6.3.42}$$

the ϕ_1 problem becomes

$$-(\psi_{0X}\psi_{1X})_X + \psi_{1YY} = 0, \tag{6.3.43}$$

$$\psi_{1X}|_{Y=0} = 0, \quad -\frac{1}{2} < X < \frac{1}{2}, \tag{6.3.44}$$

$$\psi_1 \sim Y^{\frac{8}{5}} \frac{h_0\left(\frac{AX}{Y^{\frac{4}{5}}}\right)}{A^3} + Yh_1\left(\frac{AX}{Y^{\frac{4}{5}}}\right)\frac{c_1}{c^{\frac{3}{5}}A^3}, \tag{6.3.45}$$

which is independent of z^* since $c_1 \propto c^{\frac{3}{5}}$ as can be seen from the ϕ_0 problem.

Note that this analysis applies only to that portion of the flow in front of the shock surface. Hence, we continue to find the behavior behind the shock.

In front of the shock we had (6.3.27)

$$\phi(x, \tilde{y}, \tilde{z}) = B^{\frac{2}{5}}C_0(z^*) + \phi_0(x, \tilde{y}; z^*) + B^{-\frac{6}{5}}\phi_1(x\tilde{y}; z^*) + \cdots,$$

for the inner expansion and (6.3.30),

$$\phi(x, \tilde{y}, \tilde{z}) = B^{\frac{2}{5}}\varphi_0(x^*, y^*, z^*) + \mu_1\varphi_0(x^*, y^*, z^*) + \cdots,$$

for the outer expansion. Behind the shock we assume the form

$$\phi(x, \tilde{y}, \tilde{z}) = B^{\frac{2}{5}}\tilde{C}_0 + \tilde{\phi}_0(x, \tilde{y}; z^*) + \lambda_1(B)\tilde{\phi}_1(x, \tilde{y}; z^*)$$
$$+ \lambda_2(B)\tilde{\phi}_2(x, \tilde{y}; z^*) + \cdots, \tag{6.3.46}$$

where $\lambda_2 \ll \lambda_1 \ll 1$, for the inner expansion and

$$\phi(x, \tilde{y}; z^*) = B^{\frac{2}{5}}\tilde{\varphi}_0(x^*, y^*, z^*) + \pi_1(B)\tilde{\varphi}_1(x^*, y^*, z^*) + \cdots, \tag{6.3.47}$$

where $\pi_1 \ll B^{\frac{2}{5}}$ for the outer expansion. The λ_i, π_i will be determined by matching between the inner and the outer expansion.

Substitution of (6.3.46) into the governing equation (6.3.16) and the boundary conditions (6.3.3) shows that $\tilde{\phi}_0$, $\tilde{\phi}_1$ satisfy the same equation and boundary condition at the airfoil as ϕ_0, ϕ_1 (6.3.10), (6.3.11), (6.3.12).

The shock conditions can be found identically as in Section 6.1, 6.2. The full three dimensional shock conditions are, for the shock surface given by

$$S(x, \tilde{y}, \tilde{z}) = s - g\left(\tilde{y}, \frac{\tilde{z}}{B}, B\right) = 0, \tag{6.3.48}$$

no jump of tangential velocity across the shock,

$$[\phi_s] = 0, \tag{6.3.49}$$

conservation of mass,

$$-\frac{1}{2}[\phi_x^2]_s - [\phi_{\tilde{y}}]g_{\tilde{y}} - [\phi_{\tilde{z}}]g_{\tilde{z}} = 0. \tag{6.3.50}$$

From (6.3.49), (6.3.50) the shock polar can be found,

$$-\frac{1}{2}[\phi_x^2]_s[\phi_x]_s + [\phi_{\tilde{y}}]_s^2 + [\phi_{\tilde{z}}]_s^2 = 0. \tag{6.3.51}$$

The shock jump conditions for the ϕ_i are obtained by substituting (6.3.27) into (6.3.49), (6.3.51) and expanding about the zeroth order shock locus. That is, (6.3.48) has the form

$$x = g\left(\tilde{y}, \frac{\tilde{z}}{B}; B\right) = g_0(\tilde{y}; z^*) + \theta_1(B)g_1(\tilde{y}; z^*) + \cdots . \tag{6.3.52}$$

Then for any quantity $\phi(x, \tilde{y}, \tilde{z}; B)$, with an expansion of the form (6.3.27) and (6.3.29) in front and behind the shock respectively we have

$$[\phi]_s = (\phi_{0a} - \tilde{\phi}_{0b})_{S_0} + \theta_1 g_1(\phi_{0x_a} - \tilde{\phi}_{0x_b})_{S_0} + B^{-\frac{6}{5}}(\phi_{1a} - \lambda_1\tilde{\phi}_{1b}) + \cdots , \tag{6.3.53}$$

where

$$S_0 : x = g_0(\tilde{y}; \tilde{z}^*).$$

Thus, to zeroth order we have

$$-\frac{1}{2}[\phi_{0x}^2][\phi_{0x}]_{S_0} + [\phi_{0\tilde{y}}]_{S_0}^2 = 0, \tag{6.3.54}$$

$$[\phi_0]_{S_0} = 0, \tag{6.3.55}$$

where $[\phi_0]_{S_0}$ is shorthand for

$$\tilde{\phi}_0\big(g_0(\tilde{y}; z^*), \tilde{y}, z^*\big) - \phi_0\big(g_0(\tilde{y}; z^*), \tilde{y}, z^*\big),$$

and to first order

$$[\phi_1] = \begin{cases} G_1[\phi_{0x}] & \text{if } \theta_1 = B^{-\frac{6}{5}}, \\ 0 & \text{if } \theta_1 \gg B^{-\frac{6}{5}}, \end{cases} \tag{6.3.56}$$

$$[\phi_{0x}^2][\phi_{1x}] + [\phi_{0x}\phi_{1x}][\phi_{0x}] - 2[\phi_{0\tilde{y}}][\phi_{1\tilde{y}}]$$
$$= \begin{cases} g_1\Big\{ -[\phi_{0x}]^2[\phi_{0xy}] + [\phi_{0xx}\phi_{0x}][\phi_{0x}] + 2[\phi_{0\tilde{y}\tilde{y}}][\phi_{\tilde{y}}] \Big\} & \text{if } \theta_1 = B^{-\frac{6}{5}} \\ 0 & \text{if } \theta_1 \gg B^{-\frac{6}{5}}. \end{cases} \tag{6.3.57}$$

Thus, $\tilde{\phi}_0$ is the two dimensional transonic potential. The far field expansion of $\tilde{\phi}_0$ is, (4.1.156)

$$\phi_0 \underset{\tilde{y}\to\infty}{\sim} \tilde{y}^{\frac{2}{5}}\frac{f(a\xi)}{a^3} + |\tilde{y}|^{\frac{1}{5}}\frac{f_1(a\xi)}{a^3} + \frac{\tilde{f}_2^{u,\ell}(a\xi)}{a^3} + \cdots. \tag{6.3.58}$$

We must next find the near field of φ_0. Then, by matching we will be able to identify the boundary value problems for $\tilde{\phi}_1$, $\tilde{\varphi}_1$ as well as find $\lambda_1(B)$, $\pi_1(B)$.

Substituting (6.3.30) and (6.3.47) into the shock conditions (6.3.49), (6.3.51), with the shock location given by

$$x^* = G(y^*, z^*; B) = G_0(y^*; z^*) + \cdots,$$

one finds for φ_0 the conditions

$$[\varphi_0] = 0, \tag{6.3.59}$$

$$-\frac{1}{2}[\varphi_{0x^*}^2][\varphi_{0x^*}] + [\varphi_{0y^*}]^2 + [\varphi_{0z^*}]^2 = 0, \tag{6.3.60}$$

where all the jumps are evaluated at the zeroth order shock location, $x^* = G_0(y^*, z^*)$. Also, clearly $\tilde{\varphi}_0$ satisfies the equation

$$-\tilde{\varphi}_{0x^*}\tilde{\varphi}_{0x^*x^*} + \tilde{\varphi}_{0y^*y^*} + \tilde{\varphi}_{0z^*z^*} = 0. \tag{6.3.61}$$

Our goal is to find the near field of $\tilde{\varphi}_0$.

In front of the shock (6.3.31) we had

$$\varphi_0 \sim C_0(z^*) + y^{*\frac{2}{5}} \frac{f(a\xi^*)}{a^3} + y^{*\frac{8}{5}} \frac{h(a\xi^*)}{a^3} + \cdots , \qquad (6.3.62)$$

behind the shock we expect

$$\tilde{\varphi}_0 \sim \tilde{C}_0(z^*) + y^{*\frac{2}{5}} \frac{f(a\xi^*)}{a^3} + y^{*p} \frac{\tilde{h}(a\xi^*)}{a^3} + \cdots , \qquad (6.3.63)$$

where $h = C_1(z^*)h_0(a\xi^*)$.

The shock location will have an expansion in the near field of the form

$$ax^* = y^{*\frac{4}{5}} \xi_0^{*+} + y^{*q+\frac{4}{5}} \xi_1^{*+} + \cdots ,$$

where

$$q > 0 ,$$

or

$$\xi^{*+} = \xi^{*+} + y^{*q} \xi_1^{*+} + \cdots . \qquad (6.3.64)$$

Substituting (6.3.62), (6.3.63), and (6.3.64) into the equation (6.3.61) and the shock conditions (6.3.59), (6.3.60), and using the usual expansion about the ze-roth order shock condition so that

$$[(\ )]_S = [\ \ ]_0 + [(\ )_{\xi^*}]_0 y^{*q} \xi_1^{*+} + \cdots ,$$

where 0 refers to $\xi_S^{*+} = \xi_0^{*+}$, one finds that $\tilde{h}_0$ satisfies the equation

$$\left(f' - \frac{16}{25} \xi^{*+^2} \right) \tilde{h}'' + \left(f'' + \frac{4}{5} \left(2p - \frac{9}{5} \xi^{*+} \right) \right) \tilde{h}' - p(p-1)h = \begin{cases} 0 & \text{if } p < 2, \\ -C_0'' & \text{if } p = 2. \end{cases}$$
$$(6.3.65)$$

For $p < \frac{8}{5}$, $q < 2$, the shock conditions are identical to (4.1.131), (4.1.132) with of course $\tilde{f}_1$ replaced by $\tilde{h}$, p by n_1, q by m_1. Thus there is no nontrivial solution unless $q + \frac{8}{5} = p$. Then, for $p < \frac{8}{5}$, the shock conditions are identical to (4.1.137), (4.1.139) or equivalently (4.1.152), (4.1.153). That is, all reasoning follows precisely as in the two-dimensional case. These last two conditions are equivalent to, at $\tau = \frac{5\sqrt{3}-8}{8}$,

$$S_S^S = -25p(2-5p)\tau(1+\tau)F\left(\frac{5p}{2} + 1, \frac{8-5p}{6}, \frac{3}{2}; \frac{5\sqrt{3}-8}{8} \right)$$

$$- \{(8-65p)\tau + 3(1-5p)\}F\left(\frac{5}{2}p, \frac{2-5p}{6}, \frac{1}{2}; -\frac{5\sqrt{3}-8}{8} \right) = 0, \qquad (6.3.66)$$

and

$$S_S^A = -30\tau(1+\tau)\frac{5}{18}(5p+1)(1-p)F\left(\frac{5p+3}{2}, \frac{11-5p}{6}, \frac{5}{2}, \frac{5\sqrt{3}-8}{8}\right)$$
$$+ (3(4+5p) + (65p+7)\tau)F\left(\frac{5p+1}{2}, \frac{5(1-p)}{6}, \frac{3}{2}, \frac{5\sqrt{3}-8}{8}\right) = 0. \quad (6.3.67)$$

It is easily checked, since all the hypergeometric functions have values close to one, that S_S^S and S_S^A remain > 0 for $p > \frac{2}{5}$. Thus there are no solutions of the system for $\frac{2}{5} < p < \frac{8}{5}$.

For $p = \frac{8}{5}$, we have from continuity

$$\xi_1^{*^+} = [h] = 0 \quad \text{if} \quad q - \frac{2}{5} \neq \frac{8}{5}, \quad (6.3.68)$$

where $[h] = h - \tilde{n}$, and

$$\xi_1^{*^+}[f'] - [h] = 0 \quad \text{if} \quad q - \frac{2}{5} = p, \quad (6.3.69)$$

and from the shock polar,

$$\langle h' \rangle = \frac{12}{5}\xi^{*^+}\xi^{*^+}. \quad (6.3.70)$$

Note that from continuity $[C_0] = 0$ hence since $C_0(z^*)$, $[C_0'] = [C_0''] = 0$. These conditions differ from (4.1.37), (4.1.39) because h before the shock is not zero. Hence (6.3.68) and (6.3.70) differ from (4.1.137) and (4.1.139) as conditions on $\tilde{h}$ and $\tilde{h}'$ just behind the shock in the addition of terms h and $-h'$ evaluated ahead of the shock, to the right hand side, as well as in the factor $2 \cdot \frac{12}{5}$ instead of $\frac{8}{25}$ as in (4.1.139). Eliminating $\xi_1^{*^+}$ from (6.3.68) and (6.3.70) one obtains the relationship between $\tilde{h}$, $\tilde{h}'$,

$$\tilde{h}' = -h' + \frac{\frac{24}{5}\xi_0^{*^+}(h - \tilde{h}_b)}{[f']}, \quad (6.3.71)$$

or

$$\tilde{h}' + \frac{\frac{24}{5}\xi_0^{*^+}\tilde{h}}{[f']} = -h' + \frac{24}{5}\frac{\xi_0^{*^+}}{[f']}h. \quad (6.3.72)$$

Thus this condition is only a condition on the constant multiplying the solution $\tilde{h}$ of the homogeneous equation (6.3.65). Equation (6.3.35) written in the

t variables (4.1.99) is a hypergeometric equation with $a = 4$ $b = -1$, $c = \frac{1}{2}$. Thus the solution has the form

$$\tilde{h} = C_1^{u,\ell} s^{-\frac{4}{7}}(1 - 6s) + C_2^{u,\ell}(s)^{-\frac{3}{10}} F\left(\frac{9}{2}, -\frac{1}{2}, \frac{3}{2}, \frac{3}{4}s\right), \qquad (6.3.73)$$

where the first piece of the solution was found in (6.3.17) for h_0.

Finally we must check that φ_{0x^*}, φ_{0y^*} are continuous across the wake, hence that $(\tilde{y}^{\frac{8}{5}}\tilde{h})_{y^*}$, $(\tilde{y}^{\frac{8}{5}}\tilde{h})_{x^*}$ are continuous as $\xi^* \to \infty$.

$$y^{*\frac{8}{5}}\tilde{h} \underset{s\to 0}{\sim} y^{*\frac{8}{5}}\left(C_1^{u,\ell} s^{-\frac{4}{7}} + C_2^{u,\ell} s^{-\frac{3}{10}}\right),$$

$$\underset{s\to 0}{\sim} y^{*\frac{8}{5}}\left(C_1^{u,\ell}\xi^{*\frac{10}{7}} + C_2^{u,\ell}\xi^{*\frac{3}{4}}\right).$$

So,

$$(y^{*\frac{8}{5}}\tilde{h})_{x^*} \underset{y^*\to 0}{\sim} 0$$

$$a^2(y^{*\frac{8}{5}}\tilde{h})_{y^*} \underset{y^*\to 0}{\sim} y^{*\frac{3}{5}}\left\{\frac{8}{5}C_1^{u,\ell}\xi^{*+\frac{10}{7}} + \frac{8}{5}C_2^{u,\ell}\xi^{*+\frac{3}{4}} - \frac{4}{7}C_1^{u,\ell}\xi^{*+\frac{10}{7}} - \frac{4}{5}C_2^{u,\ell}\xi^{*+\frac{3}{4}}\right\}$$

$$\sim \frac{4}{5}C_2^{u,\ell}x^{*\frac{3}{4}}.$$

Thus for continuity across the wake, $C_2^u = C_2^\ell$. Hence from the shock condition $C_1^u = C_1^\ell$, or

$$\tilde{h} = \tilde{C}_1\left\{s^{-\frac{4}{5}}(1 - 6s) + \tilde{C}_2 s^{-\frac{3}{10}} F\left(\frac{9}{2}, -\frac{1}{2}, \frac{3}{2}, \frac{3}{4}; s\right)\right\} \equiv \tilde{C}_1\tilde{h}_0,$$

where $\tilde{C}_2(\tilde{C}_1)$ as determined by the shock condition (6.3.72).

In conclusion, the expansions behind the shock are:

Inner Expansion:

$$\phi(x, \tilde{y}; \tilde{z}) = B^{\frac{2}{5}}\tilde{C}_0(z^*) + \tilde{\phi}_0(x, \tilde{y}; z^*) + \lambda_1(B)\tilde{\phi}_1(x, \tilde{y}; z^*) + \cdots,$$

where

$$\tilde{\phi}_0(x, \tilde{y}; z^*) \underset{\tilde{y}\to\infty}{\sim} \tilde{y}^{\frac{2}{5}}\frac{f(a\xi)}{a^3} + |\tilde{y}|^{\frac{1}{5}}\frac{\tilde{f}_1(a\xi)}{a^3} + \frac{\tilde{f}_2(a\xi)}{a^3} + \cdots,$$

Outer Expansion:

$$\phi(x, \tilde{y}, \tilde{z}) = B^{\frac{2}{5}} \tilde{\varphi}_0(x^*, y^*, z^*) + \pi_1(B)\tilde{\varphi}_1(x^*, y^*, z^*) + \cdots,$$

where

$$\tilde{\varphi}_0(x^*, y^*, z^*) \underset{y^* \to 0}{\sim} \tilde{C}_0(z^*) + y^{*\frac{2}{5}} \frac{f(a\xi)}{a^3} + y^{*\frac{8}{5}} \tilde{C}_1(z^*) \frac{\tilde{h}_0(a\xi^*)}{a^3} + \cdots.$$

Hence, by matching we see that

$$\pi_1(B) = B^{\frac{1}{5}}, \quad \lambda_1(B) = B^{\frac{6}{5}},$$

and

$$\tilde{\varphi}_1(x^*, y^*, z^*) \underset{y^* \to 0}{\sim} y^{*\frac{1}{5}} \frac{f_1(a\xi)}{a^3} + \cdots,$$

$$\phi_1(x, y; z^*) \underset{y^* \to \infty}{\sim} \tilde{y}^{\frac{8}{5}} \frac{h_1(a\xi)}{a^3} + \cdots.$$

Thus, the first lifting line correction behind the shock is of the same order as that in front of the shock. The correction behind the shock, $\tilde{\phi}_1$ solves the boundary problems,

$$-\tilde{\phi}_{1x}\tilde{\phi}_{1xx} + \tilde{\phi}_{1\tilde{y}\tilde{y}} = 0,$$

$$\tilde{\phi}_1 \underset{y \to \infty}{\sim} \tilde{y}^{\frac{8}{5}} \frac{\tilde{h}_1(a\xi)}{a^3},$$

with the conditions at the shock location $g_0(\tilde{y}; z^*)$ given by

$$[\phi_1] = g_1[\phi_{0x}]$$

$$[\phi_{0x}^2][\phi_{1x}] + [\phi_{0x}\phi_{1x}][\phi_{0x}] - 2[\phi_{0\tilde{y}}][\phi_{1\tilde{y}}]$$

$$= g_1\left\{ -[\phi_{0x}]^2[\phi_{0xy}] + [\phi_{0xx}\phi_{0x}][\phi_{0x}] + 2[\phi_{0\tilde{y}\tilde{y}}][\phi_{0\tilde{y}}] \right\}.$$

Here θ_1 has been chosen to be $B^{-\frac{6}{5}}$. It is the outer expansion which is affected more strongly by the finite aspect ratio behind the shock than in front of the shock.

To summarize, the expansions are:

Inner Expansion:

$$\phi(x, \tilde{y}, \tilde{z}) = B^{\frac{2}{5}} C_0(z^*) + \phi_0(x, \tilde{y}; z^*) + B^{-\frac{6}{5}} \phi_1(x, \tilde{y}; z^*) + \cdots,$$

in front of the shock, and

$$= B^{\frac{2}{5}} \tilde{C}_0(z^*) + \phi_0(x, \tilde{y}; z^*) + B^{-\frac{6}{5}} \phi_1(x, \tilde{y}; z^*) + \cdots,$$

behind the shock,

where

$$\phi_0 \underset{\tilde{y} \to \infty}{=} \tilde{y}^{\frac{2}{5}} \frac{f(a\xi)}{a^3} + c_0(z^*) + \frac{c_1(z^*) \tilde{y}^{\frac{1}{5}} h_1(a\xi)}{a^3} + O(\tilde{y}^{-\frac{2}{5}}),$$

$$\phi_1 \underset{\tilde{y} \to \infty}{=} \tilde{y}^{\frac{8}{5}} \frac{C_1(z^*) h_0(a\xi)}{a^3} + O(\tilde{y}),$$

$$\tilde{\phi}_0 \underset{\tilde{y} \to \infty}{=} \tilde{y}^{\frac{2}{5}} \frac{\tilde{f}_1(a\xi; z^*)}{a^3} + \frac{\tilde{f}_2(a\xi; z^*)}{a^3},$$

$$\tilde{\phi}_1 \underset{\tilde{y} \to \infty}{=} \tilde{y}^{\frac{8}{5}} \frac{\tilde{C}_1(z^*) \tilde{h}_0(a\xi)}{a^3} + \cdots.$$

Outer Expansion:

$$\phi(x, \tilde{y}, \tilde{z}) = B^{\frac{2}{5}} \varphi_0(x^*, y^*, z^*) + \varphi_1(x^*, y^*, z^*) + B^{-\frac{1}{5}} \varphi_2(x^*, y^*, z^*) + \cdots,$$

in front of the shock, and

$$\phi(x, \tilde{y}, \tilde{z}) = B^{\frac{2}{5}} \tilde{\varphi}_0(x^*, y^*, z^*) + B^{\frac{1}{5}} \tilde{\varphi}_1(x^*, y^*, z^*) + \tilde{\varphi}_2(x^*, y^*, z^*) + \cdots,$$

behind the shock,

where

$$\varphi_0 \underset{y^* \to 0}{=} C_0(z^*) + y^{*\frac{2}{5}} \frac{f(a\xi)}{a^3} + y^{*\frac{8}{5}} \frac{C_1(z^*) h_0(a\xi)}{a^3} + \cdots,$$

$$\varphi_1 \underset{y^* \to 0}{=} c_0(z^*) + O(y^{*\frac{8}{5}}),$$

$$\varphi_2 \underset{y^* \to 0}{=} c_1(z^*) \frac{y^{*-\frac{1}{5}} f_1(a\xi)}{a^3} + O(y^*),$$

and

$$\tilde{\varphi}_0 \underset{y^* \to 0}{=} \tilde{C}_0(z^*) + y^{*\frac{2}{5}} \frac{f(a\xi)}{a^3} + y^{*\frac{8}{5}} \frac{\tilde{C}_1(z^*) \tilde{h}_0(a\xi)}{a^3} + \cdots,$$

$$\tilde{\varphi}_1 \underset{y^* \to 0}{=} |y^*|^{\frac{1}{5}} \frac{\tilde{f}_1(a\xi; z^*)}{a^3} + \cdots,$$

$$\tilde{\varphi}_2 \underset{y^* \to 0}{=} \frac{\tilde{f}_2(a\xi; z^*)}{a^3} + \cdots.$$

References

[6.3.1] Guderley, G., On Transonic Airfoil Theory, *J. Aero. Sci.*, Oct. 1966, pp. 961–969.

[6.3.2] Cole, J. D., Cook, L. P., Ziegler, F., Finite Span Wings at Sonic Speed, *Mech. Rech. Comm.*, **1** (4), 1980, pp. 253–260.

7. Quasi-transonic Flow

Quasi-transonic flow occurs in supersonic flows when a significant component of the flow velocity is close to sonic. The most typical case occurs for a supersonic wing whose leading edge is swept close to the Mach angle. Then the flow component normal to the edge is nearly sonic and quasi-transonic effects occur. The theory of these local flows can be carried fairly far for certain simple cases. The conical case was originally discussed by Frankl and Watson [7.1.1].

7.1 Linearized Theory

The flow past thin supersonic wings is typically calculated by linearized theory as discussed in Section 2.5. Linearized theory for wings of triangular planform predicts a sharp peak in the drag coefficient for that free stream Mach such that the Mach cone lies along the leading edge. (cf. Figure 7.1). Experimental results on wave drag for the symmetrical (non-lifting) cases do not show this peak. (cf. Figure 7.2 taken from [7.1.1]). The discrepancy can be traced to quasi-transonic effects. In this section linearized theory is worked out for the simplest case. This identifies the singularity at the sonic leading edge and provides the far-field for the local quasi-transonic flow. Generalizations are discussed later.

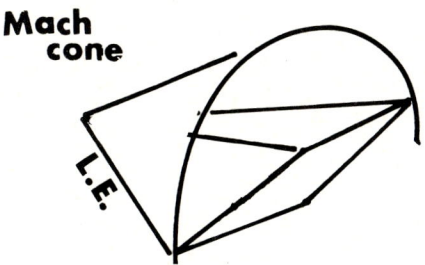

Figure 7.1

Wing with edges on Mach cone

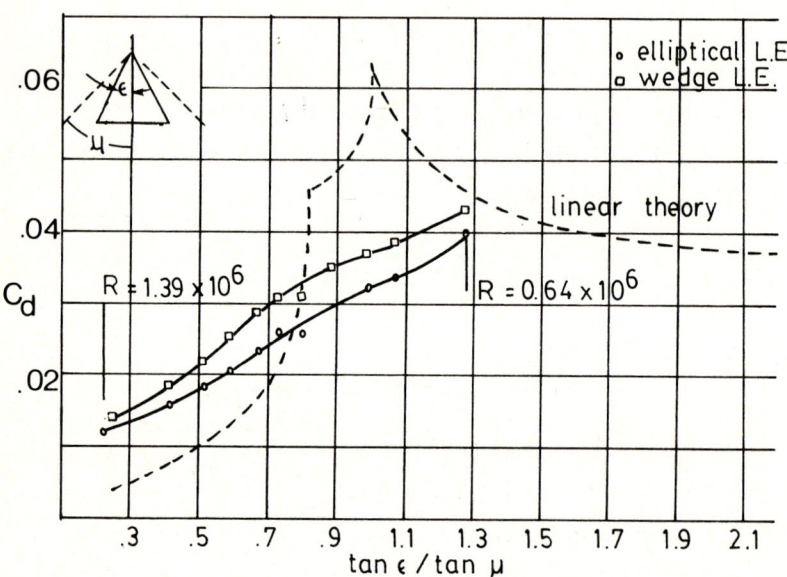

Figure 7.2

Minimum drag coefficients for 8% thick triangular wing series and
comparison with theory

Consider the wing of triangular planform with a wedge cross-section as in
Figure 7.3 where the leading edge is swept to the Mach angle $\Theta_M = \sin^{-1} \frac{1}{M_\infty}$.

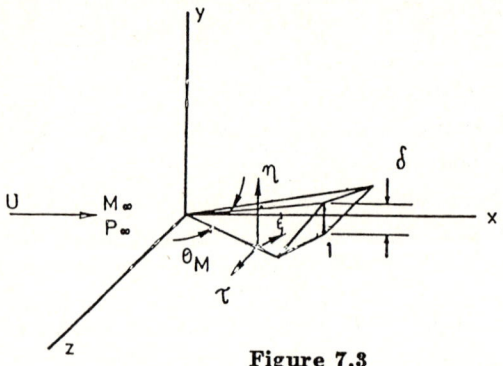

Figure 7.3

Supersonic Wing with Sonic Edges

The upper and lower surfaces are given by

$$y = \delta F_{u,\ell}(x,z) = \pm\frac{1}{2}\delta(x - \beta|z|),$$ (7.1.1)

where $\beta = \sqrt{M_\infty^2 - 1}$, δ = thickness ratio; the chord is chosen to be one.

Linearized theory calculates the perturbation potential ϕ derived from the "exact" potential Φ by the expansion

$$\Phi = U\{x + \delta\phi(x,y,z;\beta) + \cdots\}.$$ (7.1.2)

ϕ is calculated from the wave equation

$$\beta^2\phi_{xx} - (\phi_{yy} + \phi_{zz}) = 0,$$ (7.1.3)

with the boundary condition on the wing in the plane of the wing $(y = 0)$.

$$\phi_y(x,0\pm,z) = \frac{\partial F_{u,\ell}}{\partial x}(x,z).$$ (7.1.4)

The solution is represented by a supersonic source distribution in the plane of the wing $y = 0$,

$$\phi(x,y,z) = -\frac{1}{\pi}\iint_{\text{hyp.}} \frac{\phi_y(x',0+,z')}{\sqrt{(x - x')^2 - \beta^2 y^2 - \beta^2(z - z')^2}}\,dx'dz'.$$ (7.1.5)

The integral is carried out over those points in the (x,z) plane which can send a signal to (x,y,z). These points are cut out of the planform by the retrograde Mach cone and are thus bounded by the leading edges and a hyperbola. The integration is thus carried out over points on the planform for which the hyperbolic distance r_h from (x',z') to (xyz) is real, (cf. Figure 7.4)

$$r_h = \sqrt{(x - x')^2 - \beta^2 y^2 - \beta^2(z - z')^2}.$$ (7.1.6)

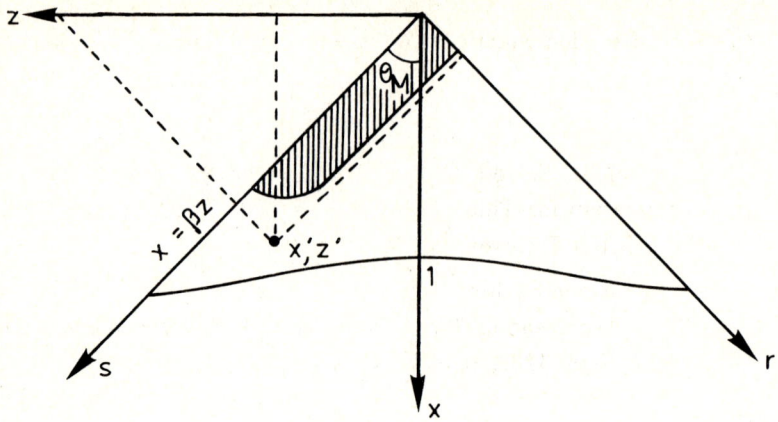

Figure 7.4

Domain of Dependence

For the special case of the wing with wedge airfoil

$$\frac{\partial F_{u,\ell}}{\partial x} = \frac{1}{2},$$ (7.1.7)

the integral (7.1.5) becomes

$$\phi = -\frac{1}{2\pi} \iint_{\text{hyp.}} \frac{dx'dz'}{\sqrt{(x-x')^2 - \beta^2 y^2 - \beta^2 (z-z')^2}}.$$ (7.1.8)

The integration is easily carried out by introducing the characteristic coordinates

$$r = x - \beta z, \quad s = x + \beta z.$$ (7.1.9)

For $y > 0$ we find

$$\phi = -\frac{1}{\pi\beta}\sqrt{x^2 - \beta^2(y^2 + z^2)} + \frac{y}{\pi}\cos^{-1}\frac{\beta y}{\sqrt{x^2 - \beta^2 z^2}}.$$ (7.1.10)

It is clear that there is a square-root behavior at the leading edge $x = \beta z$, $y = 0$ and indeed on the entire Mach cone. In particular the local pressure coefficient $C_p \sim -\phi_x$ has a square root singularity. The local integrated drag force on the leading edge is finite. However, we can not expect linearized theory to give a good approximation near the leading edge and an appreciable drag correction can result.

In the next section equations that are locally valid are derived.

References

[7.1.1] Fraenkel, L. E. and Watson, R., The Formulation of a Uniform Approximation for Thin Conical Wings with Sonic Leading Edges, *Proc. Symposium Transsonicum*, K. Oswatitsch, ed. Springer Verlag, 1962.

[7.1.2] Love, Eugene S., Investigations at Supersonic Speeds of 22 Triangular Wings Representing Two Airfoil Sections for each of 11 Apex Angles. *NACA Rept.* 1238, 1949.

7.2 Quasi-transonic Equations

Linearized theory is derived from the exact potential equation by considering the limit $\delta \to 0$ (x, y, z, M_∞ fixed). Quasi-transonic theory can be derived by considering a different limit $\delta \to 0$, M_∞ fixed, such that the representative point approaches the Mach cone. At the same time the leading edge must approach the Mach cone.

In this section we consider the case where the leading edge is exactly on the Mach cone, but for a general airfoil cross-section. Near the leading edge let

$$F_u(x, z) = a(z)(x - \beta z)^\kappa \{1 + \cdots\}, \quad z > 0, \quad 0 < \kappa \le 1. \tag{7.2.1}$$

The full potential equation (cf. Section 2.4) is

$$a^2 \{\Phi_{xx} + \Phi_{yy} + \Phi_{zz}\} = \Phi_x^2 \Phi_{xx} + 2\Phi_x \Phi_y \Phi_{xy} + \Phi_y^2 \Phi_{yy}$$
$$+ 2\Phi_y \Phi_z \Phi_{yz} + \Phi_z^2 \Phi_{zz} + 2\Phi_z \Phi_x \Phi_{xz}, \tag{7.2.2.}$$

where

$$\frac{a^2}{U^2} = \frac{1}{M_\infty^2} + \frac{\gamma - 1}{2}\left\{1 - \frac{\Phi_x^2 + \Phi_y^2 + \Phi_z^2}{U^2}\right\}.$$

The general form of the expansion near the leading edge is

$$\Phi = U\{x + \epsilon(\delta)\varphi(\xi, \eta, \varsigma) + \cdots\}, \tag{7.2.3}$$

where

$$\xi = \frac{x - \beta z}{\mu(\delta)}, \quad \eta = \frac{y}{\nu(\delta)}, \quad \tau = z; \quad \epsilon, \mu, \nu \to 0.$$

ξ is a local coordinate measuring distance downstream from the leading edge, η a local coordinate measuring vertical distance, τ spanwise location. The local scaling factors ϵ, μ, ν have to be determined so that a locally significant equation results where solutions can satisfy the boundary conditions on the wing surface and match the linearized solutions away from the leading edge. For the assumed form (7.2.3) note that

$$\frac{\Phi_x}{U} = 1 + \frac{\epsilon}{\mu}\varphi_\xi , \qquad \frac{\Phi_y}{U} = \frac{\epsilon}{\nu}\varphi_\eta , \qquad \frac{\Phi_z}{U} = -\frac{\beta\epsilon}{\mu}\varphi_\xi + \epsilon\varphi_\tau .$$

The surface boundary condition $\mathbf{q} \cdot \nabla B = 0$ on $B = y - \delta F_u(x,z) = 0$ is thus

$$\left(1 + \frac{\epsilon}{\mu}\varphi_\xi\right)(-\delta F_{u,x}) + \frac{\epsilon}{\nu}\varphi_\eta + \epsilon\varphi_\tau(-\delta F_{uz}) = 0, \quad \text{on} \quad B = 0 . \qquad (7.2.4)$$

But (7.2.1) shows

$$F_{ux} = \kappa a(x - \beta z)^{\kappa-1} + \cdots = \kappa a(\tau)\mu^{\kappa-1}\xi^{\kappa-1} + \cdots ,$$
$$F_{uz} = a'(z)(x - \beta z)^\kappa - \kappa\beta a(x - \beta z)^{\kappa-1} + \cdots = \mu^\kappa a'(z)\xi^\kappa - \kappa\beta a(\tau)\mu^{\kappa-1}\xi^{\kappa-1} .$$

The dominant terms in (7.2.4) become

$$\varphi_\eta(\xi, 0\pm, \tau) = \kappa a(\tau)\xi^{\kappa-1} , \qquad (7.2.5)$$

if

$$\boxed{\frac{\epsilon}{\nu} = \delta\mu^{\kappa-1}} . \qquad (7.2.6)$$

(7.2.6) provides one relationship for the three scalings (ϵ, μ, ν). Next we consider the full potential equations (7.2.2). Note that

$$\frac{\Phi_{xx}}{U} = \frac{\epsilon}{\mu^2}\varphi_{\xi\xi} , \qquad \frac{\Phi_{yy}}{U} = \frac{\epsilon}{\nu^2}\varphi_{\eta\eta} , \qquad \frac{\Phi_{zz}}{U} = \frac{\epsilon\beta^2}{\mu^2}\varphi_{\xi\xi} - 2\epsilon\frac{\beta}{\mu}\varphi_{\xi\tau} + \epsilon\varphi_{\tau\tau} ,$$

$$\frac{\Phi_{xy}}{U} = \frac{\epsilon}{\mu\nu}\varphi_{\xi\eta} , \qquad \frac{\Phi_{yz}}{U} = -\frac{\epsilon}{\nu}\frac{\beta}{\mu}\varphi_{\eta\xi} + \cdots , \qquad \frac{\Phi_{zx}}{U} = -\frac{\beta\epsilon}{\mu^2}\varphi_{\xi\xi} + \cdots ,$$

$$\frac{a^2}{U^2} = \frac{1}{M_\infty^2} + \frac{\gamma-1}{2}\left\{1 - \left(1 + 2\frac{\epsilon}{\mu}\varphi_\xi + \cdots\right)\right\}$$

$$= \frac{1}{M_\infty^2} - (\gamma-1)\frac{\epsilon}{\mu}\varphi_\xi + \cdots , \qquad (7.2.7)$$

$$\frac{\Phi_x^2}{U^2} = 1 + 2\frac{\epsilon}{\mu}\varphi_\xi + \cdots . \qquad (7.2.8)$$

Thus (7.2.2) becomes

$$\left(\frac{1}{M_\infty^2} - (\gamma - 1)\frac{\epsilon}{\mu}\varphi_\xi + \cdots\right)\left(\frac{\epsilon}{\mu^2}\varphi_{\xi\xi} + \frac{\epsilon}{\nu^2}\varphi_{\eta\eta} + \beta^2\frac{\epsilon}{\mu^2}\varphi_{\xi\xi} - 2\epsilon\frac{\beta}{\mu}\varphi_{\xi\tau} + \cdots\right)$$

$$= \left(1 + 2\frac{\epsilon}{\mu}\varphi_\xi + \cdots\right)\frac{\epsilon}{\mu^2}\varphi_{\xi\xi} + 2(1 + \cdots)\frac{\epsilon}{\nu}\varphi_\eta\frac{\epsilon}{\mu\nu}\varphi_{\xi\eta} + \frac{\epsilon^2}{\nu^2}\varphi_\nu^2\frac{\epsilon}{\nu^2}\varphi_{\eta\eta}$$

$$+ 2\frac{\epsilon}{\mu}\varphi_\eta\left(-\beta\frac{\epsilon}{\mu}\varphi_\xi + \cdots\right)\left(-\frac{\epsilon\beta}{\mu\nu}\varphi_{\eta\xi}\right) + \left(\frac{\beta^2\epsilon^2}{\mu^2}\varphi_\xi^2 + \cdots\right)\left(\frac{\epsilon\beta^2}{\mu^2}\varphi_{\xi\xi} + \cdots\right)$$

$$+ 2(1 + \cdots)\left(-\beta\frac{\epsilon}{\mu}\varphi_\xi + \cdots\right)\left(-\frac{\beta\epsilon}{\mu^2}\varphi_{\xi\xi} + \cdots\right), \quad \beta^2 = M_\infty^2 - 1.$$

The dominant terms of this equation $O(\frac{\epsilon}{\mu^2})$ cancel identically. The next largest terms are $O(\frac{\epsilon^2}{\mu^3}, \frac{\epsilon}{\mu}, \frac{\epsilon}{\nu^2})$. In order to have a distinguished limit all the terms must balance and this also guarantees that all variables (ξ, η, τ) appear as derivatives. If

$$\boxed{\frac{\epsilon^2}{\mu^3} = \frac{\epsilon}{\mu} = \frac{\epsilon}{\nu^2}}, \tag{7.2.9}$$

the equation above simplifies to

$$-(\gamma + 1)M_\infty^2\varphi_\xi\varphi_{\xi\xi} + \frac{1}{M_\infty^2}\varphi_{\eta\eta} - \frac{2\beta}{M_\infty^2}\varphi_{\xi\tau} = 2\varphi_\xi\varphi_{\xi\xi} + 2\beta^2\varphi_\xi\varphi_{\xi\xi},$$

or

$$\boxed{(\gamma + 1)M_\infty^4\varphi_\xi\varphi_{\xi\xi} - \varphi_{\eta\eta} + 2\beta\varphi_{\xi\tau} = 0}. \tag{7.2.10}$$

Equation (7.2.10) is the basic quasi-transonic equation. From (7.2.6), (7.2.9) the various orders can be determined

$$\nu = \sqrt{\mu}, \quad \epsilon = \mu^2, \quad \mu^{\frac{3}{2}} = \delta\mu^{\kappa-1},$$
$$\mu = \delta^{\frac{2}{5-2\kappa}}, \quad \nu = \delta^{\frac{1}{5-2\kappa}}, \quad \epsilon = \delta^{\frac{4}{5-2\kappa}}. \tag{7.2.11}$$

The quasi-transonic expansion and coordinates can be rewritten (cf. 7.2.3)

$$\Phi = U\{x + \delta^{4\sigma}\varphi(\xi, \eta, \tau) + \cdots\}, \tag{7.2.12}$$

where

$$\xi = \frac{x - \beta z}{\delta^{2\sigma}}, \quad \eta = \frac{y}{\delta^\sigma}, \quad \tau = z \quad \text{and} \quad \sigma = \frac{1}{5 - 2\kappa}.$$

The surface boundary conditions for the flow near the leading edge is (7.2.5). Disturbances should vanish at upstream infinity. Further the solution to (7.2.10) should match with the linearized theory away from the leading edge as $(\xi, \eta \to \infty,$ τ fixed). Details are given for an explicit example below.

A striking analogy appears in equation (7.2.10). This equation is formally identical to that for unsteady two-dimensional transonic flow derived in Section (3.11). For the analogy $K = 0$, $x \leftrightarrow \xi$, $\tilde{y} \leftrightarrow \eta$, $t \leftrightarrow \tau$ with some readjustment of constants. That is, each spanwise station corresponds to an instant of time in the suddenly started motion of a nose shape at sonic speed. The shock wave that develops ahead of the nose in the unsteady flow is analogous to the shock wave ahead of the leading edge of the three-dimensional wing.

Equation (7.2.10) is a conservation equation

$$\frac{\gamma+1}{2} M_\infty^4 \frac{\partial}{\partial \xi} u^2 - \frac{\partial}{\partial \eta} v + 2\beta \frac{\partial}{\partial \tau} u = 0\,, \tag{7.2.13}$$

where

$$u = \varphi_\xi\,, \quad v = \varphi_\eta\,.$$

The integrated form of (7.2.13) holds across the shocks and of course $\left[\varphi\right] = 0$.

7.3 Application to Delta Wing with Wedge Cross-Section.

Some details are provided here for the delta wing with wedge cross-section for which

$$F_u(x, z) = \frac{1}{2}(x - \beta|z|)\,, \tag{7.3.1}$$

$$\frac{\partial F_u}{\partial x} = \frac{1}{2} = \text{constant}\,. \tag{7.3.2}$$

In order for the quasi-transonic theory to be valid it is only required that (7.3.1) describe the shape near the leading edge, but for simplicity here the entire wing is considered to have that shape.

Then in the general theory (cf. 7.2.12) $\kappa = 1$, $\sigma = \frac{1}{3}$, $a(\tau) = \frac{1}{2}$, the expansion becomes $(z^* > 0$ say)

$$\Phi = U\left\{x + \delta^{\frac{4}{3}} \varphi(\xi, \eta, \tau) + \cdots\right\}\,, \tag{7.3.3}$$

where

$$\xi = \frac{x - \beta z}{\delta^{\frac{2}{3}}}\,, \quad \eta = \frac{y}{\delta^{\frac{1}{3}}}\,, \quad \tau = z\,.$$

The particular problem here is conical and this fact can be used to reduce the number of independent variables. This was the problem considered by Fraenkel and Watson [7.1.1].

The matching of (7.3.3) to the linearized solution for this wing is now discussed briefly. The inner limit of the outer (linearized) solution (7.1.10) can be obtained succinctly here by writing the outer solution in inner coordinates,

$$x^2 - \beta^2 z^2 = (x - \beta z)(x + \beta z) = \delta^{\frac{2}{3}} \xi(2\beta z + \cdots) = \delta^{\frac{2}{3}}(2\beta \xi \tau) + \cdots.$$

Thus (7.1.10) becomes

$$\phi = -\frac{1}{\beta\pi} \sqrt{\delta^{\frac{2}{3}}(2\beta\xi\tau) - \delta^{\frac{2}{3}}\beta^2\eta^2} + \delta^{\frac{1}{3}} \frac{\eta}{\pi} \cos^{-1}\left(\frac{\beta\delta^{\frac{1}{3}}\eta}{\sqrt{2\beta\delta^{\frac{2}{3}}\xi\tau}} \right). \tag{7.3.4}$$

Asymptotic matching means that $\delta\phi$ and $\delta^{\frac{4}{3}}\varphi$ should agree in form as $(\xi, \eta \to \infty,$ $x \to \beta z, y \to 0)$

$$\delta\phi \leftrightarrow \delta^{\frac{4}{3}}\varphi. \tag{7.3.5}$$

Thus, the orders in δ are correctly given for matching and (7.3.4) provides the far-field boundary condition for φ,

$$\varphi \to -\frac{1}{\beta\pi} \sqrt{2\beta\xi\tau - \beta^2\eta^2} + \frac{\eta}{\pi} \cos^{-1} \frac{\beta\eta}{\sqrt{2\beta\xi\tau}}. \tag{7.3.6}$$

This far-field applies downstream. Shock conditions are included in (7.2.10) and the proper shock location and jumps must be found.

The problem expressed by (7.2.10) with boundary condition (7.3.2) and far-field (7.3.6) has conical symmetry. The velocity is constant on rays through the apex. This conical problem is the one considered in detail by Fraenkel and Watson [7.1.1].

A conical representation is

$$\varphi = \tau\Psi(X, Y), \tag{7.3.7}$$

where the conical coordinates are

$$X = \frac{\xi}{\tau}, \quad Y = \frac{\eta}{\tau}.$$

In this flow

$$\varphi_\xi = \Psi_X, \quad \varphi_{\xi\xi} = \frac{1}{\tau}\Psi_{XX},$$

$$\varphi_\eta = \Psi_Y, \quad \varphi_{\eta\eta} = \frac{1}{\tau}\Psi_{YY},$$

$$\varphi_{\xi\tau} = \Psi_{XX}\left(-\frac{X}{\tau}\right) + \Psi_{XY}\left(-\frac{Y}{\tau}\right).$$

Thus (7.2.10) becomes

$$((\gamma+1)M_\infty^4 \Psi_X - 2\beta X)\Psi_{XX} - 2\beta Y\Psi_{XY} - \Psi_{YY} = 0. \qquad (7.3.8)$$

The far-field condition now reads

$$\Psi \rightarrow -\frac{1}{\pi}\sqrt{\frac{2X}{\beta} - Y^2} + \frac{Y}{\pi}\cos^{-1}\frac{\sqrt{\beta}Y}{\sqrt{2X}}, \qquad (7.3.9)$$

for

$$X,Y \rightarrow \infty, \quad X > \frac{\beta Y^2}{2}.$$

$X = \beta\frac{Y^2}{2}$ is the shock location (Mach cone) near infinity. The surface boundary condition is

$$\Psi_Y(X,0) = \frac{1}{2}, \quad X > 0. \qquad (7.3.10)$$

The domain of the solution is shown in Figure 7.5.

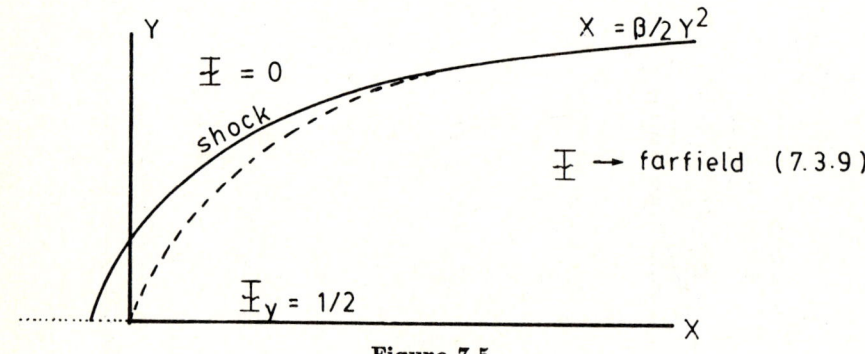

Figure 7.5

Conical Domain

We have assumed symmetry here so that $\Psi_Y(X,0) = 0$, $X < 0$. The shock location is unknown. But, shock jump conditions are specified along this locus. The shock conditions can be derived from the proper conservation form of (7.3.8). Alternately, the shock jumps for unsteady transonic flow, derived in Section 2, can be transformed to conical form. This step is omitted here since it is more convenient to transform to the coordinates used by Fraenkel and Watson [7.1.1].

These coordinates are derived from conical cylindrical coordinates centered on the x-axis. Let

$$\left\{ R = \frac{X}{\beta} - \frac{1}{2}Y^2, \quad \vartheta = Y \right\}, \tag{7.3.11}$$

and

$$\Psi(X,Y) \to \beta\psi(R,\vartheta).$$

Note that

$$\Psi_X = \psi_R, \quad \Psi_Y = \beta(\psi_\vartheta - Y\psi_R), \quad \text{etc.},$$

so that the basic conical potential equation (7.3.8) becomes

$$\left((\gamma + 1)M_\infty^4\psi_R - 2\beta^2 R\right)\psi_{RR} - \beta^2\psi_{\vartheta\vartheta} + \beta^2\psi_R = 0. \tag{7.3.12}$$

In conservation form we have

$$\left((\gamma + 1)M_\infty^4\frac{\psi_R^2}{2} - 2\beta^2 R\psi_R + 3\beta^2\psi\right)_R - \beta^2(\psi_\vartheta)_\vartheta = 0. \tag{7.3.13}$$

The boundary conditions transform correspondingly. They are indicated on Figure 7.6.

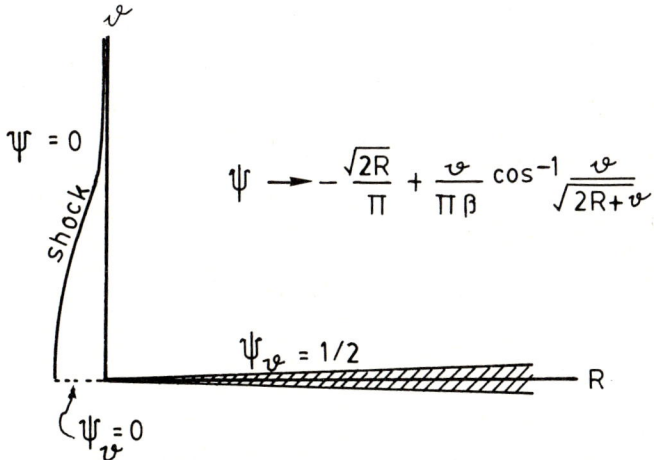

Figure 7.6
Conical Domain in (R, ϑ).

The shock jump conditions obtained by integration of (7.3.13) are

$$
\left\{
\begin{array}{c}
\left(\dfrac{\gamma+1}{2} M_\infty^4 \dfrac{\psi_R^2}{2} - 2\beta^2 R_s \psi_R \right) d\vartheta_s + \beta^2 \psi_\vartheta \, dR_s = 0 \\[2mm]
\psi_R \, dR_s + \psi_\vartheta \, d\vartheta_s = 0
\end{array}
\right\}
\tag{7.3.14}
$$

on $R = R_s$, $\vartheta = \vartheta_s$.

The characteristics of (7.3.12) are found from

$$
\frac{dR}{d\vartheta} = \pm\sqrt{\frac{(\gamma+1)}{\beta^2} M_\infty^4 \psi_R - 2R} \, .
\tag{7.3.15}
$$

Thus there are no real characteristics, the equation is elliptic in the far-field. A study of the shock jump conditions shows that the equation is also elliptic behind the detached shock and is elliptic in the domain.

A crude solution has been obtained by using a finite difference form of (cf. [7.3.2]) (7.3.13) with an approximate shock-condition. After some relaxation iterations a reasonable solution is found.

From this solution an estimate of the drag correction due to the quasi-transonic effects can be made as follows. For the upper surface

$$
\Delta D_u = \frac{\rho_\infty U^2}{2} \iint_{LE} (c_{PQT} - c_{PLIN}) \frac{\delta}{2} \, dx dz \,,
\tag{7.3.16}
$$

where

$$
c_{PQT} = \text{quasi-transonic pressure coeff.} = -2\delta^{\frac{2}{3}} \varphi_\xi \,,
$$

$$
c_{PLIN} = \text{linearized pressure coefficient} = -2\delta \phi_x \,,
$$

$$
\Delta C_{D_u} = \frac{\Delta D_u}{\frac{\rho_\infty}{2} U^2 \delta^2} = \delta^{\frac{1}{3}} \int_0^{\frac{1}{\beta}} d\tau \int_0^{\infty} d\xi \left\{ -\frac{1}{\pi}\sqrt{\frac{\tau}{2\beta\xi}} - \varphi_\xi \right\} .
$$

The singular part of the linearized solution has been used. Transforming to conical coordinates

$$
\Delta C_{D_u} = -\frac{\delta^{\frac{1}{3}}}{2\beta^2} \int_0^{\infty} \left(\frac{1}{\pi}\sqrt{\frac{1}{2\beta X}} + \Psi_X(X,0+) \right) dX \,,
$$

or

$$
\Delta C_{D_u} = -\frac{\delta^{\frac{1}{3}}}{2\beta} \int_0^{\infty} \left(\frac{1}{\pi\sqrt{2\beta^2 R}} + \psi_R(R,0+) \right) dR \,,
\tag{7.3.17}
$$

and integration by parts shows

$$\Delta C_{D_u} = \frac{\delta^{\frac{1}{3}}}{2\beta} \psi(0+,0+).$$ (7.3.18)

The numerical results for $\psi(0+,0+)$ are approximately

$$\psi(0+,0+) = -.14.$$ (7.3.19)

This gives about the correct order of magnitude for drag reduction compared with experiments. Better calculations are needed.

References

[7.3.1] Love, Eugene S., Investigations at Supersonic Speeds of 22 Triangular Wings Representing Two Airfoil Sections for each of 11 Apex Angles. *NACA Rept.* 1238, 1949.

[7.3.2] Rimbey, S., Private communication.

7.4 Remarks

The considerations of the previous paragraphs can be generalized to cases where the leading edge is sufficiently close to the Mach angle instead of exactly along it. For the wedge airfoil section the order of the difference from the Mach angle for quasi-transonic theory to be valid is $\delta^{\frac{2}{3}}$. Although the detailed singularity at the leading edge is different the behavior is essentially the same under matching.

Finally we remark that quasi-transonic theory can apply over an entire wing of suitable planform. If the chord of the wing is sufficiently small, the span sufficiently large and the sweep along the Mach angle (or close enough) the quasi-transonic theory applies over the entire surface. The conditions are sketched in Figure 7.7 below.

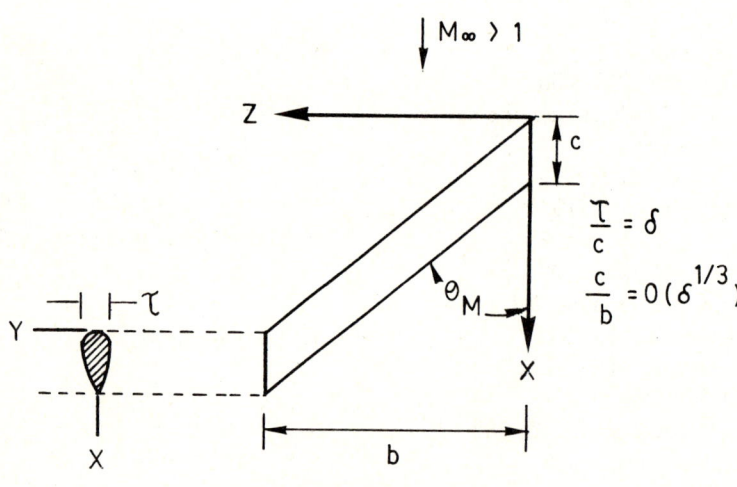

Figure 7.7

Quasi-Transonic Swept Wing

Index